Design and Analysis of Composite Structures for Automotive Applications

Automotive Series

Advanced Battery Management Technologies for Electric Vehicles
Rui Xiong, Weixiang Shen

Noise and Vibration Control in Automotive Bodies
Jian Pang

Automotive Power Transmission Systems
Yi Zhang, Chris Mi

High Speed Off-Road Vehicles: Suspensions, Tracks, Wheels and Dynamics
Bruce Maclaurin

Hybrid Electric Vehicles: Principles and Applications with Practical Perspectives, 2nd Edition
Chris Mi, M. Abul Masrur

Hybrid Electric Vehicle System Modeling and Control, 2nd Edition
Wei Liu

Thermal Management of Electric Vehicle Battery Systems
Ibrahim Dincer, Halil S. Hamut, Nader Javani

Automotive Aerodynamics
Joseph Katz

The Global Automotive Industry
Paul Nieuwenhuis, Peter Wells

Vehicle Dynamics
Martin Meywerk

Modelling, Simulation and Control of Two-Wheeled Vehicles
Mara Tanelli, Matteo Corno, Sergio Saveresi

Vehicle Gearbox Noise and Vibration: Measurement, Signal Analysis, Signal Processing and Noise Reduction Measures
Jiri Tuma

Modeling and Control of Engines and Drivelines
Lars Eriksson, Lars Nielsen

Advanced Composite Materials for Automotive Applications: Structural Integrity and Crashworthiness
Ahmed Elmarakbi

Guide to Load Analysis for Durability in Vehicle Engineering
P. Johannesson, M. Speckert

Design and Analysis of Composite Structures for Automotive Applications

Chassis and Drivetrain

Vladimir Kobelev
Department of Natural Sciences, University of Siegen, Germany

WILEY

This edition first published 2019

Registered Offices
John Wiley & Sons, Inc., 111 River Street, Hoboken, NJ 07030, USA
John Wiley & Sons Ltd, The Atrium, Southern Gate, Chichester, West Sussex, PO19 8SQ, UK

Editorial Office
The Atrium, Southern Gate, Chichester, West Sussex, PO19 8SQ, UK

For details of our global editorial offices, customer services, and more information about Wiley products visit us at www.wiley.com.

Wiley also publishes its books in a variety of electronic formats and by print-on-demand. Some content that appears in standard print versions of this book may not be available in other formats.

Library of Congress Cataloging-in-Publication Data

Names: Kobelev, Vladimir, 1959- author.
Title: Design and analysis of composite structures for automotive applications : chassis and drivetrain / Vladimir Kobelev, Department of Natural Sciences, University of Siegen, Germany.
Description: First edition. | Hoboken, NJ : Wiley, 2019. | Series: Automotive series | Includes bibliographical references and index. |
Identifiers: LCCN 2019005286 (print) | LCCN 2019011866 (ebook) | ISBN 9781119513841 (Adobe PDF) | ISBN 9781119513865 (ePub) | ISBN 9781119513858 (hardback)
Subjects: LCSH: Automobiles–Chassis. | Automobiles–Power trains. | Automobiles–Design and construction.
Classification: LCC TL255 (ebook) | LCC TL255 .K635 2019 (print) | DDC 629.2/4–dc23
LC record available at https://lccn.loc.gov/2019005286

Cover Design: Wiley
Cover Images: © Vladimir Kobelev, Background: © solarseven/ShuWerstock

Set in 10/12pt WarnockPro by SPi Global, Chennai, India

Printed and bound by CPI Group (UK) Ltd, Croydon, CR0 4YY

10 9 8 7 6 5 4 3 2 1

Contents

Foreword

From a materials science point of view, composite materials of glass and carbon fibers have a specific potential and already some practical importance in several applications under high dynamic loads. Comparing the fibers, glass fibers are the better material for spring applications because their lower modulus of elasticity compared to carbon fibers. This is favorable in terms of high strokes and deformation requirements. Due to their high specific strength and the stiffness of composite materials, it is in principle possible to achieve weight savings of 30 – 70% of the weight of a steel spring depending on application. In addition to reduce the unsprung masses for suspension, it is also possible to improve driving dynamics as well as noise, vibration and hardness behavior (NVH), since the material properties are better in some significant areas. Furthermore, due to the high corrosion resistance and resistance against other environmental influences, surface protection is not necessary in most of the applications.

However, the usage of composite materials for springs have not reached high quantities due to some limitations. Load transmission requires special designs. Considering suspension coil springs, high loads transverse to the main load direction occur. Therefore, the load transmission does not follow ideally to the fiber direction and only medium loads can act on the matrix. In addition, in the case of large-scale production and the available manufacturing processes, value adjustments must be made in comparison with units made of steel. These are currently the focus of research and development efforts throughout the world. Endless, unidirectional fiber materials, such as those used for structural elements in automotive engineering, exhibit strong anisotropic, i.e. direction-dependent, properties. The fibers used are oriented with respect to the loads that occur. Therefore, the leaf spring, where loading results almost in tension stresses of the fibers is the perfect match with composite materials. Huge weight reduction up to 75% is possible to achieve by using the material properties and the design flexibility of glass fiber reinforced composite in the best way. A single composite tension leaf spring can substitute a steel multi-leaf spring with a progressive spring load characteristic. The special design leads to a very homogenous, progressive spring characteristic and therefore, a better driving performance. Furthermore, we know already some designs for suspension steel coil springs substitution such as one-by-one substitution by composite coil spring and a meander spring design. In both case these springs do need special tools for the design and did not reach the market breakthrough due to huge different load-rate requirements within the platforms.

There are some processes existing for the production of glass fiber composite springs. Nevertheless, the prepreg process (pre-impregnated fibers) has proven itself as the best due to the realizable good properties under dynamic loads. Prepreg processes result in an optimal adhesive strength due to low porosity and allows flexibility in design, such as geometry, width and height of the spring. It is also possible to produce the elements of chassis in general and suspension particular using the resin injection process. For this resin injection process, a fiber structure is first produced from the dry reinforcement fibers, which follows the desired component geometry. If required, structural cohesion can be achieved using textile methods, such as sewing or bonding, which bond the fibers together. Such fiber structures are called preforms. The injection of the resin influences the orientation of the fibers and therefore, those springs do not reach the performance of prepreg composites due to potential ondulation.

Automotive manufacturers' requirements for carbon dioxide reduction, lower vehicle weight, the reduction of unsprung masses and the robustness of the springs, especially in the event of corrosion, will further increase in the future. The optimal application of the materials used plays a decisive role, supported by material properties, best technology and processes as well as an efficient design. Therefore, alternative materials, such as composites, may become higher importance for dynamic loaded suspension applications.

Prof. Dr. Vladimir Kobelev was born in Rostow-na Donu, Russian Federation. He studied Physical Engineering at the Moscow Institute of Physics and Technology. After his PhD at the Department of Aerophysics and Space Research (FAKI), he habilitated at the University of Siegen, Scientific-Technical Faculty. Today, Prof. Kobelev is lecturer and APL professor at the University of Siegen in the subject area of Mechanical Engineering.

In his industrial career, Prof. Kobelev is an employee at Mubea, a successful automotive supplier located near Cologne/Germany. In the Corporate Engineering Department, Prof. Kobelev is responsible for the development of calculation methods and physical modeling of Mubea components.

Joerg Neubrand
CTO, Managing Director and
Member of the Executice Board of the Mubea Group

Series Preface

Fuel efficiency continues to "drive" significant research and development in the automotive sector. In many instances, this is propelled by regulations that target reduced emissions as well as reduced fuel consumption. Even with more efficient vehicles and electric hybrid or purely electric driven systems, the need for reduced energy consumption is demanded by the market. This is due to the fact that the customer base is demanding increased efficiency as this brings better performance, lower costs and extended range of the vehicle. One clear means by which fuel efficiency can be enhanced is by reducing the weight of the vehicle. Lightweighting can be accomplished by a number of means, one of which is lighter weight material substitution. That is to say, one may substitute a lighter material for a heavier one on a vehicle component. Composites have been used to replace metal components in efforts to lightweight aircraft for decades. More recently, advances in materials, manufacture, and design have made composites cost effective and viable in the automotive sector. Two major stumbling blocks that have hindered composite use in the automotive sector are the cost of the composite components, and the ability to rapidly and economically produce such components in quantities that are needed by the automotive sector. Recently, these stumbling blocks have been overcome. However, for most commercial automotive applications, composites remain relegated to less critical elements of the vehicle system such as body panels. The use of composite for more critical vehicle applications such as suspension and drive train elements have been left to extremely demanding automotive scenarios such as Formula One. However, this scenario is about to change.

Design and Analysis of Composite Structures for Automotive Applications, provides an in-depth technical analysis of critical suspension and drive train elements with a focus on composite materials. This includes basic principles for the design and optimization of critical vehicle elements using composite materials, as well as classical concepts related to mass reduction in automotive systems. The author, Professor Kobelev, skillfully integrates concepts related to vehicle parameters such as stiffness into overall vehicle dynamics using closed form solutions that are described in exquisite detail. The discussions focus on key elements of the vehicle including suspension and powertrain. These discussions are both comprehensive as well as the first of their kind in a text book, making this text an important reference for any automotive engineer on the leading edge.

Design and Analysis of Composite Structures for Automotive Applications is part of the *Automotive Series* that addresses new and emerging technologies in automotive engineering, supporting the development of next generation vehicles using next generation technologies, as well as new design and manufacturing methodologies. The series provides technical insight into a wide range of topics that is of interest and benefit to people working in the advanced automotive engineering sector. *Design and Analysis of Composite Structures for Automotive Applications* is a welcome addition to the *Automotive Series* as it primary objective is to supply pragmatic and thematic reference and educational materials for researchers and practitioners in industry, and postgraduate/advanced undergraduates in automotive engineering. The text is a state-of-the art book written by a leading world expert in composites and its application to suspensions and is a welcome addition to the *Automotive Series.*

Thomas Kurfess
March 2019

List of Symbols and Abbreviations

E_L	Longitudinal modulus of elasticity of composite parallel to fiber direction
E_T	Transverse modulus of elasticity of composite perpendicular to the fiber direction
G_{TL}	Transverse-longitudinal shear modulus of composite
G_{TT}	Transverse shear modulus of composite
r_f	Radius of the fiber
$r_m = r_f / \sqrt{V_f}$	External radii of hypothetical matrix cylinders
V_f	Fiber volume content
V_m	Matrix volume content, $V_m = 1 - V_f$
ν_{TL}	Transverse-longitudinal Poisson ratio of a composite
ν_{TT}	Transverse Poisson's ratio of a composite
E_f	Longitudinal modulus of elasticity of fibers
$E_{f.T}$	Transverse modulus of elasticity of fibers
E_m	Modulus of elasticity of matrix (resin)
ν_m	Poisson's ratio of matrix (resin)
G_m	Shear modulus of matrix (resin)
$G_{f.TL}$	Transverse-longitudinal shear modulus of fibers
$\nu_{f.TL}$	Transverse-longitudinal Poisson's ratio of fibers
$\nu_{f.TT}$	Transverse Poisson's ratio of fibers
$\mathbf{S} = [S_{ijpq}]$	Compliance tensor of rank four, $i, j, p, q = 1, 2, 3$
$\mathbf{C} = [C_{ijpq}]$	Elasticity tensor of rank four
$\mathbf{C}^{(0)} = [c_{ijpq}]$	Elasticity tensor of rank four, in the layer coordinate system
$\mathbf{S}^{(0)} = [s_{ijpq}]$	Compliance tensor of rank four, in the layer coordinate system
$\mathbf{I} = [I_{ijpq}]$	Fourth rank identity tensor
$\boldsymbol{\sigma}$	Voigt's stress vector
$\boldsymbol{\varepsilon}$	Voigt's strain vector
$\mathbf{C}$	Voigt's elasticity matrix
$\hat{\boldsymbol{\sigma}}$	Kelvin's stress vector
$\hat{\boldsymbol{\varepsilon}}$	Kelvin's strain vector
$\hat{\mathbf{C}}$	Kelvin's elasticity tensor

$\mathbf{S}^{(0)}$	Compliance matrix in Voigt's notation in intrinsic coordinates
$\widehat{\mathbf{S}}^{(0)}$	Compliance matrix in Kelvin's notation in intrinsic coordinates
$\mathbf{t} = [t_{lk}]$	Transformation (rotation) square 3×3 or 2×2 matrix
$\mathbf{T}_\sigma$	σ-transformation (rotation) square 6×6 matrix, in Voigt's notation
$\mathbf{T}_\varepsilon$	ε-transformation (rotation) square 6×6 matrix, in Voigt's notation
$\widehat{\mathbf{T}}$	Transformation (rotation) square 6×6 matrix, in Kelvin's notation
$\mathbf{A}$	Plane modulus quadrant, square 3×3 matrix (entries for the membrane elasticity tensor) in Voigt's notation
$\mathbf{B}$	Coupling quadrant, square 3×3 matrix (entries for the coupling elasticity tensor) in Voigt's notation
$\mathbf{D}$	Plate quadrant, square 3×3 matrix (entries for the bending elasticity tensor) in Voigt's notation
$\boldsymbol{\varepsilon}^T = [\varepsilon_{11}, \varepsilon_{22}, \gamma_{12} = 2\varepsilon_{12}]$	Strain vector in Voigt's notation
$\boldsymbol{\kappa}^T = [\kappa_{11}, \kappa_{22}, 2\kappa_{12}]$	Curvature vector in Voigt's notation
$\mathbf{N}^T = [N_{11}, N_{22}, N_{12}]$	In-plane forces vector in Voigt's notation
$\mathbf{M}^T = [M_{11}, M_{22}, M_{12}]$	Bending moments vector in Voigt's notation
$\mathbf{Q}$	Reduced stiffness matrix in Voigt's notation
$\widetilde{\mathbf{T}}_\sigma$	σ-transformation (rotation), square 3×3 matrix in Voigt's notation
$\widetilde{\mathbf{T}}_\varepsilon$	ε-Transformation (rotation) square 3×3 matrix in Voigt's notation
$\widehat{\mathbf{A}}$	Plane modulus quadrant, square 3×3 matrix (entries for the membrane elasticity tensor) in Kelvin's notation
$\widehat{\mathbf{B}}$	Coupling quadrant, square 3×3 matrix (entries for the coupling elasticity tensor) in Kelvin's notation
$\widehat{\mathbf{D}}$	Plate quadrant, square 3×3 matrix (entries for the bending elasticity tensor in Kelvin's notation
$\widehat{\boldsymbol{\varepsilon}}^T = [\varepsilon_{11}, \varepsilon_{22}, \sqrt{2}\varepsilon_{12}]$	Strain vector in Kelvin's notation
$\widehat{\boldsymbol{\kappa}}^T = [\kappa_{11}, \kappa_{22}, \sqrt{2}\kappa_{12}]$	Curvature vector in Kelvin's notation
$\widehat{\mathbf{N}}^T = [N_{11}, N_{22}, \sqrt{2}N_{12}]$	In-plane reaction forces, vector in Kelvin's notation
$\widehat{\mathbf{M}}^T = [M_{11}, M_{22}, \sqrt{2}M_{12}]$	Bending moments, vector in Kelvin's notation
$\widehat{\mathbf{Q}}$	Reduced stiffness matrix in Kelvin's notation
$\widetilde{\mathbf{T}}$	Transformation matrix for rotation (3×3) in Kelvin's notation
X_t, X_c	Tensile or compressive strengths in the fiber direction
Y_t, Y_c	Tensile or compressive strengths in the transverse direction
$\widetilde{\mathbf{F}}^{(0)} = [f_{ijkl}]$	Matrix of the Mises–Hill criterion, Voigt's notation in intrinsic coordinates
F, G, H, L, M, N	Characteristic values of the Mises–Hill criterion

Λ_i	Eigenvalues of the Mises–Hill criterion, $i = 1..6$
$\widetilde{\sigma}_{11}, \widetilde{\sigma}_{22}, \widetilde{\sigma}_{33}$	Ultimate normal stresses in the Mises–Hill criterion
$\widetilde{\tau}_{23}, \widetilde{\tau}_{31}, \widetilde{\tau}_{12}$	Ultimate shear stresses in the Mises–Hill criterion
$\widehat{\mathbf{F}}^{(0)}$	Matrix of the Mises–Hill criterion, Kelvin's notation in intrinsic coordinates
Φ	Rotation angle of fibers in the plane "1–2"
$\widehat{\mathbf{F}}(\Phi)$	Matrix of the Mises–Hill criterion in the rotated axes, Kelvin's notation
$\widetilde{\mathbf{F}}_p$	Matrix of the pressure-sensitive Mises–Hill criterion, Voigt's notation in intrinsic coordinates
$\widehat{\mathbf{F}}_p$	Matrix of the pressure-sensitive Mises–Hill criterion, Kelvin's notation in intrinsic coordinates
$\mathbf{F}^{(2)}, \mathbf{F}^{(4)}, \mathbf{F}^{(6)}$	Tensors of the 2nd, 4th and 6th ranks of the Goldenblat–Kopnov tensor fracture criterion
σ_f	Axial stresses in cylindrical fibers
σ_m	Axial stresses in hollow matrix cylinders
τ	Shear stress at the fiber surface
τ_p	Yield point of the matrix
$\mathrm{p}(z), \mathrm{q}(z)$	Auxiliary functions, $\mathrm{p} = u_m - u_f$, $\mathrm{q} = u_m + u_f$.
u_m, u_f	Axial displacements of matrix and fiber cylinders
$\lambda, \mu, \widetilde{K}_3$	Parameters of the length dimension
$\widetilde{K}_1, \widetilde{K}_2$	Parameters of the inverse length dimension
l_p	Length of the plastic zone
R_f, R_f	Crack extension resistance of fibers and matrix
$\overline{E}$	Auxiliary modules
K_f, K_m	Fracture toughness of fibers and matrix
$d\mathrm{U}_e/da$	Energy release rate per thickness unit
$d\mathrm{U}_f/da$	Crack extension resistance per thickness unit
$\varphi_i(\rho, \eta)$	Potential functions, $i = 1, 2$
$a = l_c/2r_f$	Dimensionless length of adhesive or debonding region
$K_{\max}, K_{\max}$	Maximum and minimum stress intensity factor
$Y(a)$	Dimensionless parameter that reflects the geometry
$c_f = c_f(R_\sigma)$	Material constant of matrix or resin for a given stress ratio R_σ
$R_\sigma = K_{\min}/K_{\max}$	Stress ratio of cyclic load
$p_c(K) = K^{-p}$	Paris–Erdogan crack propagation function
C_T	Torsional rigidity of bar (driveshaft)
I_{b1}, I_{b2}	Moment of inertia of cross section with respect to both bending axes
M_T^{**}	Critical torque in Greenhill's problem
$W_e(\Phi)$	Density of elastic energy
$W_e^{*}(\Phi)$	Elastic energy per mass unit (specific elastic energy)
$\widetilde{W}_e^{*}$	Ultimate elastic energy per mass unit (specific ultimate elastic energy)
$\widetilde{W}_e$	Ultimate elastic energy
X_E, Y_E, Z_E	Axes of the earth-fixed coordinate system

X_V, Y_V, Y_V	Axes of the vehicle-fixed coordinate system
X, Y, Z	Axes of the horizontal coordinate system
X_W, Y_W, Z_W	Axes of the wheel coordinate system
φ	Roll angle
θ	Pitch angle
ψ	Yaw angle
β	Sideslip angle of a vehicle
$\varsigma = \psi - \beta$	Course angle
$S_{X;V}$	Circumferential slip
ω	Rotational speed of a wheel
ω_0	The rotational speed of a straight and freely rolling wheel
α	Sideslip angle of a wheel
$\mu_{X,W}$	Coefficients of circumferential force
$\mu_{Y,W}$	Coefficients of lateral force
C_α	Cornering stiffness
$C_{\alpha f}, C_{\alpha r}$	Cornering stiffness of front and rear tire
I_z	Mass moment of inertia of the vehicle around the vertical axis
m	Mass of a vehicle
L_f	Horizontal distance from front axle to center of mass
L_r	Horizontal distance from rear axle to center of mass
$L_w = L_f + L_r$	Wheel base
δ	Steer angle at the wheel (part of steer angle due to steer)
$\frac{\partial \delta}{\partial a_y}$	Steer angle gradient
δ_a	Steer angle according to Ackermann
Δ_V	External excitation in the course of steering
ω_V	Yaw circular frequency of a vehicle
D_V	Damping factor of a vehicle's yaw oscillation
L_0	Free length of a helical spring
L_{rel}	Released length of a helical spring
L_{comp}	Compressed length of a helical spring
L_c	Close up length of a compressed spring
$s = L_{rel} - L_{comp}$	Spring travel
c	Axial compression or extension spring constant
$c_{\theta F}$	Compression-twist spring rate
c_θ	Twist spring rate
U_e	Elastic energy
U_f	Work of applied forces
F	Axial force on a helical spring
M_θ	Axial torque on a helical spring
c^*	Design value for a spring constant
τ_w	Ideal stress at solid height
d_{opt}	Optimal diameter of a wire
m_{opt}	Lower boundary for spring mass
ϖ	Middle surface of an undeformed conical spring
Ω	Middle surface of a deformed conical spring

t_c	Thickness of an anisotropic conical spring
r_a, r_b	Inner and outer radius of the middle surface of a free conical spring
R_a, R_b	Outer radius of the middle surface of a deformed conical spring
$\Delta = r_b / r_a$	Ratio of the outer radius to inner radius of a conical spring
$\widehat{c}$	Inversion point of a conical spring
z_a, z_b	Heights of the inner and outer edges of a free spring
Z_a, Z_b	Heights of the inner and outer edges of a deformed spring
ε_1	Circumferential mid-surface strain of a conical spring
κ_1	Circumferential curvature changes of a conical spring
$\widetilde{E}$	Effective elastic modulus
$\overline{E}$	Effective orthotropic elastic modulus
$s_Q = s_b + s_s$	Total transversal displacement of a helical spring
Q	Shear force of equivalent column for a helical spring
M_B	Bending moment of a helical spring
m_B	External torque per unit length of a helical spring
f_Q	External load in the transverse direction of a helical spring
$C_{44} I_T$	Twist stiffness of a wire with respect to its axis
$C_{33} I_b$	Stiffness of a spring wire in the case of bending in a binormal direction
$C_{33} I_n$	Stiffness of a spring wire in the case of bending in a normal direction
$s_o(z)$	Initial transverse deflection of a helical spring
$v_o(z)$	Initial transverse velocity of a helical spring
ω_k	Circular natural frequency in the order of k of a helical spring
$\xi = L_0 / D$	Slenderness ratio of a helical spring
$\mu = L / L_0$	Dimensionless length of a helical spring
Ω_k	Dimensionless frequency of transverse oscillations in the order of k of a helical spring
$\mu_+^*(k)$	Critical deflection during compression from the free length of a helical spring
$\mu_-^*(k)$	Critical deflection during expansion from the flattened state of a helical spring
$f(\sigma_{eff}, t)$	Anisotropic stress function for creep
t	Time for creep
$\dot{e}_{ij}$	Deviatoric component of creep strain
s_{ij}	Deviatoric component for creep stress
σ_{eq}	Mises equivalent stress
c_τ	Creep constant for shear strain
c_σ	Creep constant for uniaxial strain
γ_e	Elastic component of shear strain
γ_c	Creep component of shear strain
M_T^0	Torque at the moment $t = 0$

$M_T(t)$	Torque as a function of time
F_z^0	Spring force at the moment $t = 0$
$F_z(t)$	Spring force as a function of time
M_B^0	Bending moment at the moment $t = 0$
$M_B(t)$	Bending moment as a function of time
$\Phi_T(t)$,$\Phi_B(t)$,$\Phi_H(t)$	Relaxation functions for twisting, bending and helical spring
$f' = \partial f/\partial z$, or $\partial f/\partial x_1$, or $\partial f/\partial \xi_1$	"Prime" denotes a derivative with respect to a coordinate: z, or x_1, or ξ_1
$\dot{f} = \partial f/\partial t$	"Dot" denotes a time derivative

Abbreviations

AF	Aramid fiber
AFRP	Aramid fiber reinforced plastic
Autoclave	Heated pressure tank
CE	Cyanate ester resin
CF	Carbon fiber
CFRP	Carbon fiber reinforced plastic
DSA	Driveshaft axis
EP	Epoxy resin
Fabric	Biaxially woven textile
FRP	Fiber reinforced plastic
GF	Glass fiber
GFRP	Glass fiber reinforced plastic
HM	High modulus
HT	High tensile strength
Laminate	Layer construction of cured, individual plies
Matrix	Resin in which the fibers are embedded
NVH	Noise, vibration and hardness
PA	Polyamide
PEEK	Polyetheretherketone
PF	Phenolic resin
PMMA	Polymethyl-methylacrylate
PPS	Polyphenyl sulfide
Prepreg	Preimpregnated fibres – fibers or textiles pre-saturated with resin
PU	Polyurethane resin
Roving	Soft strand of twisted, attenuated, freed of external matter fiber ready to conversion into yarn
RTM	Resin transfer molding – resin is injected into an enclosed mold in which fibers have been placed
Tg	Glass transition temperature
UD	Unidirectional fibers are oriented in only one direction
UP	Unsaturated polyester
VE	Vinyl ester resin

Introduction

Composites in Automotive Chassis and Drivetrain

In times of climate change and rising emissions in the environment, lightweight construction has found its way into almost all industries. The authorities, particularly in the automotive industry, formulate endlessly decreasing targets in emission reduction. Because increasingly stringent emissions can be minimized through weight reduction, an optimal structural design made with lightweight materials is one of the principal tendencies in contemporary development of passenger cars. Some of the most attractive improvements have been seen in the use of composites to replace parts and components traditionally manufactured from steel (Miravete 1996; Tucker and Lindsey 2002). In particular, carbon-fiber-reinforced plastics and glass reinforced plastics have great potential to reduce the weight of passenger cars. Cost of a product remains a key issue. As with any lightweight material, all further expenses, which are incurred in addition to the cost of the base material, must be accounted for. The major task to make the product attractive to customers and the market is to reduce lightweight construction and production costs as much as possible along with significant weight reduction and other extra benefits. Despite manufacturing processes being continuously improved, there is still substantial progress to be made for cost-effective mass production. Safety is another dominant criterion for passenger cars. Hence, new designs must be structurally robust enough to adhere to current and future crash safety targets.

Over recent years, car body and drivetrain have come under deep examination in the attempt to reduce mass of structure and a range of innovative concepts have been developed (Kedward 2000; Brooks 2000; O'Rourke 2000). Lightweight construction has become an optimization goal that is valid for several components of automobiles (Lu and Pilla 2014a, 2014b, 2014c; Elmarakbi and Azoti 2015; Njuguna 2016; Ishikawa et al. 2018; Hayashi 2000; Nomura 2000).

Weight reduction of the chassis has gained in importance as well. The chassis has a substantial potential of weight reduction (Neubrand 2014). Among other lightweight materials for chassis design, glass-fiber-reinforced plastic provides a good alternative to steel. Moreover, the unsprung mass can be lessened. This factor brings distinct advantages for the driving dynamics and comfort.

As previously mentioned, reducing weight of vehicles is an indispensable requirement in the automotive manufacturing sector. There are several "material factors" that are used for characterization of weight reduction (Ashby 2010). Apparently, the material density is the most trivial material factor in determining the best suited material for a certain application. The density determines the relative weights of structures, but provides no information about their strength and flexibility. Another, also simple factor is the material price of the mass unit. The specific price factor determines the material that is best suited for a price-critical application. The price factor, as a sum of raw material price and manufacturing expenses, is commercially important in material evaluation.

Another factor is the specific strength, or the strength-to-weight ratio of the material. The specific strength of an isotropic material is obviously given by the tensile or yield strength divided by the density of the material. A material with a high specific strength will be suitable for load-carrying elements. There are several secondary criteria of this kind, which depend upon the art of dominant load: uniaxial stress or bending stress. The specific strength factors distinguish whether axial stress or stress due to bending dominates. In the first case, stress is constant over the cross-section of the part. Specific strength is equal to tension load divided by the cross-sectional area and density of the material. In the second case, stress is a linear function of the thickness coordinate. The specific strength is equal to the bending moment is divided by the resistance moment and density of the material. For the fixed weight of material, the thickness of the material with a lower density is greater. The resistance moment for the material with the lower density is greater as well. Accordingly, the material with lower density possesses higher specific strength for bending even for the equal ultimate strength of both materials. This remark makes the application of lighter materials attractive for bending-dominated applications. Similar speculations are applicable for a shaft loaded by the given torque. In this load case, the shear stress depends linearly on the radius, and a material with lower density leads to lower stress on the surface if the torque and mass of the shaft are prescribed.

The specific stiffness is basically the elastic module-to-density ratio of the material (Ashby 2010). The specific strength of an isotropic material is given either by the Young's modulus divided by the density of the material for the elements predominantly in a uniaxial tension or compression state or by the shear modulus divided by the density of the material for the elements mainly in a state of shear. A material with a high specific stiffness will be suitable for elements that guarantee maximal stiffness. For example, the specific stiffness characterizes the performance of materials for structural elements that are responsible for buckling performance, dynamic and static stiffness, and for aeroelastic critical applications. Analogous to the specific strength, the specific stiffness distinguishes whether the uniaxial stress or stress due to bending dominates. For example, compare two materials in a state of bending or torsion with equal elasticity modules, but different densities. If we assume that both elements have equal mass, the specific stiffness of the lighter material is higher than the stiffness of heavier material. This occurs because the plate thickness or shaft diameter made of a material with the lower density is higher. Accordingly, the corresponding bending or torsional stiffness will be higher. These conjectures make the application of composites preferable for structural elements in bending or torsion stress states.

The elements that provide energy storage must be characterized by specific energy density. The ratio of specific energy density to mass density of a material characterizes the material for energy harvesters, different springs and flexible structural elements of an automotive chassis. These conjectures about the preferably light material in the states of torsion or bending are generally not valid for specific energy density. The argument is as follows. Consider, for example, two shafts (torsion springs) with the equal mass, equal shear modulus, equal applied torque, but two different densities. The shaft made of the lighter material possesses the higher diameter and also has the greater torsion stiffness. The twist for the given torque will be lower. Consequently, the stored energy of the lighter shaft is lower if the torque is given. But simultaneously the maximal stress will be lower, because the resistance moment of the shaft with the higher diameter is higher. The specific energy density in the case of torsion is roughly the ratio of the squared ultimate stress in the material divided by the density and elasticity modulus. The most favorable application of one or other material also depends on the load character. If the material is anisotropic, the formulas for calculation of specific factors will be somewhat complicated. These thoughts will be explained in detail later in the book.

There are other specific material parameters in the automotive praxis. Among others, the specific plastic energy release rate. This material parameter is applied for car body design. Specific plastic energy release rate indicates the suitability of the material for applications in the zones of energy adsorption. The materials with the higher specific plastic energy release rate behave preferably in the event of an accident.

Therefore, the advanced specific "performance-to-density" ratios are essential for comparison of engineering materials in engineering design. These factors include Young's modulus to density, Young's modulus to specific price, strength to density, strength to toughness, strength to elongation, strength to cost, strength to maximal service temperature, specific stiffness to specific strength, electrical resistivity to cost, recycle fraction to cost and energy content to cost (Ashby 2010). The specific factors deal commonly with the uniaxial tensile load and therefore are scalars. If the stresses are multi-axial and alternating, the scalar performance-to-density factors provide only rough estimations of design efficiency.

Moreover, the majority of "material factors" was developed for isotropic materials and takes no notice of the anisotropy of composite materials. Anisotropy is characterized by the fact that there is a shear-stretch coupling. This means that a normal stress in the longitudinal direction additionally causes a displacement. Similarly, a shear stress additionally causes an elongation. In other words, the fiber-reinforced composite materials have habitually different stiffnesses in diverse load directions. The use of scalar factors for materials with strong anisotropy and variable stress fields delivers, as a rule, an unreliable estimation of design features. The reliable optimal design of a composite material must be based on a deep structural analysis and comprehensive exertion of the specific advantages of composite materials.

Physical Properties of Composite Materials

A material is referred to as a composite if at least two diverse components are combined on the microscopic level to a new concrete mixture. The separate substances, based on their various properties, accomplish different tasks. By their nature, dissimilar materials are also frequently joined so the combination gains remarkable properties that both components cannot achieve separately. Comprehensive surveys of the physical properties and manufacturing of composites have been given (Peters 1998; Kelly and Zweben 2000; Kleinholz et al. 2010; Neitzel et al. 2014; ECSS-E-HB-32-20 2011).

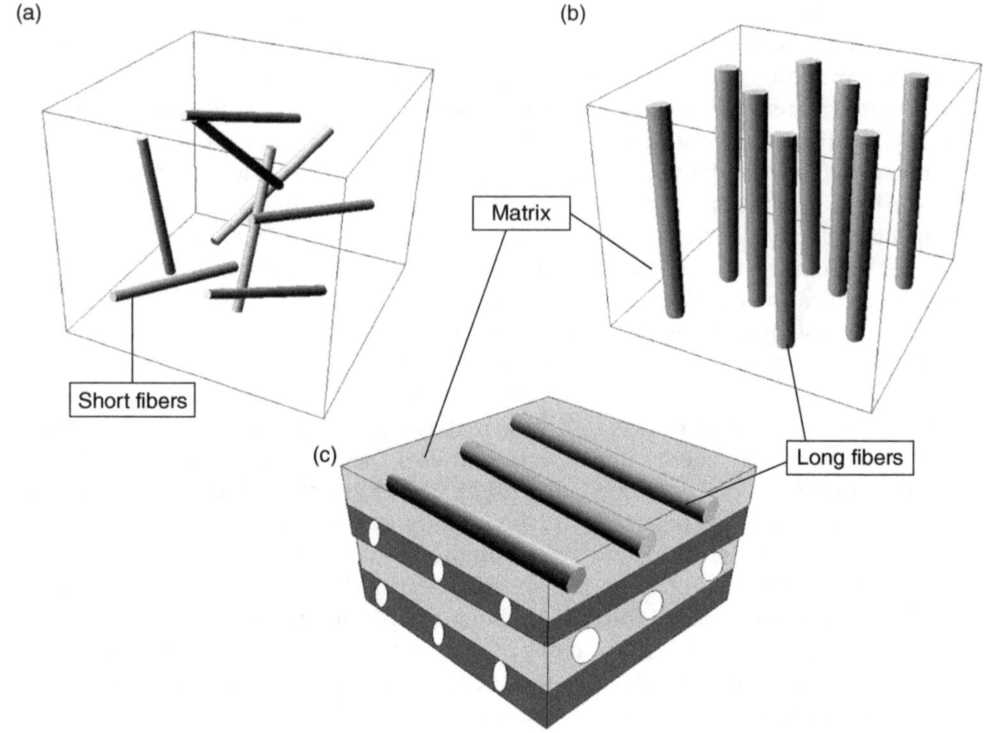

Figure 1 Types of composite material: (a) reinforcement by short fibers or whiskers; (b) unidirectional composite, reinforced by continuous fibers and (c) multilayered, laminated composite made of multiple layers.

Specifically, "composite" generally refers to solid combinations of high-strength, but brittle reinforcement fibers embedded in a weak, but ductile matrix (Figure 1). The synergy effect is that the properly synthesized composite inherits the high stiffness and load capacity from fibers and high ductility from the matrix.

Chemical industry produces artificial fibers, for example, carbon, glass or aramid fibers. These fibers possess outstanding mechanical and chemical properties. Matrix materials are being steadily developed as well. The raw materials for fibers are usually very brittle and possess only a restricted strength. However, as the fiber diameter decreases, the strength increases tremendously. The explanation of the increase in strength is comprehended in the size effect. The size effect unfolds within the remarkable features. On the one hand, the size of the flaw is limited in a thin fiber. The flaw must be many times smaller to generate an endless fiber. According to statistical considerations, the length of a flawless fiber section continues to grow for thin fibers. In other words, the thinner the fiber, the longer the flawless area (Argon 1974).

Consider a large, bulk body; for example, a glass pane. The number and size of individual defects increase with the size of the component. There are numerous dilute defects in the large volumes of the homogenous material. These initially existing defects make the homogenous material brittle. The material fails to arrest the small initial cracks. The crack grows unrestricted through the volume and finally provokes instant fracture. For the destruction of a bulk homogenous part, the principle of the weakest link is applicable. Namely, the principle of the weakest link is based on the size effect. The weakest link principle declares that a chain is only as strong as its weakest link. This means that if a part contains a defect it breaks. In application to fibers of composite materials, if a fiber contains a defect it breaks.

The picture of fracture of a heterogeneous composite material is different. Several hundred thousand fibers are present in parallel in one bundle and if one fiber breaks, the other remains intact and continues to carry the full load. Thus, the load from the failed fiber redistributes to the many fibers without failure of the bundle. The stiffness of material alters with size as well. Accordingly, a material in fiber form has significantly higher rigidity and strength than a raw material. The smaller the fiber cross-section, the higher its strength.

In fiber-reinforced plastics, the mechanical properties, such as stiffness and strength, are determined primarily by reinforcement fibers. The fibers are made of a variety of materials, and processed to form diverse semi-finished products.

For the production of fiber-reinforced plastics, mostly inorganic fibers, such as glass fibers as well as carbon fibers or aramid fibers, are used. The carbon-fiber-reinforced polymer is an evolving construction material that exhibits exceptional mechanical properties, such as strength and rigidity, with light material density at the same time. The commercial production of carbon fibers started in the 1970s. Application was primarily in the aviation and aerospace industry. The carbon fibers invaded motorsport at the beginning of the 1980s (O'Rourke 2000). Nowadays, carbon fibers are gaining in significance again due to the trend for lightweight construction and are coming increasingly into focus in the automotive industry. Prominent automobile manufacturers are already using carbon fibers with success for visible auxiliary elements and currently researching their application in principal structures.

With fiber composite materials, at least one type of fiber is "bonded" with a basic material, called a matrix, or resin. In addition to the fiber type, the matrix system plays a significant role in determination of the material properties. The new development and spreading of the composite material in modern industry is closely linked to the improvement of plastics, especially the synthetic matrix. Synthetic resins are the perfect adhesive for fiber constructions, are light, adhere to several types of fibers, are corrosion resistant and assure the flawless impregnation of the fibers. Thermosetting plastic and thermoplastic plastics are used as matrix (or resin) materials, each of which has different advantages and disadvantages when viewed individually. For moderately loaded structural and safety applications of composite components, a thermosetting plastic must be applied, as this will possess a higher resistance than thermoplastics. However, the positive features of thermoplastics are their simpler manufacturing and reusability. As already mentioned, the combination of high-strength glass fibers with a corresponding thermoset plastic matrix enables synergy effects to be optimally exploited.

Material behavior depends not only on the type of fiber and volumetric fiber content. Because fibers can chiefly support tensile loads in the longitudinal direction, material behavior depends on the arrangement of fibers within a matrix. Consequently, the composite material requires the precise alignment of reinforcing fibers. In addition, the alignment of the fibers is responsible for the anisotropic properties of a composite material. Because the composite material is created during the manufacturing process, the production method and the component geometry have a considerable influence on the resulting properties of the composite material.

The basic element of every component made of composite material is the unidirectional layer. The laminate is the assembly of several differently oriented individual layers. The properties of laminates are mostly directional and strongly dependent on the sequence of the individual plies. The composite materials are commonly built from several single layers, even though the fiber directions in all single layers can coincide. The strength and stiffness of a component can be determined by the number of layers, the fiber volume fraction, the fiber direction and the thickness of a single layer as well as the layer order.

In addition to standard strength values, such as tensile strength and yield strength, other parameters, such as interlaminar strength value and glass transition temperature, serve as characteristic values for laminated composites. As the name describes, interlaminar shear strength limits the strength of the fiber-matrix separation surface. Shear stresses occur on the fiber-matrix separation surface already at tensile/compressive loading, since the fibers and the matrix each have a different modulus of elasticity, which results in different strains and stresses in the two materials. The standardized short bending test requires relatively small test specimens (ECSS-E-HB-32-20 2011). This test permits the evaluation of interlaminar strength and is well-matched for quality control of material components.

Since the strength values are determined, a simple comparison with other material classes can be made. Despite the already mentioned high-strength of original single fibers, the yield stress of composites is consistently lower compared to steel. Due to the low density of the composite material, however, the specific material properties of composites are higher that of steel. Accordingly, glass-fiber-reinforced plastic has a higher lightweight construction potential compared to steel. Furthermore, glass-fiber-reinforced plastic has the advantage that no corrosion can occur as the material is non-conductive and chemically resistant to the environment media.

The lightweight potential dramatically increases if the structure is dominantly in the state of bending. The bending stress resistance is the third power and the flexure resistance is the second power of laminate thickness. Accordingly, the stiffness and strength of plates, and especially of sandwich structures made of composites, potentiates with their thickness without violation of the weight.

However, there are also some shortcomings of composites. These can be easily compared on the basis of the construction. For example, in comparison to helical springs made of steel, a glass-fiber-reinforced plastic meander spring requires a larger installation space. On the other hand, the volume of stressed material of fiberglass is higher as the absolute strength per unit of volume is lower in comparison to a high-grade steel alloy. Another drawback of the composite is the remaining deficiency of knowledge in some applications. It is not yet adequately known, for example, how a material behaves under a rapid load. This aspect is important for crash design of passenger cars. Finally, the quality of manufacturing turns out to be the key issue, especially for multilayered laminates and sandwich composite structures.

For the production of chassis and drivetrain components, there are different possibilities that are strongly dependent on the used resin system or whole material compound. The complex geometries of the structural elements are produced by the draping process. A flat semi-finished product, the "prepreg," is "stacked" in layers so that a blank of the finished product is formed at the end. Prepreg is a semi-finished product made of pre-impregnated continuous fibers embedded in an unhardened thermosetting plastic. For better handling, storage, transport and protection of the prepreg both surfaces are covered with foil. The production of prepregs is also very complex. In this process, the fibers are impregnated directly with the resin to simplify further processing. The fiber manufacturer bundles the individual glass fibers or glass filaments to form a strand with several thousand filaments. This strand is called "roving." In the first stage of prepreg production, the rovings are "threaded" into the plant. This is done manually by eyelets and a shearing comb, which guarantees the orientation of the roving. The roving bundle is then pressed onto the carrier film coated with a defined quantity of resin. After the roving bundle has been placed in the matrix, a short heat treatment takes place that initiates the curing reaction of the resin. A further film is then applied to protect the prepreg and the prepreg is cooled and wound.

When the prepreg is processed, the temperature and humidity are precisely controlled to ensure that the resin reaches a viscous state for better handling during the drape process. This ensures that the prepreg layers can also be draped manually and glued sufficiently for the time being. The curing process is accelerated thermally at elevated temperatures. This fact is reused during the final pressing process. The blank is heated to just under 125 °C, so that the thermoset matrix achieves a low viscosity. At the same time, the curing reaction begins. The pressing process is therefore a very important part of the glass-fiber-reinforced plastic component manufacturing process, as the final properties of the material and the component are achieved in this process, without which no firm bond between the individual prepreg layers can be achieved. In addition to the high temperature, the applied press pressure and the pressing time or curing time are also important. At the same time, the final fiber volume share of the structural element is adjusted during pressing, as a small part of the matrix is pressed out. If possible, when the viscous resin escapes, consideration must be given to achieving a uniform resin flow over the entire component. Depending on the geometry of the structural elements, the press stroke is defined by the thickness or width. With the "resin transfer molding" process, this can be achieved directly via the exact amount of resin supplied as soon as the mold is closed. The tempering process begins after the hardened structural element has been removed from the press. The structural elements are tempered in the oven at approximately 120 °C for 180 minutes. After being removed from the furnace, post-processing can be carried out. The edges must be machined in order to remove the resin.

Resin transfer molding is a frequently used process. The rovings consist of parallel dry filaments. The dry glass roving first generates a preform in the tool. The mold is then partially closed. The liquid heated plastic, in combination with a corresponding hardener, is injected into the mold. The resin spreads under pressure consistently throughout the mold. The orientation of the fibers remains unaffected. After polymerization of the matrix, the part can be removed from the mold. The advantage of this process is that the curing and pressing time is reduced by up to 50% compared to the pressing process after a draping of a similar component.

The manufacturing process is a major cost factor in the development of a series process. The cycle time of a single structural element is actually high. Manual draping requires hours, conditional on the number of layers and the size of the part. To minimize cycle times, manufacturing robotics must be developed. The robots can place the cut prepreg layers and compress them to one assembly. Robots can also perform edge processing. The hardening time of high-strength composite elements with contemporary matrix materials must be considerably reduced.

The most remarkable feature of composites is that the material and structural elements are created simultaneously and in one solitary process. Only deep understanding of this act will pave the way to creation of lightweight composite structures with their extraordinary properties in future.

Structure of the Book

The book presents the concepts of high-strength elastic elements of automotive chassis and drivetrain made of composite materials from the point of classical mechanics. The book studies the principal ideas, which are necessary for understanding of analysis and optimization of composite elements. What all considered problems have in common is that they are solved in closed form without application of finite elements or other numerical methods. All equations and formulas are derived from the primarily declared hypotheses. The building blocks of the physical background for the book are anisotropic elasticity on structural and micromechanical levels, anisotropic creep and elements of fracture mechanics.

In the book, the design principles for optimal material anisotropy of structural elements are studied from the viewpoint of stiffness, loading capacity and energy storage. For this purpose, we first evaluate the specific stiffness, strength and stored energy of elastic materials with orientation-depending mechanical properties. The basics of the anisotropic material and the fiber composite are given in Chapter 1, as they are needed in the context of this work. Based on the fiber composites and the characteristic properties of the fibers and matrix, the special features of the mechanical properties of a unidirectional single layer are described. The common calculation equations for determining the engineering constants are briefly displayed. The deformation behavior of laminates is explained. The transformation laws, based on tensor and vector notations of Voigt and Kelvin, are presented. The Kelvin form of vector notation leads to a tensor-invariant representation and, consequently, is favorable for use in the optimization problems.

Evaluation of strength for the unidirectional material as well as of the layered composite can be carried out with two main approaches: phenomenological and micromechanical. Chapter 2 exploits the traditional phenomenological approach for evaluation of strength. The traditional phenomenological approach uses local stress or average stress failure criteria. This approach is based on definite approximations of the experimental strength data. The phenomenological approach, however, requires a large amount of empirical data to perform a reliable forecast for specific materials and laminates. Chapter 2 summarizes the basic information about failure criteria. The common phenomenological approaches are presented in matrix forms, which are suited for later use in optimization problems. The main focus is fixed on the tensor-polynomial criteria because these criteria allow tensor-invariant transformations of fracture formulas between different coordinate systems. From the viewpoint of optimization, the tensor-invariant transformations are highly beneficial. The anisotropic failure criteria are presented with the traditional Voigt's notation and Kelvin's tensor notations. Further, the tensor rotation of the quadratic tensor-polynomial criteria is explained.

An alternative approach describes the failure of the composite material from the micromechanical perspective. The strength characteristics of a composite are explained in Chapter 3 as the fracture phenomena for fibers and matrix. For application of this method, the micromechanical models of relevant materials are developed. Using these models, the damage of the composite is modeled and qualitatively compared to the experiment. This micromechanical approach provides qualitative understanding of the mechanism of fracture and proposes the methods for optimization and for proper material selection. Two main effects are displayed: crack bridging cross to fiber direction and debonding of fibers in their direction. These two damage mechanisms are frequent for unidirectional fiber bundles. The parametric dependences of crack toughness are derived in closed form.

Chapter 4 presents the common principles for the optimal design of structures, which are based on optimization of the anisotropic properties of materials. Problems of maximizing the ultimate rupture load, total stiffness and stored energy capacity are formulated. The optimal orientation of the anisotropy axes in the structure elements is discussed. The method of study of this problem is based on tensor notation in space and vector and tensor notations in the six-dimensional stress space. Optimality conditions are derived using different notations of failure criteria. Special cases of the torsion and bending of shafts are considered. Bilateral achievable estimations are obtained. The conditions for achieving the upper and lower bounds agree with the necessary optimality conditions. Three main problems of the optimization for anisotropic materials are investigated in Chapter 4. The first problem studies the optimal placement of the anisotropy axes, which maximize or minimize the elastic energy for given external loads. The maximization of an elastic energy for the given loads leads to the most resilient design of the structural element, which allows for maximal flexibility. The minimization of an elastic energy for the given loads leads to the stiffest design of the structural element, with the lowest flexibility. The second problem searches the orientation of anisotropic properties to minimize the solely failure number of the structural member disrespectable to the stiffness. This optimization aim is interesting for load bearing structural elements. The third problem examines the orientation of anisotropic properties to maximize stored elastic energy in an element but limiting the failure number of the structural member. This optimization aim is interesting for the energy harvesters or spring elements.

Chapter 5 surveys the torsion and bending of moment-carrying elements of the vehicle powertrain. The solid and thin-walled beams of closed section are studied. This chapter exposes the model of the thin-walled anisotropic driveshaft with a closed profile of a general cross-sectional form. The model is based on the adoption of kinematic hypotheses and thin anisotropic shell equations. The kinematic hypotheses describe the strain state of the cross-section of the shaft. It is assumed that the profile of the normal cross-section of the thin-walled driveshaft is not distorted but is only rotated and displaced in space as a solid body. Using the accepted kinematic hypotheses, we calculate the strains and curvatures of the middle surface of the rod. The strain equations are obtained by varying the potential energy functional of the semi-momentless theory of anisotropic shells. As an example, we solve the problem of determining the critical load for a thin-walled rod in axial compression. The proposed model can be used in the design of driveshafts made of composite materials. The twist and flexure of the members are investigated for different orientation of anisotropy axes with respect to the driveshaft axis and for different forms of cross-section. The torsional stability is studied using the

Greenhill method. The results are applicable to the components of the powertrain and other composite driveshafts of ground vehicles.

Chapter 6 reviews the principal types of wheel suspensions of the vehicles. The structural elements of wheel suspensions, except of spring elements, remain in this context solid. The elementary introduction of vehicle dynamics provides the basis for subsequent accounting of flexible structural elements of wheel suspensions and study of the coupling effects. Models of dynamics for the vehicles with the solid, stiff elements of suspensions are presented in Chapter 6. Parameters are usually assumed to be known or can be estimated. The aim of modeling is to obtain a mathematical-analytical description of the dynamic behavior of a vehicle with coupling between its principal degrees of freedom, but neglecting elasticity of axle components.

Chapter 7 explains the dynamics of a vehicle, accounting for the elasticity of structural components. An important example of the highly flexural and resilient wheel suspension delivers the twist-beam axle. This type of suspension is widespread among front wheel drive cars, almost all compact and subcompact models. At the beginning, the basic designs of cars with twist-beam axles are labeled, the differentiations from other wheel suspensions are identified and the advantages and disadvantages of this type of suspension are pinpointed. The influences of the flexible axle on driving dynamics are figured. As an example, the coupling effect during cornering is given. The suspension elements made of composite materials possess some distinctive mechanical features, which impact the driving dynamics of a vehicle. First, the anisotropy provides an additional coupling between the degrees of freedom and supplementary interaction between the wheels and body. Second, the lower modulus of elasticity of material relative to its strength leads to an excessive elastic response of suspension elements. Third, the ratios of flexibilities in different directions have an effect on the steering behavior of a vehicle. The models accounting for these properties of composite materials on lateral dynamics are developed. The unidirectional layers are used for the laminate structure of the cross beam. The cross-section for the cross beam of the countersteering twist-beam axle is calculated.

Chapter 8 explains the optimal design of anisotropic helical and conical springs for automotive applications. Helical composite springs possess the form of spiral of a uniform cross-section. A distance between the successive coils of a spring is maintained so that the axis of the wire forms a helix. The design formulas for anisotropic helical springs are developed and the optimization problems are studied in closed form. In Chapter 8, it is proved that the optimal wire shape, determined from the certain equal stress condition, guarantees the lowest possible mass of spring. This mass depends only on the ultimate allowable stress for the spring material, the load at full stroke and the spring stiffness. This is an important milestone for comparison of different spring designs and spring materials.

Conical springs (also known as Belleville washers) are shallow conical rings that are subjected to axial loads. Normally, the ring thickness is constant and the applied load is evenly distributed over the upper inside edge and lower outside edge. The structural models for the anisotropic conical springs are developed and formulas for deformation of the anisotropic conical springs presented. Based on the analytical formulas, the optimization of the conical springs made of composite materials is performed.

Analysis of space-curved beams and, particularly, the helical spring is grounded on Kirchhoff's equations. In Chapter 9, the helical composite spring is substituted with a conceivable flexible beam, which coincides with the axis of the helix. The "equivalent beam" equations are significantly easier to handle than Kirchhoff's. This imaginable equivalent beam possesses identical mechanical features to the helical spring itself. Its bending, torsion and compression stiffness are equal to the corresponding stiffness of a helical spring made of anisotropic material. This beam is known also as an "equivalent column."

Chapter 10 examines the design and simulation of leaf springs that primarily withstand bending. A leaf composite spring assembly for a wheel suspension of a motor vehicle comprises a leaf spring of fiber-reinforced plastics for resiliently supporting a wheel carrier of the motor vehicle. Common types of leaf springs are studied and the optimization principles are demonstrated.

Chapter 11 overviews the corrugated or meander springs. A meander-shaped spring is the folded form of the classical elongated leaf springs. The meander spring possesses a shorter dimension in the length direction and allows for assembly in the independent type of passenger car suspensions. The meander spring is thought to be the replacement for helical springs, because the box dimensions of both spring types are comparable. On the other hand, the manufacturing of a meander spring of composite material is unpretentious in comparison to the awkward fabrication of the helical spring.

In Chapter 12, the time-dependent performance of spring elements made of composite materials, which are exposed to steady heavy loads, is studied. Common anisotropic creep laws are implemented for composite materials. For basic spring elements, the anisotropic Norton–Bailey constitutive models with anisotropic creep functions are studied. The anisotropy of the material is accounted for by the Altenbach approach, which applies the quadratic creep potential in a similar way to the common Mises–Hill plastic criterion. Analytical models are developed for the relaxation of stresses and creep under constant load. Closed-form solutions of the analytical models of creep and relaxation of composite shafts and springs in twist and bending states are presented.

Target Audience of the Book

This book was written as an accompanying script for the courses "Applied Mechanics of the Automobile," "Automotive Engineering, Chassis, II, III," "Structural Optimization in Automotive Engineering" and "Powertrain Modeling and Optimization," which have been delivered by the author at the University of Siegen, North-Rhine-Westphalia, Germany, since 2001.

This book is recommended for engineers dealing with design and development of chassis and drivetrains, graduating from automotive or mechanical engineering courses in technical high school, or in other higher engineering schools. Researchers working on elastic elements and energy harvesting equipment will also find a general review for technology and simulation of composite structural elements.

The content of this book is logically related to the work *Durability of Springs* by the same author and could be seen as a substantial extension in the direction of composite materials. That book (Kobelev 2018) exhibits the mechanics of the elastic elements made of steel alloys with a focus on metal springs for the automotive industry. Contrarily, this book investigates the distinguishing features of composites, such their anisotropy, inhomogeneity, load direction dependence, stress-coupling and stacking capabilities; all together, outstanding design possibilities on one hand, and in-born deficiencies on the other.

References

Argon, A.S. (1974). Statistical aspects of fracture. In: *Fracture and Fatigue: Composite Materials* (ed. L.J. Boutman and R.H. Krock), 153. New York: Academic Press.
Ashby, M.F. (2010). *Materials Selection in Mechanical Design*, 4e. Butterworth-Heinemann.
Brooks, R. (2000). Composites in automotive applications: design. In: *Comprehensive Composite Materials*, vol. 6. Elsevier Science Ltd. ISBN: 0-080437249.
ECSS-E-HB-32-20 (2011). *Structural Materials Handbook*. Noordwijk, The Netherlands: Space Engineering ESA-ESTEC Requirements & Standards Division.
Elmarakbi A., Azoti W.L. (2015) Novel composite materials for automotive applications: concepts and challenges for energy-efficient and safe vehicles, in: 10th International Conference on Composite Science and Technology ICCST/10, A.L. Araújo, J.R. Correia, C.M. Mota Soares, et al. (eds), IDMEC 2015.
Hayashi, T. (2000). Application of MMCS to engine cylinder blocks and brake disks. In: *Comprehensive Composite Materials*, vol. 6. Elsevier Science Ltd. ISBN: 0-080437249.
Ishikawa, T., Amaoka, K., Masubuchi, Y. et al. (2018). Overview of automotive structural composites technology developments in Japan. *Compos. Sci. Technol.* 155: 221–246.
Kedward, K.T. (2000). Generic approaches and issues for structural composite design and application. In: *Comprehensive Composite Materials*, vol. 6. Elsevier Science Ltd. ISBN: 0-080437249.
Kelly, A. and Zweben, C. (2000). *Comprehensive Composite Materials*. Pergamon: Elsevier Ltd.
Kleinholz, R., Mathes, V., and Witten, E. (eds.) (2010). *Handbuch Faserverbundkunststoffe*. Wiesbaden: Vieweg+Teubner GWV Fachverlage GmbH.
Kobelev, V. (2018). *Durability of Springs*. Wien: Springer International Publishing.
Lu, C. and Pilla, S. (2014a). *Design of Automotive Composites, PT-164*. SAE International.
Lu, C. and Pilla, S. (2014b). *CAE Design and Failure Analysis of Automotive Composites, PT-166*. SAE International.
Lu, C. and Pilla, S. (2014c). *Biocomposites in Automotive Applications, PT-165*. SAE International.
Miravete, A. (1996). *Optimisation of Composite Structures Design*. Woodhead Publishing. ISBN: 978-1-85573-208-7.
Neitzel M., Mitschang P., Breuer U. (2014). *Handbuch Verbundwerkstoffe, Werkstoffe, Verarbeitung, Anwendung*, Carl Hanser Verlag, München/Wien.
Neubrand, J. (2014). Possibilities of Coil Springs and fiber-reinforced suspension parts. In: *Encyclopedia of Automotive Engineering*. John Wiley & Sons, Ltd.
Njuguna, J. (ed.) (2016). *Lightweight Composite Structures in Transport Design, Manufacturing, Analysis and Performance*. Kidlington, Oxford: Woodhead Publishing.
Nomura, M. (2000). Automobile bumpers. In: *Comprehensive Composite Materials*, vol. 6. Elsevier Science Ltd. ISBN: 0-080437249.
O'Rourke, B.P. (2000). Formula 1 applications of composite materials. In: *Comprehensive Composite Materials*, vol. 6. Elsevier Science Ltd. ISBN: 0-080437249.
Peters, S. (ed.) (1998). *Handbook of Composites*, 2e. London: Chapman and Hall.
Tucker, N. and Lindsey, K. (eds.) (2002). *An Introduction to Automotive Composites*. Shawbury, UK: Rapra Technology Limited.

About the Companion Website

This book is accompanied by a companion website:

www.wiley.com/go/kobelev/automotive_suspensions

The website includes:

MAPLE worksheets for analytical study of the problems studied in the book. The worksheets could be used for the classes on optimization and composite materials in technical universities.

Scan this QR code to visit the companion website:

1

Elastic Anisotropic Behavior of Composite Materials

1.1 Anisotropic Elasticity of Composite Materials

1.1.1 Fourth Rank Tensor Notation of Hooke's Law

Fiber composites consist of fibers with very high stiffness and strength that are embedded in a matrix of plastic. Fibers alone can absorb high tensile forces but cannot withstand bending or compression loads. In order to achieve a desired spectrum of properties that could not be achieved individually by each component, several material components are combined in a suitable form and spatial distribution. When plastics are combined with reinforcing materials, the aim is to achieve lightweight construction of highly stressed structural parts by increasing stiffness, hardness and strength. The main problem of material optimization lies in the inadequate or missing dependencies of such parameters as loading limit, fracture toughness and critical stress intensity factor from design variables (such as fiber diameter, fiber elasticity modulus and matrix and distance between fibers). The basic task is to obtain these dependencies in an analytical form.

The components of plastics can be relatively brittle (thermosetting reaction resins) or rather flexible (thermoplastics). Only through the combination of fibers and plastics and the firm connection of the plastic matrix to the fibers can high-strength components, such as aircraft and vehicle parts, be produced. For material laws of fiber-reinforced composites, the literature provides a broad knowledge base (e.g. Moser 1992; Chou 1990; Nettles 1994; Gibson 2016). In this chapter, the necessary information for modeling and optimization of structural components in the automotive powertrain and suspension will be provided.

The most general anisotropic form of linearly elastic constitutive relations is given by the generalized Hooke's law:

$$\varepsilon_{ij} = S_{ijkl}\sigma_{kl}, \boldsymbol{\varepsilon} = \mathbf{S} \cdot \cdot \boldsymbol{\sigma}, \tag{1.1}$$

$$\sigma_{ij} = C_{ijkl}\varepsilon_{kl}, \boldsymbol{\sigma} = \mathbf{C} \cdot \cdot \boldsymbol{\varepsilon}. \tag{1.2}$$

The tensor $\mathbf{S} = [S_{ijpq}]$ is the compliance tensor of rank four, $i, j,.. = 1,.., 3$. The summation convention is applied such that the repeated indices are implicitly summed over. The elasticity tensor of rank four is $\mathbf{C} = [C_{ijpq}]$. The relationship between the two tensors is:

$$C_{ijkl}S_{klpq} = I_{ijpq}, I_{ijpq} = (\delta_{ip}\delta_{jq} + \delta_{iq}\delta_{jp})/2, \mathbf{C} \cdot \cdot \mathbf{S} = \mathbf{I}. \tag{1.3}$$

Design and Analysis of Composite Structures for Automotive Applications: Chassis and Drivetrain, First Edition. Vladimir Kobelev.

Companion website: www.wiley.com/go/kobelev/automotive_suspensions

In (1.3), $\mathbf{I} = [I_{ijpq}]$ is the fourth rank identity tensor and δ_{jq} is the Kronecker symbol. The number of independent coefficients in the elasticity tensor varies depending on the grade of a material's symmetry. In general, it describes a tensor of rank four, which contains 81 different material specific elastic coefficients. The requirement that stress components occurring in the material are symmetric reduces the number of independent coefficients from a total of 81 to 36 due to the following correlation:

$$C_{ijkl} = C_{ijlk} = C_{jikl} = C_{jilk}. \tag{1.4}$$

1.1.2 Voigt's Matrix Notation of Hooke's Law

The generalized Hookes' law in Voigt's notation is a matrix equation in which the stiffness of the material is represented in the matrix with six rows and columns, and the stresses and strains in the column vectors with six components. According to Eq. (1.2), the generalized Hooke's law in Voigt's notation looks as follows:

$$\boldsymbol{\sigma} = \mathbf{C} \cdot \boldsymbol{\varepsilon}, \text{ or } \sigma_i = C_{ij}\varepsilon_j, \tag{1.5}$$

$$\boldsymbol{\varepsilon} = \mathbf{S} \cdot \boldsymbol{\sigma}, \text{ or } \varepsilon_i = S_{ij}\sigma_j, \tag{1.6}$$

with $i, j = 1, .., 6$ and

$\boldsymbol{\sigma} = [\sigma_{11} = \sigma_1, \sigma_{22} = \sigma_2, \sigma_{33} = \sigma_3, \sigma_{13} = \sigma_4, \sigma_{32} = \sigma_5, \sigma_{21} = \sigma_6]^T$
Voigt's stress vector,
$\boldsymbol{\varepsilon} = [\varepsilon_{11} = \varepsilon_1, \varepsilon_{22} = \varepsilon_2, \varepsilon_{33} = \varepsilon_3, 2\varepsilon_{13} = \varepsilon_4, 2\varepsilon_{32} = \varepsilon_5, 2\varepsilon_{21} = \varepsilon_6]^T$
Voigt's strain vector,

C Voigt's elasticity matrix (6×6).

The elasticity matrix, which establishes the linear relationship between the stresses and distortions under a uniaxial load or stress, consists of 36 coefficients, 21 of which are independent of each other due to symmetry to the main diagonal. The structures of the elasticity and compliance matrices are:

$$\mathbf{C} = \begin{pmatrix} C_{11} & C_{12} & C_{13} & C_{14} & C_{15} & C_{16} \\ & C_{22} & C_{23} & C_{24} & C_{25} & C_{26} \\ & & C_{33} & C_{34} & C_{35} & C_{36} \\ & & & C_{44} & C_{45} & C_{46} \\ & sym. & & & C_{55} & C_{56} \\ & & & & & C_{66} \end{pmatrix} \equiv \begin{pmatrix} C_{1111} & C_{1122} & C_{1133} & C_{1123} & C_{1113} & C_{1112} \\ & C_{2222} & C_{2233} & C_{2223} & C_{2213} & C_{2212} \\ & & C_{3333} & C_{2223} & C_{3313} & C_{3312} \\ & & & C_{2323} & C_{2313} & C_{2312} \\ & sym. & & & C_{1313} & C_{1312} \\ & & & & & C_{1212} \end{pmatrix}, \tag{1.7}$$

$$\mathbf{S} = \begin{pmatrix} S_{11} & S_{12} & S_{13} & S_{14} & S_{15} & S_{16} \\ & S_{22} & S_{23} & S_{24} & S_{25} & S_{26} \\ & & S_{33} & S_{34} & S_{35} & S_{36} \\ & & & S_{44} & S_{45} & S_{46} \\ & sym. & & & S_{55} & S_{56} \\ & & & & & S_{66} \end{pmatrix} \equiv \begin{pmatrix} S_{1111} & S_{1122} & S_{1133} & 2S_{1123} & 2S_{1113} & 2S_{1112} \\ & S_{2222} & S_{2233} & 2S_{2223} & 2S_{2213} & 2S_{2212} \\ & & S_{3333} & 2S_{2223} & 2S_{3313} & 2S_{3312} \\ & & & 4S_{2323} & 4S_{2313} & 4S_{2312} \\ & sym. & & & 4S_{1313} & 4S_{1312} \\ & & & & & 4S_{1212} \end{pmatrix}. \tag{1.8}$$

There are different forms of direction-dependence or anisotropy of reinforced fiber composites.

In the most general case of elastic symmetry, there is one plane of symmetry. In this case, there are 13 elastic coefficients of each of the matrices (1.7) and (1.8). There is one additional relation between these coefficients, so 12 coefficients are independent. For example, if the symmetry plane is the plane 1–2, then the certain coefficients in both matrices (1.7) disappear:

$$\mathbf{C} = \begin{pmatrix} C_{11} & C_{12} & C_{13} & 0 & 0 & C_{16} \\ & C_{22} & C_{23} & 0 & C_{25} & C_{26} \\ & & C_{33} & 0 & 0 & C_{36} \\ & & & C_{44} & C_{45} & 0 \\ & sym. & & & C_{55} & 0 \\ & & & & & C_{66} \end{pmatrix} \equiv \begin{pmatrix} C_{1111} & C_{1122} & C_{1133} & 0 & 0 & C_{1112} \\ & C_{2222} & C_{2233} & 0 & C_{2213} & C_{2212} \\ & & C_{3333} & 0 & 0 & C_{3312} \\ & & & C_{2323} & C_{2313} & 0 \\ & sym. & & & C_{1313} & 0 \\ & & & & & C_{1212} \end{pmatrix}, \tag{1.9}$$

$$\mathbf{S} = \begin{pmatrix} S_{11} & S_{12} & S_{13} & 0 & 0 & S_{16} \\ & S_{22} & S_{23} & 0 & S_{25} & S_{26} \\ & & S_{33} & 0 & 0 & S_{36} \\ & & & S_{44} & S_{45} & 0 \\ & sym. & & & S_{55} & 0 \\ & & & & & S_{66} \end{pmatrix} \equiv \begin{pmatrix} S_{1111} & S_{1122} & S_{1133} & 0 & 0 & 2S_{1112} \\ & S_{2222} & S_{2233} & 0 & 2S_{2213} & 2S_{2212} \\ & & S_{3333} & 0 & 0 & 2S_{3312} \\ & & & 4S_{2323} & 4S_{2313} & 0 \\ & sym. & & & 4S_{1313} & 0 \\ & & & & & 4S_{1212} \end{pmatrix}. \tag{1.10}$$

Due to the prevailing symmetries, however, the dominating case is the orthotropy. Here, there are three orthogonal planes of symmetry. If the intersection lines of the underlying symmetry planes are used as a coordinate system, the shear stresses and strains are completely decoupled in this case. If the material contains symmetries, the number of independent constants is abridged. Depending on the position and number of symmetry planes, different anisotropy cases are distinguished. For orthotropy only nine independent constants are required. For example, if the distances between fibers in unidirectional composite are distinct in two directions, the material will be orthotropic (Figure 1.1). The elasticity matrix in this case is as follows (Eqs. (2.107)–(2.108) and Vannucci 2018):

$$\mathbf{S} = \begin{pmatrix} S_{11} & S_{12} & S_{13} & 0 & 0 & 0 \\ S_{12} & S_{22} & S_{23} & 0 & 0 & 0 \\ S_{13} & S_{23} & S_{33} & 0 & 0 & 0 \\ & & & S_{44} & 0 & 0 \\ & sym & & & S_{55} & 0 \\ & & & & & S_{66} \end{pmatrix} = \mathbf{C}^{-1}, \tag{1.11}$$

$$\mathbf{C} = \begin{pmatrix} C_{11} & C_{12} & C_{13} & 0 & 0 & 0 \\ C_{12} & C_{22} & C_{23} & 0 & 0 & 0 \\ C_{13} & C_{23} & C_{33} & 0 & 0 & 0 \\ & & & C_{44} & 0 & 0 \\ & sym & & & C_{55} & 0 \\ & & & & & C_{66} \end{pmatrix},$$

$$\mathbf{C} \equiv \begin{pmatrix} \frac{S_{22}S_{33}-S_{23}^2}{S} & \frac{S_{13}S_{32}-S_{12}S_{33}}{S} & \frac{S_{12}S_{23}-S_{13}S_{22}}{S} & 0 & 0 & 0 \\ & \frac{S_{11}S_{33}-S_{13}^2}{S} & \frac{S_{21}S_{13}-S_{23}S_{11}}{S} & 0 & 0 & 0 \\ & & \frac{S_{11}S_{22}-S_{12}^2}{S} & 0 & 0 & 0 \\ & & & S_{44}^{-1} & 0 & 0 \\ & sym. & & & S_{55}^{-1} & 0 \\ & & & & & S_{66}^{-1} \end{pmatrix},$$

$$S = S_{11}S_{22}S_{33} - S_{11}S_{23}^2 - S_{22}S_{13}^2 - S_{33}S_{12}^2 + 2S_{13}S_{23}S_{31}. \tag{1.12}$$

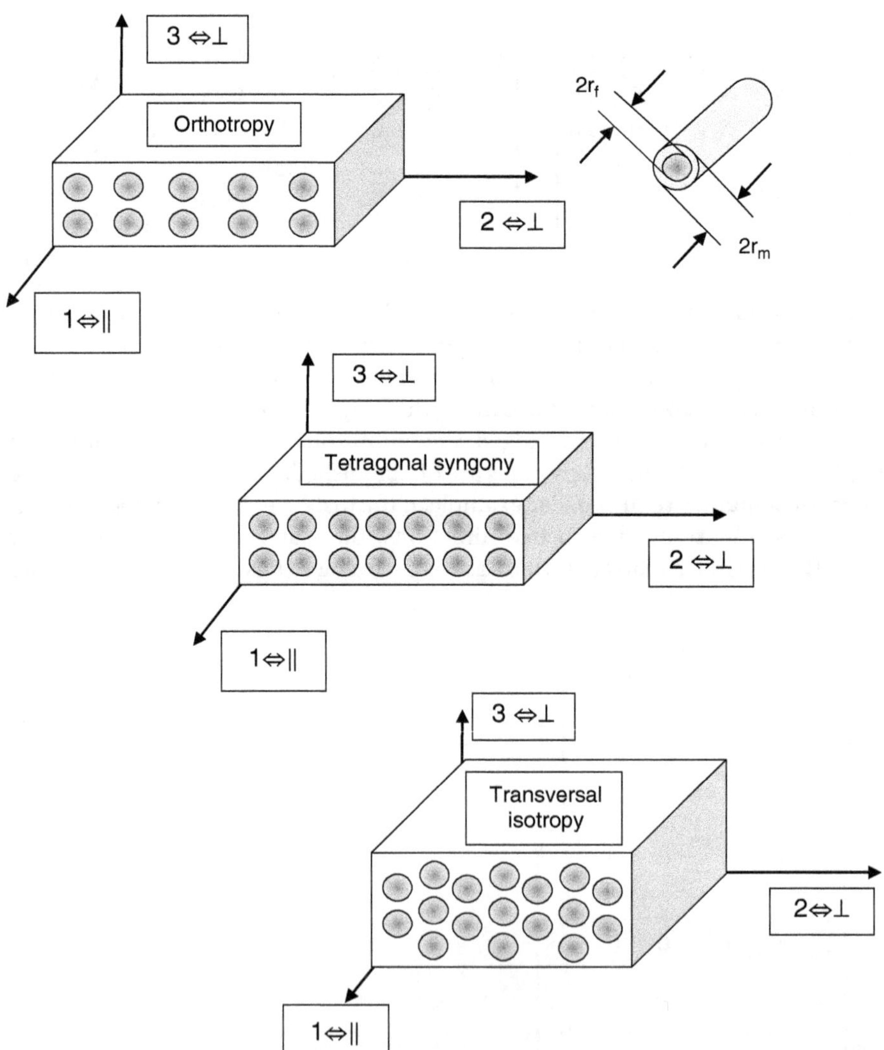

Figure 1.1 Coordinate system and elastic symmetry.

The special case of orthotropy is the tetragonal elastic syngony with six elastic constants:

$$\mathbf{S} = \begin{pmatrix} S_{11} & S_{12} & S_{13} & 0 & 0 & 0 \\ S_{12} & S_{11} & S_{13} & 0 & 0 & 0 \\ S_{13} & S_{13} & S_{33} & 0 & 0 & 0 \\ & & & S_{44} & 0 & 0 \\ & sym & & & S_{44} & 0 \\ & & & & & S_{66} \end{pmatrix} = \mathbf{C}^{-1}, \mathbf{C} = \begin{pmatrix} C_{11} & C_{12} & C_{13} & 0 & 0 & 0 \\ C_{12} & C_{11} & C_{13} & 0 & 0 & 0 \\ C_{13} & C_{13} & C_{33} & 0 & 0 & 0 \\ & & & C_{44} & 0 & 0 \\ & sym & & & C_{44} & 0 \\ & & & & & C_{66} \end{pmatrix}. \quad (1.13)$$

Such symmetry appears in composites if the fibers arranged in a preferable orthogonal mosaic pattern with the equal distances between fibers in both directions (Figure 1.1).

The specific case of orthotropy is the transversal isotropy. Transversal isotropy presents when an infinite number of symmetries exist perpendicular to an isotropic plane. In the unidirectional material, the isotropic plane is normal to fiber direction. For example, the pattern of fibers allows threefold rotational symmetry. Each fiber after a rotation of 120° comes into the place of another fiber. Otherwise, if the fibers are chaotically arranged in the isotropic plane, directional dependence disappears as well. In these cases, the number of elastic constants reduces to five, because the additional constraint appears due to symmetry along axis 3:

$$C_{66} = C_{11} - C_{12}.$$

To complete the information given here, we will briefly discuss the elasticity tensor **C** for isotropic materials with no directional dependence. The isotropic material requires only the modulus of elasticity E and the contraction number ν:

$$S_{11} = 1/E, S_{12} = -\nu/E, S_{44} = 1/G = 2(1+\nu)/E.$$

1.1.3 Kelvin's Matrix Notation of Hooke's Law

Kelvin's notation was proposed by Kelvin (1856, 1878). The components of fourth rank elasticity tensors in three dimensions can be arranged into a matrix of six dimensions:

$$\hat{\boldsymbol{\sigma}} = \hat{\mathbf{C}} \cdot \hat{\boldsymbol{\varepsilon}}, \text{or } \hat{\sigma}_i = \hat{C}_{ij}\hat{\varepsilon}_j, \quad (1.14)$$

$$\hat{\boldsymbol{\varepsilon}} = \hat{\mathbf{S}} \cdot \hat{\boldsymbol{\sigma}}, \text{or } \hat{\varepsilon}_j = \hat{S}_{ij}\hat{\sigma}_i, i,j = 1,..,6, \quad (1.15)$$

with

$\hat{\boldsymbol{\sigma}} = [\sigma_{11} = \hat{\sigma}_1, \sigma_{22} = \hat{\sigma}_2, \sigma_{33} = \hat{\sigma}_3, \sqrt{2}\sigma_{13} = \hat{\sigma}_4, \sqrt{2}\sigma_{32} = \hat{\sigma}_5, \sqrt{2}\sigma_{21} = \hat{\sigma}_6]^T$ Kelvin's stress vector,

$\hat{\boldsymbol{\varepsilon}} = [\varepsilon_{11} = \hat{\varepsilon}_1, \varepsilon_{22} = \hat{\varepsilon}_2, \varepsilon_{33} = \hat{\varepsilon}_3, \sqrt{2}\varepsilon_{13} = \hat{\varepsilon}_4, \sqrt{2}\varepsilon_{32} = \hat{\varepsilon}_5, \sqrt{2}\varepsilon_{21} = \hat{\varepsilon}_6]^T$ Kelvin's strain vector,

$\hat{\mathbf{C}}$ Kelvin's elasticity tensor (6×6):

$$\hat{\mathbf{C}} = \begin{pmatrix} \hat{C}_{11} & \hat{C}_{12} & \hat{C}_{13} & \hat{C}_{14} & \hat{C}_{15} & \hat{C}_{16} \\ & \hat{C}_{22} & \hat{C}_{23} & \hat{C}_{24} & \hat{C}_{25} & \hat{C}_{26} \\ & & \hat{C}_{33} & \hat{C}_{34} & \hat{C}_{35} & \hat{C}_{36} \\ & & & \hat{C}_{44} & \hat{C}_{45} & \hat{C}_{46} \\ & sym. & & & \hat{C}_{55} & \hat{C}_{56} \\ & & & & & \hat{C}_{66} \end{pmatrix},$$

$$\hat{\mathbf{C}} \equiv \begin{pmatrix} C_{1111} & C_{1122} & C_{1133} & \sqrt{2}C_{1123} & \sqrt{2}C_{1113} & \sqrt{2}C_{1112} \\ & C_{2222} & C_{2233} & \sqrt{2}C_{2223} & \sqrt{2}C_{2213} & \sqrt{2}C_{2212} \\ & & C_{3333} & \sqrt{2}C_{2223} & \sqrt{2}C_{3313} & \sqrt{2}C_{3312} \\ & & & 2C_{2323} & 2C_{2313} & 2C_{2312} \\ & sym. & & & 2C_{1313} & 2C_{1312} \\ & & & & & 2C_{1212} \end{pmatrix}, \tag{1.16}$$

$\hat{\mathbf{S}}$ Kelvin's compliance tensor (6×6):

$$\hat{\mathbf{S}} = \begin{pmatrix} \hat{S}_{11} & \hat{S}_{12} & \hat{S}_{13} & \hat{S}_{14} & \hat{S}_{15} & \hat{S}_{16} \\ & \hat{S}_{22} & \hat{S}_{23} & \hat{S}_{24} & \hat{S}_{25} & \hat{S}_{26} \\ & & \hat{S}_{33} & \hat{S}_{34} & \hat{S}_{35} & \hat{S}_{36} \\ & & & \hat{S}_{44} & \hat{S}_{45} & \hat{S}_{46} \\ & sym. & & & \hat{S}_{55} & \hat{S}_{56} \\ & & & & & \hat{S}_{66} \end{pmatrix} \equiv \begin{pmatrix} S_{1111} & S_{1122} & S_{1133} & \sqrt{2}S_{1123} & \sqrt{2}S_{1113} & \sqrt{2}S_{1112} \\ & S_{2222} & S_{2233} & \sqrt{2}S_{2223} & \sqrt{2}S_{2213} & \sqrt{2}S_{2212} \\ & & S_{3333} & \sqrt{2}S_{2223} & \sqrt{2}S_{3313} & \sqrt{2}S_{3312} \\ & & & 2S_{2323} & 2S_{2313} & 2S_{2312} \\ & sym. & & & 2S_{1313} & 2S_{1312} \\ & & & & & 2S_{1212} \end{pmatrix}. \tag{1.17}$$

The components of Kelvin's matrices of the elasticity and compliance will change under a change of coordinate system following the tensor transformation rule. The components of Kelvin's matrices change as the components of the second-rank tensor change and this notable feature is significant for optimization methods, as recognized in Mehrabadi and Cowin (1990).

The symmetry of Kelvin's matrices for orthotropic, tetragonal syngony and transversal isotropy is analogous to those of Voigt's matrices (Table 1.1).

Table 1.1 Coefficients of Voigt's and Kelvin's matrices for orthotropic materials.

	Kelvin's notation	Voigt's notation
$i=1,2,3$	$\hat{S}_{ii} = S_{ii} = \frac{1}{E_i}$, $\sigma_{11} = \hat{\sigma}_1 = \sigma_1$, $\sigma_{22} = \hat{\sigma}_2 = \sigma_2$, $\sigma_{33} = \hat{\sigma}_3 = \sigma_3$,	$\varepsilon_{11} = \hat{\varepsilon}_1 = \varepsilon_1$, $\varepsilon_{22} = \hat{\varepsilon}_2 = \varepsilon_2$, $\varepsilon_{33} = \hat{\varepsilon}_3 = \varepsilon_3$,
$i,j = 1,2,3,$ $i \neq j$	$\hat{S}_{ji} = S_{ji} = -\frac{\nu_{ij}}{E_i}$,	
$i,j = 1,2,3,$ $i \neq j,$ $k = 4,5,6$	$\hat{S}_{kk} = \frac{1}{2G_{ij}}$ $\sqrt{2}\sigma_{13} = \hat{\sigma}_4, \sqrt{2}\varepsilon_{13} = \hat{\varepsilon}_4,$ $\sqrt{2}\sigma_{32} = \hat{\sigma}_5, \sqrt{2}\varepsilon_{32} = \hat{\varepsilon}_5,$ $\sqrt{2}\sigma_{21} = \hat{\sigma}_6, \sqrt{2}\varepsilon_{21} = \hat{\varepsilon}_6.$	$S_{kk} = \frac{1}{G_{ij}}$, $\sigma_{13} = \sigma_4, 2\varepsilon_{13} = \varepsilon_4,$ $\sigma_{32} = \sigma_5, 2\varepsilon_{32} = \varepsilon_5,$ $\sigma_{21} = \sigma_6, 2\varepsilon_{21} = \varepsilon_6.$

1.2 Unidirectional Fiber Bundle

1.2.1 Components of a Unidirectional Fiber Bundle

The anisotropy of a composite material can be defined as the difference in physical properties, such as the modulus and Poisson's ratio, when the material is loaded in different directions. The modern and comprehensive discussion of this topic is given by Vannucci (2018).

The properties of the composite material are particularly dependent on the proportion of fiber and matrix material in the end product. The fiber volume content V_f describes the ratio of the volume fraction of the fiber material to the total volume of the composite. The achievable fiber volume ratio and the fiber orientation are strongly dependent on the manufacturing process used.

The radius of the fibers r_f and the radius r_m of the surrounding cylinders made of matrix, or resin, material. The imaginary radius r_m is determined from the volume proportions of the fibers, such that the volume content of fiber and matrix materials to be preserved. Explicitly, the volume fractions of fibers V_f and matrix V_m must be directly proportional to the cross-sectional areas of fibers and matrix, respectively:

$$s_f/s_m = V_f/V_m, V_m = 1 - V_f, \tag{1.18}$$

$$s_f = \pi\, r_f^2, s_m = \pi(r_m^2 - r_f^2). \tag{1.19}$$

There is the following relation between the radii of the fibers r_f and the radii of the matrix cylinders r_m:

$$r_m = r_f/\sqrt{V_f}. \tag{1.20}$$

For the unidirectional layer, some assumptions for idealization are made that are required for the calculation of the mechanical properties. It is assumed that the fibers are ideally straight and parallel in one direction without interruption. Furthermore, the fibers are evenly distributed over the cross-section and adhere ideally to the matrix; that is, no shifts at the fiber-matrix interface occur during loading.

1.2.2 Elastic Properties of a Unidirectional Fiber Bundle

With knowledge of the special properties of fiber and matrix material and the structure of the unidirectional layer, it is easy to imagine that the unidirectional layer has anisotropic, that is, direction-dependent, mechanical properties. The stiffness of the unidirectional layer in the longitudinal direction of the fibers under tensile load is naturally many times higher than in the transverse direction. In the longitudinal direction of the fibers, the mechanical properties also depend on the direction of loading; that is, the strength properties are not as good under compressive loading as under tensile loading.

A unidirectional reinforced material with macroscopically homogeneous modules of elasticity is transversal isotropic. If such a fiber composite material also has an even distribution of reinforcing fibers, this results in a direction-independent modulus of elasticity normal to the fiber direction. For the transverse isotropy, five material constants are sufficient to characterize the stiffness behavior of the composite. With a unidirectional layer, the isotropic plane is normal to the longitudinal direction of the fiber; that is, all sections normal to the isotropic plane have the same properties. The fiber composite material is modeled from a series of elastic round cylinders of unlimited length, which are located in a cylindrical, tubular, elastic matrix (Figure 1.1).

For a complete description of the linear elasticity law for the unidirectional layer in space, only five independent constants are necessary. The engineering constants are (Figure 1.1):

E_L	longitudinal modulus of elasticity in the "1" direction,
E_T	transverse modulus of elasticity in the "2" and "3" directions,
G_{TL}	shear modulus in ("1–2") and ("1–3") planes,
G_{TT}	shear modulus in the ("2–3") plane,
ν_{TL}	Poisson's ratio in ("1–2") and ("1–3") planes,
ν_{TT}	Poisson's ratio in the ("2–3") plane.

The engineering constants are also known as basic elasticity values. The different experimental methods are available for determining the basic elasticity values. For our purposes, as well as for the introductory design of components, estimation of basic elasticity values could be done by means of analytical formulas as follows.

For a spatial unidirectional layer with known engineering constants, the law of elasticity is formulated with engineering constants; that is, the modulus of elasticity, shear modulus and transverse contraction numbers provide the compliance matrix in Voigt's notation:

$$\mathbf{S}^{(0)} = \begin{pmatrix} 1/E_L & -\nu_{TL}/E_L & -\nu_{TL}/E_L & & & \\ -\nu_{TL}/E_L & 1/E_T & -\nu_{TT}/E_T & & & \\ -\nu_{TL}/E_L & -\nu_{TT}/E_T & 1/E_T & & & \\ & & & 1/G_{TT} & & \\ & & & & 1/G_{TL} & \\ & & & & & 1/G_{TL} \end{pmatrix}. \tag{1.21}$$

The compliance matrix in Kelvin's notation reads:

$$\hat{\mathbf{S}}^{(0)} = \begin{pmatrix} 1/E_L & -\nu_{TL}/E_L & -\nu_{TL}/E_L & & & \\ -\nu_{TL}/E_L & 1/E_T & -\nu_{TT}/E_T & & & \\ -\nu_{TL}/E_L & -\nu_{TT}/E_T & 1/E_T & & & \\ & & & 1/2G_{TT} & & \\ & & & & 1/2G_{TL} & \\ & & & & & 1/2G_{TL} \end{pmatrix}. \tag{1.22}$$

1.2.3 Effective Elastic Constants of Unidirectional Composites

There are several approximate relationships between the modules of fibers and matrix and the homogenized, effective modules of elasticity of the composite material (Hill 1963; ECSS-E-HB-32-20 2011, Schürmann 2007; Younes et al. 2012). The formulas

Table 1.2 Effective modules of unidirectional composite material (Schürmann 2007).

Effective modules of a unidirectional composite material	
Modulus of elasticity	$E_L = E_f V_f + E_m V_m$
	$E_T = \frac{E_m}{1-\nu_m^2} \cdot \frac{1+0.85 \cdot V_f^2}{V_m^{1.25} + \frac{E_m}{E_{f.T}} \cdot \frac{V_f}{1-\nu_m^2}}$
Shear modulus	$G_{TL} = \frac{G_m(1+0.4\sqrt{V_f})}{V_m^{1.45} + V_f G_m / G_{f.TL}}$
	$G_{TT} = \frac{E_T}{2+2\nu_{TT}}$
Poisson coefficient	$\nu_{\mathrm{TL}} = V_f \cdot \nu_{fTL} + V_m \cdot \nu_m$
	$\nu_{\mathrm{TT}} = \nu_{fTT} V_f + \nu_m V_m \frac{1+\nu_m - \nu_{TL} E_m / E_L}{1-\nu_m^2 + \nu_m \nu_{TL} E_m / E_L}$

Table 1.3 Input values for calculation of effective modules.

$\nu_{f.\mathrm{TL}}$	Transverse, longitudinal, transverse Poisson' ratio of fibers
ν_{m}	Poisson' ratio of matrix (resin)
$\nu_{f.TT}$	Transverse Poisson' ratio of fibers
$\nu_{m,\,eff}$	Effective matrix cross Poisson' ratio
ν_{LT}	Poisson's ratio
G_m	Shear modulus of the matrix (resin)
$G_{f.\,TL}$	Transverse longitudinal shear modulus of the fiber
$E_{f.\,T}$	Transverse modulus of elasticity of the fiber
E_f	Longitudinal modulus of elasticity of the fiber
E_m	Modulus of elasticity of the matrix (resin)
V_f	Fiber volume content
$V_m = 1 - V_f$	Matrix volume content

(Schürmann 2007) provide estimations of effective elastic constants, especially for a high volume concentration of fibers. These expressions are given in Tables 1.2 and 1.3.

For the fiber volume content of $V_f = 0.6$, the effective technical constants of the unidirectional layer:

$$\mathrm{E_L} = 45680 MPa, \mathrm{E_T} = 13698 MPa, \mathrm{G_{TT}} = 5101 MPa,$$

$$\mathrm{G_{TL}} = 5384 MPa, \nu_{\mathrm{TL}} = 0.272, \nu_{\mathrm{TT}} = 0.3427.$$

Table 1.4 Elastic constants and densities of matrix and fibers of UD fiberglass.

E_m	3200	E_f	74 000	MPa
G_m	1185	G_f	30 327	MPA
ν_m	0.35	ν_f	0.22	
ρ_m	1800	ρ_f	2000	kg m^{-3}

For this example (Table 1.4), the elasticity and compliance matrices from Eqs. (1.10) and (1.12) of the unidirectional layer in Voigt's notation read:

$$\mathbf{S}^{(0)} = \begin{pmatrix} 21.89 & -5.95 & -5.95 & & & \\ -5.95 & 73.0 & -25.02 & & & \\ -5.95 & -25.02 & 73.0 & & & \\ & & & 196.04 & & \\ & & & & 185.7 & \\ & & & & & 185.7 \end{pmatrix} [10^{-12}\text{Pa}^{-1}] \tag{1.23}$$

$$\mathbf{C}^{(0)} = \begin{pmatrix} 48987 & 6079 & 6079 & & & \\ 6079 & 16276 & 6074 & & & \\ 6079 & 6074 & 16276 & & & \\ & & & 5100 & & \\ & & & & 5384 & \\ & & & & & 5384 \end{pmatrix} [\text{MPa}]. \tag{1.24}$$

1.3 Rotational Transformations of Material Laws, Stress and Strain

The layered view and the directional dependency of the mechanical properties of laminates means that when designing composite material components at least two coordinate systems must be distinguished; the component coordinate system xyz and the fiber coordinate system $x_f y_f z_f$ (Figure 1.2). The direction of a fiber coordinate system coincides with fiber direction, and the z- and z_f-axes are normal to the layer direction.

The direction in the longitudinal direction of a fiber of a unidirectional layer is marked with "x_f." The direction is transverse to the longitudinal direction of the fiber and the

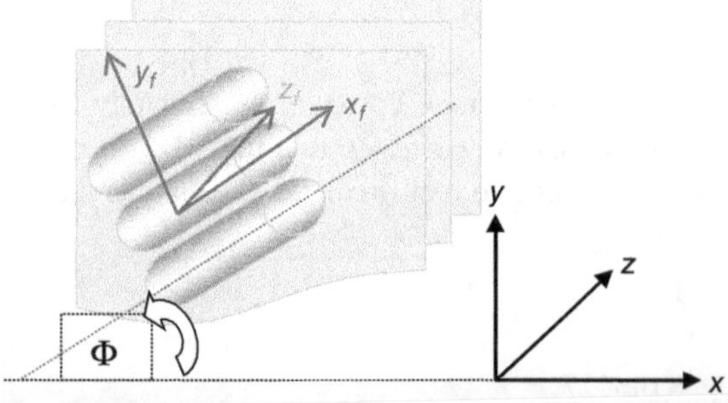

Figure 1.2 Coordinate system associated with fiber direction and rotated coordinate system of the structural component.

layer plane is described using the symbol "y_f." The third spatial direction, z, which is also transverse to the fiber direction but also perpendicular to the layer plane, has to be considered and is called the normal plane. The thickness direction of the laminate is the z-coordinate of the component coordinate system. The fiber angle Φ is defined as the mathematically positive angle between the x-axis and x_f-axes. The rotation of the component coordinate system around the z-axis is described by the rotation angle Φ. For further details, see Vannucci (2018).

1.3.1 Rotation of Fourth Rank Elasticity Tensors

The elasticity and compliance properties for a general anisotropic material are usually expressed in an arbitrarily chosen coordinate system. To exclude the orientational arbitrariness of a coordinate system, a special coordinate system based on a material's intrinsic orientation is needed. For example, the intrinsic coordinate system is associated with the direction of fibers of a unidirectional composite (layer coordinate system). The elasticity and compliance tensor in the layer (intrinsic) coordinate system possess the components:

$$\mathbf{C}^{(0)} = \mathbf{C}(\Phi = 0) = [c_{pqns}], \mathbf{S}^{(0)} = \mathbf{S}(\Phi = 0) = [s_{pqns}]. \tag{1.25}$$

The rotation coordinate system of an anisotropic material modifies the components of elasticity and compliance tensors. The tensors of the fourth rank $\mathbf{S}(\Phi)$ and $\mathbf{C}(\Phi)$ in the rotated coordinate system possess the components:

$$S_{ijkl} = S_{ijkl}(\Phi), C_{ijkl} = C_{ijkl}(\Phi), i, j, k, l = 1, 2, 3. \tag{1.26}$$

The components depend on the orientation of the principal axes of anisotropy, associated with the intrinsic layer coordinate system, relative to the axes of rotated coordinate system. The relations between the tensors in the intrinsic (1.25) and rotated coordinate system (1.26) are:

$$\begin{gathered}\mathbf{S} = \mathbf{t}\cdot\mathbf{t}\cdot\mathbf{S}^{(0)}\cdot\mathbf{t}^T\cdot\mathbf{t}^T, \mathbf{C} = \mathbf{t}\cdot\mathbf{t}\cdot\mathbf{C}^{(0)}\cdot\mathbf{t}^T\cdot\mathbf{t}^T, \\ S_{ijkl} = s_{pqns}t_{ip}t_{jq}t_{kn}t_{ls}, C_{ijkl} = c_{pqns}t_{ip}t_{jq}t_{kn}t_{ls}.\end{gathered} \tag{1.27}$$

The rotation matrix $\mathbf{t} = [t_{lk}]$ is orthogonal:

$$\mathbf{t}\cdot\mathbf{t}^T = \mathbf{I}, t_{sp}t_{pl} = \delta_{sl}. \tag{1.28}$$

For the rotation along the z-axis with the angle Φ:

$$\mathbf{t}(\varphi) = [t_{ij}] = \begin{bmatrix} c & s & 0 \\ -s & c & 0 \\ 0 & 0 & 1 \end{bmatrix}, \text{with } c = \cos\Phi, s = \sin\Phi.$$

The angle Φ can differ from point to point of the structural element. The general rotations in space could be studied using the established methods (Altmann 1986).

1.3.2 Rotation of Elasticity Matrices in Voigt's Notation

It is not possible to perform the rotation of the elasticity matrices with one orthogonal rotation matrix, as required by the tensor transformation law. Two transformation matrices $\mathbf{T}_\sigma$, $\mathbf{T}_\varepsilon$ are required for the rotation of the elasticity matrices in Voigt's notation. These matrices are used also for rotation of stress and strain vectors:

$$\mathbf{C} = \mathbf{T}_\sigma \cdot \mathbf{C}^{(0)} \cdot \mathbf{T}_\sigma^T, \quad \mathbf{S} = \mathbf{T}_\varepsilon \cdot \mathbf{S}^{(0)} \cdot \mathbf{T}_\varepsilon^T, \tag{1.29}$$

In the Eq. (1.29) the following values are used:

$\mathbf{T}_\sigma$	σ-transformation matrix,
$\boldsymbol{\sigma} = \mathbf{T}_\sigma \cdot \boldsymbol{\sigma}^{(0)}$	stress vector in a rotated coordinate system,
$\boldsymbol{\sigma}^{(0)}$	stress vector in an intrinsic coordinate system,
$\mathbf{T}_\varepsilon$	ε-transformation matrix,
$\boldsymbol{\varepsilon} = \mathbf{T}_\varepsilon \cdot \boldsymbol{\varepsilon}^{(0)}$	strain vector in a rotated coordinate system,
$\boldsymbol{\varepsilon}^{(0)}$	strain vector in an intrinsic coordinate system,
$\mathbf{C} = [C_{ij}]$	elasticity matrix in a rotated coordinate system,
$\mathbf{C}^{(0)} = [c_{ij}]$	elasticity matrix in an intrinsic coordinate system,
$\mathbf{S} = [S_{ij}]$	compliance matrix in a rotated coordinate system,
$\mathbf{S}^{(0)} = [s_{ij}]$	compliance matrix in an intrinsic coordinate system.

The matrices $\mathbf{T}_\sigma$, $\mathbf{T}_\varepsilon$ in Eq. (1.29) obey the rules:

$$\mathbf{T}_\sigma^{-1} = \mathbf{T}_\varepsilon^{\mathrm{T}}, \mathbf{T}_\varepsilon^{-1} = \mathbf{T}_\sigma^{\mathrm{T}}.$$

Their components read:

$$\mathbf{T}_\sigma = \begin{pmatrix} c^2 & s^2 & 0 & 0 & 0 & 2cs \\ s^2 & c^2 & 0 & 0 & 0 & -2cs \\ 0 & 0 & 1 & 0 & 0 & 0 \\ 0 & 0 & 0 & c & -s & 0 \\ 0 & 0 & 0 & s & c & 0 \\ -sc & sc & 0 & 0 & 0 & c^2 - s^2 \end{pmatrix}, \tag{1.30}$$

$$\mathbf{T}_\varepsilon = \begin{pmatrix} c^2 & s^2 & 0 & 0 & 0 & cs \\ s^2 & c^2 & 0 & 0 & 0 & -cs \\ 0 & 0 & 1 & 0 & 0 & 0 \\ 0 & 0 & 0 & c & -s & 0 \\ 0 & 0 & 0 & s & c & 0 \\ -2sc & 2sc & 0 & 0 & 0 & c^2 - s^2 \end{pmatrix}. \tag{1.31}$$

We substitute of the relations

$$\boldsymbol{\sigma}^T = \boldsymbol{\sigma}^{(0)T} \cdot \mathbf{T}_\sigma{}^T = \boldsymbol{\sigma}^{(0)T} \cdot \mathbf{T}_\varepsilon{}^{-1}, \boldsymbol{\sigma} = \mathbf{T}_\sigma \cdot \boldsymbol{\sigma}^{(0)} = (\mathbf{T}_\varepsilon{}^T)^{-1} \cdot \boldsymbol{\sigma}^{(0)}.$$

and (1.29) into the expression for elastic energy density $2W_\sigma = \boldsymbol{\sigma}^T \cdot \mathbf{S} \cdot \boldsymbol{\sigma}$. The substitution assures that the elastic energy density remains invariant during rotation:

$$\begin{cases} 2W_\sigma = \boldsymbol{\sigma}^T \cdot \mathbf{S} \cdot \boldsymbol{\sigma} = \boldsymbol{\sigma}^{(0)T} \cdot \mathbf{S}^{(0)} \cdot \boldsymbol{\sigma}^{(0)}, \\ 2W_\varepsilon = \boldsymbol{\varepsilon}^T \cdot \mathbf{C} \cdot \boldsymbol{\varepsilon} = \boldsymbol{\varepsilon}^{(0)T} \cdot \mathbf{C}^{(0)} \cdot \boldsymbol{\varepsilon}^{(0)}. \end{cases} \tag{1.32}$$

The formulas for the components of compliance matrix $\mathbf{S}$ in rotated coordinate system in terms of the components of compliance matrices in intrinsic coordinate system $\mathbf{S}^{(0)}$ are:

$$\begin{aligned}
S_{11} &= s_{11}c^4 + 2(s_{12} + s_{66})c^2s^2 + s_{22}s^4,\\
S_{22} &= s_{22}c^4 + 2(s_{12} + s_{66})c^2s^2 + s_{11}s^4\\
S_{12} &= s_{12}(c^4 + s^4) + (s_{11} + s_{22} - 4s_{66})c^2s^2,\\
S_{66} &= s_{66}(c^2 - s^2)^2 + (s_{11} + s_{22} - 2s_{12})c^2s^2,\\
S_{16} &= -cs(s_{22}c^2 - s_{11}s^2 - (s^2 - c^2)(s_{12} + 2s_{66})),\\
S_{26} &= -cs(s_{11}c^2 - s_{22}s^2 + (s^2 - c^2)(s_{12} + 2s_{66})).
\end{aligned} \tag{1.33}$$

1.3.3 Rotation of Elasticity Matrices in Kelvin's Notation

The rotation of the elasticity and compliance matrices in the Kelvin's notation uses the tensor transformation rule with the rotation matrix $\widehat{\mathbf{T}}(\Phi)$ (Mehrabadi and Cowin 1990):

$$\widehat{\mathbf{C}} = \widehat{\mathbf{T}} \cdot \widehat{\mathbf{C}}^{(0)} \cdot \widehat{\mathbf{T}}^{-1}, \widehat{\mathbf{S}} = \widehat{\mathbf{T}} \cdot \widehat{\mathbf{S}}^{(0)} \cdot \widehat{\mathbf{T}}^{-1}. \tag{1.34}$$

The six-dimensional orthogonal rotation matrix rotates the components of the strain and stress vectors in Kelvin's notation as well:

$$\widehat{\mathbf{T}}(\Phi) = \begin{bmatrix} c^2 & s^2 & 0 & 0 & 0 & \sqrt{2}sc \\ s^2 & c^2 & 0 & 0 & 0 & -\sqrt{2}sc \\ 0 & 0 & 1 & 0 & 0 & 0 \\ 0 & 0 & 0 & c & -s & 0 \\ 0 & 0 & 0 & s & c & 0 \\ -\sqrt{2}sc & \sqrt{2}sc & 0 & 0 & 0 & c^2 - s^2 \end{bmatrix}, \widehat{\mathbf{T}}^{-1}(\Phi) = \widehat{\mathbf{T}}^T(\Phi), \tag{1.35}$$

$\widehat{\boldsymbol{\sigma}} = \widehat{\mathbf{T}} \cdot \widehat{\boldsymbol{\sigma}}^{(0)}$	stress vector in a rotated coordinate system,
$\boldsymbol{\sigma}^{(0)}$	stress vector in an intrinsic coordinate system,
$\widehat{\boldsymbol{\varepsilon}} = \widehat{\mathbf{T}} \cdot \widehat{\boldsymbol{\varepsilon}}^{(0)}$	strain vector in a rotated coordinate system,
$\widehat{\boldsymbol{\varepsilon}}^{(0)}$	strain vector in an intrinsic coordinate system,
$\widehat{\mathbf{C}} = [\widehat{C}_{ij}]$	elasticity matrix in a rotated coordinate system,
$\widehat{\mathbf{C}}^{(0)} = [\widehat{c}_{ij}]$	elasticity matrix in an intrinsic coordinate system,
$\widehat{\mathbf{S}} = [\widehat{S}_{ij}]$	compliance matrix in a rotated coordinate system,
$\widehat{\mathbf{S}}^{(0)} = [\widehat{s}_{ij}]$	compliance matrix in an intrinsic coordinate system.

The elastic energy density (1.32) remains invariant during rotation:

$$\begin{cases} 2W_\sigma = \widehat{\boldsymbol{\sigma}}^T \cdot \widehat{\mathbf{S}} \cdot \widehat{\boldsymbol{\sigma}} = \widehat{\boldsymbol{\sigma}}^{(0)T} \cdot \widehat{\mathbf{S}}^{(0)} \cdot \widehat{\boldsymbol{\sigma}}^{(0)}, \\ 2W_\varepsilon = \widehat{\boldsymbol{\varepsilon}}^T \cdot \widehat{\mathbf{C}} \cdot \widehat{\boldsymbol{\varepsilon}} = \widehat{\boldsymbol{\varepsilon}}^{(0)T} \cdot \widehat{\mathbf{C}}^{(0)} \cdot \widehat{\boldsymbol{\varepsilon}}^{(0)}. \end{cases} \tag{1.36}$$

1.4 Elasticity Matrices for Laminated Plates

1.4.1 Voigt's Matrix Notation for Anisotropic Plates

Lightweight structures are typically thin-walled and flat and must transmit forces and torques from different directions. The fibers are therefore arranged according to load in order to make use of the advantages of the composite material. This is only possible by stacking different layers, which results in the so-called laminate or multi-layer composite. However, it is possible not only to stack unidirectional layers, but also to use individual layers whose fibers are at an angle of 0° or 90° to each other and are therefore referred to as cross-laminated layers.

With the information about the laminate structure, that is, layer thickness and layer sequence as well as the orientation of the layers, the constitutive law for laminates reads Vannucci (2018) and Toorani and Lakis (2000):

$$\begin{bmatrix} \mathbf{N} \\ \mathbf{M} \end{bmatrix} = \begin{bmatrix} \mathbf{A} & \mathbf{B} \\ \mathbf{B} & \mathbf{D} \end{bmatrix} \begin{bmatrix} \boldsymbol{\varepsilon} \\ \boldsymbol{\kappa} \end{bmatrix}. \tag{1.37}$$

The elasticity matrix in Voigt's notation of a dimension of 6×6 determines the in-plane deformation and bending of the laminated plate. Its minors **A,B,D** of the third rank are:

$\mathbf{A} = \int_{-h/2}^{h/2} \mathbf{Q}(z)dz$	in-plane quadrant, square 3×3 matrix,
$\mathbf{B} = \int_{-h/2}^{h/2} \mathbf{Q}(z)zdz$	coupling quadrant, square 3×3 matrix,
$\mathbf{D} = \int_{-h/2}^{h/2} \mathbf{Q}(z)z^2dz$	bending quadrant, square 3×3 matrix,
$\boldsymbol{\varepsilon}^T = [\varepsilon_{11}, \varepsilon_{22}, 2\varepsilon_{12}]$	strain vector in Voigt's notation,
$\boldsymbol{\kappa}^T = [\kappa_{11}, \kappa_{22}, 2\kappa_{12}]$	curvature vector in Voigt's notation,
$\mathbf{N}^T = [N_{11}, N_{22}, N_{12}]$	in-plane force vector in Voigt's notation,
$\mathbf{M}^T = [M_{11}, M_{22}, M_{12}]$	bending moment vector in Voigt's notation.

This form of vector representation is common in engineering, although it requires different transformation matrices for all relevant elastic quantities, such as stress and strain vector and elasticity matrices.

For derivation, the stiffness relations for each singular layer are:

$$\begin{bmatrix} \sigma_{11} \\ \sigma_{22} \\ \sigma_{12} \end{bmatrix} = \mathbf{Q} \begin{bmatrix} \varepsilon_{11} \\ \varepsilon_{22} \\ 2\varepsilon_{12} \end{bmatrix}, \mathbf{Q}(z) = \begin{bmatrix} Q_{11} & Q_{12} & Q_{16} \\ Q_{12} & Q_{22} & Q_{26} \\ Q_{16} & Q_{26} & Q_{66} \end{bmatrix}. \tag{1.38}$$

The coefficients of the stiffness matrix (1.38) read:

$$Q_{ij} = Q_{ij}(z) = C_{ij} - C_{i3}C_{i3}/C_{33}, i, j = 1, 2, 6.$$

For a unidirectional layer, the reduced stiffness matrix is:

$$\mathbf{Q} = \begin{bmatrix} \frac{\mathrm{E_L}}{1 - \nu_{\mathrm{TL}} \cdot \nu_{LT}} & \frac{\nu_{LT} \cdot \mathrm{E_L}}{1 - \nu_{\mathrm{TL}} \cdot \nu_{LT}} & 0 \\ \frac{\nu_{\mathrm{LT}} \cdot \mathrm{E_L}}{1 - \nu_{\mathrm{TL}} \cdot \nu_{LT}} & \frac{\mathrm{E}_T}{1 - \nu_{\mathrm{TL}} \cdot \nu_{LT}} & 0 \\ 0 & 0 & \mathrm{G_{TL}} \end{bmatrix}.$$

The reactions and moments in Voigt's notation are expressed through the in-plane stress in singular layers:

$$\mathbf{N}^T = [N_{11}, N_{22}, N_{12}] = \int_{-h/2}^{h/2} [\sigma_{11}, \sigma_{22}, \sigma_{12}]dz, \tag{1.39}$$

$$\mathbf{M}^T = [M_{11}, M_{22}, M_{12}] = \int_{-h/2}^{h/2} z^2[\sigma_{11}, \sigma_{22}, \sigma_{12}]dz. \tag{1.40}$$

The stiffness evaluation and rotation of anisotropy orientation are also appropriate to thin shell models.

1.4.2 Rotation of Matrices in Voigt's Notation

For the rotation of each minor in Voigt's matrix notation, two rotation matrices $\widetilde{\mathbf{T}}_\sigma$ and $\widetilde{\mathbf{T}}_\varepsilon$ are used:

$$\begin{aligned} \{\mathbf{A}, \mathbf{B}, \mathbf{D}\} &= \widetilde{\mathbf{T}}_\sigma \cdot \{\mathbf{A}^{(0)}, \mathbf{B}^{(0)}, \mathbf{D}^{(0)}\} \cdot \widetilde{\mathbf{T}}_\sigma^T, \\ \{\mathbf{A}^{-1}, \mathbf{B}^{-1}, \mathbf{D}^{-1}\} &= \widetilde{\mathbf{T}}_\varepsilon \cdot \{\mathbf{A}^{(0)-1}, \mathbf{B}^{(0)-1}, \mathbf{D}^{(0)-1}\} \cdot \widetilde{\mathbf{T}}_\varepsilon^T. \end{aligned} \tag{1.41}$$

$\widetilde{\mathbf{T}}_\sigma$	σ-transformation matrix for plates (3 × 3),
$\widetilde{\mathbf{T}}_\varepsilon$	ϵ-Transformation matrix for plates (3 × 3),
$\{\mathbf{A}, \mathbf{B}, \mathbf{D}\}$	minors of elasticity matrix in a rotated coordinate system,
$\{\mathbf{A}^{(0)}, \mathbf{B}^{(0)}, \mathbf{D}^{(0)}\}$	minors of elasticity matrix in an intrinsic coordinate system,
$\{\mathbf{A}^{-1}, \mathbf{B}^{-1}, \mathbf{D}^{-1}\}$	minors of compliance matrix in a rotated coordinate system,
$\{\mathbf{A}^{(0)^{-1}}, \mathbf{B}^{(0)^{-1}}, \mathbf{D}^{(0)^{-1}}\}$	minors of compliance matrix in an intrinsic coordinate system.

The rotation matrices obey the rules $\widetilde{\mathbf{T}}_\sigma^{-1} = \widetilde{\mathbf{T}}_\varepsilon^{\mathrm{T}}$, $\widetilde{\mathbf{T}}_\varepsilon^{-1} = \widetilde{\mathbf{T}}_\sigma^{\mathrm{T}}$ and read:

$$\widetilde{\mathbf{T}}_\sigma = \begin{pmatrix} c^2 & s^2 & 2cs \\ s^2 & s^2 & -2cs \\ -sc & sc & c^2 - s^2 \end{pmatrix}, \tag{1.42}$$

$$\widetilde{\mathbf{T}}_\varepsilon = \begin{pmatrix} c^2 & s^2 & cs \\ s^2 & s^2 & -cs \\ -2sc & 2sc & c^2 - s^2 \end{pmatrix}. \tag{1.43}$$

1.4.3 Kelvin's Matrix Notation for Anisotropic Plates

The constitutive law for laminates in Kelvin's notation reads (Vannucci 2018):

$$\begin{bmatrix} \widehat{\mathbf{N}} \\ \widehat{\mathbf{M}} \end{bmatrix} = \begin{bmatrix} \widehat{\mathbf{A}} & \widehat{\mathbf{B}} \\ \widehat{\mathbf{B}} & \widehat{\mathbf{D}} \end{bmatrix} \begin{bmatrix} \widehat{\boldsymbol{\varepsilon}} \\ \widehat{\boldsymbol{\kappa}} \end{bmatrix}. \tag{1.44}$$

The elasticity matrix in Kelvin's notation of the second rank determines the in-plane deformation and bending. Its minors $\widehat{\mathbf{A}}, \widehat{\mathbf{B}}, \widehat{\mathbf{D}}$ of the third rank are:

$\widehat{\mathbf{A}} = \int_{-h/2}^{h/2} \widehat{\mathbf{Q}}(z)dz$	in-plane quadrant, square 3 × 3 matrix,
$\widehat{\mathbf{B}} = \int_{-h/2}^{h/2} \widehat{\mathbf{Q}}(z)zdz$	coupling quadrant, square 3 × 3 matrix,
$\widehat{\mathbf{D}} = \int_{-h/2}^{h/2} \widehat{\mathbf{Q}}(z)z^2dz$	bending quadrant, square 3 × 3 matrix,
$\widehat{\boldsymbol{\varepsilon}}^T = [\varepsilon_{11}, \varepsilon_{22}, \sqrt{2}\varepsilon_{12}]$	strain vector,
$\widehat{\boldsymbol{\kappa}}^T = [\kappa_{11}, \kappa_{22}, \sqrt{2}\kappa_{12}]$	curvature vector,
$\widehat{\mathbf{N}}^T = [N_{11}, N_{22}, \sqrt{2}N_{12}]$	in-plane force vector,
$\widehat{\mathbf{M}}^T = [M_{11}, M_{22}, \sqrt{2}M_{12}]$	bending moment vector.

Kelvin's notation was derived earlier (Kelvin 1856) than Voigt's (Voigt 1910); it uses the solitary orthogonal transformation matrix and obeys the tensor transformation law, but Kelvin's notation is not used habitually in engineering.

For derivation, the stiffness laws for singular layers are:

$$\begin{bmatrix} \sigma_{11} \\ \sigma_{22} \\ \sqrt{2}\sigma_{12} \end{bmatrix} = \widehat{\mathbf{Q}} \begin{bmatrix} \varepsilon_{11} \\ \varepsilon_{22} \\ \sqrt{2}\varepsilon_{12} \end{bmatrix}, \widehat{\mathbf{Q}}(z) = \begin{bmatrix} Q_{11} & Q_{12} & Q_{16}/\sqrt{2} \\ Q_{12} & Q_{22} & Q_{26}/\sqrt{2} \\ Q_{16}/\sqrt{2} & Q_{26}/\sqrt{2} & Q_{66}/2 \end{bmatrix}. \tag{1.45}$$

The reactions and moments in Kelvin's notation are expressed through in-plane stress in singular layers:

$$\widehat{\mathbf{N}}^T = [N_{11}, N_{22}, \sqrt{2}N_{12}] = \int_{-h/2}^{h/2} [\sigma_{11}, \sigma_{22}, \sqrt{2}\sigma_{12}]dz, \tag{1.46}$$

$$\widehat{\mathbf{M}}^T = [M_{11}, M_{22}, \sqrt{2}M_{12}] = \int_{-h/2}^{h/2} z^2[\sigma_{11}, \sigma_{22}, \sqrt{2}\sigma_{12}]dz. \tag{1.47}$$

1.4.4 Rotation of Matrices in Kelvin's Notation

The rotation of the elasticity and compliance minors in Kelvin's notation (Vannucci 2018) is performed with the orthogonal transformation matrix $\widetilde{\mathbf{T}}$:

$$\begin{aligned} \{\widehat{\mathbf{A}}, \widehat{\mathbf{B}}, \widehat{\mathbf{D}}\} &= \widetilde{\mathbf{T}} \cdot \{\widehat{\mathbf{A}}^{(0)}, \widehat{\mathbf{B}}^{(0)}, \widehat{\mathbf{D}}^{(0)}\} \cdot \widetilde{\mathbf{T}}^{-1}, \\ \{\widehat{\mathbf{A}}^{-1}, \widehat{\mathbf{B}}^{-1}, \widehat{\mathbf{D}}^{-1}\} &= \widetilde{\mathbf{T}} \cdot \{\widehat{\mathbf{A}}^{(0)-1}, \widehat{\mathbf{B}}^{(0)-1}, \widehat{\mathbf{D}}^{(0)-1}\} \cdot \widetilde{\mathbf{T}}^{-1}. \end{aligned} \tag{1.48}$$

$\widetilde{\mathbf{T}}$	transformation matrix for plates tensor (3 × 3),
$\{\widehat{\mathbf{A}}, \widehat{\mathbf{B}}, \widehat{\mathbf{D}}\}$	minors of elasticity matrix in a rotated coordinate system,
$\{\widehat{\mathbf{A}}^{(0)}, \widehat{\mathbf{B}}^{(0)}, \widehat{\mathbf{D}}^{(0)}\}$	minors of elasticity matrix in an intrinsic coordinate system,
$\{\widehat{\mathbf{A}}^{-1}, \widehat{\mathbf{B}}^{-1}, \widehat{\mathbf{D}}^{-1}\}$	minors of compliance matrix in a rotated coordinate system,
$\{\widehat{\mathbf{A}}^{(0)^{-1}}, \widehat{\mathbf{B}}^{(0)^{-1}}, \widehat{\mathbf{D}}^{(0)^{-1}}\}$	minors of compliance matrix in an intrinsic coordinate system.

The three-dimensional orthogonal transformation matrix is used for rotation:

$$\widetilde{\mathbf{T}}(\theta) = \begin{pmatrix} c^2 & s^2 & \sqrt{2}sc \\ s^2 & c^2 & -\sqrt{2}sc \\ -\sqrt{2}sc & \sqrt{2}sc & c^2 - s^2 \end{pmatrix}, \widetilde{\mathbf{T}}^T = \widetilde{\mathbf{T}}^{-1}. \tag{1.49}$$

1.5 Coupling Effects of Anisotropic Laminates

1.5.1 Orthotropic Laminate Without Coupling

For common structural components, the usual aim is to eliminate coupling between the in-plane quadrant **A** and the bending quadrant **D.** Depending on how the quadrants are occupied, different coupling properties of the laminate can take place. If all components of matrix $\mathbf{Q}(z)$ are symmetric, $\mathbf{Q}(z) = \mathbf{Q}(-z)$, the coupling quadrant **B** disappears and they assume the diagonal form:

$$\mathbf{A} = \begin{bmatrix} A_{11} & A_{12} & 0 \\ A_{12} & A_{22} & 0 \\ 0 & 0 & A_{66} \end{bmatrix}, \mathbf{B} = \begin{bmatrix} 0 & 0 & 0 \\ 0 & 0 & 0 \\ 0 & 0 & 0 \end{bmatrix}, \mathbf{D} = \begin{bmatrix} D_{11} & D_{12} & 0 \\ D_{12} & D_{22} & 0 \\ 0 & 0 & D_{66} \end{bmatrix}. \tag{1.50}$$

Moreover, the plate behaves as an orthotropic material; that is, the shear stress in the principal coordinates does not affect the directional stresses. Analogously, pure warping leads to no bending of a plate in principal coordinates. For example, this laminate structure occurs if the laminate type is $[0^\circ/90^\circ/90^\circ, 0^\circ]$ or $[0^\circ/\pm 45^\circ/../\pm 45^\circ, 0^\circ]$. This stacking leads to a fully symmetric structure with decoupling between in-plane deformations and bending.

1.5.2 Anisotropic Laminate Without Coupling

The coupling between bending and in-plane deformation also disappears if the matrices assume the form:

$$\mathbf{A} = \begin{bmatrix} A_{11} & A_{12} & A_{16} \\ A_{12} & A_{22} & A_{26} \\ A_{16} & A_{26} & A_{66} \end{bmatrix}, \mathbf{B} = \begin{bmatrix} 0 & 0 & 0 \\ 0 & 0 & 0 \\ 0 & 0 & 0 \end{bmatrix}, \mathbf{D} = \begin{bmatrix} D_{11} & D_{12} & D_{16} \\ D_{12} & D_{22} & D_{26} \\ D_{16} & D_{26} & D_{66} \end{bmatrix}. \tag{1.51}$$

However, the plate behaves as an anisotropic material; that is, the shear stress in the principal coordinates has an effect on the directional stresses. Analogously, pure warping causes bending of a plate in principal coordinates.

1.5.3 Anisotropic Laminate With Coupling

For some structural applications coupling is desired because this provides an additional grade of control. The smart adjustment of the additional control grade allows elimination of some undesired properties of a structure that can present in an uncoupled structure (Reddy and Miravete 1995).

The coupling between bending and in-plane deformation presents as (if the matrices are in the fully populated form):

$$\mathbf{A} = \begin{bmatrix} A_{11} & A_{12} & A_{16} \\ A_{12} & A_{22} & A_{26} \\ A_{16} & A_{26} & A_{66} \end{bmatrix}, \mathbf{B} = \begin{bmatrix} B_{11} & B_{12} & B_{16} \\ B_{12} & B_{22} & B_{26} \\ B_{16} & B_{26} & B_{66} \end{bmatrix}, \mathbf{D} = \begin{bmatrix} D_{11} & D_{12} & D_{16} \\ D_{12} & D_{22} & D_{26} \\ D_{16} & D_{26} & D_{66} \end{bmatrix}. \tag{1.52}$$

The plate behaves as an anisotropic material; that is, the shear stress in the principal coordinates has an effect on the directional stresses. Pure warping causes bending of a plate in principal coordinates. Additionally, the bending and in-plane deformations are also coupled.

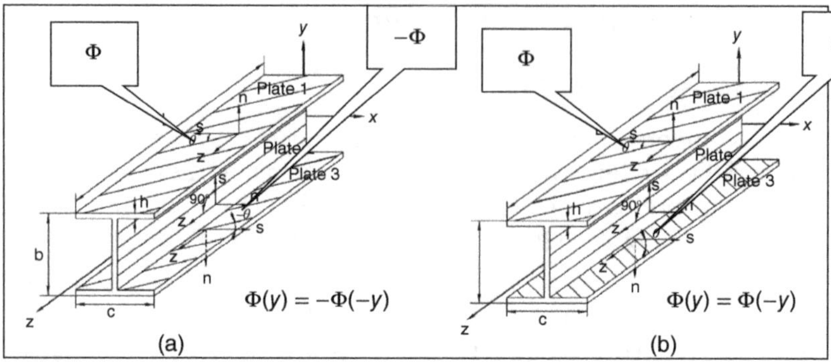

Figure 1.3 (a) Circumferentially asymmetric stiffness configuration (CAS), (b) circumferentially uniform stiffness configuration (CUS) (Librescu and Song 2006, Figure. 13.1.1) The local coordinate system *xyz* is associated with the twist-beam *z*-axis. Source: University of Siegen.

1.5.4 Coupling Effects in Laminated Thin-Walled Sections

The theory of an elastic coupling of the composite thin-walled beams with open cross-sections was developed by Chandra et al. (1990). This theory is based on the model by Vlasov and Gjelsvik, and includes the effects of transverse-shear deformation and coupling between laminate reactions and moments. Figure 1.3 shows the laminate structure for a flexure-torsion and tensile-torsion coupling for a fiber composite component with a rectangular cross-section. The beams that possess the circumferentially asymmetric stiffness (CAS) configuration (see Librescu and Song 2006; Rehfield and Atilgan 1989 and Vo and Lee 2008) are shown in Figure 1.3. This structural configuration is attainable when the ply-angle distributions are equal in the top and bottom walls as well as the lateral walls. The equations that describe coupling between bending and torsion are derived in Section 4.4.4 (Librescu and Song 2006). The circumferentially uniform stiffness (CUS) and CAS configurations are known as antisymmetric and symmetric configurations, respectively (Smith and Chopra 1991).

1.6 Conclusions

This chapter compiles the equations of the anisotropic elastic behavior of fiber composites. The common tensor and Voigt's vector notations are used for the derivation of elasticity equations of an anisotropic medium. Besides these, Kelvin's notations are also presented because this form of vector notation leads to a tensor-invariant representation and, consequently, is favorable for use in optimization problems.

For further reading, the following reference works about anisotropic elasticity and composite materials are recommended: Halpin (1992), Tsai (1968), Hill (1950), Barbero (1999), Hull and Clyne (1996), Altenbach et al. (2004), Tin (1996), Gay et al. (2013), Peters (1998), Kollár and Springer (2003), Tschoegl et al. (2002), Lakes and Wineman (2006), Hertz et al. (1981), Abramovich and Livshitz (1994), Kaiser (1999), Kaiser and Francescatti (1996), Kroker (2013), Dugas (2002), Kindervater (2008), Santiuste et al. (2008) and Foye (1972).

References

Abramovich, H. and Livshitz, A. (1994). Free vibrations free vibrations of non-symmetric cross-ply laminated composite beams. *J. Sound Vib.* 176 (5): 597–612.

Altenbach, H., Altenbach, J., and Kissing, W. (2004). *Mechanics of Composite Struktural Elements*. Springer-Verlag Berlin Heidelberg.

Altmann, S.L. (1986). *Rotations, Quaternions, and Double Groups*. Oxford: Oxford University Press.

Barbero, E. (1999). *Introduction to Composite Materials Design*. New York: Taylor and Francis.

Chandra, R., Stemple, A.D., and Chopra, I. (1990). Thin-walled composite beams under bending, torsional, and extensional loads. *J. Aircraft* 27 (7): 619–626.

Chou, T.-W. (1992). *Microstructural Design of Fiber Composites*. Cambridge: Cambridge University Press.

Dugas M. (2002) *Ein Beitrag zur Auslegung von Faserverbundtragflügeln im Vorentwurf*, Institut für Flugzeugbau der Universität Stuttgart.

ECSS-E-HB-32-20 Part 2A (2011). Space engineering, structural materials handbook. In: *Part 2: Design Calculation Methods and General Design Aspects. ESA-ESTEC, Requirements & Standards Division*, 434. Noordwijk, The Netherlands.

Foye, R.L. (1972). The transverse poisson's ratio of composites. *J. Compos. Mater.* 6: 293–295.

Gay, D., Hoa, S.V., and Tsai, S.W. (2013). *Composite Materials, Design and Applications*. Boca Raton, FL: CRC Press LLC.

Gibson, R.F. (2016). *Principles of Composite Material Mechanics*. Boca Raton: CRC Press.

Halpin, J.C. (1992). *Primer on Composite Materials Analysis*. Taylor & Francis Group, LLC.

Hill, R. (1950). *The Mathematical Theory of Plasticity*. London: Oxford University Press.

Hill, R. (1963). Elastic properties of reinforced solids: some theoretical principles. *J. Mech. Phys. Solids* 11: 357–372.

Hull, D. and Clyne, T.W. (1996). *An Introduction to Composite Materials*, 2e. Cambridge: Cambridge University Press.

Kaiser, C. (1999) *Elastische Kopplungen in aktiven und passiven Laminaten sowie dünnwandigen Faserverbundbalken*, Dissertation, TUM, München.

Kaiser C., Francescatti D. (1996) Theoretical and experimental analysis of composite beams with elastic couplings, *ICAS-96-5.4.4, ICAS Congress*, 1996, Italy, Sorrento.

Kelvin, W.T. (1856). Elements of a mathematical theory of elasticity. *Philos. Trans. R. Soc.* 146: 481–498.

Kelvin, W.T. (1878). Mathematical theory of elasticity. *Encycl. Br.* 7: 819–825.

Kindervater Ch. (2008) *Technologie – und Dimensionierungsgrundlagen für Bauteile aus Faserkunststoffverbund (FKV)*. Vorlesungsmanuskript, DLR, Institut für Bauweisen – und Konstruktionsforschung, Stuttgart.

Kollár, L.P. and Springer, G.S. (2003). *Mechanics of Composite Structures*. Cambridge University Press.

Kroker A.M. (2013) *Strukturmechanische Modellierung und Analyse des Tragverhaltens von dünnwandigen hochbelasteten Composite-Biege-und Querkraftträgern*, Darmstadt 2013.

Lakes, R. and Wineman, A. (2006). On poisson's ratio in linearly viscoelastic solids. *J. Elast.* 85 (1): 45–63.

Librescu, L. and Song, O. (2006). *Thin-Walled Composite Beams, Theory and Application, Solid Mechanics and its Applications*, vol. 131. Dordrecht, The Netherlands: Springer.
Mehrabadi, M.M. and Cowin, S.C. (1990). Eigentensors of linear anisotropic elastic materials. *Q. J. Mech. Appl. Math.* 43: 15–41.
Moser, K. (1992). *Faser-Kunststoff-Verbund: Entwurfs- und Berechnungsgrundlagen.* Springer-Verlag, Berlin/Heidelberg.
Nettles A.T. (1994) *Basic Mechanics of Laminated Composite Plates*, NASA Reference Publication 135, Marshall Space Flight Center.
Peters, S.T. (ed.) (1998). *Handbook of Composites*, 2e. London: Chapman and Hall.
Hertz, T.J., Shirk, M.H., Ricketts, R.H., and Weisshaar, T.A. (1981). *Aeroelastic Tailoring with Composites Applied to Forward Swept Wings.* Air Force Wright Aeronautical Laboratories, NASA Langley Research Center, and Purdue University.
Reddy, J.N. and Miravete, A. (1995). *Practical Analysis of Composite Laminates.* Boca Raton: CRC Press.
Rehfield, L.W. and Atilgan, A.R. (1989). Toward understanding the tailoring mechanisms for thin-walled composite tubular beams. In: *Proceedings of the First USSR–U.S. Symposium on Mechanics of Composite Materials, 23–26 May, Riga, Latvia SSR* (ed. S.W. Tsai, J.M. Whitney, T.W. Chou and R.M. Jones), 187–196. ASME Publ. House.
Santiuste, C., Sánchez-Sáez, S., and Barbero, E. (2008). Dynamic analysis of bending–torsion coupled composite beams using the flexibility influence function method. *Int. J. Mech. Sci.* 50 (12): 1611–1618.
Schürmann, H. (2007). *Konstruieren mit Faser-Kunststoff-Verbunden.* Berlin Heidelberg: Springer-Verlag.
Smith, E.C. and Chopra, I. (1991). Formulation and evaluation of an analytical model for composite box-beams. *J. Am. Helicopter Soc.* 36 (3): 23–35.
Tin, T.C.T.G. (1996). *Anisotropic Elasticity, Theory and Applications.* Oxford and New York: Oxford University Press.
Toorani, M.H. and Lakis, A.A. (2000). General equations of anisotropic plates and shells including transverse shear deformations, rotary inertia and initial curvature effects. *J. Sound Vib.* 237 (4): 561–615. https://doi.org/10.1006/jsvi.2000.3073.
Tsai, S.W. (1968). *Fundamental Aspects of Fiber Reinforced Plastic Composites.* New York: Wiley.
Tschoegl, N.W., Knauss, W.G., and Emri, I. (2002). Poisson's ratio in linear viscoelasticity – a critical review. *Mech. Time-Depend. Mater.* 6 (1): 3–51.
Vannucci, P. (2018). *Anisotropic Elasticity.* Springer Nature, Singapore Pte Ltd.
Vo, T.P. and Lee, J. (2008). Flexural–torsional behavior of thin-walled composite box beams using shear-deformable beam theory. *Eng. Struct.* 30 (7): 1958–1968.
Voigt W. (1910) *Lehrbuch der Kristallphysik: mit Ausschluß d. Kristalloptik.* Teubner, Leipzig u. a.
Younes, R., Hallal, A., Fardoun, F., and Chehade, F.H. (2012). Comparative review study on elastic properties modeling for unidirectional composite materials. In: *Composites and Their Properties* (ed. N. Hu). London: IntechOpen.

2

Phenomenological Failure Criteria of Composites

2.1 Phenomenological Failure Criteria

2.1.1 Criteria for Static Failure Behavior

In fiber-reinforced composites, both the failure behavior, which is either ductile or brittle, as well as the type of failure (fiber or matrix fracture), are strongly dependent on the components of the applied stress, type and compositions of components, the fiber arrangement and manufacturing procedures. For design purposes, most attention is paid to the geometric details of an internal structure composite. The other composition and manufacturing factors must be fixed.

Among others, the highest loading capacity for a uniaxial stress state provides unidirectional fiber-reinforced composites (Sun 2000). Unidirectional fiber-reinforced materials are best suited to chassis and drivetrain applications because the stresses in such systems can be extraordinarily high. Some failure modes of a unidirectional single layer with extensive photographs of crack surfaces in experiments are exposed in Agarwal et al. (2006). The majority of problems studied in this book focus on this type of composite.

2.1.2 Stress Failure Criteria for Isotropic Homogenous Materials

In the case of metallic materials, static strength verification is often based on the ratio of a comparative stress (e.g. von Mises stress) to a strength value obtained from a uniaxial tensile test.

The main normal stress hypothesis has established itself in the evaluation of brittle materials under tensile loading. It is assumed that the fracture of the component occurs in the plane in which the greatest normal stress prevails. The ultimate value of first principal stress can be used as the failure criterion. The Tresca shear stress hypothesis is used for a predominant compressive load.

The plastic flow through shear stresses causes the failure of ductile materials. The elastic energy, associated with changes in shape or distortion, predicts a failure of the structure in this case. According to the von Mises criterion, an isotropic material yields if the distortion energy per unit volume in that material reaches or exceeds a certain critical value.

Design and Analysis of Composite Structures for Automotive Applications: Chassis and Drivetrain, First Edition. Vladimir Kobelev.

Companion website: www.wiley.com/go/kobelev/automotive_suspensions

2.1.3 Phenomenological Failure Criteria for Composites

Failure criteria for fiber-plastic composites serve to differentiate whether an external load in a fiber-plastic composite component (see also laminate) leads to failure or not. The term "failure criterion" is more common for composites. The "breakage criterion" for the fiber, the matrix, the fiber-matrix interface and the delamination of single layers is included in this generic term. Due to the structural anisotropy of fiber-plastic composites, a distinction is made between different types of failure. There are a number of criteria whose applications depend on the problem to be calculated (Hart-Smith 2000; Agarwal et al. 2006; Talreja and Varna 2016). Accordingly, the results of the application of different criteria regularly fluctuate considerably. To avoid this uncertainty, the VDI guideline 2014 (VDI 2006) introduces a uniform calculation method for fiber-reinforced composites.

The evaluation of the critical deformation of a material is based on the comparison of the scalar function of the invariants of a stress tensor with a *reference* stress. This determination is only possible if the material can be characterized by means of a solitary strength parameter. The distinction between tensile and compressive stresses must be included in the strength parameter for fiber-reinforced materials because the fracture mechanism of these materials is sensitive to the sign of directional stress. The strengths of fiber-reinforced materials vary in spatial directions (anisotropy) and have also to be included in the strength parameter. Due to more complex failure behavior, it is not yet possible to calculate the ultimate fracture stress with the same accuracy as metallic materials. Moreover, the ultimate strengths in fiber-reinforced composites respond very sensitively to production quality. Environmental parameters, such as fiber volume, temperature, humidity and so on, also play an important role. In a fiber-reinforced material, micromechanical residual stresses mostly affect the cyclic failure resistance.

The criteria for fiber-plastic composites are subdivided according to the type of failure. The failure criteria refer to the beginning (first ply failure), degradation (gradual failure) or the (catastrophic total) failure. The overview of strength criteria for fiber-plastic composites is provided in Table 2.1. The failure criteria are grouped into three principal classes: phenomenological criteria without stress coupling; equivalent stress and tensor

Table 2.1 Overview of strength criteria for fiber-plastic composites.

		Different classes of failure criteria			
	Class of failure criteria	Phenomenological criteria without stress coupling	Equivalent stress and tensor polynomials for stress coupling criteria	Differentiating criteria and physically based criteria	
	Examples for the criteria	Maximum stresses, Maximum strain	Mises–Hill, Goldenblat–Kopnov Tsai–Wu,	Hashin	Puck, Cuntze
Character features of failure criteria	Stress interactions	no	yes	yes	yes
	Prediction of the type of failure	yes	no	yes	yes
	Physically founded	no	no	no	yes
	Micro-mechanic modeling	no	no	no	yes

polynomials for stress coupling and differentiating criteria and physically based criteria. These three classes of failure criteria describe the character of stress interactions in a different way; otherwise they predict the type of failure, are more or less physically founded and some of them are capable of modeling the micro-mechanics of a fracture.

2.1.4 Phenomenological Criteria Without Stress Coupling

2.1.4.1 Criterion of Maximum Averaged Stresses

With this criterion, the acting averaged stresses in an entire laminate are simply compared with the strength values of the material:

$$\begin{array}{llll} \sigma_1/X_t \leq 1, & \sigma_2/Y_t \leq 1 & \textit{for} & \sigma_1 > 0, \sigma_2 > 0 \\ |\sigma_1|/X_c \leq 1, & |\sigma_2|/Y_c \leq 1 & \textit{for} & \sigma_1 < 0, \sigma_2 < 0 \\ -1 \leq \tau/T_c \leq 1. & & & \end{array} \tag{2.1}$$

Here

X_t, X_c. are tensile or compressive strengths in the fiber direction,
Y_t, Y_c. are tensile or compressive strengths in the transverse direction,
T_c. is the shear strength.

Failure occurs if one of the five inequalities (2.1) is violated. The prediction accuracy under multi-axial loading can be investigated with uniaxial loading tests on unidirectional reinforced specimens whose fiber direction deviates from the loading direction. In case of the uniaxial load, only the stress component $\sigma_{11}^{(0)} = \sigma_M$ is present in the global coordinates. The deviation angle between loading direction and fiber direction is Φ. The stresses along of the local material coordinates follow from the transformation:

$$\boldsymbol{\sigma} = \mathbf{t}^T \cdot \boldsymbol{\sigma}^{(0)} \cdot \mathbf{t}. \tag{2.2}$$

If the Eq. (2.2) is resolved, the range for allowable stresses is:

$$\begin{aligned} &-X_c \leq \sigma_M \cos^2\Phi \leq X_t, \\ &-X_c \leq \sigma_M \sin^2\Phi \leq X_t, \\ &-T_c \leq \sigma_M \cos\Phi \sin\Phi \leq T_c. \end{aligned} \tag{2.3}$$

At an angle of about 25° under tensile load, this criterion (2.3) delivers strength predictions that are too optimistic.

Under compressive stress along the fiber, the failure behavior depends strongly on fiber volume content (ratio of fiber volume to total volume) and matrix properties. With a small fiber volume content, a so-called "out-of-phase" microscopic instability (microburst) can be detected in the fibers. With a high fiber volume content, which is more common in practice, an "in-phase" microburst usually occurs. Both types of failure assume that the tensile strength of the matrix material is sufficiently high and therefore no matrix failure has occurred yet. Another type of failure due to buckling of fibers appears under the action of longitudinal compressive stress.

Boundary layer failure between fiber and matrix and matrix fracture under shear tensile load are the most common types of failure. A tensile failure of fibers in their transverse direction only occurs with very precisely aligned fibers and very low transverse tensile fiber strengths. Transverse compressive stress often causes a shear failure of the matrix material, which is caused by fiber fractures and boundary layer failure. These failure processes reflect the possible intra-laminar failure of a unidirectional reinforced single layer.

2.1.4.2 Criterion of Maximum Averaged Strains

The criterion of the maximum averaged strains formulates analogously. For tensile loading in the fiber direction, the ratio of fiber elongation at break (ε_f) to matrix elongation at break (ε_m) is a decisive factor. For a case where elongation at the break of the fiber is greater than the matrix elongation at break ($\varepsilon_f > \varepsilon_m$), it is possible that a matrix breach is expected. On the other hand, with $\varepsilon_f < \varepsilon_m$, a fiber break occurs. These conditions could be written in the form of Eq. (2.1) after the replacement of stresses to the corresponding strains.

2.1.5 Phenomenological Criteria with Stress Coupling

Tensor and polynomial failure criteria can predict failure and non-failure. They make no statement about the type of fracture that occurs (fracture mode). The common criteria of this type are:

- *Mises–Hill criterion* (von Mises 1928; Hill 1950)
- *Tsai–Wu Criterion* (Tsai 1968; Tsai and Wu 1971)
- *Goldenblat–Kopnov criterion* (Goldenblat and Kopnov 1965)

The principal advantage of the tensor and polynomial criteria is their suitability for optimization purposes due to their tensor form.

2.1.5.1 Mises–Hill Anisotropic Failure Criterion

von Mises proposed initially a flow criterion (1928) to evaluate the stress of a plastic material with orthotropic elongation. The flow criterion is based on the von Mises equivalent stress of isotropic medium. This theory was adopted by Hill and is broadly used to define the breakdown response of anisotropic, originally metallic, materials (Hill 1950). In Voigt's notation, it reads in the intrinsic axes as:

$$\boldsymbol{\sigma} \bullet \widetilde{\mathbf{F}}^{(0)} \bullet \boldsymbol{\sigma} \le 1, \quad \widetilde{\mathbf{F}}^{(0)} = [f_{ijkl}], \tag{2.4}$$

$$\widetilde{\mathbf{F}}^{(0)} = \begin{bmatrix} G+H & -H & -G & 0 & 0 & 0 \\ -H & F+H & -F & 0 & 0 & 0 \\ -G & -F & F+G & 0 & 0 & 0 \\ 0 & 0 & 0 & 2L & 0 & 0 \\ 0 & 0 & 0 & 0 & 2M & 0 \\ 0 & 0 & 0 & 0 & 0 & 2N \end{bmatrix}, \boldsymbol{\sigma} = \begin{bmatrix} \sigma_{11} \\ \sigma_{22} \\ \sigma_{33} \\ \sigma_{23} \\ \sigma_{31} \\ \sigma_{12} \end{bmatrix} = \begin{bmatrix} \sigma_1 \\ \sigma_2 \\ \sigma_3 \\ \tau_4 \\ \tau_5 \\ \tau_6 \end{bmatrix}.$$

The Mises–Hill criterion in the intrinsic axes is the function of stress:

$$\boldsymbol{\sigma} \bullet \widetilde{\mathbf{F}}^{(0)} \bullet \boldsymbol{\sigma} = F(\sigma_2 - \sigma_3)^2 + G(\sigma_3 - \sigma_1)^2 + H(\sigma_1 - \sigma_2)^2 + \\ 2L\tau_4^2 + 2M\tau_5^2 + 2N\tau_6 \le 1.$$

Intrinsic axes of the unidirectional fiber-reinforced material are shown on Figure 1.1 (Chapter 1). The characteristic values F, G, H, L, M, N in (2.4) are those that characterize the anisotropy and are positive dimensionless constants that have to be determined

Table 2.2 Coefficients of the matrix and tensor Mises–Hill criterion.

$\tilde{f}_{11}$	f_{1111}	$G+H=\tilde{\sigma}_1^{-2}$
$\tilde{f}_{22}$	f_{2222}	$H+F=\tilde{\sigma}_2^{-2}$
$\tilde{f}_{33}$	f_{3333}	$F+G=\tilde{\sigma}_3^{-2}$
$\tilde{f}_{12}$	f_{1122}	$2F=\tilde{\sigma}_2^{-2}+\tilde{\sigma}_3^{-2}-\tilde{\sigma}_1^{-2}$
$\tilde{f}_{13}$	f_{1133}	$2G=\tilde{\sigma}_3^{-2}+\tilde{\sigma}_1^{-2}-\tilde{\sigma}_2^{-2}$
$\tilde{f}_{23}$	$f_{2233}.$	$2H=\tilde{\sigma}_1^{-2}+\tilde{\sigma}_2^{-2}-\tilde{\sigma}_3^{-2}$
$\tilde{f}_{44}$	f_{2323}	$2L=\tilde{\tau}_4^{-2}$
$\tilde{f}_{55}$	f_{1313}	$2M=\tilde{\tau}_5^{-2}$
$\tilde{f}_{66}$	f_{1212}	$2N=\tilde{\tau}_6^{-2}$

Table 2.3 Eigenvalues of the matrix and tensor Mises–Hill criterion.

Λ_1	0
Λ_2, Λ_3	$F+G+H \pm \sqrt{F(F-G)+G(G-H)+H(H-F)}$
$\Lambda_4, \Lambda_5, \Lambda_6$	$2L, 2M, 2N$

experimentally. The coefficients of the matrix and tensor in the intrinsic axes are shown in Tables 2.2 and 2.3.

The coefficients depend on the ultimate normal stresses $\tilde{\sigma}_1, \tilde{\sigma}_2, \tilde{\sigma}_3$ and the failure stresses and on the ultimate shear stresses $\tilde{\tau}_4, \tilde{\tau}_5, \tilde{\tau}_6$. The ultimate normal stresses and ultimate shear stresses in the Mises–Hill criterion are independent of stress sign. This assumption is normally applied for metallic materials; however, it is inappropriate for most fiber composites.

If, for example:

$$\tilde{\sigma}_1 \geq \tilde{\sigma}_2 \geq \tilde{\sigma}_3 > 0, \text{then } F \geq G \geq H > 0.$$

The quadratic form of the Mises–Hill criterion (2.4) must be positive definite. The condition of positive definiteness of the quadratic form is the possessiveness of all its eigenvalues. Except for the third eigenvalue, all others are obviously positive definite. The confirmation of the possessiveness of a third eigenvalue Λ_3 is proved based on the inequality:

$$(F+G+H)^2 = F(F-G)+G(G-H)+H(H-F)+$$
$$3FG+3GH+3HF > F(F-G)+G(G-H)+H(H-F).$$

For a transversally isotropic unidirectional composite with the fibers in the "1" direction, the ultimate stresses are equal to:

$$\tilde{\sigma}_1 = \tilde{\sigma}_L, \tilde{\sigma}_2 = \tilde{\sigma}_3 = \tilde{\sigma}_T, \tilde{\tau}_4 = \tilde{\tau}_{TT}, \tilde{\tau}_5 = \tilde{\tau}_6 = \tilde{\tau}_{LT}.$$

Consequently, the coefficients (2.4) for the transversally isotropic unidirectional composite reduce to:

$$2F = 2\tilde{\sigma}_T^{-2} - \tilde{\sigma}_L^{-2},\ 2G = 2H = \tilde{\sigma}_L^{-2},$$
$$2L = \tilde{\tau}_{TT}^{-2},\quad 2M = 2N = \tilde{\tau}_{LT}^{-2}.$$

As certain components of stress vanish ($\sigma_3 = 0$, $\tau_4 = 0$, $\tau_5 = 0$), the Hill yield criterion reads:

$$F\sigma_2{}^2 + G\sigma_1{}^2 + H(\sigma_1 - \sigma_2)^2 + 2N\tau_6^2 \leq 1.$$

Using Kelvin's tensor notation (Kelvin 1856, 1878; Mehrabadi and Cowin 1990), Eq. (2.4) transforms to:

$$\widehat{\boldsymbol{\sigma}} \bullet \widehat{\mathbf{F}}^{(0)} \bullet \widehat{\boldsymbol{\sigma}} \leq 1, \tag{2.5}$$

$$\widehat{\mathbf{F}}^{(0)} = \begin{bmatrix} G+H & -H & -G & 0 & 0 & 0 \\ -H & F+H & -F & 0 & 0 & 0 \\ -G & -F & F+G & 0 & 0 & 0 \\ 0 & 0 & 0 & L & 0 & 0 \\ 0 & 0 & 0 & 0 & M & 0 \\ 0 & 0 & 0 & 0 & 0 & N \end{bmatrix}, \widehat{\boldsymbol{\sigma}} = \begin{bmatrix} \sigma_{11} \\ \sigma_{22} \\ \sigma_{33} \\ \sqrt{2}\sigma_{23} \\ \sqrt{2}\sigma_{31} \\ \sqrt{2}\sigma_{12} \end{bmatrix} = \begin{bmatrix} \widehat{\sigma}_1 \\ \widehat{\sigma}_2 \\ \widehat{\sigma}_3 \\ \widehat{\tau}_4 \\ \widehat{\tau}_5 \\ \widehat{\tau}_6 \end{bmatrix}.$$

The Mises–Hill criterion is the tensor invariant function of stress:

$$\widehat{\boldsymbol{\sigma}} \bullet \widehat{\mathbf{F}}^{(0)} \bullet \widehat{\boldsymbol{\sigma}} = F(\widehat{\sigma}_2 - \widehat{\sigma}_3)^2 + G(\widehat{\sigma}_3 - \widehat{\sigma}_1)^2 + H(\widehat{\sigma}_1 - \widehat{\sigma}_2)^2 + L\widehat{\tau}_4^2 + M\widehat{\tau}_5^2 + N\widehat{\tau}_6^2 \leq 1.$$

The greatest advantage of the tensor notation (2.5) is its tensor invariance. The Mises–Hill criterion permits the application of tensor rules for coordinate transformations. This makes the Mises–Hill criterion in Kelvin's tensor notation applicable for optimization purposes.

2.1.5.2 Pressure-Sensitive Mises–Hill Anisotropic Failure Criterion

The Mises–Hill criterion also allows for taking the pressure sensitivity of the material into account. In Voigt's notation, the pressure-sensitive Mises–Hill criterion in the principal axes reads:

$$\boldsymbol{\sigma} \bullet \widetilde{\mathbf{F}}^{(0)} \bullet \boldsymbol{\sigma} + P\mathrm{Tr}^2(\boldsymbol{\sigma}) \equiv \boldsymbol{\sigma} \bullet \widetilde{\mathbf{F}}_P \bullet \boldsymbol{\sigma} \leq 1, \quad \mathrm{Tr}(\boldsymbol{\sigma}) = \sigma_1 + \sigma_2 + \sigma_3, \tag{2.6}$$

$$\widetilde{\mathbf{F}}_P = \begin{bmatrix} G+H+P & -H+P & -G+P & 0 & 0 & 0 \\ -H+P & F+H+P & -F+P & 0 & 0 & 0 \\ -G+P & -F+P & F+G+P & 0 & 0 & 0 \\ 0 & 0 & 0 & 2L & 0 & 0 \\ 0 & 0 & 0 & 0 & 2M & 0 \\ 0 & 0 & 0 & 0 & 0 & 2N \end{bmatrix}.$$

The constant P in (2.6) characterizes the pressure sensitivity. This value is a dimensionless constant that has to be determined experimentally. The eigenvalues of the matrix $\widetilde{\mathbf{F}}_P$ are listed in Table 2.4. The eigenvalues are all positive if the coupling constant P is positive. In this case, the surface in six-dimensional space that represents the failure criterion is convex. Namely, the convexity of the failure surface in the six-dimensional stress space is indispensable for the physical sense of any failure criterion.

Using Kelvin's tensor notation, the pressure-sensitive Mises–Hill criterion becomes:

$$\widehat{\boldsymbol{\sigma}} \bullet \widehat{\mathbf{F}}^{(0)} \bullet \widehat{\boldsymbol{\sigma}} + P\mathrm{Tr}^2(\boldsymbol{\sigma}) \equiv \boldsymbol{\sigma} \bullet \widehat{\mathbf{F}}_P \bullet \boldsymbol{\sigma} \leq 1. \tag{2.7}$$

Table 2.4 Eigenvalues of the matrix $\widetilde{\mathbf{F}}_P$ for the pressure-sensitive Mises–Hill criterion.

Λ_1	$3P$
Λ_2, Λ_3	$F+G+H \pm \sqrt{F(F-G)+G(G-H)+H(H-F)}$
$\Lambda_4, \Lambda_5, \Lambda_6$	$2L, 2M, 2N$

$$\widehat{\mathbf{F}}_P = \begin{bmatrix} G+H+P & -H+P & -G+P & 0 & 0 & 0 \\ -H+P & F+H+P & -F+P & 0 & 0 & 0 \\ -G+P & -F+P & F+G+P & 0 & 0 & 0 \\ 0 & 0 & 0 & L & 0 & 0 \\ 0 & 0 & 0 & 0 & M & 0 \\ 0 & 0 & 0 & 0 & 0 & N \end{bmatrix}.$$

The pressure-sensitive Mises–Hill criterion is the tensor invariant function of stress.

2.1.5.3 Tensor-Polynomial Failure Criterion

Criteria with a larger number of parameters offer the possibility of a valid formulation, regardless of the sign of acting stresses. The best known is the Goldenblat–Kopnov tensor-polynomial fracture criterion in tensor notation (Goldenblat and Kopnov 1965):

$$(F^{(2)}_{ij}\sigma_{ij})^a + (F^{(4)}_{ijkl}\sigma_{ij}\sigma_{kl})^b + (F^{(6)}_{ijklmn}\sigma_{ij}\sigma_{kl}\sigma_{mn})^c \leq 1. \tag{2.8}$$

The scalar quantities a, b, c are the shape factors. The tensors are the strength tensors of the second, fourth and sixth ranks, respectively:

$$\mathbf{F}^{(2)} = \left[F^{(2)}_{ij}\right], \mathbf{F}^{(4)} = [F^{(4)}_{ijkl}], \mathbf{F}^{(6)} = [F^{(6)}_{ijklmn}].$$

Since the stress tensor is symmetric $\sigma_{ij} = \sigma_{ji}$, the components of the tensors are symmetric:

$$F^{(2)}_{ij} = F^{(2)}_{ji}, F^{(4)}_{ijkl} = F^{(4)}_{ijlk} = F^{(4)}_{jikl} = F^{(4)}_{klij}, \ldots$$

Because of the tensor representation, the criterion (2.8) is valid for all coordinate systems. The fracture criterion accounts for the dependence of the shear strength on the directions of applied shear stresses as well the difference between tensile and compressive strengths.

Goldenblat and Kopnov examined in detail the first case where the shape factors are $a = 1, b = 1/2, c = -\infty$:

$$\overline{F}^{(2)}_{ij}\sigma_{ij} + \sqrt{\overline{F}^{(4)}_{ijkl}\sigma_{ij}\sigma_{kl}} \leq 1 \text{ or } \sqrt{\boldsymbol{\sigma} \bullet \widetilde{\mathbf{F}}^{(4)} \bullet \boldsymbol{\sigma}} + \widetilde{\mathbf{F}}^{(2)} \bullet \boldsymbol{\sigma} \leq 1. \tag{2.9}$$

Consequently, there are 15 unknown values in the criterion (2.8): six components of the symmetric tensor $\overline{F}^{(2)}_{ij}$ and nine components of the symmetric tensor $\overline{F}^{(4)}_{ijkl}$. Both vector Voigt's and Kelvin's notations for stress components could be used for the Goldenblat–Kopnov tensor fracture criterion.

In another important case of the shape factors $a = 1, b = 1, c = -\infty$, the tensor failure criterion of the second rank reads:

$$\overline{\overline{F}}^{(2)}_{ij}\sigma_{ij} + \overline{\overline{F}}^{(4)}_{ijkl}\sigma_{ij}\sigma_{kl} \leq 1, \text{or } \boldsymbol{\sigma} \bullet \widetilde{\mathbf{F}}^{(4)} \bullet \boldsymbol{\sigma} + \widetilde{\mathbf{F}}^{(2)} \bullet \boldsymbol{\sigma} \leq 1. \tag{2.10}$$

We use the symbols $\overline{F}$ and $\overline{\overline{F}}$ to differentiate between the components of the tensors in both frequent cases.

Consider a material with three orthogonal planes of symmetry. The planes of symmetry of the material are referred to as P_1, P_2, P_3. An important example for such material delivers the unidirectional composite. The normal to one plane in unidirectional composites is in the fiber direction (1) and the two other directions (2 and 3) are orthogonal to the fiber axis. Assume that the directions of principal stresses are orthogonal to planes P_1, P_2, P_3. For an orthotropic material, where the tensor-polynomial criterion in the coordinate system $Ox_fy_fz_f$, which is associated with planes of symmetry, the shear stresses vanish (Figure 1.2, Chapter 1). In the coordinate system $Ox_fy_fz_f$ the criterion with two positive constants a, b reads:

$$f(\sigma_1,\sigma_2,\sigma_3) = (F_{11}^{(2)}\sigma_1 + F_{22}^{(2)}\sigma_2 + F_{33}^{(2)}\sigma_3)^a + \\ (F_{1111}^{(4)}\sigma_1{}^2 + F_{2222}^{(4)}\sigma_2{}^2 + F_{2222}^{(4)}\sigma_3{}^2 + 2F_{1122}^{(4)}\sigma_1\sigma_2 + 2F_{2233}^{(4)}\sigma_2\sigma_3 + 2F_{3311}^{(4)}\sigma_3\sigma_1)^b \le 1. \tag{2.11}$$

There are nine coefficients in (2.11), which must be determined from at least nine experiments. In particular, if the shape factors are $a=1$, $b=1/2$, $c=-\infty$ or $a=1$, $b=1$, $c=-\infty$, the domain (2.11) in three-dimensional space (σ_1, σ_2, σ_3) is a triaxle ellipsoid. Its position in space is determined by three coordinates of the centrum. Additionally, three Euler angles determine the orientation and the three axes' length its shape. Similarly, in two-dimensional space, the tensor-polynomial criterion defines an ellipse. Its position determines two coordinates of centrum, two axes and one angle; so, in sum, five constants. Thus, exactly five experimental points are required for the determination of tensor coefficients in a two-dimensional case.

Notably, the tensor-polynomial criterion in both "dialects" by Goldenblat–Kopnov and Tsai–Wu leads to the same ellipsoid in a three-dimensional stress space (σ_1, σ_2, σ_3). To show this, the term with the square root in the Goldenblat–Kopnov criterion (2.9) should be isolated and then both sides squared. Then compare the squared Goldenblat–Kopnov criterion (2.9) with the Tsai–Wu criterion (2.10):

$$\overline{F}_{ijkl}^{(4)}\sigma_{ij}\sigma_{kl} \le (1 - \overline{F}_{ij}^{(2)}\sigma_{ij})^2 \text{ and } \overline{\overline{F}}_{ijkl}^{(4)}\sigma_{ij}\sigma_{kl} \le 1 - \overline{\overline{F}}_{ij}^{(2)}\sigma_{ij}. \tag{2.12}$$

Collect the same powers of stress components in both inequalities (2.12). The desired relations between the coefficients of both "dialects" of the unique tensor-polynomial criterion of the second order are the following:

$$\overline{F}_{iiii}^{(4)} - (\overline{F}_{ii}^{(2)})^2 = \overline{\overline{F}}_{iiii}^{(4)}, \\ \overline{F}_{ijij}^{(4)} - \overline{F}_{ii}^{(2)}\overline{F}_{jj}^{(2)} = \overline{\overline{F}}_{ijij}^{(4)}, \\ 2\overline{F}_{ii}^{(2)} = \overline{\overline{F}}_{ii}^{(2)}. \tag{2.13}$$

Both forms transform to each other after recalculation of their coefficients. However, the criterion pronounces much higher safety factors in the safe region. For comparison of both these failure criteria, consider an example. The three-dimensional graphs in Figures 2.1 and 2.2 of both criteria have been calculated using the values from Table 2.5.

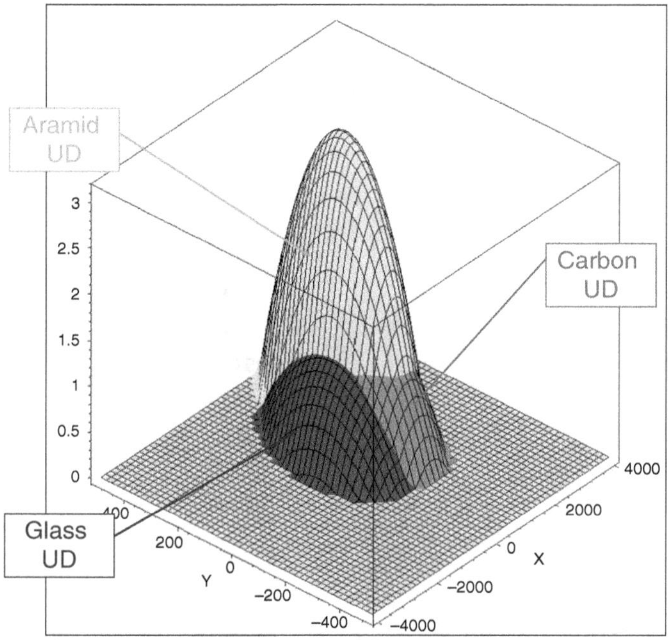

Figure 2.1 The surface of the safety factors for the Tsai–Wu criterion ($\alpha = 1, \beta = 1, \gamma = -\infty$), material properties from Table 2.5.

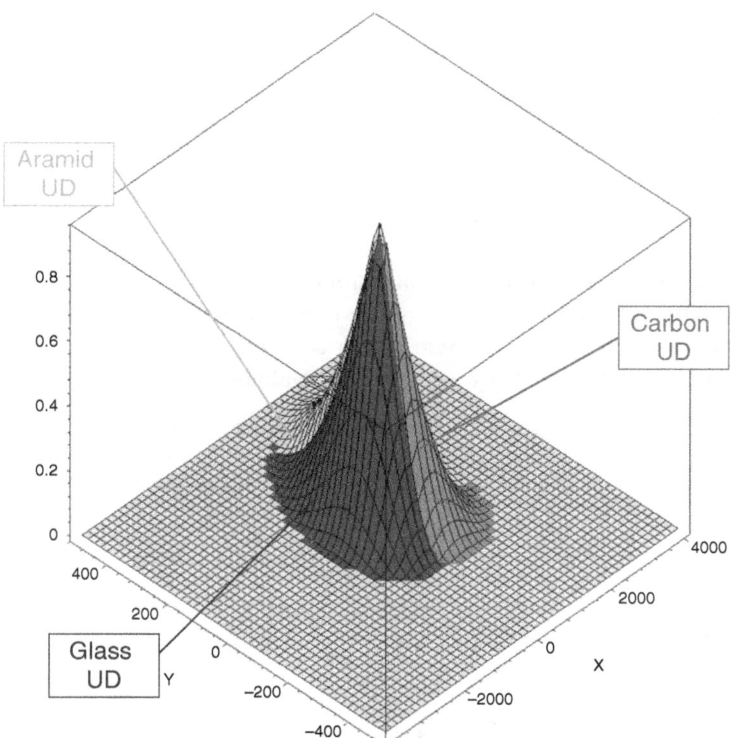

Figure 2.2 The surface of the safety factors for the Goldenblat–Kopnov criterion ($\alpha = 1, \beta = 1/2, \gamma = -\infty$), material properties from Table 2.5.

Table 2.5 Material parameters for failure criteria (Vasilev et al. 1990).

	Carbon UD	Glass UD	Aramid UD
σ_1^T, MPa	2100	1600	2200
σ_1^C, MPa	660	1108	310
$\sigma_2^T = \sigma_3^T$, MPa	210	350	280
$\sigma_2^C = \sigma_3^C$, MPa	170	89	112
$\tau_{12}^T = \tau_{12}^C$, MPa	35	35	30
$\tau_{23}^T = \tau_{31}^T$, MPa	191	310	200
$\tau_{23}^C = \tau_{31}^C$, MPa	228	935	414
E_L, MPa	70 000	142 000	95 000
E_T, MPa	4500	9130	5500
ν_{LT}	0.02	0.023	0.021
G_{LT}, MPa	2100	5490	2100
ρ, kg m^{-3}	2200	1550	1380

2.1.5.4 Tsai–Wu Criterion

The flow criterion developed by Mises and Hill was independently introduced by Tsai for orthotropic laminates and is commonly used for fracture evaluation. In particular, Tsai (1968) substituted the strengths of fiber-reinforced composites for the anisotropic flow limits of the Mises–Hill criterion. A similar criterion was referred to as the Zacharov criterion (Zacharov 1961) in the book by Goldenblat and Kopnov (1968, Eqs [57, 58] pp. 66–67). The Tsai–Wu criterion (Tsai and Wu 1971) could be considered to be a special case of the Goldenblat–Kopnov criterion for the following choice of the exponents: $a = 1$, $b = 1$, $c = -\infty$. The Tsai–Wu criterion provides much better predictions for unidirectional specimens with fiber directions deviating from the direction of loading in comparison to the criteria for maximum stresses, maximum strains or Mises–Hill. The Tsai–Wu criterion is also faultlessly suited to optimization purposes.

2.1.5.5 Assessment of Coefficients in Tensor-Polynomial Criteria

For experimental evaluation of the coefficients of tensor-polynomial criteria, a series of the experiments must be performed. For the assessment of the unknown coefficients, the experimentally acquired failure stresses are commonly evaluated:

a) three ultimate uniaxial tension vectors in the directions normal to planes P_i, $i = 1, 2, 3$:

$$\{\sigma_1^T, 0, 0\}, \{0, \sigma_2^T, 0\}, \{0, 0, \sigma_3^T\};$$

b) three ultimate uniaxial compression vectors in the directions normal to planes P_i, $i = 1, 2, 3$:

$$\{-\sigma_1^C, 0, 0\}, \{0, -\sigma_2^C, 0\}, \{0, 0, -\sigma_3^C\};$$

c) three ultimate shear stress vectors:

$$\{-\tau_{12}^T, \tau_{12}^T, 0\}, \{0, -\tau_{23}^T, \tau_{23}^T\}, \{\tau_{31}^T, 0, -\tau_{31}^T\};$$

d) three ultimate shear stress vectors:

$$\{\tau_{12}^C, -\tau_{12}^C, 0\}, \{0, \tau_{23}^C, -\tau_{23}^C\}, \{-\tau_{31}^C, 0, \tau_{31}^C\}.$$

The stresses τ_{ij}^T cause the tension $|\tau_{ij}^T|$ in fibers and equal valued compression $-|\tau_{ij}^T|$ in the normal to fiber directions. Correspondingly, the stresses $\tau_{ij}^C, i \neq j$ cause the compression $-|\tau_{ij}^C|$ in fibers and equal valued tension $|\tau_{ij}^C|$ in the normal to fiber directions.

After the substitution of the stresses (a) and (b) in the function (2.11), the six equations appear:

$$\begin{aligned} f({\sigma_1}^T, 0, 0) &= (F_{11}^{(2)} {\sigma_1}^T)^a + (F_{1111}^{(4)} {\sigma_1}^{T2})^b = 1, \\ f(0, {\sigma_2}^T, 0) &= (F_{22}^{(2)} {\sigma_2}^T)^a + (F_{2222}^{(4)} {\sigma_2}^{T2})^b = 1, \\ f(0, 0, {\sigma_3}^T) &= (F_{22}^{(2)} {\sigma_3}^T)^a + (F_{3333}^{(4)} {\sigma_3}^{T2})^b = 1, \end{aligned} \tag{2.14}$$

$$\begin{aligned} f(-{\sigma_1}^C, 0, 0) &= (-F_{11}^{(2)} {\sigma_1}^C)^a + (F_{1111}^{(4)} {\sigma_1}^{C2})^b = 1, \\ f(0, -{\sigma_2}^C, 0) &= (-F_{22}^{(2)} {\sigma_2}^C)^a + (F_{2222}^{(4)} {\sigma_2}^{C2})^b = 1, \\ f(0, 0, -{\sigma_3}^C) &= (-F_{22}^{(2)} {\sigma_3}^C)^a + (F_{3333}^{(4)} {\sigma_3}^{C2})^b = 1. \end{aligned} \tag{2.15}$$

The solution of simultaneous algebraic Eqs. (2.14) and (2.15) gives the expressions for the diagonal coefficients:

$$\begin{aligned} F_{ii}^{(2)} &= \sqrt[a]{\frac{({\sigma_i}^T)^{2b} - ({\sigma_i}^C)^{2b}}{({\sigma_i}^T)^{2b}(-{\sigma_i}^C)^a - ({\sigma_i}^C)^{2b}({\sigma_i}^T)^a}}, \\ F_{iiii}^{(4)} &= \sqrt[b]{\frac{(-{\sigma_i}^C)^a - ({\sigma_i}^T)^a}{({\sigma_i}^T)^{2b}(-{\sigma_i}^C)^a - ({\sigma_i}^C)^{2b}({\sigma_i}^T)^a}}, \quad i = 1, 2, 3. \end{aligned} \tag{2.16}$$

Substitution of stresses (c) into (2.11) leads to:

$$\begin{aligned} f(-{\tau_{12}}^T, {\tau_{12}}^T, 0) &= (-F_{11}^{(2)} + F_{22}^{(2)})^a ({\tau_{12}}^T)^a + ({\tau_{12}}^T)^{2b} (F_{1111}^{(4)} + F_{2222}^{(4)} - 2F_{1122}^{(4)})^b = 1, \\ f(0, -{\tau_{23}}^T, {\tau_{23}}^T) &= (-F_{22}^{(2)} + F_{33}^{(2)})^a ({\tau_{23}}^T)^a + ({\tau_{23}}^T)^{2b} (F_{2222}^{(4)} + F_{3333}^{(4)} - 2F_{2233}^{(4)})^b = 1, \\ f({\tau_{31}}^T, 0, -{\tau_{31}}^T) &= (-F_{33}^{(2)} + F_{11}^{(2)})^a ({\tau_{31}}^T)^a + ({\tau_{31}}^T)^{2b} (F_{3333}^{(4)} + F_{1111}^{(4)} - 2F_{3311}^{(4)})^b = 1. \end{aligned} \tag{2.17}$$

From the Eq. (2.17), follow the expressions of the off-diagonal coefficients $F_{1122}^{(4)}$, $F_{2233}^{(4)}$ and $F_{3311}^{(4)}$ in terms of ultimate shear stresses and the already defined diagonal coefficients (2.16):

$$F_{iijj}^{(4)} = \frac{F_{iiii}^{(4)} + F_{jjjj}^{(4)}}{2} - \frac{\sqrt[b]{1 - ({\tau_{ij}}^T)^a (F_{ii}^{(2)} - F_{jj}^{(2)})^a}}{2({\tau_{ij}}^T)^2}, i \neq j. \tag{2.18}$$

Substitution of stresses (d) into (2.11) leads to three more conditions:

$$\begin{aligned} f({\tau_{12}}^C, -{\tau_{12}}^C, 0) &= (F_{11}^{(2)} - F_{22}^{(2)})^a ({\tau_{12}}^C)^a + ({\tau_{12}}^C)^{2b} (F_{1111}^{(4)} + F_{2222}^{(4)} - 2F_{1122}^{(4)})^b = 1, \\ f(0, {\tau_{23}}^C, -{\tau_{23}}^C) &= (F_{22}^{(2)} - F_{33}^{(2)})^a ({\tau_{23}}^C)^a + ({\tau_{23}}^C)^{2b} (F_{2222}^{(4)} + F_{3333}^{(4)} - 2F_{2233}^{(4)})^b = 1, \\ f(-{\tau_{31}}^C, 0, {\tau_{31}}^C) &= (F_{33}^{(2)} - F_{11}^{(2)})^a ({\tau_{31}}^C)^a + ({\tau_{31}}^C)^{2b} (F_{3333}^{(4)} + F_{1111}^{(4)} - 2F_{3311}^{(4)})^b = 1. \end{aligned} \tag{2.19}$$

From Eq. (2.19) follow different expressions of the off-diagonal coefficients $F_{1122}^{(4)}$, $F_{2233}^{(4)}$, $F_{3311}^{(4)}$ in terms of ultimate shear stresses and the already defined diagonal coefficients (2.14):

$$F_{iijj}^{(4)} = \frac{F_{iiii}^{(4)} + F_{jjjj}^{(4)}}{2} - \frac{\sqrt[b]{1 - (-\tau_{ij}{}^{C})^{a}(F_{ii}^{(2)} - F_{jj}^{(2)})^{a}}}{2(-\tau_{ij}{}^{C})^{2}}, i \neq j. \tag{2.20}$$

The explanation is as follows. As mentioned before, the position and form of a triaxial ellipsoid in space is determined by exactly nine stresses (a), (b) and (c). The same is valid for a generalized ellipsoid with the shape factors *a, b*. Three parameters define the center of the ellipsoid, three its orientation in space and three the axes. The system of stresses (d), however, delivers three additional conditions. Generally speaking, for arbitrary values of stresses $\tau_{ij}{}^{C}$ the conditions (2.19) are superfluous and cannot be satisfied. The superfluous conditions could be identically fulfilled using so called "compatibility conditions for stresses" (Goldenblat and Kopnov 1965). The solutions (2.18) and (2.20) must provide the identical solution for all off-diagonal coefficients $F_{iijj}^{(4)}$. Equating (2.18) and (2.20), the compatibility conditions for stresses $\tau_{ij}{}^{T}$ and $\tau_{ij}{}^{C}$ appear as:

$$\sqrt[b]{\frac{1 - (\tau_{ij}{}^{T})^{a}(F_{ii}^{(2)} - F_{jj}^{(2)})^{a}}{1 - (\tau_{ij}{}^{C})^{a}(F_{jj}^{(2)} - F_{ii}^{(2)})^{a}}} = \left(\frac{\tau_{ij}{}^{T}}{\tau_{ij}{}^{C}}\right)^{2}, i \neq j. \tag{2.21}$$

The ellipsoid, which represents the tensor-polynomial criterion in a three-dimensional principal stress space ($\sigma_1, \sigma_2, \sigma_3$), requires nine values for explicit positioning. If there are more than nine stress states acquired experimentally, the position of the ellipsoid must be determined using data fitting. For example, the mean square distance from the ellipsoid to the experimentally acquired points could be minimized. However, the resulting expressions are too bulky for the closed form analysis and numerical data processing is required.

Two usual settings for shape factors, $a = 1$, $b = 1/2$ and $a = 1$, $b = 1$ correspond to the Goldenblat–Kopnov and Tsai–Wu criteria. For the tensor-polynomial criterion in form by Goldenblat–Kopnov with $a = 1$, $b = 1/2$, the coefficients (2.16) reduce to:

$$\overline{F}_{ii}^{(2)} \stackrel{def}{=} \frac{1}{2\sigma_i{}^{T}} - \frac{1}{2\sigma_i{}^{C}}, \quad \overline{F}_{iiii}^{(4)} \stackrel{def}{=} \left(\frac{1}{2\sigma_i{}^{T}} + \frac{1}{2\sigma_i{}^{C}}\right)^{2}. \tag{2.22}$$

With $a = 1$, $b = 1$, the solution (2.16) gives the expressions for the diagonal coefficients the tensor-polynomial criterion in the form by Tsai–Wu:

$$\overline{\overline{F}}_{ii}^{(2)} \stackrel{def}{=} \frac{1}{\sigma_i{}^{T}} - \frac{1}{\sigma_i{}^{C}}, \quad \overline{\overline{F}}_{iiii}^{(4)} \stackrel{def}{=} \frac{1}{\sigma_i{}^{T}\sigma_i{}^{C}}. \tag{2.23}$$

The compatibility conditions (2.21) for stresses for both forms of the tensor-polynomial criterion are identical:

$$\frac{1}{\sigma_{ii}^{T}} - \frac{1}{\sigma_{ii}^{C}} - \frac{1}{\sigma_{jj}^{T}} + \frac{1}{\sigma_{jj}^{C}} = \frac{1}{\tau_{ij}^{C}} - \frac{1}{\tau_{ij}^{T}}, \quad for\ i \neq j. \tag{2.24}$$

2.2 Differentiating Criteria

2.2.1 Fiber and Intermediate Break Criteria

Differentiating criteria apply only to a certain type of failure. A prediction of different failure forms (failure modes) is possible.

The differentiating criteria are closely linked to the so-called classical laminate theory, since global loads must first be converted in their effect on the individual layer of a laminate. The criteria are then related to the local stress state in the individual layer.

Compared to the general failure criteria mentioned before, it should be noted that the calculation effort for several criteria, possibly in several iterations, is significantly higher. The further parameters, the so-called curve parameters, which are not based on classical material properties, must be determined.

The breakage of fibers leads to immediate failure of an overall composite material. The fiber break criteria are:

- *Simple fiber break criterion*
- *Fiber break criterion according to Puck*
- *ultimate fiber strain criterion*
- *ultimate fiber stress criterion*

An intermediate break is a break or crack in the matrix (cohesive failure) or the interface between fiber and matrix (adhesive failure). For this type of failure, there are the following criteria:

- *Inter-fiber breakage criterion according to Puck (*1996*)*
- *Inter-fiber break criterion (Hashin* 1980*)*
- *Inter-fiber break criterion (Cuntze* 2006*)*
- *Inter-fiber break criterion (Ladevèze and LeDantec* 1992*).*

2.2.2 Hashin Strength Criterion

The strength criterion (Hashin 1980) characterizes the convex volume (failure body) in six-dimensional space of stress components. Each face of the failure body is valid for a specific failure mode. Due to the transverse isotropy of the composite materials, inter-fiber breakage can in the most general case only be a function of stress variations with respect to rotation around the fiber direction. For characterization of failure of a unidirectional lamina, Hashin applies the stress invariants. In particular, the stress σ_1 is invariant in the fiber direction "1." The criterion is invariant with respect to rotation along the fiber direction axis "1."

The most general quadratic function of components of the stress tensor:

$$I_1 = \sigma_1, I_2 = \sigma_2 + \sigma_3, I_3 = \tau_4^2 - \sigma_2\sigma_3, I_4 = \tau_6^2 + \tau_5^2$$

reads:

$$f(I_1, I_2, I_3, I_4) =$$
$$A_1 I_1 + B_1 {I_1}^2 + A_2 I_2 + B_2 {I_2}^2 + C_{12} I_1 I_2 + A_3 I_3 + A_4 I_4. \tag{2.25}$$

The scalar values I_1, I_2, I_3, I_4 are invariant with respect to rotation along the fiber direction axis "1." That is, the inequality can be used as the failure criterion for transversely isotropic lamina:

$$f(I_1, I_2, I_3, I_4) \leq 1.$$

The coefficients A_1, B_1, A_2, B_2, C_{12}, A_3, A_4 are to be determined experimentally. For the determination of the unknown coefficients, the following rules apply (Table 2.6). The stress σ_{11} causes fiber failure. Thus, the fiber failure is affected only by the invariant I_1. The matrix failure could be affected by the transverse stresses σ_2, σ_3, τ_5 only. Therefore, the matrix failure is affected only by the invariants I_2, I_3. The shear stress τ_5 and τ_6 influence both the fiber and the matrix failure modes; the corresponding shear intensity is

$$I_4 = \tau^2 \equiv \tau_5^2 + \tau_6^2. \quad (2.26)$$

Each face of the failure body could be represented separately.

From this general quadratic form, the expressions for failure criterion for the special cases result.

In the case of a single layer in-plane stress state (Table 2.7):

$$I_1 = \sigma_1, I_2 = \sigma_2, I_3 = 0, I_4 = \tau_6{}^2. \quad (2.27)$$

Table 2.6 Coefficients of quadratic form of the Hashin strength criterion in a general stress state.

$\sigma_1 > 0, I_1 > 0$	Fiber tensile failure	$\frac{I_1}{X_1^T} = 1$ or $\left(\frac{I_1}{X_1^T}\right)^2 + \frac{I_4}{X_A{}^2} = 1$
$\sigma_1 < 0, I_1 < 0$	Fiber compressive failure	$\frac{I_1}{X_1^C} = 1$
$\sigma_2 + \sigma_3 > 0, I_2 > 0$	Inter-fiber (matrix) tensile failure	$\left(\frac{I_2}{X_2^T}\right)^2 + \frac{I_3}{X_T^2} + \frac{I_4}{X_A^2} = 1$
$\sigma_2 + \sigma_3 < 0, I_2 < 0$	Inter-fiber (matrix) compressive failure	$\frac{I_2}{X_2^C}\left[\left(\frac{X_2^C}{2X^T}\right)^2 - 1\right] + \left(\frac{I_2}{2X_T}\right)^2 + \frac{I_3}{X_T^2} + \frac{I_4}{X_A^2} = 1$

Table 2.7 Coefficients of quadratic form of the Hashin strength criterion in a plane stress state.

$\sigma_1 > 0, I_1 > 0$	Fiber tensile failure	$\frac{\sigma_1}{X_1^T} = 1$ or $\left(\frac{\sigma_1}{X_1^T}\right)^2 + \left(\frac{\tau_6}{X_A}\right)^2 = 1$
$\sigma_1 < 0, I_1 < 0$	Fiber compressive failure	$\frac{\sigma_1}{X_1^C} = 1$
$\sigma_2 + \sigma_3 > 0, I_2 > 0$	Inter-fiber (matrix) tensile failure	$\left(\frac{\sigma_2}{X_2^T}\right)^2 + \left(\frac{\tau_6}{X_A}\right)^2 = 1$
$\sigma_2 + \sigma_3 < 0, I_2 < 0$	Inter-fiber (matrix) compressive failure	$\frac{\sigma_2}{X_2^C}\left[\left(\frac{X_2^C}{2X^T}\right)^2 - 1\right] + \left(\frac{\sigma_2}{2X_T}\right)^2 + \left(\frac{\tau_6}{X_A}\right)^2 = 1$

Table 2.8 Coefficients of quadratic form of the Hashin strength criterion in case of pure shear.

$I_4 = 0$, $I_3 \neq 0$.	Transverse matrix failure	$\frac{I_3}{X_T^2} = 1$ or $\frac{\tau_4^2}{X_T^2} = 1$
$I_3 = 0$, $I_4 \neq 0$.	Axial shear matrix failure	$\frac{I_4}{X_A^2} = 1$ or $\frac{\tau_5^2 + \tau_6^2}{X_A^2} = 1$

In the case of pure shear loading, the only components of stress are τ_4, τ_5, τ_6 (Table 2.8).

2.2.3 Delamination Criteria

Delamination is the separation of two layers in a layered fiber-plastic composite (Sridharan 2008; Turon et al. 2006; Jun et al. 2014, 2015; Pascoe et al. 2013). Delamination can be regarded as a special case of intermediate fiber breakage. The classification is strongly oriented toward fracture mechanics and accordingly divides the loads leading to failure into three categories:

- *Normal load on layer level (Mode I)*
- *Shear load in the layer plane, parallel to the fiber orientation (Mode II)*
- *Shear load in the layer plane, perpendicular to the fiber orientation (Mode III)*

The loads naturally do not occur in their pure form in the real component. Accordingly, the experimental technique for characterizing delamination behavior is oriented not only to the pure loads but also to the combinations:

- *Mode I tested by DCB (double cantilever beam)*
- *Mode II tested by ENF (end notch fixture)*
- *Mode I + II tested by MMFR (mixed mode fixed ratio)*

2.3 Physically Based Failure Criteria

2.3.1 Puck Criterion

Since the composite elements of automotive suspensions have different, physically decoupled failure modes, the approach of considering fiber composites with the help of physically justified failure criteria has been developed. The break criterion with a serious physical background was developed by Puck (1996) and Puck and Schürmann (1998). The Puck criterion looks at the fiber breakage and intermediate fiber breakage of a laminate layer separately. While a simple brittle failure criterion based on the uniaxial loading of the fibers is used for the direction of the fibers, a theory based on the failure resistance of the active planes has been developed for the intermediate fiber failure. In the general three-dimensional case, the failure planes must be determined iteratively, which is disadvantageous for an application, especially when calculating a large number of values. In a single layer of a laminate, however, the evaluation of the flat stress state is usually possible without an iterative process. The disadvantage of the increased effort involved in determining the effort using a physically justified criterion is only problematic when used in manual calculation. In practice, stress calculation and further evaluation are usually computer-aided. With regard to the inexpensive computing power available, a more complex determination of the effort is only disadvantageous in the context of a computationally intensive optimization.

The interaction of different stress components in this criterion is combined with a material stress. This criterion predicts whether fiber or matrix failure will occur first. The criterion of maximum stress predicts the type of failure without taking stress interactions into account.

The basis of Puck's failure criterion is that it is not the UD layer stresses that are considered, but the stresses in a possible plane of fracture, the so-called plane of action, which are evaluated. The failure criterion is based on the three basic hypotheses of Mohr and Hashin for a brittle fracture:

- the failure limit of a material is determined by the stresses in the fracture plane;
- an intermediate fiber break always occurs in a fiber-parallel cutting plane;
- for an intermediate fiber breakage, only the normal and shear stresses are decisive.

Puck uses a seven-parameter model to calculate the angle-dependent "effective stress." If the stress is greater than an ultimate value, the fiber composite under consideration theoretically fails. Puck distinguishes two states of stress:

- *Case of tension load.* In the case of tension occurs the intermediate fiber break that supported by the normal stress and the delamination as the result of shear stress in matrix;
- *Case of compression load.* In this state the possible failure impaired by normal stress.

2.3.2 Cuntze Criterion

Based on Puck's approach, Cuntze developed an invariant approach. The principle of strictly assigning a failure criterion to each failure mode, which is based on an associated strength, was pursued (Cuntze 2006). The possible failure modes include fiber breakage and intermediate fiber breakage. The advantage here is that this criterion can be calculated without iterations.

The Cuntze (Cuntze and Freund 2004; Cuntze 2006) failure criterion is based on the so-called failure mode concept. The stresses are described by invariants with the aim of achieving a coordinate system that is as generally valid as possible and independent of failure criterion for material behavior.

The failure mode concept offers several advantages:

- Each failure condition theoretically represents an independent failure mechanism and forms part of the complete failure surface of the failure body.
- Each failure mechanism is determined by strength.
- Each invariant term represents a physical mechanism in case of failure: a volume change, a change of shape or the internal friction of the affected material element.

The five failure modes occur both in the two-dimensional and in the three-dimensional stress case. Adaptation of the failure curve to test data is only permitted in areas where are defined by a unique failure mode. In the interaction area between adjacent failure modes, the adaptation of the Failure area possible due to a simple, Weibull-based row formation. The use of invariants enables a scalar formulation that automatically takes the material symmetries into account. The invariants are formed from homogeneous stress terms with the lowest possible degree. The equations allow fast calculation, which has proven to be numerically very robust. From the results the failure modes with the lowest or highest reserve factor or numerically more advantageous by its inverse, the effort can be determined. This identifies the critical areas for structural optimization of a composite component under consideration. If necessary, the design of the component can then be adapted in such a way that it is neither undersized nor oversized.

With this failure criterion it is possible to account for non-linear material curves based on the real stress–strain curve. However, there are still too few test data available to make a statement on the quality of process in the non-linear material range. The material behavior "hardening" and "softening" can be recorded. The support effect by embedding fiber composite layers in a fiber composite could be estimated. The possible changes in transverse contraction coefficients with different load states can be also implemented.

The concept has the significant advantage that the occurrence of failure in an entire UD layer is expressed by only one equation. A case distinction with regard to tensile and compressive stresses, fiber breakage and intermediate fiber breakage is not necessary.

2.4 Rotational Transformation of Anisotropic Failure Criteria

All criteria mentioned previously are formulated in the intrinsic or local coordinate system, which is associated with a preferred direction. For example, the formulas of Hashin criterion are valid if the preferred direction x is the fiber direction "1." If the fiber direction "1" and preferred x direction are inclined to each other and the axis z coincides with the axis "3," then the stresses are to be recalculated.

Consider the rotational transformation of the Mises–Hill anisotropic failure criterion. After substitution of the corresponding expressions for stresses, the criteria depend on the inclination angle Φ. We use the following formulas:

$$\hat{\boldsymbol{\sigma}} = \hat{\mathbf{T}} \cdot \hat{\boldsymbol{\sigma}}^{(0)}, \hat{\boldsymbol{\sigma}}^{(0)} = \hat{\mathbf{T}}^{-1} \cdot \hat{\boldsymbol{\sigma}} = \hat{\mathbf{T}}^{T} \cdot \hat{\boldsymbol{\sigma}}.$$

With the substitution into (2.5) we get the Mises–Hill anisotropic failure criterion in the rotated axes:

$$\hat{\boldsymbol{\sigma}}^{(0)T} \bullet \hat{\mathbf{F}}^{(0)} \bullet \hat{\boldsymbol{\sigma}}^{(0)} = \hat{\boldsymbol{\sigma}}^{T} \bullet \hat{\mathbf{T}} \bullet \hat{\mathbf{F}}^{(0)} \bullet \hat{\mathbf{T}}^{T} \bullet \hat{\boldsymbol{\sigma}} = \hat{\boldsymbol{\sigma}}^{T} \bullet \hat{\mathbf{F}}(\Phi) \bullet \hat{\boldsymbol{\sigma}} \leq 1, \tag{2.28}$$

where in the last term the rotation of the stiffness tensor $\hat{\mathbf{F}}(\Phi)$ in the Kelvin's notation (Kelvin 1856, 1878) around the fiber angle Φ is used:

$$\hat{\mathbf{F}}(\Phi) = \hat{\mathbf{T}} \bullet \hat{\mathbf{F}}^{(0)} \bullet \hat{\mathbf{T}}^{T}. \tag{2.29}$$

Correspondingly, the matrix for the pressure-sensitive Mises–Hill anisotropic failure criterion reads in a rotated coordinate system:

$$\hat{\mathbf{F}}(\Phi) = \hat{\mathbf{T}} \cdot \hat{\mathbf{F}}_p \cdot \hat{\mathbf{T}}^{-1}, \ \hat{\mathbf{T}}^{-1} = \mathbf{T}^{T}.$$

Thus, in Voigt's notation, the criterion is given by Eq. (2.4):

$$\boldsymbol{\sigma}^{T} = \boldsymbol{\sigma}^{(0)T} \cdot \mathbf{T}_{\sigma}{}^{T} = \boldsymbol{\sigma}^{(0)T} \cdot \mathbf{T}_{\varepsilon}{}^{-1},$$

$$\boldsymbol{\sigma} = \mathbf{T}_{\sigma} \cdot \boldsymbol{\sigma}^{(0)} = (\mathbf{T}_{\varepsilon}{}^{T})^{-1} \cdot \boldsymbol{\sigma}^{(0)}.$$

In Voigt's notation, the criterion reads:

$$\boldsymbol{\sigma}^T \bullet \widetilde{\mathbf{F}}(\Phi) \bullet \boldsymbol{\sigma} = \boldsymbol{\sigma}^{(0)T} \cdot \mathbf{T}_{\varepsilon}^{-1} \bullet \widetilde{\mathbf{F}}(\Phi) \bullet (\mathbf{T}_{\varepsilon}^{T})^{-1} \cdot \boldsymbol{\sigma}^{(0)} =$$
$$\boldsymbol{\sigma}^{(0)T} \cdot \widetilde{\mathbf{F}}^{(0)} \cdot \boldsymbol{\sigma}^{(0)} \leq 1. \tag{2.30}$$

Thus, the matrix for the Mises–Hill anisotropic failure criterion in a rotated coordinate system reads as:

$$\widetilde{\mathbf{F}}^{(0)} = \mathbf{T}_{\varepsilon}^{-1} \bullet \widetilde{\mathbf{F}}(\Phi) \bullet (\mathbf{T}_{\varepsilon}^{T})^{-1}, \text{ or } \widetilde{\mathbf{F}}(\Phi) = \mathbf{T}_{\varepsilon} \bullet \widetilde{\mathbf{F}}^{(0)} \bullet \mathbf{T}_{\varepsilon}^{T}. \tag{2.31}$$

The same transformation laws are valid for the pressure-sensitive criterion:

$$\widetilde{\mathbf{F}}_P(\Phi) = \mathbf{T}_{\varepsilon} \bullet \widetilde{\mathbf{F}}_P^{(0)} \bullet \mathbf{T}_{\varepsilon}^{T}. \tag{2.32}$$

The coefficients of both the ordinary and pressure-sensitive Mises–Hill anisotropic failure criterion in Voigt's notation rotate similarly to the components of a compliance matrix (Eq. 1.29 in Chapter 1).

The non-zero components of the matrix (2.25) of the Mises–Hill criterion in the inclined coordinate system are presented in Table 2.9.

Because of the tensor invariance in the notation (Mehrabadi and Cowin 1990), the eigenvalues of the matrix $\|\widehat{\mathbf{F}}(\Phi)\|$ do not depend on the inclination angle. For the pressure-independent Mises–Hill criterion, the coefficient P in the Table 2.9 should be set to zero.

The rotational transformation of the components for tensor-polynomial criteria (Goldenblat–Kopnov, Tsai–Wu) could be obtained analogously:

$$F_{ijkl}^{(4)}(\Phi) = t_{ip}t_{jq}t_{kn}t_{ls}f_{pqns}^{(4)}, \quad F_{ij}^{(2)}(\Phi) = t_{ip}t_{jq}f_{pq}^{(2)}. \tag{2.33}$$

The effect of in-plane rotation is demonstrated Figures 2.3–2.5.

Table 2.9 Coefficients of the pressure-sensitive Mises–Hill criterion in the inclined coordinate system $c = \cos\Phi$, $s = \sin\Phi$.

$\widehat{F}_{11}$.	$(G-2N+F+4H)c^4+(-2F-4H+2N)c^2+H+F+P$
$\widehat{F}_{12} = \widehat{F}_{21}$.	$(-4H-G-F+2N)c^4+(G-2N+F+4H)c^2-H+P$
$\widehat{F}_{13} = \widehat{F}_{31}$.	$(-G+F)c^2-F+P$
$\widehat{F}_{16} = \widehat{F}_{61}$.	$-s(G-2N+F+4H)\sqrt{2}c^3-s(-2H-F+N)\sqrt{2}c$
$\widehat{F}_{22}$.	$(G-2N+F+4H)c^4+(2N-4H-2G)c^2+H+G$
$\widehat{F}_{23} = \widehat{F}_{32}$	$(-F+G)c^2-G+P$
$\widehat{F}_{26} = \widehat{F}_{62}$	$s(G-2N+F+4H)\sqrt{2}c^3+s(-G-2H+N)\sqrt{2}c$
$\widehat{F}_{33}$	$F+G+P$
$\widehat{F}_{36} = \widehat{F}_{63}$	$-sc\sqrt{2}(-G+F)$
$\widehat{F}_{44}$	$(-M+L)c^2+M$
$\widehat{F}_{45} = \widehat{F}_{54}$	$(-M+L)cs$
$\widehat{F}_{55}$	$(M-L)c^2+L$
$\widehat{F}_{66}$	$(4N-2F-2G-8H)c^4+(-4N+8H+2F+2G)c^2+N$

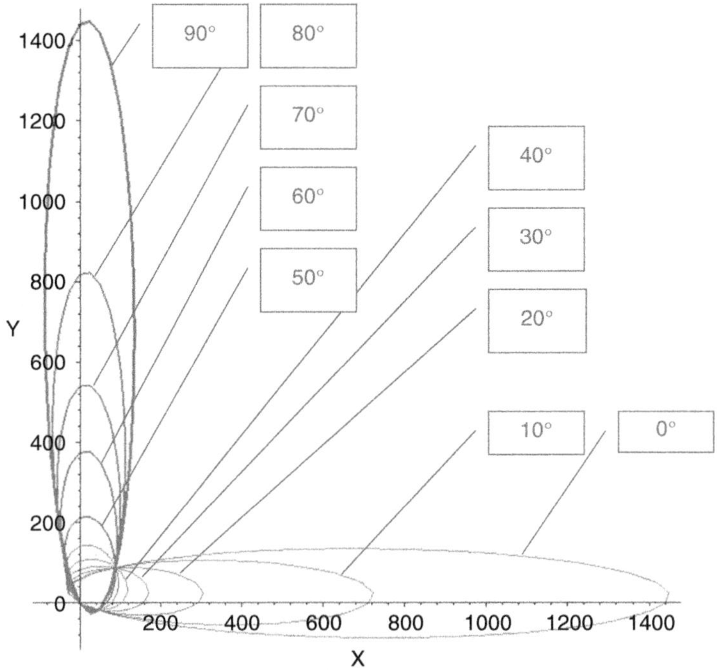

Figure 2.3 Transformation of tensor-polynomial failure criteria for a unidirectional fiberglass composite due to rotation of the fiber axis (Carbon UD, Table 2.5).

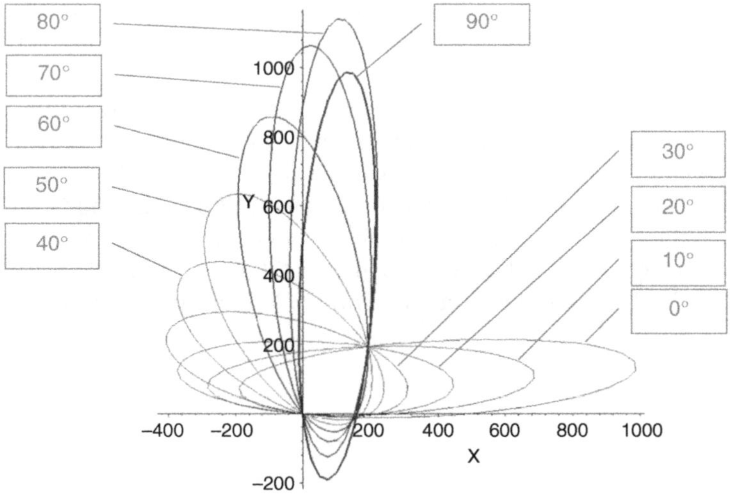

Figure 2.4 Transformation of tensor-polynomial failure criteria for a unidirectional fiberglass composite due to rotation of the fiber axis (Glass UD, Table 2.5).

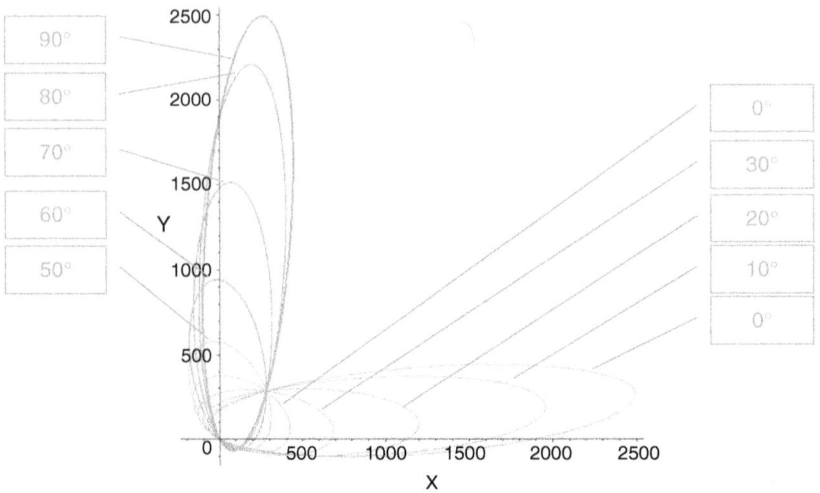

Figure 2.5 Transformation of tensor-polynomial failure criteria for a unidirectional fiberglass composite due to rotation of the fiber axis (Aramid UD, Table 2.5).

2.5 Conclusions

Chapter 2 summarizes the information about the failure criteria required for optimization purposes. The phenomenological approaches are presented in the matrix notation, which are suited for optimization problems. The main focus is on tensor-polynomial criteria because these allow invariant coordinate transformations, including rotation of coordinate systems. From the viewpoint of optimization, the invariant properties are advantageous. The traditional Voigt's notation and tensor Kelvin's notations are discussed. The rotation matrices for the quadratic tensor-polynomial criteria are derived.

Several aspects of anisotropy and strength of composite materials have been intensively studied in the following references: Elanchezhian et al. (2014), Daniel (2014), Daniel et al. (2011), Sevenois and Van (2015), Kaminski et al. (2015), Chou (1992), Timoshenko (1948), Goldenblat et al. (1977), Narayanaswami and Adelmann (1977), Reifsnider and Masters (1987), Knops (2008), Kuksenko and Tamuzs (1981), Hull and Clyne (1996), Altenbach et al. (2004), Gay et al. (2013), Handbook of Composites (1998), Kollár and Springer (2003), Liu and Nairn (1990), Robinson et al. (2012), Bitterli (1987) and Vassilopoulos and Keller (2011). These studies are recommended for further reading.

References

Agarwal, B.D., Broutman, L.-J., and Chandrashekhara, K. (2006). *Analysis and Performance of Fiber Composites*, 3e. Wiley.

Altenbach, H., Altenbach, J., and Kissing, W. (2004). *Mechanics of Composite Structural Elements*. Springer-Verlag, Berlin Heidelberg.

Bitterli, K. (1987). Ermüdungsverhalten eines glasfaserverstärkten Epoxidharzes mit besonderer Berücksichtigung von Schubeinflüssen. *Diss. Techn. Wiss. ETH Zürich*, 8407: 2599.

Chou, T.-W. (1992). *Microstructural Design of Fiber Composites*. Cambridge: Cambridge University Press.

Cuntze, R.G. (2006). Efficient 3D and 2D failure conditions for UD laminae and their application within the verification of the laminate design. *Compos. Sci. Technol.* 66: 1081–1096.

Cuntze, R.G. and Freund, A. (2004). The predictive capability of failure mode concept-based strength criteria for multidirectional laminates. *Compos. Sci. Technol.* 64: 343–377.

Daniel I.M. (2014) Failure of composite materials under multi-axial static and dynamic loading, International Symposium on Dynamic Response and Failure of Composite Materials, *DRaF2014, Procedia Engineering* 88, 10–17.

Daniel, I.M., Werner, B.T., and Fenner, J.S. (2011). Strain-rate-dependent failure criteria for composites. *Compos. Sci. Technol.* 71: 357–364.

Elanchezhian C., Ramnath B.V., Hemalatha J. (2014) Mechanical behaviour of glass and carbon fiber reinforced composites at varying strain rates and temperatures, *3rd International Conference on Materials Processing and Characterisation (ICMPC 2014). Proc. Mat. Sci.*, 6, 1405–1418.

Gay, D., Hoa, S.V., and Tsai, S.W. (2013). *Composite Materials, Design and Applications*. Boca Raton, FL: CRC Press LLC.

Goldenblat, I.I. and Kopnov, V.A. (1965). Strength of glass-reinforced plastics in the complex stress state. *Mech. Polym.* 2: 70.

Goldenblat, I.I. and Kopnov, V.A. (1968). *Criteria of Strength and Plasticity of Structural Materials*. Moscow: "Mashinostroenie" Publishing House.

Goldenblat, I.I., Bazhanov, V.L., and Kopnov, V.A. (1977). *Long-Term Strength in Mechanical Engineering*. Moscow: Mashinostroyeniye.

Hart-Smith, L.J. (2000). Strength of Unnotched laminates under multiaxial stress. In: *Comprehensive Composite Materials*, vol. 1 (ed. A. Kelly), 667–681. ISBN: 0-080437192.

Hashin, Z. (1980). Failure criteria for unidirectional Fiber composites. *J. Compos. Mater.* 46 (6): 329–334.

Hill, R. (1950). *Mathematical Theory of Plasticity*, 355. Oxford: Clarendon Press.

Hull, D. and Clyne, T.W. (1996). *An Introduction to Composite Materials, Second Edition*. Cambridge University Press.

Jun, L., Guiqiong, J., Bo, W. et al. (2014). Damage characteristics and constitutive modeling of the 2D C/SiC composite: Part I – experiment and analysis. *Chin. J. Aeronaut.* 27 (6): 1586–1597.

Jun, L., Guiqiong, J., Bo, W. et al. (2015). Damage characteristics and constitutive modeling of the 2D C/SiC composite: Part II – material model and numerical implementation. *Chin. J. Aeronaut.* 28 (1): 314–326.

Kaminski, M., Laurin, F., Maire, J.F. et al. (2015). Fatigue damage modeling of composite structures: the onera viewpoint. *Aerospace Lab* 1–12. https://doi.org/10.12762/2015.AL09-06.

Kelvin, W.T. (1856). Elements of a mathematical theory of elasticity. *Philos. Trans. R. Soc.* 146: 481–498.

Kelvin, W.T. (1878). Mathematical theory of elasticity. *Encycl. Br.* 7: 819–825.

Knops, M. (2008). *Analysis of Failure in Fiber Polymer Laminates – The Theory of Alfred Puck*. Springer. ISBN: 978-3-540-75764-1.

Kollár, L.P. and Springer, G.S. (2003). *Mechanics of Composite Structures*. Cambridge University Press.

Kuksenko, V. and Tamuzs, V. (1981). *Fracture Micromechanics of Polymer Materials*. Dordrecht: Springer Science+Business Media.

Lad'eveze, P. and LeDantec, E. (1992). Damage modelling of the elementary ply for laminated composites. *Compos. Sci. Technol.* 43 (3): 257–267.

Liu S., Nairn J.A. (1990) Fracture Mechanics Analysis of Composite Microcracking: Experimental Results in Fatigue, *Proceedings of the 5th Technical Conference on Composite Materials, American Society of Composites*, East Lansing, Michigan, June 11–14, 1990.

Mehrabadi, M.M. and Cowin, S.C. (1990). Eigentensors of linear anisotropic elastic materials. *Q. J. Mech. Appl. Math.* 43: 15–41.

Mises, R.V. (1928). Mechanik der plastischen Formänderung von Kristallen. *Zeits. f. angew. Math. u. Mech.*, 8 (3): 161–185.

Narayanaswami, R.H. and Adelmann, H. (1977). Evaluation of the tensor polynomial and Hoffmann strength theories for composite Materials. *J. Compos. Mater.* 11: 366.

Pascoe, J.A., Alderliesten, R.C., and Benedictus, R. (2013). Methods for the prediction of fatigue delamination growth in Compositesand adhesive bonds – a critical review. *Eng. Fract. Mech.* 112–113: 72–96.

Peters, S.T. (1998). *Handbook of Composites*, 2e. London: Chapman and Hall.

Puck, A. (1996). *Festigkeitsanalyse von Faser-Matrix-Laminates (Modelle für die Praxis)*. München, Wien: Hanser.

Puck, A. and Schürmann, H. (1998). Failure analysis of frp laminates by means of physically based phenomenological models. *Compos. Sci. Technol.* 58 (7): 1045–1067.

Reifsnider K.L., Masters J.L. (1987) Investigation of Characteristic Damage States in Composite Materials. *ASME Paper* No. 78-Aero-4, 3.

Robinson, P., Greenhalgh, E., and Pinho, S. (eds.) (2012). *Failure Mechanisms in Polymer Matrix Composites, Criteria, Testing and Industrial Applications*, Woodhead Publishing Series in Composites Science and Engineering. Woodhead Publishing.

Sevenois, R.D.B. and Van, P.W. (2015). Fatigue damage modeling techniques for textile composites: review and comparison with unidirectional composite modeling techniques. *Appl. Mech. Rev., ASME* 67 (2): 020802-1. https://doi.org/10.1115/1.4029691.

Sridharan, S. (ed.) (2008). *Delamination Behaviour of Composites*. Woodhead Publishing.

Sun, C.T. (2000). Strength analysis of unidirectional composites and laminates. In: *Comprehensive Composite Materials*, vol. 1 (ed. A. Kelly), 641–666. ISBN: 0-080437192.

Talreja, R. and Varna, J. (eds.) (2016). *Modeling Damage, Fatigue and Failure of Composite Materials*, Woodhead Publishing Series in Composites, Science and Engineering: Number 65. Cambridge: Elsevier Ltd.

Timoshenko, S. (1948). *Strength of Materials*. Toronto, New York, London: Van Nostrand.

Tsai, S.W. (1968). *Fundamental Aspects of Fiber Reinforced Plastic Composites*. New York: Wiley Interscience.

Tsai, S.W. and Wu, E.M. (1971). A general theory of strength for anisotropic materials. *J. Compos. Mater.* 5: 58–80.

Turon A., Costa J., Camanho P.P., Dávila C.G. (2006). Simulation of Delamination Propagation in Composites under High-Cycle Fatigue by means of Cohesive-Zone Models, *NASA/TM-2006-214532*.

Vasilev, V.V., Protasov, V.D., and Bolotin, V.V. (1990). *Composite Materials, Handbook*, 512. Moscow: Maschinostroenie.

Vassilopoulos, A.P. and Keller, T. (2011). *Fatigue of Fiber-reinforced Composites*. London: Springer-Verlag.

VDI (2006) *Development of FRP components (fibre-reinforced plastics) Analysis*, VDI 2014, Part 3.

Zacharow, K.V. (1961). Strength criterion for plastic materials. *Plastic Masses* 30 (6): 48–51.

3

Micromechanical Failure Criteria of Composites

3.1 Pullout of Fibers from the Elastic-Plastic Matrix

3.1.1 Axial Tension of Fiber and Matrix

This chapter deals with composite materials in which straight fibers are located as continuous fibers in a plastic embedding resin (matrix). The basic element of a fiber composite laminate, a so-called fiber multilayer composite, is the unidirectional single layer (monolaminate, monolayer). The component is given a mechanical anisotropy by a predetermined orientation of the fibers. This allows the strength to be specifically adapted to the stresses being applied, prevents crack propagation and significantly reduces brittleness.

The models of fracture mechanics of fiber composite materials were introduced in earlier work (Chamis 1969; Chamis et al. 1971; Dow and Rosen 1965). In the fracture mechanics of fiber-reinforced composites, a distinction is made between violent fracture, fatigue fracture and hot fracture (Figure 3.1). The most frequent fracture forms for a fiber composite material are shown in simplified form in Figure 3.2.

Usually, the mechanical properties of fiber composites are investigated by considering them as an effective homogeneous medium. If, however, the values of the stresses applied are comparable to the yield strength, the material can no longer be considered homogeneous. As long as the stresses are lower than the breaking stress, the deformation is almost uniform. As the stresses increase, the stress concentration in the fracture zones increases. The resulting non-linear deformation is concentrated in the areas of crack tips, local imperfections, fiber breaks and other material defects. It is therefore impossible to consider the material as effectively homogeneous. Thus, an investigation of micromechanical behavior is necessary. Microanalysis proves to be efficient when considering the strength, brittleness, crack propagation and fracture toughness of fiber composites. In addition, only microanalysis can provide information on the dependence of fracture properties on parameters, which is of great importance for strength optimization.

The basic mechanisms of the various types of failure of fiber composite materials (fiber breakage, matrix failure, delamination etc.) have not yet been precisely clarified in some cases and in several cases have only been described qualitatively. Nevertheless, there are a number of failure criteria that, within certain limits, make it possible to estimate the load-bearing capacity of the materials (Sun 2000; Hart-Smith 2000).

Design and Analysis of Composite Structures for Automotive Applications: Chassis and Drivetrain, First Edition. Vladimir Kobelev.

Companion website: www.wiley.com/go/kobelev/automotive_suspensions

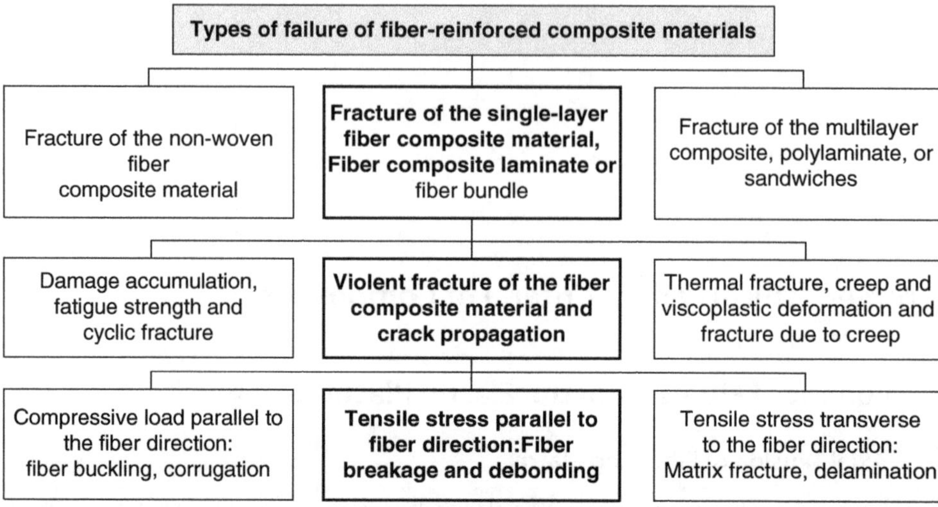

Figure 3.1 Types of failure for fiber-reinforced composite materials.

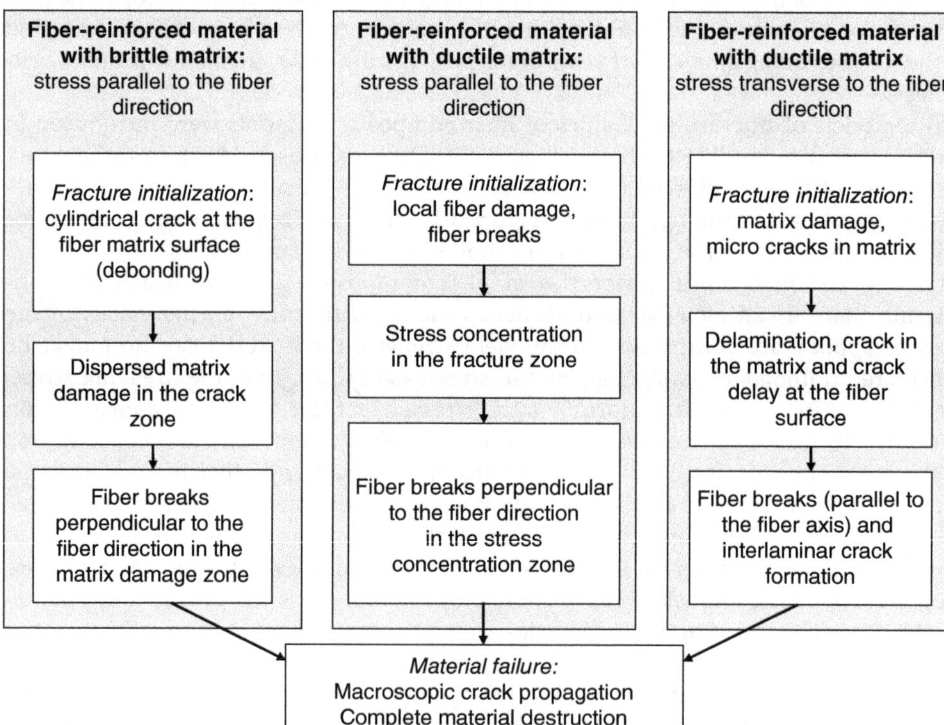

Figure 3.2 Common types of fracture mechanism of fiber-reinforced composite materials under tension load.

Violent fractures are caused by mechanical overload under moderately rapid to abrupt loading (Tsai et al. 1966; Argon 2000).

The characteristic feature of a fatigue fracture is that it does not occur as a result of a single extreme stress like a violent fracture, but rather in the course of time under the operating stress acting in an oscillating manner. Fatigue is thus caused by mechanical alternating stresses. The formation of a cyclic fracture surface indicates three phases. The phase of an initially submicroscopic and then microscopic crack formation (fiber fracture and matrix fracture) passes into the phase of macroscopic crack propagation. After damage accumulation for the formation of cracks, the cyclic crack grows gradually until the remaining material cross-section fails due to the constantly increasing stress caused by violent fracture. The detailed analysis of the violent fracture in FVWS can be found in studies by Argon (2000), Hancock (1970), Owen and Bishop (1974), Owen and Morris (1971), Kuksenko and Tamuzs (1981), Barbero (1999) and Hull and Clyne (1996).

Thermal fracture is caused by combined mechanical and thermal stress. Increased temperature and simultaneously acting mechanical stresses lead to changes in material properties, creep deformation and formation of pores and microcracks. Halpin and Meinecke (1969), Schapery (1972) and Sims (1972) investigate these events.

Violent fractures occurring during tensile loading of unidirectional fiber composites are investigated in this chapter. The material is subjected to tensile stress parallel to the fiber direction. The essential microscopic features and phenomenology of similar failures are described by Fichter (1969), Lifshitz and Rotem (1970) and Lockett (1970).

The other frequent fracture phenomena in composites are known, such as fiber curvature due to compressive stress parallel to fiber direction (Foye 1966; Greszczuk 1972, 1973) and crack propagation transverse to the fiber direction (Beaumont 1974; Koeneman 1969; Wu and Reuter 1965).

The present work deals with the fracture problem and failure criteria for fiber composites with ductile long fibers. The fibers are embedded in a brittle or ductile matrix material. Only single-fiber layers or fiber bundles are considered.

The fracture models of fiber-reinforced composites are usually based on exact solutions to idealized problems that can be illustrated by various aspects of the fracture process. The most important terms and modern models of fracture in fiber composite materials have been developed by Herrmann and Ferber (1995), Jones (1975), Kelly and Zweben (2000), Piggott (1980) and Sih (1991).

One of the first studies on micromechanical modeling of the crack in a fiber composite material can be found at (Hedgepeth and Van Dyke 1967), in which the equilibrium of a layer continuum with regularly changing elastic planes of two types is examined. Hedgepeth and Van Dyke investigated local defects and analytical calculation of stress distributions in plane fiber composites using the shear-lag method. This method requires that the fiber resists elongation and the matrix resists shear forces. The unidirectional single layer extends along the layer plane and contains a discontinuity. It considers displacements and shear stresses in the matrix and normal stresses in fibers and calculates stresses in the fibers adjacent to torn fibers. The stress concentration depends on one or more damaged layers in a plane fiber composite. It was found that elongation in the vicinity of a broken fiber exceeds infinite elongation by one-third and the so-called stress factor is equal to one-third. The original method was extended to describe the non-planar fiber arrangement and the plastic flow of the matrix.

Within this model, Dharani et al. (1983) and Goree and Gross (1979) solved the problems of fiber breakage and matrix decomposition. In addition, the stresses of the surrounding discontinuity were calculated.

Mikhailov (1973) used the Hedgepeth–Van Dyke shear model to treat the fracture of the fiber-reinforced plastic with parallel fiber orientation. The equations of equilibrium of the monolayer are solved under the following boundary conditions: the displacements of the undamaged fibers on the crack surface are zero; the stresses of the torn fibers on the crack surface are equal to zero.

The interaction of a fiber with the matrix was the subject of studies for failure assessment of fiber-reinforced composites (Aveston et al. 1971; Zhou et al. 1992; Meda and Steif. 1994; Xia et al. 1994).

In the work by Korczynskij et al. (1981) the conditions of crack propagation in a brittle, fiber-reinforced matrix were investigated, applying the energy hypothesis by Griffiths (1920). In the relaxation zone around the crack, the connection between the fibers and the matrix is broken. At the separation surface, constant shear stresses act between the matrix and the fibers; in the relaxation zone, the strain energy release rate is calculated. As a result, the strength of the composite material is a function of the fiber concentration. In the work by Hannant et al. (1983) it was assumed that fibers influence the size of the relaxation zone, in contrast to Korczynskij et al. (1981).

Selvadurai (1983) considered a problem of disk cracking in an infinite, fiber-reinforced, single-force loaded matrix. The presence of fibers bridging the crack means that there are certain normal stresses on the crack surface. In the crack environment, the fibers detach from the matrix (debonding). The stresses on the crack surface and the crack opening depend on the elongation of the detached fibers. The problem is summarized by a second kind of Fredholm integral equation and the dependence of the stress intensity factor on the elastic constants of the matrix and the fibers is calculated.

Budianski et al. (1986) studied an infinitely long crack in a fiber-reinforced matrix under the effect of homogeneous elongation in the fiber direction. The model accounted for the elastic contacts between fibers and matrix. The possibility of separation of fiber and matrix was also considered. Based on the hypothesis of the shape change energy, a comparison of the tensile strength of different fiber composites in undamaged fibers was carried out.

Stang (1987) developed a double inclusion model for a flat round crack. The crack in the matrix was modeled as an elliptical inclusion with zero stiffness in an unlimited tensile loaded matrix. The fibers bridging the crack were replaced by a fictitious, homogeneous, elliptical body. The separation of the fibers from the matrix was considered, resulting in a stress-free plane. Based on the strain energy hypothesis, the tensile strength of the fiber composite was calculated.

Tirosh (1973) described a model of a crack that propagated in the matrix parallel to the fiber layers. The fiber composite material was stretched in the fiber direction. Based on experimental data, the plastic zone in the crack tip area was simulated as a thin strip perpendicular to the discontinuity. The size of the plastic zone and the stresses at the boundaries were obtained by calculating the J-integral around the crack tip.

Interface debonding and frictional sliding associated with the fiber pullout process have been regarded as two significant mechanisms to increase the toughness of unidirectional fiber-reinforced composites (Liu et al. 1994; Zhang et al. 1999).

Hutchinson and Jensen (1990) studied the debonding and pullout after the fiber breaks that were complemented by frictional sliding. The fiber-matrix interaction was modeled by a cylindrical cell with two sets of boundary conditions. One condition modeled an isolated fiber-matrix unit and the other a matrix containing an array of unidirectional fibers in an infinite elastic cylindrical matrix.

Slaughter and Sanders (1991) investigated load-diffusion from axially loaded fibers embedded in elastic matrices. For the solution, fundamental elastostatic solutions were used for a point force and a point dilatation in either a fully or semi-infinite space. Tangential tractions across the fiber-matrix interface were modeled based on these solutions. The solutions of axial-symmetric elasticity problems are the three-dimensional analogs of Melan's first problem and Reissner's problem.

Meng and Wang (2015) studied the failure mechanism of interfacial debonding in fiber-reinforced composites. The stress and displacement fields of fiber and matrix were derived. The relation between the pullout length and debonding length of fiber was obtained. An energy release rate criterion for interfacial debonding was given. Parameter studies of the theoretical model were carried out for a glass fiber-reinforced epoxy composite.

In the publication by Newaz and Yung (1988), this model was used to study the behavior of a crack between layers under multiple loading of the pattern. In the work by Hollen et al. (1969) some models were provided to describe mechanical behavior and strength of fiber composites.

Specific failure modes of composites are debonding and delamination. The causes for the occurrence of delamination are manifold. In addition to poor production of the laminate, stresses (mainly in the thickness direction) at the free edges are of great importance. This was determined and investigated very early by Pipes and Pagano (1970), Pagano (1978) and Wang and Choi (1982). The "Free-Edge-Effect" (Achenbach 1975) causes delamination of layered plates starting directly from the edge.

The dynamic stress on the laminate due to impact is another cause of delamination because high transverse shear and normal stresses can occur (Blackman and Williams 1999).

The fiber-matrix separation or matrix cracks, which appear during the fabrication of a material, are also able to cause delamination. It is therefore often difficult to clearly identify the origin of failure (Corigliano 2007). The possible interactions of the individual failure types at the fiber-matrix level, the single layer level and the laminate level were reviewed by Ochoa and Reddy (1992) and Yin (1998) gives a clear summary of the delamination process in layered laminates and a detailed list of.

The linear-elastic solution for the crack tip fields in a general orthotropic medium has been solved by Sih et al. (1965). Knowledge of the stress and displacement fields then leads to the energy release rate.

An attempt to link the energy release rate to micromechanical characteristics of a component is demonstrated in the paper by Liu and Nairn (1990), which describes using the new energy release rate analysis for a fracture mechanics-based interpretation of microcrack formation during fatigue loading.

The terms "base laminate" and "sub-laminate" are habitually used in the context of delamination. The "base laminate" is the cross-sectional part that has the larger dimension in the thickness direction at the delamination point. The "sub-laminate" designates the remaining thinner cross-sectional part. Both "base laminate" and "sub-laminate" can consist of several individual layers. In particular, delamination is of great importance with regard to a possible stability failure due to buckling of the "base laminate" or "sub-laminate." An additional problem is that the local delamination is invisible from the outside and is difficult to detect. The numerous publications on the metrological recording of delamination clearly prove this (Davies et al. 1998). When composites are processed into multidirectional laminates, many different damage events will usually occur (Robinson et al. 2012; Schürmann, 2007). Some of this damage may initially appear uncritical but can impair future fatigue strength. Although damage and local failure are caused by processes in microscopic dimensions, we will take a simplified look at the macroscopic damage forms of fiber breakage and intermediate fiber breakage. Although the complete failure of multidirectional laminates under a quasi-static load is usually caused by fiber breakage, until this load-bearing capacity limit is reached after a large number of previous damage events, especially in the form of a matrix, there are breaks between the fibers. These inter-fiber breaks can already occur at loads far below the actual failure load in the presence of residual temperature or moisture stresses but have little effect on the quasi-static load-bearing capacity. Their prediction therefore requires a simulation of damage development under increasing mechanical load, whereby the effect of any damage occurring on the redistribution of stresses in the laminate must be modeled. Inter-fiber breaks produce stress concentrations at the boundary layers of the adjacent layers with local interlaminar stresses, which provoke delamination under fatigue load and separation of the layers from each other. From a certain size, the delamination can reduce the stability of the laminate to such an extent that instability failure with subsequent material failure occurs under pressure loads. The formation and spread of delamination under fatigue load are not completely understood as very complex processes. The simulation of already known mechanisms is also the subject of research rather than a means of practical interpretation. That is why you stay on the safe side when you measure a component you are looking at against the occurrence of the first layer. The calculation methods presented in the following are based on the plane stress state, as it can be calculated using multilayer theory. The formation and spread of delamination are caused by interlaminar stresses that are not covered by multilayer theory.

In this section, the approach of an infinite elastic cylindrical single-fiber debonding from elastic-plastic matrix material under the tension load along its axis is applied (Cox and Marshall 1994).

Due to the shear stresses at the separation surface, the fibers and matrix extend in the direction of the fiber axis, $u_f(z)$, $u_m(z)$. On the cylindrical interface surface between the fibers and the matrix, the shear stresses transmit the axial load between both components. The axial tensile stresses present in the fiber and the matrix. On the other hand, the shear stresses dominate in the matrix such that the shear stresses could be neglected in the thin fibers. Roving is a bundle, strand or multifilament yarn of parallel filaments (continuous filaments). After final polymerization of the matrix, the roving forms the continuous mechanical reinforcement. The reinforcement possesses much higher stiffness and strength than the surrounding matrix. There is a clear visible separation surface between the reinforcement and the matrix. The roving-reinforcements could be also modeled as a homogeneous single-fiber with the appropriate effective mechanical properties.

Consider a single-fiber surrounded by an elastic matrix. Both matrix and fibers are assumed to be isotopic with elasticity modules E_m and E_f, correspondingly. The fiber is an infinite solid cylinder with the radius r_f. The matrix is assumed to be an infinite hollow cylinder with the inner radius r_f and outer radius r_m. The volume concentration of fibers V_f:

$$r_m = r_f / \sqrt{V_f}. \tag{3.1}$$

The cross-section areas of hollow matrix and solid fiber, correspondingly, are:

$$s_m = \pi(r_m^2 - r_f^2), \quad s_f = \pi r_f^2. \tag{3.2}$$

The axial tensile stresses in the cylindrical fibers and in the hollow matrix cylinders read:

$$\sigma_m = E_m du_m / dz, \quad \sigma_f = E_f du_f / dz. \tag{3.3}$$

with these expressions, the axial forces in fibers and in matrix correspondingly turn into:

$$F_m = s_m\, \sigma_m, \qquad F_f = s_f\, \sigma_f. \tag{3.4}$$

The axial forces in matrix and fibers are in static equilibrium. The derivatives of the axial forces in the cylinders of matrix F_m and fibers F_f are equal to the shear forces f_s at the fiber surface:

$$dF_m / dz = f_s, \qquad dF_f / dz = -f_s. \tag{3.5}$$

The shear force f_s at the interface surface of matrix and fibers per length unit is a product of the shear stress at the fiber surface $\tau(z)$ and the circumference of the fiber cylinders $2\,\pi\, r_f$:

$$f_s = 2\,\pi\, r_f\, \tau. \tag{3.6}$$

The substitution of the shear force (3.6) in the equilibrium equation (3.5) leads to:

$$dF_m / dz = 2\,\pi\, r_f\, \tau, \quad dF_f / dz = -2\,\pi\, r_f\, \tau, \tag{3.7}$$

$$s_m d\sigma_m / dz = 2\pi r_f \tau, \quad s_f d\sigma_f / dz = -2\,\pi\, r_f\, \tau. \tag{3.8}$$

The axial extension of fibers and matrix remain linear-elastic until breakage. Consequently, the substitution of (3.3) into (3.8) leads to the equations (see Figure 3.3):

$$E_m s_m d^2 u_m / dz^2 = 2\pi\, r_f \tau, \quad E_f s_f d^2 u_f / dz^2 = -2\pi\, r_f \tau. \tag{3.9}$$

3.1.2 Shear Stresses in Matrix Cylinders

We consider cross-breakage, which splits either resin or fiber cylinders or both components into two parts. The upper broken part is in the upper half space $z > 0$ and the lower is in the lower half space $z < 0$. The plane crack also divides the components of the composite material along the rupture plane $z = 0$. The lower part is shown in Figure 3.4.

The axial displacement in the cross-section of fibers does not depend upon radius and is equal across the fiber cross-section. Thus, the shear in fibers disappears.

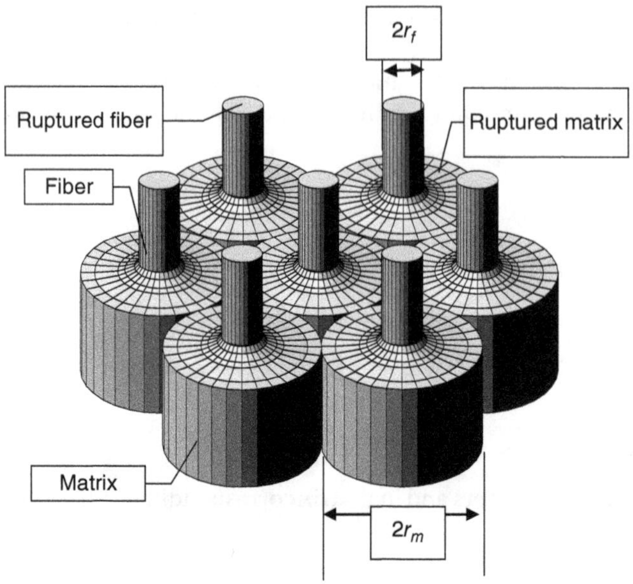

Figure 3.3 Pullout of fibers and rupture surfaces of fibers and matrix on the plane $z = 0$.

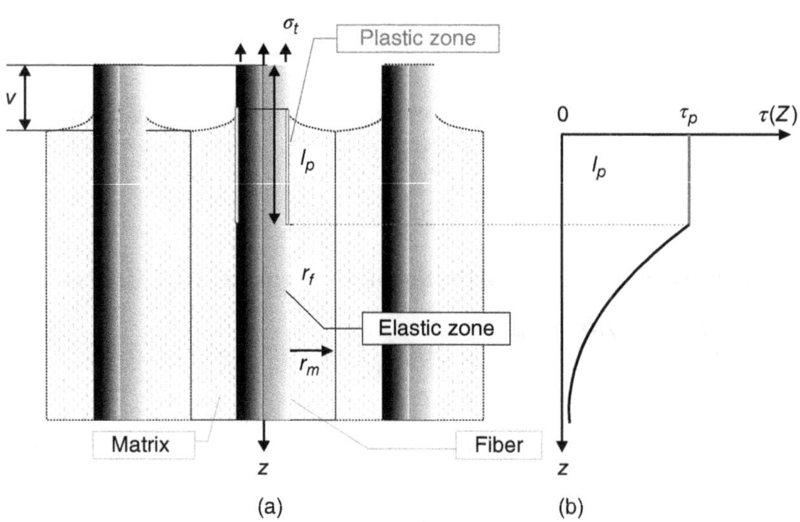

Figure 3.4 Zones along the pullout fibers. (a) Pulling the fibers from the elastic-plastic matrix. (b) Shear stresses at the fiber-matrix interface.

Otherwise, there is the shear stress $\tau_{rz}(r, z)$ in matrix (resin). The shear stress in matrix far from breakage $z > l_p$ is elastic. The shear stress increases in the vicinity of the separation surface in matrix $2\,\pi\, r_f$. If the shear stress at the interface exceeds the yield point of the matrix τ_p, in the vicinity of breakage $0 < z < l_p$ the shear deformation will be plastic. In the plastic area, the fiber glides on the matrix. The shear stress on the separation surface is:

$$\tau_{rz}(r_f, z) = \tau(z) \quad \text{for} \quad r = r_f. \tag{3.10}$$

Notably, the plastic sliding occurs only on the interface surface between fibers and matrix, because the shear stress in the outer zone of matrix $r > r_f$ gradually reduces as the inverse of radius, $\sim 1/r$. Therefore, the shear stress is insufficient to cause the plastic flow in the resin material far from the separation surface.

The axial displacement of the matrix $u_a(r, z)$ can be obtained by solving the boundary-value problem (Kobelev 2001):

$$\left\{ \tau_{rz} = G_m \frac{\partial u_a}{\partial r}, \quad \frac{\partial \tau_{rz}}{\partial r} + \frac{\tau_{rz}}{r} = 0 \quad \text{for } r_f < r < r_m, \right. \tag{3.11}$$

The solutions of Eq. (3.11) for shear stress $\tau_{rz}(r, z)$ and axial displacement $u_a(r, z)$ are:

$$\tau_{rz}(r, z) = \widetilde{C}_1(z)/r \tag{3.12}$$

and

$$u_a(r, z) = \widetilde{C}_1(z) \ \ln \ r + \widetilde{C}_2(z). \tag{3.13}$$

The boundary condition on the interface surface $r = r_f$ is:

$$u_a(r_f, z) = u_f(z). \tag{3.14}$$

The boundary condition on the outer surface of the matrix $r = r_m$ is equal to the mean axial displacement of matrix material:

$$u_a(r_m, z) = u_m(z). \tag{3.15}$$

The shear stress at the interface $\tau(z)$ depends upon the character of matrix deformation. The matrix deforms plastically in the region $0 < z < l_p$ near the rupture plane $z = 0$. Far from the separation surface $z > l_p$, the matrix deforms elastically. On the separation surface, the shear stress in the matrix material is the following function of the distance along the fiber axis:

$$\tau = \begin{cases} \tau_p & \textit{for } 0 < z < l_p, \\ \dfrac{G_m(u_m - u_f)}{r_f \ln(r_m/r_f)} & \textit{for} \quad z > l_p. \end{cases} \tag{3.16}$$

3.1.3 Coupled Elongation of Fibers and Matrix

The axial displacements of the fibers and the matrix are coupled together. We introduce two new auxiliary functions p(z), q(z). The auxiliary functions in terms of mean axial displacements of matrix and fiber cylinders are:

$$\mathrm{p} = u_m - u_f, \quad \mathrm{q} = u_m + u_f, \tag{3.17}$$

Accordingly, the mean axial displacements of matrix and fiber cylinders depend on the auxiliary functions p(z), q(z):

$$u_m = \frac{1}{2}\mathrm{p} + \frac{1}{2}\mathrm{q}, \quad u_f = -\frac{1}{2}\mathrm{p} + \frac{1}{2}\mathrm{q}. \tag{3.18}$$

After the substitution of the shear stress on the interface surface $\tau(z)$ (3.16) into Eq. (3.9), the equations appear for the axial displacements of the fibers and the matrix in the elastic region in terms of auxiliary functions:

$$\begin{cases} \lambda^2 d^2\mathrm{p}/dz^2 = \mathrm{p}, \\ \mu^2 d^2\mathrm{q}/dz^2 = \mathrm{p}, \end{cases} \quad \text{for } z > l_p. \tag{3.19}$$

In Eq. (3.19) we use the following parameters for length:

$$\lambda = \sqrt{\frac{r_m^2 E_T}{2G_m} \ln\left(\frac{r_m}{r_f}\right)}, \tag{3.20}$$

$$\mu = \sqrt{\left(\frac{1}{E_m V_m} - \frac{1}{E_f V_f}\right)^{-1} \frac{r_m^2}{2G_m} \ln\left(\frac{r_m}{r_f}\right)}. \tag{3.21}$$

Correspondingly, in the plastic region, the equations of the axial displacements of the fibers and the matrix could be expressed with two new constants $\widetilde{K}_1$, $\widetilde{K}_2$:

$$\begin{cases} d^2u_m/dz^2 = \widetilde{K}_1, \\ d^2u_f/dz^2 = -\widetilde{K}_2, \end{cases} \quad \text{for } 0 < z < l_p. \tag{3.22}$$

The parameters in (3.22) possess the dimension of inverse length:

$$\widetilde{K}_1 = 2\frac{\tau_p r_f}{V_m E_m r_m^2}, \quad \widetilde{K}_2 = 2\frac{\tau_p r_f}{V_f E_f r_m^2}. \tag{3.23}$$

3.1.4 Failures in Matrix and Fibers

3.1.4.1 Equations for Mean Axial Displacements of Fibers and Matrix

Now we study the three cases of failures of matrix and of fibers, as mentioned before. If both fibers and matrix are completely broken over the rupture surface $z = 0$, the crack in the composite material appears. Axial stresses in both components of the material vanish on the rupture surface:

$$E_f \frac{du_f}{dz}\bigg|_{z=0} = 0, \quad E_m \frac{du_m}{dz}\bigg|_{z=0} = 0.$$

This case was studied in detail by Sih and Chen (1981).

Otherwise, if only matrix separation at the rupture surface occurs, the fiber remains intact (Hutchinson and Jensen 1990). Accordingly, the boundary conditions for $z = 0$ express the absence of the fiber displacement and of axial stress in matrix:

$$u_f|_{z=0} = 0, \quad E_m \frac{du_m}{dz}\bigg|_{z=0} = 0. \tag{3.24}$$

Far from the breakage surface, the strains in fibers and matrix match:

$$\frac{d\,u_m}{dz}\bigg|_{z\to\infty} = \frac{d\,u_f}{dz}\bigg|_{z\to\infty} \leftrightarrow \mathrm{p}|_{z\to\infty} = 0.$$

Correspondingly, the average stress in the composite material tends to a given constant σ. The boundary conditions as $z \to \infty$ are:

$$\mathrm{p}|_{z\to\infty} = 0, \quad \left(V_m E_m \frac{d\,u_m}{dz} + V_f E_f \frac{d\,u_f}{dz}\right)\bigg|_{z\to\infty} = \sigma. \tag{3.25}$$

The second condition (3.25) expresses the average axial stress in composite material in terms of the axial stress in matrix and fiber multiplied by their corresponding cross-sections:

$$(V_m\,\sigma_m + V_f\sigma_f)|_{z\to\infty} = \sigma. \tag{3.26}$$

In the elastic range, the stresses in fibers and matrix decline exponentially fast toward the corresponding mean values:

$$E_m\frac{d\,u_m}{dz}\bigg|_{z\to\infty} = \sigma_m|_{z\to\infty},\quad E_f\frac{d\,u_f}{dz}\bigg|_{z\to\infty} = \sigma_f|_{z\to\infty}. \tag{3.27}$$

The solution to the boundary-value problem (3.19), (3.25) in the elastic region far from breakage is:

$$\begin{cases} \mathrm{p} = \widetilde{C}_1 e^{-\frac{z}{\lambda}} + \widetilde{C}_2 e^{\frac{z}{\lambda}}, \\ \mathrm{q} = \widetilde{C}_3 + \widetilde{C}_4 z + \dfrac{\lambda^2}{\mu^2}\left[\widetilde{C}_1 e^{-\frac{z}{\lambda}} + \widetilde{C}_2 e^{\frac{z}{\lambda}}\right], \end{cases} \quad \textit{for}\quad z > l_p. \tag{3.28}$$

In the region of the vicinity of breakage, the resin flows plastically along the surface of fibers. In the region of plastic flow of matrix, the solution of (3.22) reads:

$$\begin{cases} \mathrm{p} = \widetilde{D}_1 + \widetilde{D}_2 z + (\widetilde{K}_1 + \widetilde{K}_2)z^2/2, \\ \mathrm{q} = \widetilde{D}_3 + \widetilde{D}_4 z + (\widetilde{K}_1 - \widetilde{K}_2)z^2/2, \end{cases} \quad \textit{for}\quad 0 < z < l_p. \tag{3.29}$$

The quantities $\widetilde{C}_i, \widetilde{D}_i, i = 1.4$ are integration constants. The integration constants are determined by inserting the relationships (3.28) and (3.29) in (3.24) and (3.25).

To fulfill the boundary condition: $\mathrm{p}|_{z\to\infty} = 0$, we set $\widetilde{C}_2 = 0$.

The axial displacements in fibers and matrix $\mathrm{p}(z), \mathrm{q}(z), u_f(z), u_m(z)$ must be continuous and smooth functions:

$$\lim_{\varepsilon\to 0+}\mathrm{p}(l_p+\varepsilon) = \lim_{\varepsilon\to 0+}\mathrm{p}(l_p-\varepsilon),\quad \lim_{\varepsilon\to 0+}\frac{d\mathrm{p}}{dz}(l_p+\varepsilon) = \lim_{\varepsilon\to 0+}\frac{d\mathrm{p}}{dz}(l_p-\varepsilon), \tag{3.30}$$

$$\lim_{\varepsilon\to 0+}\mathrm{q}(l_p+\varepsilon) = \lim_{\varepsilon\to 0+}\mathrm{q}(l_p-\varepsilon),\quad \lim_{\varepsilon\to 0+}\frac{d\mathrm{q}}{dz}(l_p+\varepsilon) = \lim_{\varepsilon\to 0+}\frac{d\mathrm{q}}{dz}(l_p-\varepsilon). \tag{3.31}$$

The general solution of the ordinary differential equations (3.28) and (3.29) contains arbitrary integration constants $\widetilde{C}_i, \widetilde{D}_i, i = 1.4$ and a parameter l_p, which characterizes the length of the plastic zone on the inner surface of matrix cylinder. In the point $z = l_p$ the shear stress on the inner surface is equal to the plastic stress of matrix material τ_p:

$$\frac{G_m(u_m - u_f)}{r_f\ln(r_m/r_f)}\bigg|_{z=l} = \tau_p. \tag{3.32}$$

The resolution of the Eq. (3.32) over the difference of the axial displacements leads to:

$$u_m - u_f = \widetilde{K}_3 \qquad \textit{for}\quad z = l_p. \tag{3.33}$$

The parameter in the Eq. (3.33) is the parameter of the length dimension:

$$\widetilde{K}_3 = \frac{\tau_p r_f}{G_m}\ln\left(\frac{r_m}{r_f}\right). \tag{3.34}$$

3.1.4.2 Solutions of Equations for Mean Axial Displacements of Fibers and Matrix

The expressions (3.28) and (3.29) must be used in the boundary conditions (3.24), (3.25), (3.30), (3.31) and (3.33). After substitution, eight variables must satisfy the following eight equations:

$$
\begin{aligned}
&\widetilde{D}_4 + \widetilde{D}_2 = 0, \quad \widetilde{D}_3 - \widetilde{D}_1 = 0,\\
&(V_m E_m + V_f E_f)\widetilde{C}_4 - 2\sigma = 0,\\
&\widetilde{D}_1 + \widetilde{D}_2 l_p + \frac{1}{2}(\widetilde{K}_1 + \widetilde{K}_2){l_p}^2 - \widetilde{C}_1 \exp\left(-\frac{l_p}{\lambda}\right) = 0,\\
&\widetilde{D}_2 l_p + (\widetilde{K}_1 + \widetilde{K}_2) l_p + \frac{1}{\lambda}\widetilde{C}_1 \exp\left(-\frac{l_p}{\lambda}\right) = 0,\\
&\widetilde{C}_2 + \widetilde{C}_4 l_p + \frac{\lambda^2}{\mu^2} C_1 \exp\left(-\frac{l_p}{\lambda}\right) = \widetilde{D}_3 + \widetilde{D}_4 l_p + \frac{1}{2}(\widetilde{K}_1 - \widetilde{K}_2){l_p}^2,\\
&\widetilde{C}_4 - \frac{\lambda^2}{\mu^2}\widetilde{C}_1 \exp\left(-\frac{l_p}{\lambda}\right) = \widetilde{D}_4 + (\widetilde{K}_1 - \widetilde{K}_2) l_p,\\
&\widetilde{D}_1 + \widetilde{D}_2 l_p + \frac{1}{2}(\widetilde{K}_1 + \widetilde{K}_2){l_p}^2 = \widetilde{K}_3.
\end{aligned}
\tag{3.35}
$$

The first three equations (3.35) follow from conditions (3.24) and (3.25), the last four equations satisfy (3.31) and (3.33). The solution of the linear simultaneous equations (3.35) delivers the integration constants $\widetilde{C}_i, \widetilde{D}_i, i = 1.4$:

$$
\begin{aligned}
&\widetilde{C}_1 = 2\exp\left(\frac{l_p}{\lambda}\right)\frac{(\varepsilon - l_p\widetilde{K}_1)}{\lambda^2 + \mu^2}\mu^2\lambda,\\
&\widetilde{C}_3 = \widetilde{K}_2{l_p}^2 + 2\frac{\lambda^2 - \mu^2}{\lambda^2 + \mu^2}(l_p + \lambda)(\widetilde{K}_1 l_p - \varepsilon),\\
&\widetilde{C}_4 = 2\varepsilon \equiv 2{E_L}^{-1}\sigma,\\
&\widetilde{D}_2 = -\widetilde{D}_4 = \frac{\mu^2 - \lambda^2}{\lambda^2 + \mu^2} l_p\widetilde{K}_1 - l_p\widetilde{K}_2 - \frac{2\varepsilon\mu^2}{\lambda^2 + \mu^2},\\
&\widetilde{D}_1 = \widetilde{D}_3 = \frac{1}{2}{l_p}^2\widetilde{K}_2 + 2\frac{\mu^2\varepsilon}{\lambda^2 + \mu^2}(l_p + \lambda) - \frac{\widetilde{K}_1}{\lambda^2 + \mu^2}\left[2\mu^2\lambda l_p + \frac{1}{2}(3\mu^2 - \lambda^2){l_p}^2\right].
\end{aligned}
\tag{3.36}
$$

Finally, the length of the plastic zone results in:

$$
l_p = \frac{1}{2} r_m^2 E_T \left(\frac{\sigma}{r_f \tau_p E_f V_f} - \frac{1}{\lambda G_m}\ln\frac{r_m}{r_f}\right). \tag{3.37}
$$

with the formulas (3.35), the displacement of the matrix at the breakage surface is:

$$
u_{m.0} \equiv u_m(z = 0) = \frac{\mathrm{p}(0) + \mathrm{q}(0)}{2} = \frac{\lambda + l_p}{E_f V_f}\sigma - \frac{\tau_p r_f l_p}{E_T r_m^2}(l_p + 2\lambda). \tag{3.38}
$$

3.1.5 Rupture of Matrix and Pullout of Fibers from Crack Edges in a Matrix

First, consider the case of the matrix breakage on the surface $z = 0$ (Figure 3.5a). The fibers remain intact and connect both crack edges. There are three possible regions of fiber pullout depending on the applied stress in the material. This model of the fiber composite material is particularly suitable for a material with a brittle matrix. The strength of the matrix for such materials is exhausted earlier than the strength of the fibers. The crack propagation in the matrix dominates the failure of the entire material.

3.1.5.1 Elastic Elongation (Case I)

The average stress in unidirectional material σ is considerably lower than σ_p:

$$\sigma < \sigma_p = \frac{r_f \tau_p E_f V_f}{\lambda G_m} \ln \frac{r_m}{r_f}. \tag{3.39}$$

In this case, the stress in the matrix does not exceed the value τ_p and the matrix deforms elastically. Therefore, the plastic zone disappears: $l_p = 0$.

From (3.9) and (3.49) the elastic displacement in the moment of breakage is the following:

$$u_R(\sigma) = \frac{E_m V_m}{E_L} u_{m.0} = \frac{E_m V_m}{E_f V_f} \frac{\lambda}{E_L} \sigma. \tag{3.40}$$

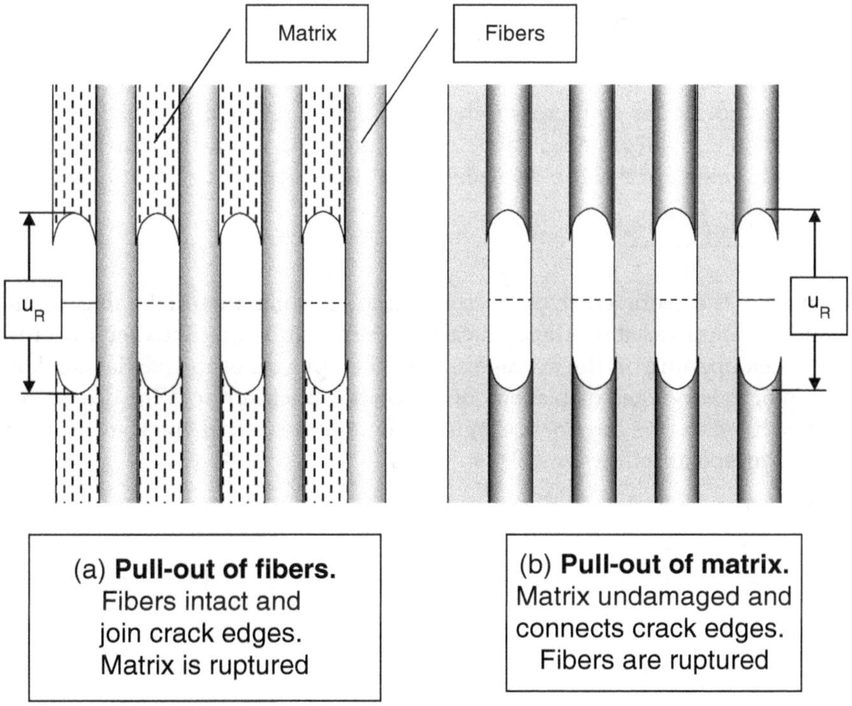

Figure 3.5 Bridging of crack surfaces by fibers (a) and by matrix (b).

The magnitude of elastic displacement in this case is directly proportional to average stress in a unidirectional material σ:

$$\sigma(u_R) = k_0 u_R, \quad \text{with } k_0 = \frac{E_f V_f}{E_m V_m} \frac{E_L}{\lambda}. \tag{3.41}$$

3.1.5.2 Plastic Sliding on the Fiber Surface (Case II)

If the shear stress in the matrix exceeds the plastic limit τ_p, the plastic flow on the separation surface between matrix and fiber occurs. The fiber slides over the surface of matrix in the vicinity of breakage. The plastic displacement $u_R(\sigma)$ on the breakage surface of fiber is the quadratic function of the average stress in the unidirectional material σ:

$$u_R(\sigma) = \frac{\sigma^2 + k_2}{k_1}, \tag{3.42}$$

$$k_1 = \frac{4r_f \tau_p E_f^2 V_f^2}{V_m E_m r_m^2} \frac{E_L}{E_T}, \quad k_2 = \frac{2(V_f \tau_p E_m r_f)^2}{E_T G_m r_m{}^2} \ln \frac{r_m}{r_f}. \tag{3.43}$$

Correspondingly, the stress on fibers is the square root of the axial displacement on the breakage:

$$\sigma(u_R) = \sqrt{k_1 u_R - k_2}. \tag{3.44}$$

The plastic flow on the interface surface occurs if the displacement on the breakage surface exceeds the ultimate value:

$$u_R \geq u_p = \frac{\tau_p E_m c_m r_f}{E_L G_m} \ln \frac{r_m}{r_f}. \tag{3.45}$$

If the displacement on breakage is low, $u_R < u_p$, pure elastic pullout of fibers happens.

3.1.5.3 Fiber Breakage (Case III)

If the average stress continues to increase, the condition is reached when the tensile strength of the fibers σ_{fb} will be exhausted. At the moment when the tensile stress achieves its ultimate value $\sigma = \sigma_{fb}$, the displacement at the fracture surface is equal to:

$$u_R = u_b \equiv \frac{(E_L \sigma_{fb}/E_f)^2 + k_2}{k_1}. \tag{3.46}$$

For the simulation, the averaged stress σ in the unidirectional material will be considered as an independent variable. The application of simulation methods requires the dependence of crack opening on the average stress. The opening of the partial crack is equal to the jump of the averaged displacements u_R on the crack surface. The direct and inverse functions that describe the fiber bridging effect on the partial crack surface are correspondingly denoted as follows:

$$u_R = f(\sigma), \quad \sigma = f^{-1}(u_R).$$

The relationship of the averaged stress in the composite material σ and the jump of the displacement u_R in all three regions, until to the total break of fibers occurs, reads (Figure 3.6):

$$u_R = f(\sigma) = \begin{cases} \sigma/k_0 & \mathit{for} \quad \sigma < \sigma_p, \\ \dfrac{\sigma^2 + k_2}{k_1} & \mathit{for} \ \sigma_p < \sigma < \sigma_b, \\ \infty & \mathit{for} \quad \sigma > \sigma_b, \end{cases} \tag{3.47}$$

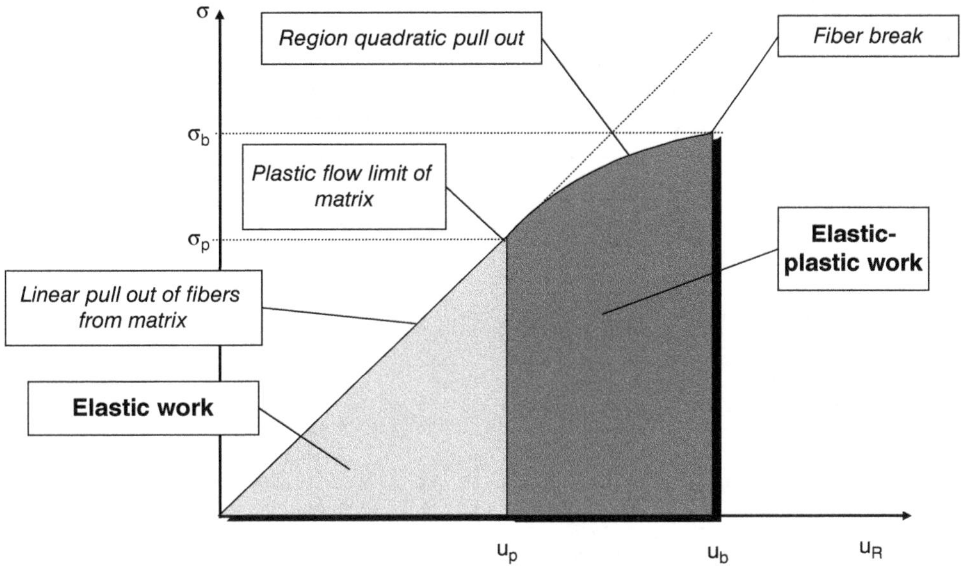

Figure 3.6 Dependence of the average stress upon the displacement of the matrix.

with

σ_b the ultimate fracture stress of fibers, $u_b = f(\sigma_b)$
σ_p the proportionality limit of matrix flow in the case of ruptured matrix cylinders, $u_p = f(\sigma_p)$.

3.1.6 Rupture of Fibers, Matrix Joints and Crack Edges

However, if the matrix shows a tough behavior during fracture, the fracture phenomenon changes completely. In this case, the fibers break practically independently of each other first. The fibers are ruptured over the surface, $z = 0$, but the matrix still joins both flaw edges. This rare case occurs when the matrix is extensible and ductile and the fibers are weak and brittle. The direct and inverse functions that describe the matrix bridging effect on the partial crack surface are correspondingly denoted as follows:

$$u_R = \widetilde{f}(\sigma), \quad \sigma = \widetilde{f}^{-1}(u_R).$$

The relationship of the average stress σ and the jump of the displacement on the crack surface u_R, until final disruption of matrix happens, reads:

$$u_R = \widetilde{f}(\sigma) = \begin{cases} \sigma/\widetilde{k}_0 & for \quad \sigma < \widetilde{\sigma}_p, \\ \dfrac{\sigma^2 + \widetilde{k}_2}{\widetilde{k}_1} & for \ \widetilde{\sigma}_p < \sigma < \widetilde{\sigma}_b, \\ \infty & for \quad \sigma > \widetilde{\sigma}_b. \end{cases} \tag{3.48}$$

The constants in the Eq. (3.48) are:

$\widetilde{\sigma}_b$ ultimate fracture stress of matrix, $\widetilde{u}_b = \widetilde{f}(\widetilde{\sigma}_b)$,
$\widetilde{\sigma}_p$ proportionality limit for matrix flow (fibers detached), $\widetilde{u}_p = \widetilde{f}(\widetilde{\sigma}_p)$.

Table 3.1 Coefficients of Eqs. (3.47) and (3.48).

	Fibers intact, matrix detached, Eq. (3.47)	Matrix intact, fibers ruptured, Eq. (3.48)
Elastic constant	$k_0 = \frac{E_f V_f}{E_m V_m} \frac{E_L}{\lambda}$	$\tilde{k}_0 = \frac{E_m V_m}{E_f V_f} \frac{E_L}{\lambda}$
Plastic constants	$k_1 = \frac{4 r_f \tau_p E_f^2 V_f^2}{V_m E_m r_m^2} \frac{E_L}{E_T}$	$\tilde{k}_1 = \frac{4 r_f \tau_p E_f^2 V_f^2}{V_m E_m r_m^2} \frac{E_L}{E_T}$
	$k_2 = \frac{2(V_f \tau_p E_m)^2}{E_T G_m} \ln \frac{r_m}{r_f}$	$\tilde{k}_2 = \frac{2(V_f \tau_p E_m r_f)^2}{E_T G_m r_m{}^2} \ln \frac{r_m}{r_f}$
plastic flow limit stress	$\sigma_p = \frac{r_f \tau_p E_f V_f}{\lambda G_m} \ln \frac{r_m}{r_f}$	$\tilde{\sigma}_p = \frac{r_m \tau_p E_m c_m}{\lambda G_m} \ln \frac{r_m}{r_f}$
Displacement to plastic flow	$u_p = \frac{\tau_p E_m V_m r_f}{E_L G_m} \ln \frac{r_m}{r_f}$	$\tilde{u}_p = \frac{\tau_p E_f V_f r_m}{E_L G_m} \ln \frac{r_m}{r_f}$
Displacement to break	$u_b = \frac{(E_L \sigma_{fb}/E_m)^2 + k_2}{k_1}$ Final fiber rupture	$\tilde{u}_b = \frac{(E_L \sigma_{mb}/E_m)^2 + \tilde{k}_2}{\tilde{k}_1}$ Final matrix rupture

In Eq. (3.47) the elastic constant of the matrix must be replaced by that of the fiber and vice versa. The coefficients of Eq. (3.47) and (3.48) are shown in the Table 3.1. Here σ_{mb} is the tensile strength of the matrix.

3.2 Crack Bridging in Elastic-Plastic Unidirectional Composites

3.2.1 Crack Bridging in Unidirectional Fiber-Reinforced Composites

In order to extend the crack in a plate of unit thickness under the effect of transverse tensile stress, a crack length increase must occur at its tips (see Figure 3.7). A portion of the elastic energy of the body is released and converted on one hand into work U_f to overcome resistance where the body opposes the crack extension, and on the other hand into kinetic energy U_k of the moving body parts. If the elastic energy in the cracked body before the crack extension U_0 and the energy released during the crack extension is U_e valid for the total energy of the system U_1 at any moment:

$$U_1 = U_0 - U_e. \tag{3.49}$$

All energy values are taken as a unit of thickness of the material. The crack becomes unstable and spreads in the material if U_k becomes positive. This is equivalent to the

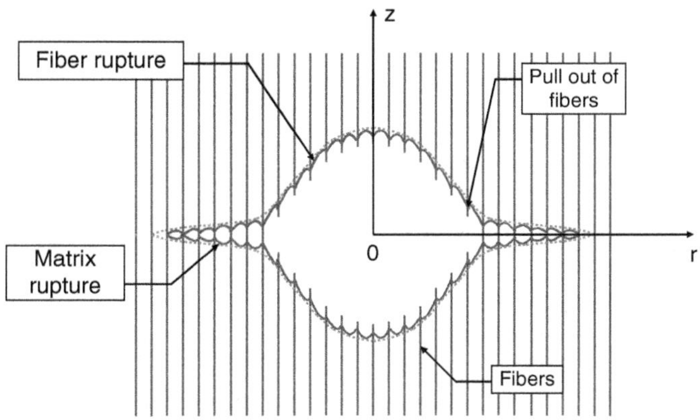

Figure 3.7 Crack in the unidirectional material.

condition:

$$dU_e/da > dU_f/da. \tag{3.50}$$

In the Eq. (3.50), dU_e/da is the energy release rate per thickness unit. The value dU_f/da in (3.50) is the energy necessary to the crack extension per thickness unit or, equivalently, the crack extension resistance R. This energy is required in the formation of new surfaces and the pullout of fibers (or matrix) from new surfaces. The condition (3.50) states that unstable crack extension is always possible if more released elastic energy is available than energy consumed for crack propagation. The relation between crack extension resistance R and fracture toughness K was given by Sih and Liebowitz (1968) and Poe (1984).

The crack extension resistance of both components embedded in the composite material is given in terms of the fracture toughness:

$$\mathrm{R}_f = K_f{}^2/\overline{E}, \quad \mathrm{R}_m = K_m{}^2/\overline{E}, \tag{3.51}$$

where (Sih and Liebowitz 1968):

$$\overline{E} = \sqrt{\frac{2\,E_L\,E_T}{(E_T/E_L)^2 + E_L/(2G_{LT}) - \nu_{LT}}}. \tag{3.52}$$

3.2.2 Matrix Crack Growth

If matrix rupture happens (Figure 3.5a), the specific crack extension resistance is the resistance force for the matrix:

$$\mathrm{R}_m = 2\,\gamma_m\,V_m. \tag{3.53}$$

For the average stress in material σ, the energy release rate:

$$\frac{dU_e}{da} = \int_0^{\sigma} \sigma du_R =$$

$$\begin{cases} \dfrac{\sigma^2}{2k_0}, & \sigma < \sigma_p, \\ \dfrac{\sigma^3 + 3k_2\sigma - \sigma_p{}^3 - 3k_2\sigma_p}{3k_1} + \dfrac{\sigma_p{}^2}{2k_0}, & \sigma_p < \sigma < \sigma_b, \\ \dfrac{\sigma_b{}^3 + 3k_2\sigma_b - \sigma_p{}^3 - 3k_2\sigma_p}{3k_1} + \dfrac{\sigma_p{}^2}{2k_0}, & \sigma > \sigma_b. \end{cases} \tag{3.54}$$

Unstable crack prolongation therefore begins when the energy release rate is greater than the crack extension resistance. If the energy release rate (3.54) for any averaged stress in the composite material σ will be equal to the crack extension resistance in matrix (3.53), the matrix crack will spread without restraint. In a pure elastic region, the stress that provokes the unabridged matrix crack growth is equal to:

$$\sigma_{\mathrm{ME}} = \sqrt{4k_0\gamma_m V_m} = K_m\sqrt{2k_0/\overline{E}}. \tag{3.55}$$

3.2.3 Fiber Crack Growth

If the fiber rupture happens (Figure 3.5b), the specific crack extension resistance is the resistance force for the fibers:

$$\mathrm{R}_f = 2\,\gamma_f\, V_f. \tag{3.56}$$

Similarly, for the average stress in material σ, the energy release rate in case of only fiber separation reads:

$$\frac{dU_e}{da} = \int_0^{\sigma} \sigma du_R =$$

$$\begin{cases} \dfrac{\sigma^2}{2\widetilde{k}_0}, & \sigma < \widetilde{\sigma}_p, \\ \dfrac{\sigma^3 + 3\widetilde{k}_2\sigma - \sigma_p{}^3 - 3\widetilde{k}_2\sigma_p}{3\widetilde{k}_1} + \dfrac{\sigma_p{}^2}{2\widetilde{k}_0}, & \widetilde{\sigma}_p < \sigma < \widetilde{\sigma}_b, \\ \dfrac{\sigma_b{}^3 + 3\widetilde{k}_2\sigma_b - \sigma_p{}^3 - 3\widetilde{k}_2\sigma_p}{3\widetilde{k}_1} + \dfrac{\sigma_p{}^2}{2\widetilde{k}_0}, & \widetilde{\sigma} > \widetilde{\sigma}_b. \end{cases} \tag{3.57}$$

The crack, which only disrupts fibers, grows if the energy release rate (3.56) for a stress σ matches the crack extension resistance in the fibers (3.53). In the pure elastic region, the stress, which leads to the unrestricted fiber crack growth, is equal to:

$$\sigma_{\mathrm{FE}} = \sqrt{4\widetilde{k}_0\gamma_f\, V_f} = K_f\sqrt{2\widetilde{k}_0/\overline{E}}. \tag{3.58}$$

with (3.54)–(3.57) the dependence of the stress that leads to both forms of damage in material is determined as the function of fiber volume fraction. We estimate the stress

Table 3.2 Properties of fiber and matrix for calculations.

Matrix EP	$E_m = 4.2GPa,$ $v_m = 0.35$	$\sigma_{mb} = 100MPa$ $\tau_p = 72MPa$	$K_m = 4MNm^{-3/2}$
Fiber Carbon $r_f = 0.1mm$	$E_f = 275GPa$ $v_f = 0.2$	$\sigma_{fb} = 3.5GPa$	$K_f = 40MNm^{-3/2}$

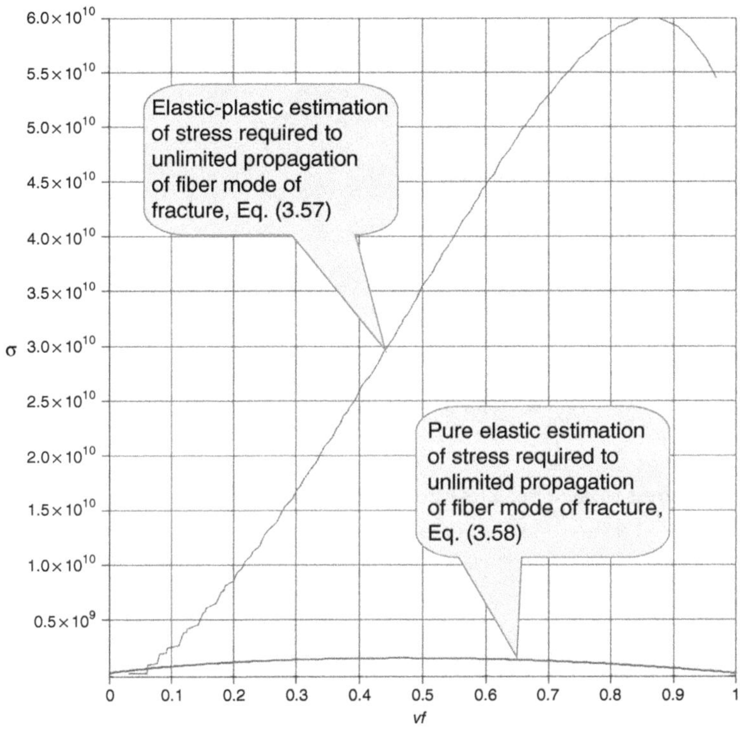

Figure 3.8 Dependence of the stress to unlimited break in fibers as function of fiber volume fraction.

that leads to both forms of damage in a material for a composite with the polymeric matrix. For calculation, the values from the Table 3.2 are assigned.

Figure 3.8 displays the curves of critical stress that provokes the emergence of a crack that destroys the fibers without violating the matrix. This mode is shown in Figure 3.5b. The pure elastic estimation diverges too far from the elastic-plastic calculation and cannot counted, as the values are obviously too low. The elastic-plastic solution presents values that are approximately equal to the critical stress of fiber alone.

Figure 3.9 shows the curves of critical stress that lead to the appearance of a flaw that destroys the matrix without damage in the fibers, the mode pictured in Figure 3.5a. The pure elastic estimation almost delivers the values of the elastic-plastic calculation.

Figure 3.10 presents the comparison of elastic-plastic stress solutions for flaws that lead to destruction of the matrix without damage to the fibers (Figure 3.5a, light gray curve) and damages the fibers without damage to the matrix (Figure 3.5a, dark gray

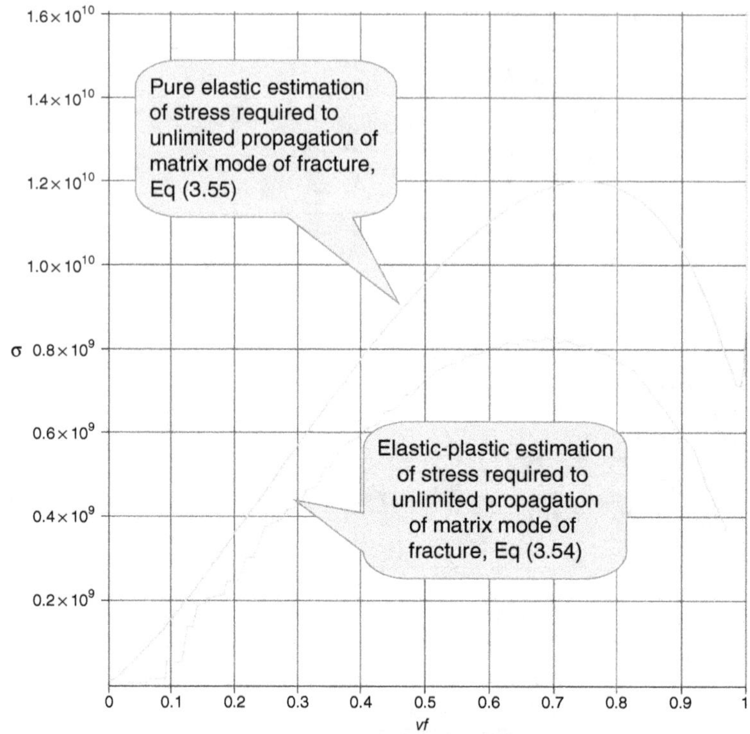

Figure 3.9 Dependence of the stress to unlimited break in matrix as function of fiber volume fraction.

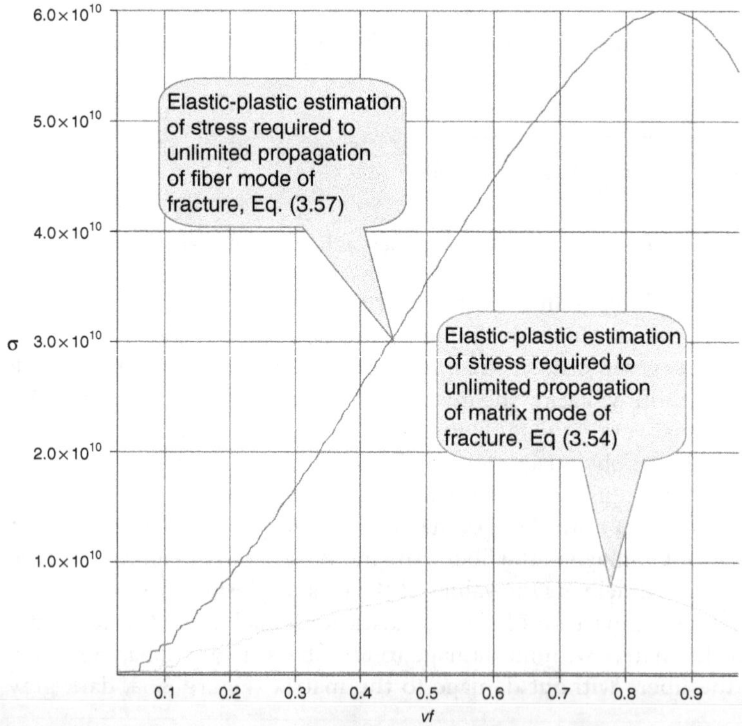

Figure 3.10 Dependence of the stress to unlimited break in fibers and matrix as function of fiber volume fraction for pure elastic pullout.

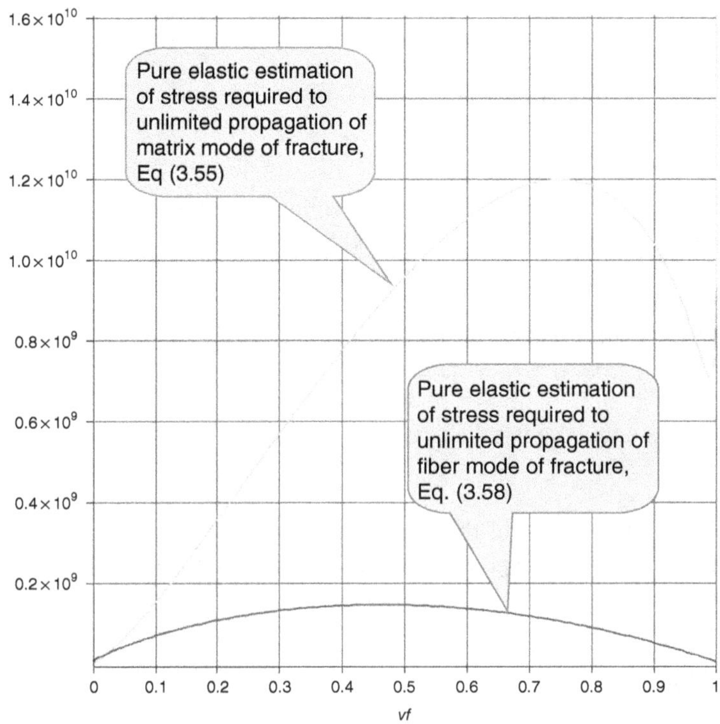

Figure 3.11 Dependence of the stress to unlimited break in fibers and matrix as function of fiber volume fraction for elastic-plastic pullout.

curve). The fiber damage stresses are much higher than the stresses that destroy the matrix.

Figure 3.11 displays the comparison of pure elastic stress solutions for flaws that lead to destruction of the matrix without damage to the fibers (Figure 3.5a, light gray curve) and damage the fibers without damage to the matrix (Figure 3.5a, dark gray curve). These curves do not provide a realistic estimation of critical stresses.

3.2.4 Penny-Shaped Crack

3.2.4.1 Crack in a Transversal-Isotropic Medium

Let's consider the problem of a plane microscopically small crack in the fiber composite with parallel fiber orientation whose surface should be perpendicular to the fibers. The diameter of the crack is larger than the mean distance between the fibers. The homogeneous tensile stresses act in the infinite direction of the fibers.

We want to replace the fiber composite material that has parallel fiber alignment around the crack with an effective, transversal-isotropic medium. In rotationally symmetric deformation, its elastic properties are characterized by five constants C_{11}, C_{33}, C_{44}, C_{13}, C_{12}, which are determined by the following effective elastic properties of the components of the material (Lekhnitskii 1981; Vannucci 2018):

$$S_{11} = \frac{1}{E_L}, S_{13} = -\frac{\nu_{LT}}{E_L}, S_{44} = \frac{1}{G_{LT}}, S_{12} = -\frac{\nu_{TT}}{E_L}, S_{33} = \frac{1}{E_T}, \tag{3.59}$$

and

$$C_{11} = \frac{S_{11}S_{33} - S_{13}{}^2}{\widetilde{m}(S_{11} - S_{12})}, C_{12} = \frac{-S_{12}S_{33} + S_{13}{}^2}{\widetilde{m}(S_{11} - S_{12})}, C_{13} = -\frac{S_{13}}{\widetilde{m}},$$
$$C_{44} = \frac{1}{S_{44}}, C_{33} = \frac{S_{11} + S_{12}}{\widetilde{m}}, \widetilde{m} = (S_{11} + S_{12})S_{33} - 2S_{13}{}^2. \tag{3.60}$$

We now introduce a cylindrical coordinate system (r, θ, z) into a transversal-isotropic medium in such a way that the center of the disk crack lies in the surface $z = 0$ at the origin of the coordinate system. The radius of the crack is r_c. The crack then occupies a flat surface with polar coordinates $r \le r_c$, $z = 0$.

For solution, dimensionless cylindrical coordinates are used:

$$\rho = r/r_c, \quad \eta = z/r_c. \tag{3.61}$$

The penny-shaped crack occupies the flat disk cavity in dimensionless coordinates:

$$\rho \le 1, \quad \eta = 0. \tag{3.62}$$

We label the components of the tensor of the effective stresses with σ_η, $\tau_{\rho\eta}$ and the components of the vector of effective displacements in the direction η with u_η. Correspondingly, the values of the averaged displacement u and the average stress σ in material at $z = 0$ are:

$$u_\eta(\rho, 0) = u, \qquad \sigma_\eta(\rho, 0) = \sigma. \tag{3.63}$$

3.2.4.2 Mechanisms of the Fracture Process

Fiber bridging plays a considerable role in fatigue delamination growth in composite materials. Fiber bridging decreases crack growth rate significantly with an increase of delamination length (Yao et al. 2016; Xu et al. 1998; Landis et al. 2000).

Crack propagation in a fiber composite material takes place through matrix and fiber breakage. We consider two mechanisms of the fracture process:

- the decomposition of the matrix (with debonding or decohesion at the fiber-matrix interface) when the stress intensity coefficient exceeds the critical value, and
- fiber breakage when the tension in the fibers reaches its tensile strength.

The greatest elongation is absorbed by the fibers in the central area of the crack. Therefore, it is assumed (and this is confirmed by numerical modeling) that, in the general case, the disk crack in fiber composites is divided into three annular zones (Figure 3.12):

- in the outer zone ($\rho_2 < \rho < 1$) the fibers and the matrix deform elastically;
- in the middle zone ($\rho_1 < \rho < \rho_2$) the matrix becomes plastic along the Fa-ser matrix interface, and the fibers detach themselves from the matrix. Here, plastic sliding and debonding (decohesion) take place at the fiber-matrix boundary;
- in the central zone ($\rho_1 > \rho$) the fibers are ruptured.

3.2.4.3 Crack Bridging in an Orthotropic Body With Disk Crack

In this section, the boundary-value problem for the disk crack is formulated. Due to the symmetry in relation to the plane $\eta = 0$, the problem can be considered in terms of the load on the half space $\eta \ge 0$. The boundary conditions are set at the boundary of the

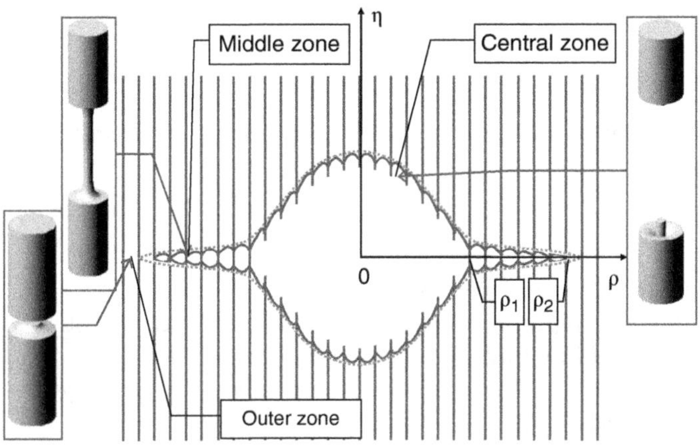

Figure 3.12 Crack section and zone structure.

half space. To solve the problem, we use the super-position principle, which only applies to constant stresses along the barrel direction.

The material is stretched in infinite distance from the crack along the fastening direction (axis η) by a homogeneous stress field: $\sigma_\eta = P$.

On the crack surface ($\eta = 0,\ \rho \le 1$) the shear stress is zero $\tau_{\rho\eta}(\rho, 0) = 0$. In the middle and outer zone ($\eta = 0,\ \rho_1 \le \rho \le 1$) the unbroken fibers connect both fissures. The normal stress $\sigma_\eta(\eta = 0) = \sigma$ at the crack banks is a known function of the opening of the crack: $\sigma = f^{-1}(u_\eta)$ with the inverse function, see Eq. (3.47).

According to (3.37) the normal stress in the outer zone is linear and in the middle zone non-linearly dependent on the opening.

In the central zone $\rho_1 > \rho$, there is no connection between the crack surfaces and the normal stress at the crack edges vanishes:

$$\sigma_\eta(\eta = 0) = 0. \tag{3.64}$$

To transform the boundary conditions in the middle and outer zone ($\eta = 0,\ \rho_1 \le \rho \le 1$), a homogeneous stress field overloads the initial stress field:

$$\sigma_\eta = -P. \tag{3.65}$$

The new boundary-value problem provides the conditions for normal stresses:

$$\begin{cases} \sigma_\eta(\eta \to \infty) = 0 & \text{in infinity, } \eta \to \infty \\ \sigma_\eta(\eta = 0) = \sigma - P & \text{on the crack edges} \end{cases} \tag{3.66}$$

In the plane $\eta = 0$, $\rho > 1$, the half spaces are coupled to each other. The boundary condition in the plane $\eta = 0$ means that the displacements on the extension of the crack surface ($\rho > 1$) are equal to zero:

$$u_\eta(\rho, 0) = 0, \quad \text{for } \rho > 1. \tag{3.67}$$

Summarizing these conditions, the mixed boundary values appear as:

$$\tau_{\rho\eta}(\rho, 0) = 0. \quad \text{for } \rho \ge 0. \tag{3.68}$$

$$\sigma_\eta(\rho, \eta = 0) = f^{-1}[u_\eta(\rho, 0)] - P \quad \text{for } \rho \le 1, \tag{3.69}$$

$$u_\eta(\rho, 0) = 0. \quad \text{for } \rho > 1. \tag{3.70}$$

In (3.69), $\sigma = f^{-1}(u_\eta(\rho, 0))$ is the inverse function, see Eq. (3.47). The normal and shear stresses are fading away as $\rho^2 + \eta^2 \to \infty$:

$$\tau_{\rho\eta} \to 0, \qquad \sigma_\eta \to 0. \tag{3.71}$$

3.2.4.4 Solution to an Axially Symmetric Crack Problem

To solve the axisymmetric equilibrium problem (Figure 3.13) for a transversal-isotropic medium, we introduce the potential functions $\varphi_1(\rho, \eta)$, $\varphi_2(\rho, \eta)$ according to Sneddon and Lowengrub (1969). The stresses and displacements identically satisfy the equilibrium conditions and can be expressed by potential functions:

$$\sigma_\eta = \frac{1}{a^2}\left(\alpha_1 \frac{\partial^2 \varphi_1}{\partial \eta^2} + \alpha_2 \frac{\partial^2 \varphi_2}{\partial \eta^2}\right), \tag{3.72}$$

$$\tau_{\rho\eta} = \frac{1}{a^2}\left(\beta_1 \frac{\partial^2 \varphi_1}{\partial \rho \partial \eta} + \beta_2 \frac{\partial^2 \varphi_2}{\partial \rho \partial \eta}\right), \tag{3.73}$$

$$u_\eta = \frac{1}{a}\left(\gamma_1 \frac{\partial \varphi_1}{\partial \eta} + \gamma_2 \frac{\partial \varphi_2}{\partial \eta}\right). \tag{3.74}$$

To convert the problem (3.68)–(3.74) into an integral equation, we represent the potential functions as Hankel transformations of auxiliary functions:

$$\varphi_i^*(\xi) = \mathrm{H}_0\{\, \mathbf{L}^R[\varphi_i(\rho, 0)],\ \rho \to \xi\} = \int_0^\infty \varphi_i(\rho, 0)\rho \mathrm{J}_0(\xi\rho)d\rho, \tag{3.75}$$

$$\varphi_i(\rho, \eta) = \int_0^\infty \xi\, \varphi_i^*(\xi)\mathrm{J}_0(\xi\, \rho)\exp(-\xi\, \eta/\omega_i)d\xi, \quad q_i = 1/\omega_i. \tag{3.76}$$

Condition (3.68) returns the relationship between the functions $\varphi_1^*(\xi)$ and $\varphi_2^*(\xi)$:

$$\omega_2\beta_1\varphi_1^*(\xi) + \omega_1\beta_2\varphi_2^*(\xi) = 0. \tag{3.77}$$

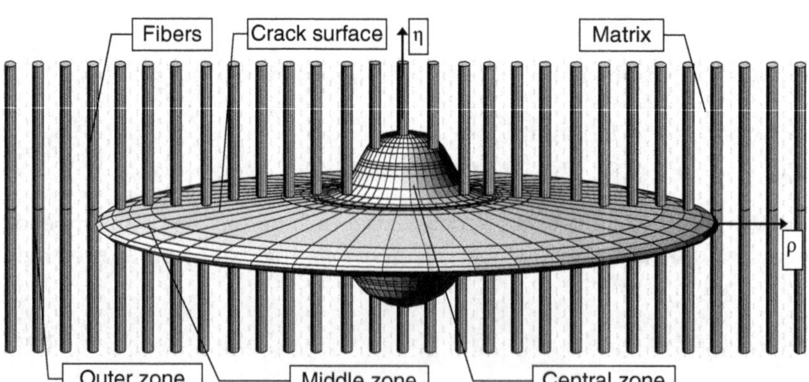

Figure 3.13 Zones at penny-shaped crack.

This condition is used to eliminate the function $\varphi_2^*(\xi)$:

$$\begin{cases} \sigma_\eta(\rho,\eta=0) = \left(\dfrac{\alpha_1}{a^2\omega_1^2} - \dfrac{\alpha_2\beta_1}{a^2\omega_1\omega_2\beta_2}\right)\int_0^\infty \xi^3\,\varphi_1^*(\xi)\mathrm{J}_0(\xi\,\rho)d\xi \\ u_\eta(\rho,\eta=0) = \left(\dfrac{\gamma_1}{a\omega_1} - \dfrac{\gamma_2\,\beta_1}{a\omega_1\,\beta_2}\right)\int_0^\infty \xi^2\,\varphi_1^*(\xi)\mathrm{J}_0(\xi\,\rho)d\xi \end{cases} \tag{3.78}$$

with the formula:

$$\varphi(\xi) = \xi^2\frac{\beta_1}{a^2\omega_1}\left(\frac{\alpha_1}{\beta_1\omega_1} - \frac{\alpha_2}{\omega_2\beta_2}\right)\varphi_1^*(\xi) \tag{3.79}$$

we get the expressions for the crack opening and pullout stress on the crack surface:

$$\begin{cases} u_\eta(\rho,\eta=0) = \Omega_R\int_0^\infty \varphi(\xi)\mathrm{J}_0(\xi\,\rho)d\xi, \\ \sigma_\eta(\rho,\eta=0) = \int_0^\infty \xi\,\varphi(\xi)\,\mathrm{J}_0(\xi\,\rho)d\xi. \end{cases} \tag{3.80}$$

In the first expression the constant reads:

$$\Omega_R = a(\gamma_1\beta_2 - \gamma_2\beta_1)\left(\frac{\alpha_1\beta_2}{\omega_1} - \frac{\alpha_2\beta_1}{\omega_2}\right)^{-1}.$$

From Eqs. (3.49), (3.69) and (3.70), an integral equation pair with respect to an unknown function $\varphi(\xi)$ appears:

$$\int_0^\infty \xi\,\varphi(\xi)\,\mathrm{J}_0(\xi\,\rho)\,d\xi + f^{-1}\left(\Omega_R\int_0^\infty \varphi(\xi)\mathrm{J}_0(\xi\,\rho)\,d\xi\right) = P \quad \text{for } \rho < 1, \tag{3.81}$$

$$\int_0^\infty \varphi(\xi)\mathrm{J}_0(\xi\,\rho)\,d\xi = 0 \quad \text{for } \rho \geq 1. \tag{3.82}$$

Equation (3.82) is fulfilled identically by substitution:

$$\varphi(\xi) = \int_0^1 g(t)\sin(\xi\,t)dt, \quad g(0) = 0. \tag{3.83}$$

where $g = g(t)$, $0 < t < 1$ – is the new potential function. The formula (3.83) follows from the relation:

$$\int_0^\infty \varphi(\xi)\mathrm{J}_0(\xi\,\rho)d\xi = \int_0^1 g(t)dt\int_0^\infty \sin(\xi\,t)\mathrm{J}_0(\xi\,\rho)d\xi =$$

$$= \begin{cases} \int_\rho^1 \dfrac{g(t)dt}{\sqrt{t^2-\rho^2}} & \textit{for } 0 \leq \rho < 1 \\ 0 & \textit{for } \quad \rho > 1 \end{cases}. \tag{3.84}$$

with

$$\int_0^\infty \sin(\xi\,t)\mathrm{J}_0(\xi\,\rho)d\xi = \frac{H(t-\rho)}{\sqrt{t^2-\rho^2}}. \tag{3.85}$$

The function $H(x)$ in (3.85) is the Heaviside function:

$$H(x) = 0 \quad \text{for } x \leq 0 \tag{3.86}$$

and

$$H(x) = 1 \quad \text{for } x > 0. \tag{3.87}$$

By substitution of $\varphi(\xi)$ in (3.49), Eq. (3.71) can be given as a non-linear integral-differential equation:

$$\int_0^\rho \frac{dg(t)}{dt}\frac{dt}{\sqrt{\rho^2 - t^2}} + f^{-1}\left(\Omega_R \int_\rho^1 \frac{g(t)dt}{\sqrt{t^2 - \rho^2}}\right) = P \quad \text{with } g(0) = 0. \tag{3.88}$$

For the derivation of Eq. (3.88), we perform the stepwise evaluation of the integral:

$$\int_0^\infty \xi\varphi(\xi)J_0(\xi\,\rho)d\xi = \int_0^\infty \xi J_0(\xi\,\rho)\left[-\frac{1}{\xi}g(1)\cos(\xi) + \frac{1}{\xi}\int_0^1 g'(t)\cos(\xi t)dt\right]d\xi =$$

$$= -g(1)\int_0^1 \cos(\xi)J_0(\xi\,\rho)d\xi + \int_0^1 g'(t)dt\int_0^\infty \cos(\xi\,t)J_0(\xi\,\rho)d\xi =$$

$$= \begin{cases} \displaystyle\int_0^\rho \frac{g'(t)dt}{\sqrt{\rho^2 - t^2}} & \textit{for } \ 0 \le \rho < 1, \\ \displaystyle -\frac{g(1)}{\sqrt{\rho^2 - 1}} + \int_0^1 \frac{g'(t)dt}{\sqrt{\rho^2 - t^2}} & \textit{for } \quad \rho > 1. \end{cases} \tag{3.89}$$

The stresses in the crack surface $\eta = 0$ are equal:

$$\sigma_\eta(\rho, 0) = \begin{cases} \displaystyle -\int_0^\rho \frac{g'(t)dt}{\sqrt{\rho^2 - t^2}} & \textit{for } \ 0 \le \rho < 1, \\ \displaystyle \frac{g(1)}{\sqrt{\rho^2 - 1}} - \int_0^1 \frac{g'(t)dt}{\sqrt{\rho^2 - t^2}} & \textit{for } \quad \rho > 1. \end{cases} \tag{3.90}$$

The stress behavior at the crack tip is considered decisive for crack propagation. The stresses are reduced to zero at a large distance from the crack tip. At the crack tip, all stress components show the same $1/\sqrt{r}$ – singularity:

$$\rho \to 1 + \varepsilon, \quad r = \rho - 1 > 0.$$

The level of the stress components is determined by the variable:

$$K_I = A_{44}g(1)\sqrt{r_c}. \tag{3.91}$$

The quantity K_I is the "stress intensity factor." The stress intensity factor depends on the external stress, the crack geometry and beyond that on the material structure.

The crack opening is:

$$u_\eta(\rho, 0) = \Omega_R \int_\rho^1 \frac{g(t)dt}{\sqrt{t^2 - \rho^2}} \quad \text{for } \ 0 \le \rho < 1. \tag{3.92}$$

At a given tension, the crack edges open parabolically along the radius of the crack ρ ($\rho << 1$). The maximum value of the crack opening depends linearly on the "stress intensity factor" K_I.

With the application of the operator:

$$\mathbf{I}[F(\xi), \xi \to t] = \frac{2}{\pi}\int_0^t \frac{F(\xi)\xi d\xi}{\sqrt{t^2-\xi^2}}, \tag{3.93}$$

the non-linear integral equation for the function $g(t)$ follows from (3.88):

$$g(t) + \frac{2}{\pi}\int_0^t f^{-1}\left(\Omega_R \int_y^1 \frac{g(t)dt}{\sqrt{t^2-\xi^2}}\right)\frac{\xi d\xi}{\sqrt{t^2-\xi^2}} = \frac{2}{\pi}Pt \text{ with } g(0) = 0. \tag{3.94}$$

According to (3.49), in the case of the elastic withdrawal of the fibers from the matrix the surface stress linearly depends on the crack opening:

$$\widetilde{f}(t) = \widetilde{k}_0\, t,$$

where $\widetilde{k}_0 = \frac{E_f V_f}{E_m(1-V_f)}\frac{E_L}{\lambda}$.

In this case the non-linear integral equation (3.94) is transformed into a second Fredholm linear integral equation:

$$g(t) + \frac{2\kappa_0}{\pi}\int_0^t g(\xi)\ln\frac{t+\xi}{t-\xi}d\xi = \frac{2}{\pi}Pt, \text{ with } g(0) = 0, \kappa_0 = \widetilde{k}_0\Omega_R. \tag{3.95}$$

The Eq. (3.95) could be solved using the numerical method. The solution is searched in the form of power series:

$$g(t) = \sum_{n=1}^{\infty} C_n t^n. \tag{3.96}$$

For the solution, the following identity is used:

$$\chi_n(t) \stackrel{def}{=} \int_0^t \xi^n \ln\frac{t+\xi}{t-\xi}d\xi, \tag{3.97}$$

$$\chi_n(t) \equiv \frac{t^{n+1}}{2n\,(n+1)}\left[n\,\Psi\left(\frac{n}{2}\right) + 2\,\gamma\, n - n\,\Psi\left(\frac{n+1}{2}\right) + 2n\ln 2 + 2n\,\Psi(n) + 4\right].$$

In Eq. (3.97) are:

$\Psi(n) = d\ln\Gamma(x)/dx$ the digamma function,
γ the Euler constant.

The values of the function $\chi_n(t)$ are shown in Table 3.3. Substitution of the power series (3.96) into Eq. (3.95) and combination of equal powers of t leads to the recurrent linear algebraic equations with respect to C_n. The solution of the recurrent linear algebraic equations for coefficients (3.97) is:

$$C_n = \frac{2(-\kappa_0)^{n-1}}{\pi^n}\prod_{k=1}^{n-1}\frac{k\,\Psi\left(\frac{k}{2}\right) + 2\,\gamma\, k - k\,\Psi\left(\frac{k+1}{2}\right) + 2k\ln 2 + 2k\,\Psi(k) + 4}{(1+k)k}. \tag{3.98}$$

The first six coefficients are shown in Table 3.3.

With (3.98) the stress intensity factor is the function of the factor $\kappa_0 = \widetilde{k}_0\Omega_R$ only and reads:

$$K_I = C_{44}g(1)\sqrt{r_c} = C_{44}\sqrt{r_c}\sum_{n=1}^{\infty} C_n. \tag{3.99}$$

Table 3.3 Expressions for integral $\chi_n(t)$ and solutions for coefficients C_n.

n	$\chi_n(t)$	C_n
1	t^2	$2/\pi$
2	$t^3(1+2\ln 2)/3$	$-4\kappa_0/\pi^2$
3	$2t^4/3$	$8\kappa_0{}^2(1+2\ln 2)/3\pi^3$
4	$t^5(3+4\ln 2)/10$	$-32\kappa_0{}^3(1+2\ln 2)/9\pi^4$
5	$23t^6/45$	$32\kappa_0{}^4(1+2\ln 2)(3+4\ln 2)/45\pi^5$
6	$t^7(11+12\ln 2)/42$	$-1472\kappa_0{}^5(1+2\ln 2)(3+4\ln 2)/2025\pi^6$

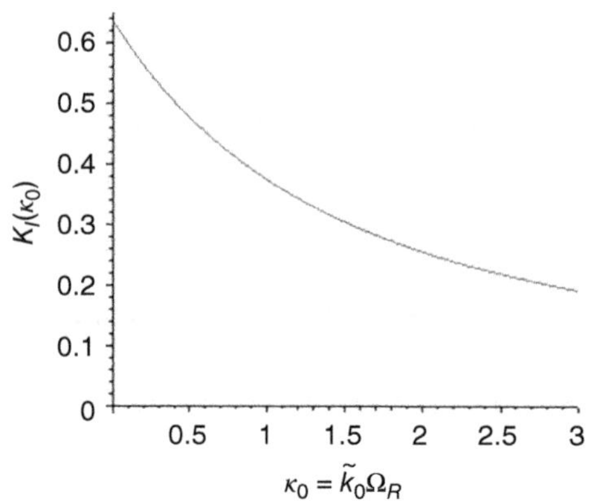

Figure 3.14 Stress intensity factor K_I as the function of $\widetilde{k}_0\Omega_R$.

The graph of the function $K_I(\kappa_0) \equiv K_I(\widetilde{k}_0\Omega_R)$ was calculated with 20 terms in the sum and is plotted in Figure 3.14.

3.2.5 Plane Crack Problem

3.2.5.1 Equations of the Plane Crack Problem

The stresses and displacements fulfill the equilibria of the plane crack problem (Figure 3.12) and can be expressed by means of potential functions (Eq. C.25, Appendix C):

$$\sigma_\eta = -\frac{1}{a^2}\left(\frac{1}{p_1^2}\frac{\partial^2\psi_1}{\partial\eta^2} + \frac{1}{p_2^2}\frac{\partial^2\psi_2}{\partial\eta^2}\right), \tag{3.100}$$

$$\tau_{\eta\zeta} = -\frac{1}{a^2}\left(\frac{\partial^2\psi_1}{\partial\eta\,\partial\zeta} + \frac{\partial^2\psi_2}{\partial\eta\,\partial\zeta}\right), \tag{3.101}$$

$$u_\eta = \frac{1}{a}\left(\delta_1\frac{\partial\psi_1}{\partial\eta} + \delta_2\frac{\partial\psi_2}{\partial\eta}\right). \tag{3.102}$$

To express the boundary-value problem (3.68)–(3.70) in the form of an integral equation, the potential functions $\psi_i(\zeta,\ \eta)$ expressed as Fourier transformations of auxiliary functions $\psi_i^*(\xi)$are:

$$\psi_i(\zeta,\eta) = \mathfrak{F}_C\{\psi_i^*(\xi)e^{-\xi\eta p_i}, \xi \to \zeta\} = \sqrt{\frac{2}{\pi}}\int_0^\infty \psi_i^*(\xi)e^{-\xi\eta p_i}\cos(\zeta\xi)d\xi. \tag{3.103}$$

There are the relationships between the stresses and displacements and the auxiliary functions $\psi_i^*(\xi)$:

$$\sigma_\eta(\zeta,\eta=0)=-\frac{1}{a^2}\sqrt{\frac{2}{\pi}}\int_0^\infty [p_1^2\psi_1^*(\xi)+p_2^2\psi_2^*(\xi)]\xi^2\cos(\zeta\xi)d\xi, \tag{3.104}$$

$$\tau_{\eta\zeta}(\zeta,\eta=0)=-\frac{1}{a^2}\sqrt{\frac{2}{\pi}}\int_0^\infty [p_1\psi_1^*(\xi)+p_2\psi_2^*(\xi)]\sin(\zeta\xi)d\xi, \tag{3.105}$$

$$u_\eta(\zeta,\eta=0)=\frac{1}{a}\sqrt{\frac{2}{\pi}}\int_0^\infty [\delta_1 p_1\psi_1^*(\xi)+\delta_2 p_2\psi_2^*(\xi)]\xi\cos(\zeta\xi)d\xi. \tag{3.106}$$

Condition (3.68) returns the relationship between the functions $\psi_1^*(\xi)$, $\psi_2^*(\xi)$:

$$p_1\psi_1^*(\xi)+p_2\psi_2^*(\xi)=0. \tag{3.107}$$

This relationship allows us to eliminate $\psi_1^*(\xi)$, the function from (3.104) to (3.106):

$$\psi(\xi)=\frac{1}{a^2}\sqrt{\frac{2}{\pi}}(p_1-p_2)p_1\psi_1^*(\xi)\xi, \tag{3.108}$$

$$\psi_1^*(\xi)=\sqrt{\frac{\pi}{2}}\frac{a^2\psi(\xi)}{\xi(p_1-p_2)p_1}, \tag{3.109}$$

$$\psi_2^*(\xi)=-\sqrt{\frac{\pi}{2}}\frac{a^2\psi(\xi)}{\xi(p_1-p_2)p_2}. \tag{3.110}$$

From the Eqs. (3.108)–(3.110) follow the representations of normal stress and displacement in terms of the sole function $\psi(\xi)$:

$$u_\eta(\zeta,\eta=0)=\Omega_E\int_0^\infty \psi(\xi)\cos(\xi\,\zeta)d\xi,\quad \Omega_E=a\frac{\delta_1-\delta_2}{p_1-p_2}, \tag{3.111}$$

$$\sigma_\eta(\zeta,\eta=0)=\int_0^\infty \xi\psi(\xi)\cos(\xi\,\zeta)d\xi=\frac{d}{d\zeta}\int_0^\infty \psi(\xi)\sin(\xi\,\zeta)d\xi. \tag{3.112}$$

In turn, the Eq. (3.82) is identically fulfilled by the substitution:

$$\psi(\xi)=\int_0^1 g(t)\mathrm{J}_0(t\xi)dt.$$

The function $\psi(\xi)$ could be replaced by the other potential function $g(t)$. The simplification of expressions for the stresses and displacements is carried out with the integral formulas:

$$\int_0^\zeta \psi(\xi)\cos(\zeta\xi)d\xi=\int_\zeta^1\frac{g(t)dt}{\sqrt{t^2-\zeta^2}}\ \text{for}-1\le\zeta<1, \tag{3.113}$$

$$\int_0^\zeta \psi(\xi)\sin(\zeta\xi)d\xi=\begin{cases}\int_0^\zeta \frac{g(t)dt}{\sqrt{\zeta^2-t^2}} & \text{for}\ \ \zeta<1,\\ \int_0^1 \frac{g(t)dt}{\sqrt{\zeta^2-t^2}} & \text{for}\ \ \zeta\ge 1\end{cases}. \tag{3.114}$$

With (3.114), the stresses in the crack surface $\eta = 0$ are expressed as

$$\sigma_\eta(\zeta, 0) = \begin{cases} \dfrac{d}{d\zeta}\displaystyle\int_0^\zeta \frac{g(t)dt}{\sqrt{\zeta^2 - t^2}} & \text{for } \zeta < 1, \\ \dfrac{d}{d\zeta}\displaystyle\int_0^1 \frac{g(t)dt}{\sqrt{\zeta^2 - t^2}} & \text{for } \zeta \geq 1. \end{cases} \tag{3.115}$$

The Eq. (3.113) leads to the formula for crack opening:

$$u_\eta(\zeta, 0) = \Omega_E \int_\zeta^1 \frac{g(t)dt}{\sqrt{t^2 - \zeta^2}}. - 1 \leq \zeta < 1. \tag{3.116}$$

3.2.5.2 Solution to the Plane Crack Problem

The substitution for $\psi(\xi)$ allows Eq. (3.81) to be expressed as a non-linear integral-differential equation:

$$\frac{d}{d\zeta}\int_0^\zeta \frac{g(t)dt}{\sqrt{\zeta^2 - t^2}} + f^{-1}\left(\Omega_E \int_\zeta^1 \frac{g(t)dt}{\sqrt{t^2 - \zeta^2}}\right) = P \text{ with } g(0) = 0. \tag{3.117}$$

The Eq. (3.117) could be solved numerically (Kobelev 2001). In the central zone, the broken tips of the fibers remain above the matrix surface, detached from the matrix (see Figure 3.15): It can be seen that the length of the detached part of the fibers does not

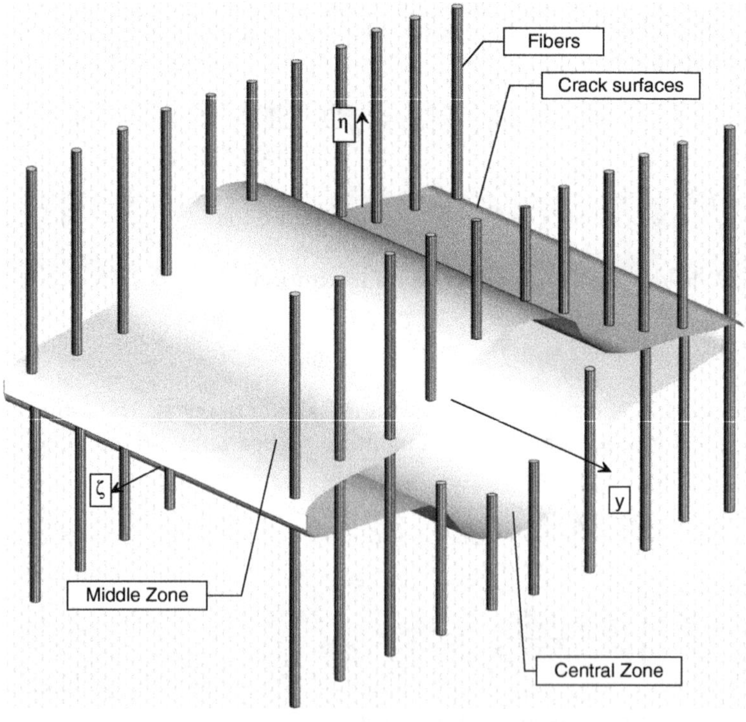

Figure 3.15 Plane crack problem.

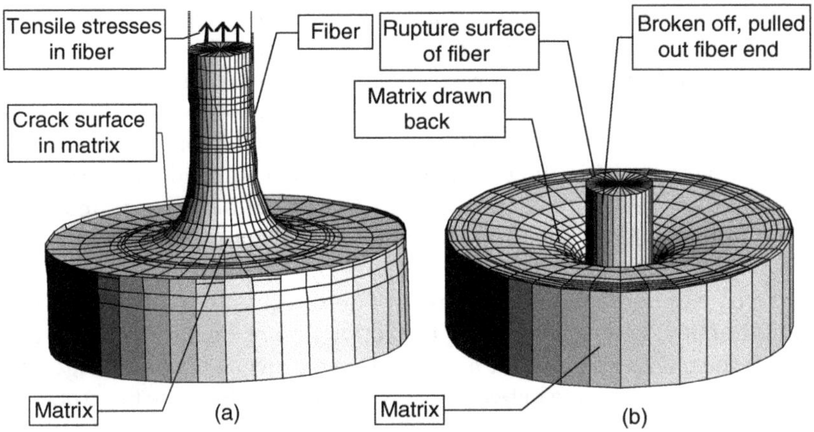

Figure 3.16 Elastic-plastic pullout of the fiber from the matrix (a) and the spring-back of the fiber after the fiber rupture (b).

depend on the applied load, but exclusively on the elastic properties of the components of the fiber composite material. The maximum size of the plastic draw-off of the fiber up to the break results in (Figure 3.16a):

$$l_b = \frac{E_L\, u_b}{E_m\, c_m}.$$

After rupture, an elastic relief takes place and the fiber pulls itself into the matrix (Figure 3.16b). When the load is relieved, the fiber sinks into the matrix by the size of:

$$l_p = \frac{E_L\, u_p}{E_m\, c_m}.$$

Above the surface, the peeling tips of the torn fibers are removed from the remaining length:

$$l_r = l_b - l_p.$$

The appearance of the microcrack predicted by the model corresponds to the experimental data (Hannant et al. 1983). A figure in Hannant et al. (1983) shows a micrograph of the crack in the fiber composite consisting of a tungsten fiber-reinforced copper matrix; the zones of the crack in the matrix and the fiber break with partially released fibers can be seen (see Figure 3.16).

3.3 Debonding of Fibers in Unidirectional Composites

3.3.1 Axial Deformation of Unidirectional Fiber Composites

In this section, we consider the problem of delamination, or debonding, of a matrix on the boundary of fibers. Debonding occurs when an adhesive breaks off adhering to a substrate material. Debonding happens if the forces that hold the bond together are broken, perhaps by an external load or environmental interference. Delamination is failure in a laminated material, in our case a composite, which leads to separation of the layers of reinforcement or plies. Delamination failure can be one of quite a few types. Delamination fracture occurs within the adhesive or matrix, fracture within the reinforcement or debonding of the resin matrix from the reinforcement fiber. In this final case, it is the debonding that leads to delamination. There is a faint distinction between debonding and delamination; namely, debonding – when two materials lose adhesion to each other and delamination – when a laminated material becomes definitely separated, perhaps induced by poor processing during production, impact in service or environmental influence.

One of the most frequent occurrences of fracture of unidirectional composites is the debonding of a unidirectional composite under the action of tensional force parallel to the fiber direction. If the tensile stress is not high enough to provoke fiber breakage, the fiber remains intact but the debonding between fibers and matrix along the fiber-matrix boundary occurs. The debonding arises in a form of a cylindrical crack twinned around the fiber. The prominent debonding occurs during the cyclic load. If an initial cylindrical crack appears, its size increases with the number of cycles. If the concentration of the debonding defects is substantial, the extent of the cylindrical crack influences the average material properties. With the elongation of the loosened contact region, the average material stiffness declines. This effect is known as the stiffness degradation of the composite material. The debonding model that makes an exact estimation of the process parameters and cyclic extension of the twinned cylindrical crack on the boundary is introduced in this section.

The typical picture of fatigue failure of a unidirectional fiberglass-epoxy composite is shown in Figure 3.17. The meander spring was exposed to 5000 full stroke cycles. The debonding of fibers is clearly visible, but no significant force reduction is measurable. Similar results are reported by Schürmann (2015).

The method is based on the model of a cylindrical fiber partially embedded in an infinite elastic matrix. Both matrix and fiber are assumed to be isotropic for briefness, but the transversal isotropy of both materials could be accounted for without alternation of the mathematical method.

An infinite cylindrical fiber is embedded in matrix such that its axis coincides with the direction of the z-axis of a cylindrical coordinate system $\{r, z\}$. In the stress-free state, the shear modulus and Poisson coefficient of the fiber are, correspondingly, G_f, μ_f and its diameter is $2r_f$. In the dimensionless cylindrical coordinates

$$\rho = r/r_f, \quad \eta = z/r_f,$$

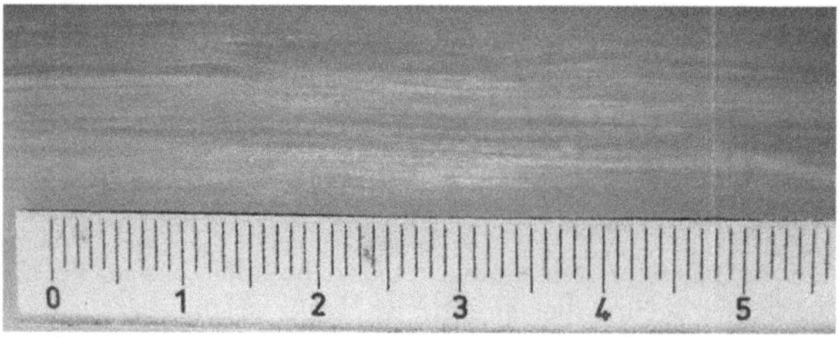

Figure 3.17 Fatigue failure of unidirectional fiberglass-epoxy composite. Visible debonding of fibers.

the infinite cylinder, occupied by the fiber material, is:

$$\rho \leq 1, \quad -\infty < \eta < \infty;$$

the infinite volume with a cylindrical cavity filled by the matrix volume is:

$$\rho \geq 1, \quad -\infty < \eta < \infty.$$

The matrix is modeled as a body of the outer diameter $2r_m$ with the cylindrical cavity. The cavity possesses an infinite extension in the direction of fiber. The outer radius of the matrix is

$$r_m = r_f / \sqrt{V_f}.$$

This choice satisfies the equation for volume content of fibers. The cavity is held in reserve for the fiber. The shear modulus and Poisson coefficient of the matrix are, correspondingly, G_m, μ_m and the diameter of the cylindrical cavity in matrix is also $2r_f$. There is a perfect adhesion on the fiber-matrix boundary in the stress-free state . The stress-free state (A) is depicted in Figure 3.18. In this state, the perfect adhesion and ideal

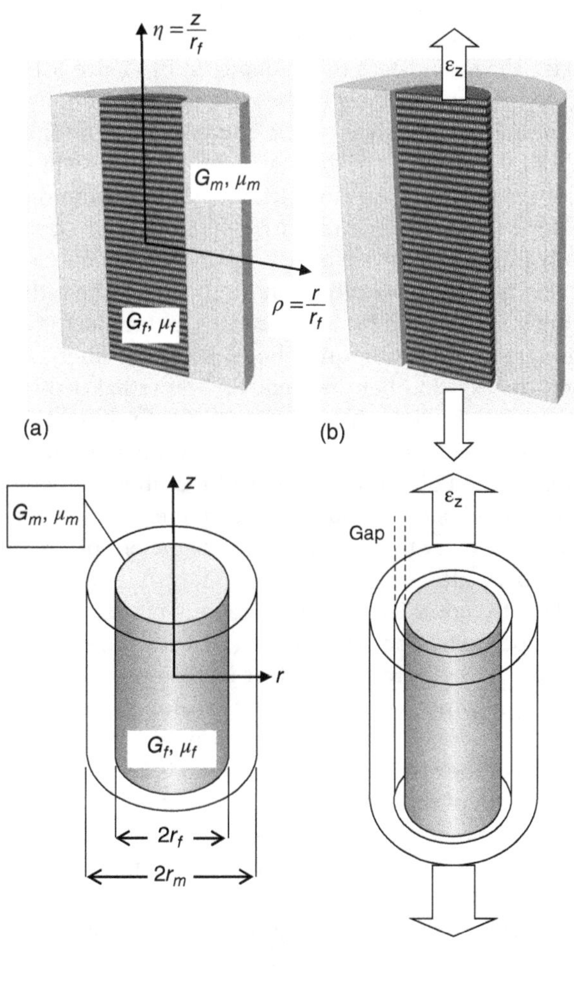

Figure 3.18 Fiber-matrix debonding model. (a) Undeformed stress state, perfect adhesion. (b) Deformed stress state, total debonding along the full length.

elastic contact along the full contact boundary are assumed. The fiber fills the cavity in the matrix and entirely fits in it in an undeformed state.

Second, consider the homogenous elastic elongation of the compound body along the fiber axis. The stress state (B) is depicted in Figure 3.18. There is no contact on the fiber-matrix boundary in the state of the homogenous elastic elongation. In this state, the matrix and fiber deform independently. The axial elongation of the compound medium is accompanied by the radial deformation of fiber and matrix. If the Poisson coefficients of fiber and matrix match, the only stress arising due to the homogenous elastic elongation stress component is stress in the fiber direction. In this case, the radial deformation of fiber and matrix is homogeneous and the same in both materials. However, the Poisson coefficients of fiber and matrix are usually different (see, for example, Appendix A). In this case the homogenous elastic elongation of the compound body along the fiber axis is accompanied by the different radial deformation in fiber and matrix. The gap on the contact boundary of fiber and matrix occurs. If the Poisson coefficient of the fiber is higher than the Poisson coefficient of matrix, the positive gap on the contact boundary of fiber and matrix occurs. If the Poisson coefficient of fiber is lower than the Poisson coefficient of matrix, the negative gap (overlapping) on the contact boundary of fiber and matrix takes place.

Third, consider the homogenous elastic elongation of a compound body along the fiber axis. The stress state (C) is depicted in Figure 3.19. There is a perfect adhesion on the fiber-matrix boundary in the state of homogenous elastic elongation. In this state, ideal adhesion and faultless elastic contact along the full contact boundary are assumed. The fiber fills the cavity in the matrix and flawlessly fits in it in a state of homogenous elastic elongation. The axial elongation of the compound medium is accompanied by the radial deformation of fiber and matrix. If the Poisson coefficients of fiber and matrix match, the only phenomenon arising due to the homogenous elastic elongation stress component is stress in the fiber direction. In this case, the radial deformation of fiber and matrix is homogeneous and the same in both materials. In a general case, the homogenous elastic elongation of a compound body along the fiber axis is accompanied by different radial deformation in fiber and matrix. The radial displacements of fiber and matrix exactly match on the boundary due to ideal adhesion. The homogeneous axial deformation is accompanied by the appearance of a radially inhomogeneous stress field in matrix and fiber. If the Poisson coefficient of the fiber is higher than the Poisson coefficient of the matrix, the positive radial stress in fiber and matrix occurs. If the Poisson coefficient of a fiber is lower than the Poisson coefficient of the matrix, the compression of fiber and matrix results.

Fourth, consider the homogenous elastic elongation of the compound body along the fiber axis. The stress state (D) with the separation in the periphery is depicted in Figure 3.19. There is a perfect adhesion on the fiber-matrix boundary in the state of the homogenous elastic elongation along a part of the boundary and the abandonment of contact (debonding) along a limited cylindrical region of contact surface. The total length of the contact adhesive region is $2ar_f$. The dimensionless variable a characterizes the length of the adhesive region divided by the diameter of fiber $2r_f$, $a = l_c/2r_f$.

The homogeneous axial deformation is accompanied by the appearance of axially and radially inhomogeneous stress field in matrix and fiber. If the Poisson coefficient of a fiber is higher than the Poisson coefficient of matrix, positive radial stress in fiber and matrix along the contact region occurs. Along the debonding zone, a gap arises. This case

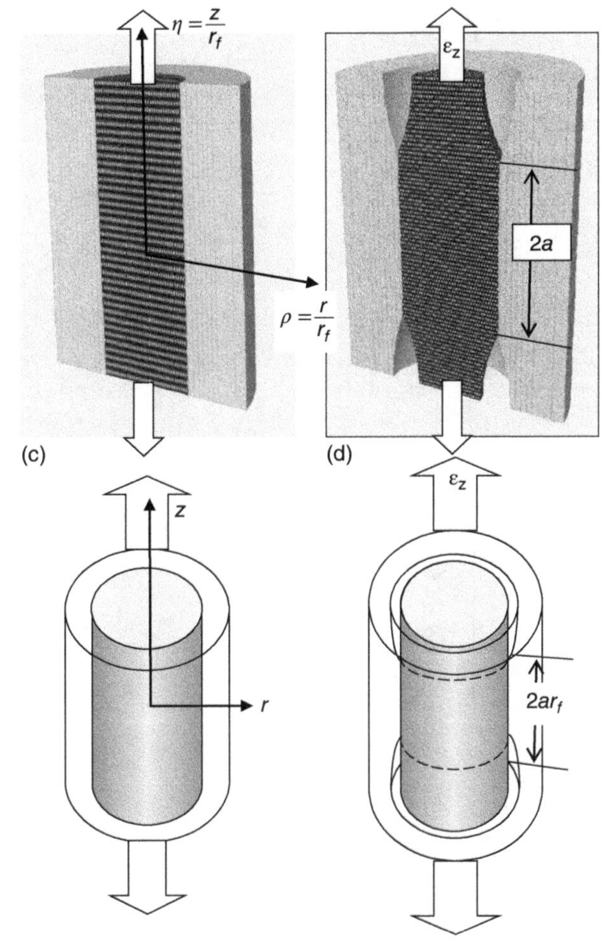

Figure 3.19 Fiber-matrix debonding model. (C) Deformed stress state, perfect adhesion along the full length. (D) Deformed stress state, partial debonding.

is interesting for investigation. Stress concentration on the tips of cylindrical flaw occurs if Poisson coefficient of a fiber is higher than the Poisson coefficient of the matrix. Otherwise, if the Poisson coefficient of fiber is lower than the Poisson coefficient of matrix, the homogeneous compression of fiber and matrix along the total, infinite contact surface takes place (like in case C).

Analogously, the breakage with separation in the central region (E) could be considered (Figure 3.20). The total length of the debonding region in this case is $2ar_f$, where a is the dimensionless ratio of the length of the debonding region to the diameter of the fiber.

3.3.2 Stresses in Unidirectional Composite in Cases of Ideal Debonding or Adhesion

3.3.2.1 Equations of an Axially Loaded Unidirectional Compound Medium (A)

To start, we evaluate the stress-deformation states in the cases (B) and (C). The mathematical derivation is performed in the dimensioned cylindrical coordinates $\{r, z\}$. The

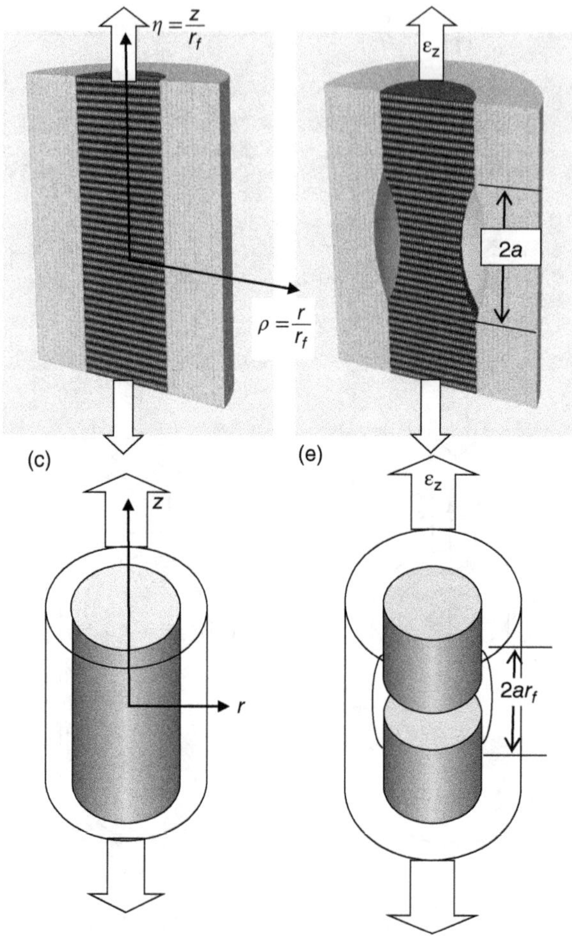

Figure 3.20 Fiber-matrix debonding model. (A) Undeformed stress state, perfect adhesion. (E) Deformed stress state, partial debonding.

consideration is performed in the cylindrical coordinate system associated with the direction of the fiber axis. The shear stresses and strains vanish because of cylindrical symmetry. The diagonal components of strain in fibers (f) and matrix (m) in vector notation are:

$$\boldsymbol{\varepsilon}_{(f)} = \left[\frac{du_{(f)}}{dr}, \frac{u_{(f)}}{r}, \varepsilon_z\right], \quad \boldsymbol{\varepsilon}_{(m)} = \left[\frac{du_{(m)}}{dr}, \frac{u_{(m)}}{r}, \varepsilon_z\right]. \tag{3.118}$$

Similarly, the diagonal components of stress in fibers (f) and matrix (m) in vector notation are:

$$\boldsymbol{\sigma}_{(f)} = [\sigma_{(f)r}, \sigma_{(f)\theta}, \sigma_{(f)z}], \quad \boldsymbol{\sigma}_{(m)} = [\sigma_{(m)r}, \sigma_{(m)\theta}, \sigma_{(m)z}]. \tag{3.119}$$

The compliance matrices for fiber and matrix correspondingly read:

$$\mathbf{S}_{(f)} = \begin{bmatrix} 1/E_f & -\nu_f/E_f & -\nu_f/E_f \\ -\nu_f/E_f & 1/E_f & -\nu_f/E_f \\ -\nu_f/E_f & -\nu_f/E_f & 1/E_f \end{bmatrix}, \tag{3.120}$$

$$\mathbf{S}_{(m)} = \begin{bmatrix} 1/E_m & -\nu_m/E_m & -\nu_m/E_m \\ -\nu_m/E_m & 1/E_m & -\nu_m/E_m \\ -\nu_m/E_m & -\nu_m/E_m & 1/E_m \end{bmatrix}, \tag{3.121}$$

$$\boldsymbol{\varepsilon}_{(f)} = \mathbf{S}_{(f)} \cdot \boldsymbol{\sigma}_{(f)}, \quad \boldsymbol{\varepsilon}_{(m)} = \mathbf{S}_{(m)} \cdot \boldsymbol{\sigma}_{(m)}. \tag{3.122}$$

The tangential strains in fiber and matrix from (3.122) follow as:

$$\frac{u_{(f)}}{r} = -\frac{\nu_f}{E_f}\sigma_{(f)r} + \frac{1}{E_f}\sigma_{(f)\theta} - \frac{\nu_f}{E_f}\sigma_{(f)z}, \tag{3.123}$$

$$\frac{u_{(m)}}{r} = -\frac{\nu_m}{E_m}\sigma_{(m)r} + \frac{1}{E_m}\sigma_{(m)\theta} - \frac{\nu_m}{E_m}\sigma_{(m)z}. \tag{3.124}$$

The equations (3.123) and (3.124) express the radial components of the displacements in fiber and matrix. The radial strains in fiber and matrix follow from (3.122) as:

$$\frac{du_{(f)}}{dr} = \frac{1}{E_f}\sigma_{(f)r} - \frac{\nu_f}{E_f}\sigma_{(f)\theta} - \frac{\nu_f}{E_f}\sigma_{(f)z}, \tag{3.125}$$

$$\frac{du_{(m)}}{dr} = \frac{1}{E_m}\sigma_{(m)r} - \frac{\nu_m}{E_m}\sigma_{(m)\theta} - \frac{\nu_m}{E_m}\sigma_{(m)z}. \tag{3.126}$$

The axial strains in fiber and matrix are the given constants. From (3.122), these constants reduce to:

$$\varepsilon_z = -\frac{\nu_f}{E_f}\sigma_{(f)r} - \frac{\nu_f}{E_f}\sigma_{(f)\theta} + \frac{1}{E_f}\sigma_{(f)z}, \tag{3.127}$$

$$\varepsilon_z = -\frac{\nu_m}{E_m}\sigma_{(m)r} - \frac{\nu_m}{E_m}\sigma_{(m)\theta} + \frac{1}{E_m}\sigma_{(m)z}. \tag{3.128}$$

Our aim is to eliminate the variables

$$u_{(f)}, u_{(m)}, \sigma_{(f)\theta}\ \sigma_{(m)\theta}, \sigma_{(f)z}\ \sigma_{(m)z}$$

from the equations (3.123)–(3.128) and obtain the equilibrium equations in terms of unknown functions $\sigma_{(f)r}$ and $\sigma_{(m)r}$, and the given constant ε_z. From Eqs. (3.123)–(3.128) the variables are expressed in terms of unknown functions $\sigma_{(f)r}$ and $\sigma_{(m)r}$ as:

$$\sigma_{(f)\theta} = -\frac{1}{\nu_f}(E_f\varepsilon_z + \nu_f\sigma_{(f)r} - \sigma_{(f)z}),$$

$$\sigma_{(m)\theta} = -\frac{1}{\nu_m}(E_m\varepsilon_z + \nu_m\sigma_{(m)r} - \sigma_{(m)z}),$$

$$\sigma_{(f)z} = \frac{d(\nu_f\sigma_{(f)r})}{dr} + 2\nu_f\sigma_{(f)r} + E_f\varepsilon_z,$$

$$\sigma_{(m)z} = \frac{d(\nu_m\sigma_{(m)r})}{dr} + 2\nu_m\sigma_{(m)r} + E_m\varepsilon_z,$$

$$u_{(f)} = -\frac{\nu_f\sigma_{(f)r} - \sigma_{(f)\theta} + \nu_f r\sigma_{(f)z}}{E_f},$$

$$u_{(m)} = -\frac{\nu_m\sigma_{(m)r} - \sigma_{(m)\theta} + \nu_m r\sigma_{(m)z}}{E_m}.$$

Equilibrium equations in the fiber and matrix material are:

$$\frac{d(r\sigma_{(f)r})}{dr} - \sigma_{(f)\theta} = 0, \quad \frac{d(r\sigma_{(m)r})}{dr} - \sigma_{(m)\theta} = 0. \tag{3.129}$$

The equilibrium equations in terms of functions $\sigma_{(f)r}$ and $\sigma_{(m)r}$ reduce to:

$$\nu_f r \frac{d(\sigma_{(f)r})}{dr} + 2\nu_f \sigma_{(f)r} + E_f \varepsilon_z = 0, \tag{3.130}$$

$$\nu_m r \frac{d(\sigma_{(m)r})}{dr} + 2\nu_m \sigma_{(m)r} + E_m \varepsilon_z = 0. \tag{3.131}$$

From the solution of Eqs. (3.130) and (3.131), we get the expressions for stresses and radial displacements in fiber and matrix with four integration constants $c_{11}, c_{12}, c_{21}, c_{22}$:

$$\sigma_{(f)r} = c_{11} + \frac{c_{21}}{r^2}, \quad \sigma_{(m)r} = c_{12} + \frac{c_{22}}{r^2}, \tag{3.132}$$

$$\sigma_{(f)\theta} = c_{11} - \frac{c_{21}}{r^2}, \quad \sigma_{(m)\theta} = c_{12} - \frac{c_{22}}{r^2}, \tag{3.133}$$

$$\sigma_{(f)z} = 2\nu_f c_{11} + E_f \varepsilon_z, \quad \sigma_{(m)z} = 2\nu_m c_{12} + E_m \varepsilon_z, \tag{3.134}$$

$$u_{(f)} = -\nu_f \varepsilon_z r + \frac{1 - \nu_f - 2\nu_f^{\,2}}{E_f} c_{11} r - \frac{1 + \nu_f}{E_f} \frac{c_{21}}{r}, \tag{3.135}$$

$$u_{(m)} = -\nu_m \varepsilon_z r + \frac{1 - \nu_m - 2\nu_m^{\,2}}{E_m} c_{12} r - \frac{1 + \nu_m}{E_m} \frac{c_{22}}{r}. \tag{3.136}$$

3.3.2.2 Total Debonding (B)

Now consider two special cases that were briefly discussed before. First, study case B, namely the deformed stress state with total debonding along the full length. In this case, the following conditions are valid:

$$\begin{aligned} &\sigma_{(f)r}|_{r=r_f} = 0, \quad \sigma_{(f)r}|_{r\to 0} = 0, \\ &\sigma_{(m)r}|_{r=r_f} = 0, \quad \sigma_{(m)r}|_{r=r_0} = 0, \end{aligned} \tag{3.137}$$

$$u_{(f)} = -\nu_f \varepsilon_z r, \quad u_{(m)} = -\nu_m \varepsilon_z r. \tag{3.138}$$

with the trivial solution for integration constants:

$$c_{11} = c_{12} = c_{21} = c_{22} = 0.$$

The gap between the surfaces of fiber and matrix is

$$\delta_0 = (\nu_f - \nu_m)\varepsilon_z r.$$

Thus, if $\nu_f - \nu_m > 0$ for positive deformation ε_z (tension of the media), the reduction of the cross-section of fibers is greater than the matrix and a gap occurs. Contrarily, if $\nu_f - \nu_m < 0$ for positive tensile deformation ε_z, the reduction of the cross-section of a matrix is greater than the fiber and an overlap with, consequently, pressure on the boundary takes place.

3.3.2.3 Ideal Adhesion (C)

Second, revise case C, namely deformed stress state with perfect adhesion along the full length. In this case, the following conditions are valid:

$$\begin{aligned} &\sigma_{(f)r}|_{r=r_f} = \sigma_{(m)r}|_{r=r_f}, \quad u_{(f)}|_{r=r_f} = u_{(m)}|_{r=r_f}, \\ &\sigma_{(f)r}|_{r\to 0} = 0, \quad \sigma_{(m)r}|_{r=r_0} = 0. \end{aligned} \tag{3.139}$$

The solution for the integration constants in this case reads as:

$$\begin{aligned} &c_{22} = E_f E_m (\nu_m - \nu_f) {r_m}^2 {r_f}^2 \varepsilon_z \Delta_d^{-1}, \\ &c_{21} = 0, \\ &c_{11} = E_f E_m (\nu_m - \nu_f)({r_m}^2 - {r_f}^2)\varepsilon_z \Delta_d^{-1}, \\ &c_{12} = E_f E_m (\nu_m - \nu_f) {r_f}^2 \varepsilon_z \Delta_d^{-1}, \\ &\Delta_d = {r_f}^2 [E_f(\nu_m - 1 + 2\nu_m^2) - E_m(\nu_f - 1 + 2\nu_f^2)] + \\ &\qquad {r_m}^2[-E_f(\nu_m + 1) + E_m(\nu_f - 1 + 2\nu_f^2)]. \end{aligned} \tag{3.140}$$

The stresses on the interface boundary fiber to matrix result in:

$$\begin{aligned} &\sigma_{(m)\theta}|_{r=r_f} = -E_f E_m(\nu_m - \nu_f)({r_m}^2 + {r_f}^2)\varepsilon_z \Delta_d^{-1}, \\ &\sigma_{(f)r}|_{r=r_f} = \sigma_{(m)r}|_{r=r_f} = \sigma_{(f)\theta}|_{r=r_f} = E_f E_m (\nu_m - \nu_f)({r_m}^2 - {r_f}^2)\varepsilon_z \Delta_d^{-1} \end{aligned} \tag{3.141}$$

$$\begin{aligned} \sigma_{(f)z}|_{r=r_f} = {} & {r_f}^2 E_f [E_f(\nu_m - 1 + 2\nu_m^2) + E_m(-\nu_f + 1 - 2\nu_f^2)]\varepsilon_z \Delta_d^{-1} \\ & + {r_m}^2 E_f[-E_f(\nu_m + 1) + E_m(\nu_f - 1 + 2\nu_m \nu_f)]\varepsilon_z \Delta_d^{-1}. \end{aligned} \tag{3.142}$$

$$\begin{aligned} \sigma_{(m)z}|_{r=r_f} = {} & {r_f}^2 E_m [E_f(\nu_m - 1 + 2\nu_f \nu_m) + E_m(-\nu_f + 1 - 2\nu_f^2)]\varepsilon_z \Delta_d^{-1} \\ & + {r_m}^2 E_m[-E_f(\nu_m + 1) + E_m(\nu_f - 1 + 2\nu_f^2)]\varepsilon_z \Delta_d^{-1}. \end{aligned} \tag{3.143}$$

with the equations (3.141), the radial, circumferential and axial stresses on the interface are determined:

- if the Poisson coefficients of matrix and fiber are equal, $\nu_m = \nu_f$, the radial and circumferential stresses vanish;
- if $\nu_m > \nu_f$ the radial stresses on the interface boundary are positive. The circumferential stress in fiber is equal to the radial stress and is also positive. The circumferential stress in matrix is negative;
- if $\nu_m < \nu_f$ the radial stresses on the interface boundary are negative and no separation occurs even if adhesion fails. The circumferential stress in fiber is equal to the radial stress and is also negative. The circumferential stress in matrix is positive. This could provoke the radial breakage of matrix, giving rise to star-formed radial cracks surrounding the fiber.

In this section, the ideal debonding and the ideal adhesive contact problems were solved for a uniformly tensioned unidirectional composite. The solution is simplified by the cylindrical symmetry of the problem, resulting in an easy to solve problem with only one radial independent variable.

3.3.3 Stresses in a Unidirectional Composite in a Case of Partial Debonding

The partial debonding problem incorporates the wide class of contact problems of elasticity theory. The description of a partial debonding of a cylindrical fiber in a cylindrical cavity in an elastic matrix transforms to an axial-symmetric contact problem. The principal equations of the contact problem are based on the solution of the axisymmetric boundary-value problem of the theory of elasticity for a finite circular cylinder with an arbitrary loaded curved surface (Filon 1901, 1902). The classical Filon problem concerns the uniformly distributed tangential load applied along two rings at the curved surface. The equilibrium of an infinite circular cylinder under the action of normal loads has been considered in detail, for example, in Lur'e (1964) and Meleshko (2003).

The static axisymmetric contact problems of an infinite linear-elastic, isotropic and homogeneous cylinder and the static axisymmetric problem of a space with an infinite cylindrical cavity were considered by Alexandrov and Pozharskii (2001). We apply their method for solution of contact problems to the contact problem of debonding of a matrix from a fiber in a composite material.

Two auxiliary boundary problems are to be examined for the solution of the debonding problem. The mathematical derivation is performed in the dimensionless cylindrical coordinates

$$\{\rho = r/r_f, \eta = z/r_f\}.$$

3.3.3.1 Partial Radial Load on the Fiber Surface

The first auxiliary boundary problem is the deformation of an elastic fiber that is pressed over a finite region at its outer surface. For this purpose, the fiber is represented as infinite cylinder radius r_f. The length of the cylindrical crack, which warps over the fiber is l_c. The fiber material occupies the space $\{\rho \leq 1, -\infty < \eta < \infty\}$, with the variable η along the axis of cylinder and ρ in the radial direction. The dimensionless length of the crack is $a = l_c/2r_f$.

The boundary conditions on the outer surface of cylinder $\rho = 1$ read:

$$\sigma_{(f)r}(\eta)|_{\rho=1} = -\widetilde{q}(\eta), \quad \tau_{rz}(\eta)|_{\rho=1} = 0,$$
$$\widetilde{q}(\eta) = \begin{cases} q(\eta), & |\eta| < a, \\ 0, & |\eta| \geq a. \end{cases} \tag{3.144}$$

The variable $\widetilde{q}$ represents the pressure on the outer surface. The shear stress τ_{rz} on the surface is absent. The solution of (3.144) is given by Alexandrov and Pozharskii (2001, Eq. (3.16)–(3.18)):

$$u_{(f)}(\eta)|_{\rho=1} = -\frac{1}{\pi \mu_f} \int_{-a}^{a} q(t) K_f(t-\eta)dt,$$
$$K_f(t) = \int_0^\infty \xi^{-1} \cos(\xi t) L_f(\xi) d\xi. \tag{3.145}$$

The coefficients of the integral equation (3.145) are shown in Table 3.4.

3.3.3.2 Partial Radial Load on the Matrix Cavity Surface

The second auxiliary boundary problem is the static problem for a matrix material in an infinite space, which contains an infinite cylindrical cavity of radius r_f. In the

Table 3.4 Kernel of the integral equation for elastic cylinder, Eq. (3.145).

Kernel of integral	$L_f(u) = \dfrac{u}{u^2(\omega_0^2(u) - 1) - 2 + 2\nu_f}$
Coefficient of kernel	$\omega_0(u) = \mathrm{I}_0(u)/\mathrm{I}_1(u)$, $\mathrm{I}_0(u)$,$\mathrm{I}_1(u)$ Bessel functions
Elastic constant	$\mu_f = \dfrac{G_f}{1 - \nu_f}$
Asymptotic of kernel	$\dfrac{L_f(u)}{\mu_f} = \begin{cases} A_f u + o(u), & u \to 0 \\ B_f + o(1/u), & u \to \infty \end{cases}$
Asymptotic coefficient $u \to 0$	$A_f = \dfrac{1 - \nu_f}{2(1 + \nu_f)G_f}$
Coefficient of limit value $u \to \infty$	$B_f = \mu_f^{-1} \equiv \dfrac{1 - \nu_f}{G_f}$

dimensionless cylindrical coordinates, the elastic matrix medium occupies the space $\{\rho \geq 1, -\infty < \eta < \infty\}$. The boundary conditions on the surface $\rho = 1$ of an infinite cavity are:

$$\sigma_{(m)r}(\eta)|_{\rho=1} = -\widetilde{q}(\eta), \quad \tau_{rz}(\eta)|_{\rho=1} = 0,$$
$$\widetilde{q}(\eta) = \begin{cases} q(\eta), & |\eta| < a, \\ 0 & |\eta| \geq a. \end{cases} \tag{3.146}$$

The condition of boundedness of the stresses as $\rho \to \infty$ must be satisfied as well. The pressure $\widetilde{q}(\eta)$ expands the cavity in the matrix, at $\rho = a$. Once again, the shear stress on the outer surface of matrix cylinder vanishes, at $\rho = 1$.

The solution of (3.146) reads (Alexandrov and Pozharskii 2001, Eq. (3.22)–(3.24)):

$$u_{(m)}(\eta)|_{\rho=1} = \frac{1}{\pi\mu_m}\int_{-a}^{a} q(t)K_m(t - \eta)dt,$$
$$K_m(t) = \int_0^\infty \xi^{-1}\cos(\xi t)L_m(\xi)d\xi. \tag{3.147}$$

Table 3.5 exhibits the coefficients of the integral equation (3.147).

3.3.3.3 Partial Debonding With Central Adhesion Region (D)

The boundary conditions of the cylindrical interface surface for contact problem with the given contact gap $\delta_0(z)$. The conditions on the contact cylindrical surface $\rho = 1$ are the following:

$$[u_{(f)}(\eta) - u_{(m)}(\eta)]_{\rho=1} = -\delta_0(\eta), \qquad |\eta| < a,$$
$$\sigma_{(m)r}(\eta)|_{\rho=1} = 0, \quad \sigma_{(f)r}(\eta)|_{\rho=1} = 0, \quad |\eta| \geq a,$$
$$\tau_{rz}(\eta)|_{\rho=1} = 0. \tag{3.148}$$

The sense of the boundary conditions (3.148) is as follows. Consider, for example, the case of tension of the unidirectional composite material in the direction of fibers. The

Table 3.5 Kernel of the integral equation for elastic cavity, Eq. (3.147).

Kernel of integral	$L_m(u) = \dfrac{u}{-u^2(\Omega_0^2(u) - 1) + 2 - 2\nu_m}$
Coefficient of kernel	$\Omega_0(u) = K_0(u)/K_1(u)$, $K_0(u), K_1(u)$ Bessel functions
Elastic constant	$\mu_m = \dfrac{G_m}{1 - \nu_m}$
Asymptotic of kernel	$\dfrac{L_m(u)}{\mu_m} = \begin{cases} A_m u + o(u), & u \to 0 \\ B_m + o(1/u), & u \to \infty \end{cases}$
Asymptotic coefficient $u \to 0$	$A_m = \dfrac{1}{2G_m}$
Coefficient of limit value $u \to \infty$	$B_m = \mu_m^{-1} \equiv \dfrac{1 - \nu_m}{G_m}$

separation of the surfaces of the fiber and matrix takes place if the increase of the diameter of cavity in matrix is greater than the increase of the fiber. In this case, the Poisson coefficient of the matrix is greater than the Poisson coefficient of the fiber. If the gap is positive and on the part of the boundary $|\eta| < a$, the separation is prevented by the adhesion of both materials, tension (formally a negative pressure) $p(\eta)$ arises in the adhesion zone. The other part of boundary remains free and the distance between the surfaces of fiber and matrix gradually increases from end points, $|\eta| > a$. The gap asymptotically reaches the limit value. This limit value of gap far from the adhesion region corresponds to the gap in the case of perfect separation. This was calculated in the previous section. The solution of (3.148) transforms to the solution of an integral equation:

$$\delta_0(\eta) = \frac{2}{\pi \hat{m}} \int_{-a}^{a} q(t) K_c(t - \eta) dt,$$

$$K_c(t) = \int_0^{\infty} \xi^{-1} \cos(\xi\, t) L_c(\xi) d\xi. \tag{3.149}$$

with respect to the unknown contact pressure $p(\eta)$ on the surface $\rho = 1$ (Alexandrov and Pozharskii 2001, Eq. (3.22)–(3.24)):

$$\sigma_{(m)r}(\eta)|_{\rho=1} = -p(\eta), \quad \sigma_{(f)r}(\eta)|_{\rho=1} = p(\eta), \quad |\eta| < a. \tag{3.150}$$

The coefficients of the integral equation (3.149) are displayed in Table 3.6.

For the subsequent study of the integral equation, the study of the asymptotic behavior of the kernel is indispensable. The kernel of the integral equation (3.149) possesses the asymptotic behavior as listed in Table 3.4. Unfortunately, no closed form solution of the integral equation with the original kernel $L_c(u)$ was discovered. Opportunely, a powerful method was discovered by Alexandrov and Pozharskii (2001). The referenced method employs the suitable approximation for the kernel of the integral equation, which leads to a closed form solution. This closed form solution permits the comprehensive study of parameters and asymptotic behavior of the solution. Following the referenced method,

Table 3.6 Kernel of the integral equation for contact problem of elastic cylinder and cavity, Eq. (3.149).

Kernel of integral	$L_c(u) = \frac{L_f(u)}{2\mu_f} + \frac{L_m(u)}{2\mu_m}$
Ratio of the elastic constants	$\widehat{m} = \frac{\mu_f}{\mu_m} = \frac{G_f}{G_m}\frac{1-\nu_m}{1-\nu_f}$
Asymptotic of kernel	$L_c(u) = \begin{cases} Au + o(u), & u \to 0 \\ B + o(1/u), & u \to \infty \end{cases}$
Asymptotic coefficient $u \to 0$	$A = \frac{(1-\nu_f)G_m + (1-\nu_m)G_f}{4(1+\nu_f)G_m G_f}$
Coefficient of limit value $u \to \infty$	$B = \frac{(1-\nu_f)G_m + (1-\nu_m)G_f}{2G_m G_f}$

we replace the exact kernel $L_c(u)$ with an auxiliary kernel that behaves congruently at the points $u \to 0$ and $u \to \infty$, as the original kernel:

$$l_c(u) = B\tanh\left(\frac{Au}{B}\right) = \begin{cases} Au + o(u), & u \to 0, \\ B + o(1/u), & u \to \infty. \end{cases} \tag{3.151}$$

The original formulation of method assumed that the materials of fiber and matrix were equal (Alexandrov and Pozharskii 2001, Eq. (3.143)). This hypothesis is a strong limitation for the actual problem. For our purposes, it is principally significant that the elastic constants of the fiber and matrix substantially differ. Consequently, the assumption of equality of elastic constants should be dropped.

The integrals for the exact and approximated kernels are shown here:

$$\begin{aligned} K_c(\eta) &= \int_0^\infty \cos(\xi\,\eta) L_c(\xi)\frac{d\xi}{\xi}, \\ k_c(\eta) &= \int_0^\infty \cos(\xi\,\eta) l_c(\xi)\frac{d\xi}{\xi} = -B\ln\left(\tanh\left(\frac{\pi\,\eta\,B}{4A}\right)\right). \end{aligned} \tag{3.152}$$

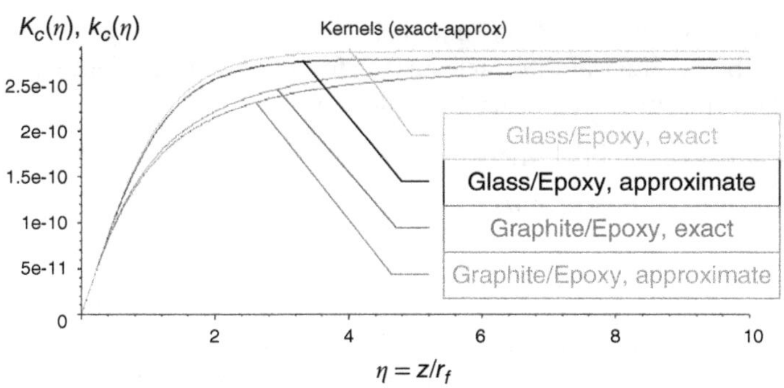

Figure 3.21 Exact and approximated kernels $K_c(\eta)$, $k_c(\eta)$ of the integral equation of the contact problem.

Table 3.7 Material data for debonding problem (Barbero 1999).

	Glass	Carbon T300
E_f, GPa	73	230
ν_f	0.22	0.2
E_m, GPa	3.12	3.12
ν_m	0.38	0.38

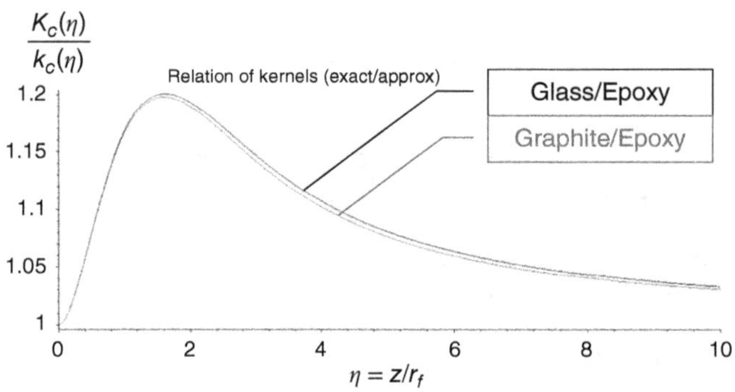

Figure 3.22 Relation of exact to the approximate kernel $K_c(\eta)/k_c(\eta)$.

The first integral contains the Bessel functions. Exact and approximated kernels $K_c(\eta)$, $k_c(\eta)$ of the integral equation of the contact problem are plotted in Figure 3.21. Exact and approximated kernels of the first integral equation for the contact problem could only be evaluated numerically, but the second can be taken in closed form. The material data of fiberglass and carbon composites with epoxy resin used for calculations are listed in Table 3.7. The shaded lines depict the exact and approximated kernels for glass fibers in an epoxy matrix, respectively. The lines show the exact and approximated kernels for graphite fibers in an epoxy matrix. The relations of exact to approximate kernel $K_c(\eta)/k_c(\eta)$ are shown for these materials in Figure 3.22. The maximal difference for kernels is about 20%.

3.3.3.4 Partial Debonding With Central Debonding Region (E)

The boundary conditions of the cylindrical interface surface for a contact problem with the given contact gap $\delta_0(\eta)$ on the surface $\rho = 1$are the following:

$$[u_{(f)}(\eta) - u_{(m)}(\eta)]_{\rho=1} = -\delta_0(\eta), \quad |\eta| \geq a;$$
$$\sigma_{(m)r}(\eta)|_{\rho=1} = 0, \quad \sigma_{(f)r}(\eta)|_{\rho=1} = 0, \quad |\eta| < a;$$
$$\tau_{rz}(\eta)|_{\rho=1} = 0.$$

The sense of the boundary conditions is as follows. The separation of the surfaces of the fiber and matrix takes place if the increase of the diameter of cavity in matrix is greater than the increase of the fiber. If the gap is positive, and on the periphery part of the boundary $|\eta| > a$ separation is prevented by the adhesion of both materials, the tension (formally a negative pressure) $p(\eta)$ arises in the adhesion zone. The central part of the interface border $|\eta| \leq a$ remains free and the distance between the surfaces of fiber and matrix gradually increases from the end points.

3.3.3.5 Semi-Infinite Debonding With Central Debonding Region (F)

The boundary conditions of the cylindrical interface surface for a contact problem with the given contact gap $\delta_0(\eta)$ on the surface $\rho = 1$ are the following:

$$
\begin{aligned}
&[u_{(f)}(\eta) - u_{(m)}(\eta)]_{\rho=1} = -\delta_0(\eta), \quad \eta \geq 0, \\
&\sigma_{(m)r}(\eta)|_{\rho=1} = 0, \;\; \sigma_{(f)r}(\eta)|_{\rho=1} = 0, \;\; \eta < 0, \\
&\tau_{rz}(\eta)|_{\rho=1} = 0.
\end{aligned} \tag{3.153}
$$

The boundary condition (3.153) expresses the following. If the gap is positive, for the positive values of the axial coordinate $\eta \geq a$ separation is prevented by the adhesion of both materials, tension (formally a negative pressure) $p(\eta)$ arises in the adhesion zone. For the positive values of δ_0, the axial coordinates on the interface border $\eta < a$ remain free.

3.3.4 Contact Problem for a Finite Adhesion Region

The solution of the problem for partial debonding with central adhesion region (D) is given in closed form. The contact problem (D) reduces to the singular integral equation with the kernel (3.152):

$$
\int_{-a}^{a} g(t)k_c(|\eta - t|)dt = f(\eta), \;\; for \;\; -a \leq \eta \leq a, \tag{3.154}
$$

or

$$
\int_{-a}^{a} g(t) \ln \left|\coth \frac{\eta - t}{4}\right| dt = f(\eta) \;\; for \;\; -a \leq \eta \leq a. \tag{3.155}
$$

The solution of the integral equation (3.155) was given by Gohberg and Krein (1967) and reads according to Polyanin and Manzhirov (2008):

$$
\begin{aligned}
&g(\eta) = g_1(\eta) + g_2(\eta) + g_3(\eta), \\
&g_1(\eta) = \frac{w(\eta, a)}{2M'(a)} \left[\frac{d}{da} \int_{-a}^{a} f(t)w(t, a)dt\right], \\
&g_2(\eta) = -\frac{1}{2} \int_{|\eta|}^{a} w(\eta, \xi)\frac{d}{d\xi} \left[\frac{1}{M'(\xi)} \frac{d}{d\xi} \int_{-\xi}^{\xi} f(t)w(t, \xi)dt\right] d\xi, \\
&g_3(\eta) = -\frac{1}{2} \frac{d}{d\eta} \int_{|\eta|}^{a} \frac{w(\eta, \xi)}{M'(\xi)} \left[\int_{-\xi}^{\xi} f'(t)w(t, \xi)dt\right] d\xi.
\end{aligned} \tag{3.156}
$$

The functions in the Eq. (3.156) are following:

$$
M(\xi) = \frac{\mathbf{P}_{-1/2}(\cosh \xi)}{\mathbf{Q}_{-1/2}(\cosh \xi)}, \tag{3.157}
$$

$$
w(\eta, \xi) = \frac{1}{\pi \mathbf{Q}_{-1/2}(\cosh \xi)\sqrt{2\cosh \xi - 2\cosh \eta}}, \tag{3.158}
$$

where:

$\mathbf{P}_{-1/2}(\cosh\xi)$ is the Legendre function of the first kind;
$\mathbf{Q}_{-1/2}(\cosh\xi)$ is the Legendre function of the second kind.

For a constant right side in the interval $-a \le \eta \le a$, the right side reads:

$$f(\eta) = \begin{cases} 1, & -a \le \eta \le a, \\ f_a(\eta) & -\eta \le a \text{ and } a \le \eta. \end{cases}$$

The solution (3.156) for the function $f(\eta) = 1$ in $-a \le \eta \le a$ compresses to:

$$g_1(\eta) = \frac{c_g(a)}{\sqrt{2\cosh a - 2\cosh \eta}} \quad \mathit{for} \quad -a \le \eta \le a. \tag{3.159}$$

The normalized solution (3.159) is given by the equation:

$$\frac{g_1(\eta)}{g_1(0)} = \frac{\sqrt{2\cosh a}}{\sqrt{2\cosh a - 2\cosh \eta}}.$$

This formula for the dimensionless lengths of contact region of $a = 1, 2$ and 4 is shown in Figure 3.23.

The solution (3.159) identically satisfies the integral equation:

$$\int_{-a}^{a} g_1(t) \ln\left|\coth\frac{\eta - t}{4}\right| dt = 1 \quad \mathit{for} \quad -a \le \eta \le a. \tag{3.160}$$

It is worth mentioning that the method of solution of the singular integral equation (Gohberg and Krein 1967) satisfies only the right side inside the interval $-a \le \eta \le a$.

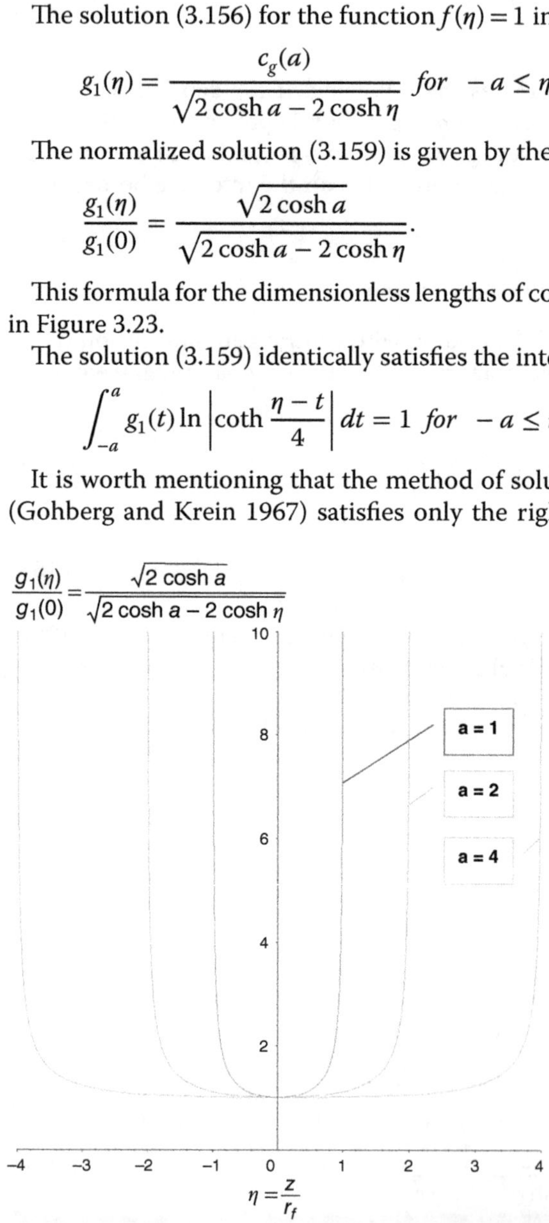

Figure 3.23 Normalized solution $g_1(\eta)/g_1(0)$ (3.159) for the dimensionless length of contact region $a = 1, 2, 4$.

Outside this interval the right side $f(\eta)$ is not defined and no equation for it must be fulfilled. In other words, the function $f_a(\eta)$ is arbitrary on $-\eta \leq a$ and $a \leq \eta$. This function could be calculated with the obtained exact solution $g_1(t)$. We show that this function exponentially fades with the distance from endpoints of the interval. Strictly speaking, this means that the nullifying of the normal stresses (3.148) outside the adhesion interval $-\eta \leq a$ and $a \leq \eta$ is not identically fulfilled with the closed form solution (3.158) or (3.159). The acquired formulas are acceptable for the practical evaluation of debonding effects and parameter study. Right sides of the equation $f(\eta)$ for different dimensionless length of contact region are shown on the Figure 3.24.

The dimensionless constant in Eq. (3.159) depends on the dimensionless length of the contact zone and is represented as:

$$c_g(a) = \frac{1+\exp(a)}{2\pi^2}\exp\left(-\frac{a}{2}\right)\mathbf{E}\left(\frac{\exp(a)-1}{\exp(a)+1}\right) - \frac{1}{(1+\exp(a))\pi^2}\left(1+\frac{\mathbf{Q}_{1/2}(\cosh a)}{\mathbf{Q}_{-1/2}(\cosh a)}\right)\exp\left(\frac{a}{2}\right)\mathbf{K}\left(\frac{\exp(a)-1}{\exp(a)+1}\right). \tag{3.161}$$

In the formula (3.161) the following special functions are presented:

$\mathbf{K}(x)$ is the complete elliptic integral of the first kind,
$\mathbf{E}(x)$ is the complete elliptic integral of the second kind.

Once again, the dimensionless variable a characterizes the length of the adhesive region divided by the diameter of fiber $2r_f$.

For large values of the dimensionless length a, the asymptotic behavior of the formula (3.161) is as follows:

$$c_g(a) \underset{a\to\infty}{\to} \frac{1}{2\pi^2}\exp(a/2). \tag{3.162}$$

That is, for sufficiently large values of the dimensionless length a, the following approximate formula could be used:

$$g_1(\eta) \underset{a>>1}{\approx} \frac{1}{2\pi^2}\frac{\exp(a/2)}{\sqrt{2\cosh a - 2\cosh \eta}} \; for - a \leq \eta \leq a, \quad g_1(0) \underset{a>>1}{\approx} \frac{1}{2\pi^2}. \tag{3.163}$$

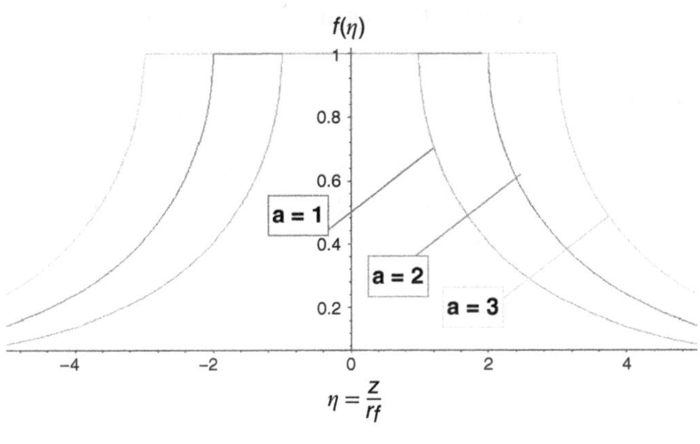

Figure 3.24 Right side of the equation $f(\eta)$ for different dimensionless length of contact region $a = 1$; 2; 3.

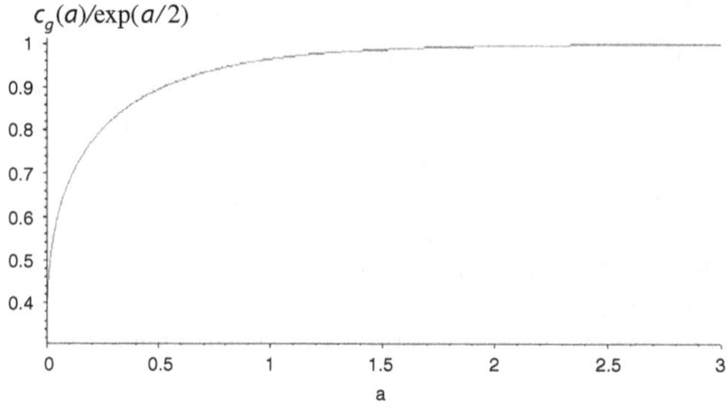

Figure 3.25 The coefficient $c_g(a)/\exp(a/2)$ as a function of debonding length a.

The behavior of the function $c_g(a)$ for large values of a are illustrated in Figure 3.25. The asymptotic behavior for $\eta^* \to 0$ of the solution near the crack tip $\eta = a - \eta^*$ is the following:

$$\bar{g}_1(\eta^*) \overset{def}{=} \frac{g_1(a-\eta^*)}{g_1(0)}$$

$$= \sqrt{\frac{\exp(a)-1}{\exp(a)+1}}\left(1+\frac{\exp(2a)+1}{\exp(2a)-1}\frac{\eta^*}{4}\right)\frac{1}{\sqrt{\eta^*}} + o(\sqrt{\eta^*}). \tag{3.164}$$

The value η^* is the dimensionless distance to the tip of debonding. The dimensioned distance to the tip is $\eta^* r_f$. The formula (3.164) demonstrates that the function $\bar{g}_1(\eta^*)$ follows the inverse-square root law η^* near the tip of debonding:

$$\bar{g}_1(\eta^*) \approx \sqrt{\frac{\exp(a)-1}{\exp(a)+1}}\frac{1}{\sqrt{\eta^*}}, \quad g_1(\eta^*) \approx \sqrt{\frac{\exp(a)-1}{\exp(a)+1}}\frac{\sqrt{2\cosh a}}{\sqrt{\eta^*}} = \frac{Y(a)}{\sqrt{\eta^*}}, \tag{3.165}$$

where is the dimensionless factor that accounts for the geometry of the crack:

$$Y(a) = \sqrt{\frac{\exp(a)-1}{\exp(a)+1}}\sqrt{2\cosh a}.$$

The coefficient in (3.165) appears in the stress intensity factor for an finite debonding region as Y the geometrical factor: $K = Y(a)\Delta\sigma_r\sqrt{r_f}$. The graphs of the exact expression for $\bar{g}_1(\eta)$ and asymptotical expression (3.163) of the solution near the tip of debonding are shown for $a = 2$ on Figure 3.26.

The solution of the problem for partial debonding with peripheral adhesion region (E) could be found in closed form as well.

With the equations (3.159)–(3.163), the solution of the integral equation for the partial contact region is complete.

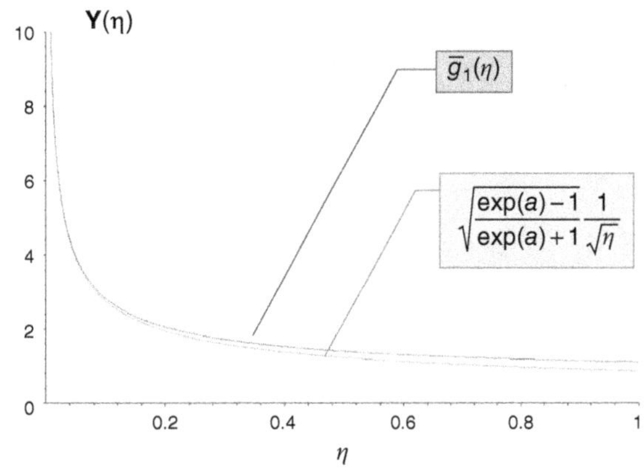

Figure 3.26 Graphs of the exact and asymptotical expressions of the solution near the tip of debonding.

3.3.5 Debonding of a Semi-Infinite Adhesion Region

For the semi-infinite debonding region $-\infty \le \eta < 0$, the integral equation of contact problem (F) reads

$$\int_0^\infty g_2(t) \ln \left| \coth \frac{\eta - t}{4} \right| dt = f_0(\eta) \ for \ 0 \le \eta < \infty. \tag{3.166}$$

For the typical debonding problem case, the gap is constant. Namely, in this case, the right side of the integral equation (3.166) will be also constant. For a constant right side $f(\eta) = 1$, the solution (3.159) reduces in the case of a semi-infinite contact region (F) to:

$$g_2(\eta) = \frac{1}{2\pi^2\sqrt{1 - \exp(-\eta)}} \underset{\eta\to\infty}{\to} \frac{1}{2\pi^2}. \tag{3.167}$$

In other words, the solution (3.167) identically satisfies the integral equation:

$$\int_0^\infty g_2(t) \ln \left| \coth \frac{\eta - t}{4} \right| dt = 1 \ for \ 0 \le \eta < \infty. \tag{3.168}$$

The solution (3.167) is plotted in Figure 3.27.

The asymptotical behavior for $\eta \to 0$ of the solution near the crack tip is the following:

$$\bar{g}_2(\eta) \overset{def}{=} \frac{g_2(\eta)}{\lim\limits_{\eta\to\infty} g_2(\eta)} = \frac{1}{\sqrt{1 - \exp(-\eta)}} = \frac{1}{\sqrt{\eta}} + \frac{\sqrt{\eta}}{4} + o(\sqrt{\eta}). \tag{3.169}$$

The value η is the dimensionless distance to the tip of debonding. The formula (3.169) shows that the function $\bar{g}_2(\eta)$ follows – similar to the function $\bar{g}_1(\eta)$ – the inverse-square root law $1/\sqrt{\eta}$ near the tip of debonding:

$$\bar{g}_2(\eta) \approx \frac{1}{\sqrt{\eta}}, \quad g_2(\eta) \approx \frac{1}{2\pi^2\sqrt{\eta}} = \frac{Y_\infty}{\sqrt{\eta}}, \quad Y(a) \to_{a\to\infty} Y_\infty = \frac{1}{2\pi^2}. \tag{3.170}$$

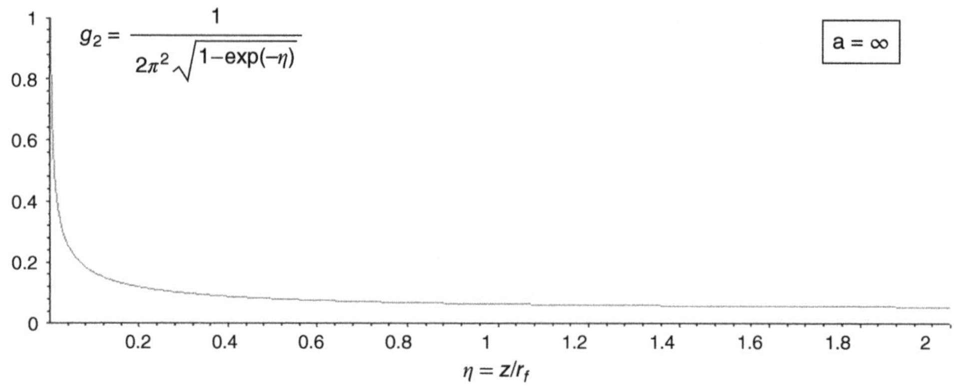

Figure 3.27 Solution $g_2(t)$ (3.167) for a semi-infinite contact region $a = \infty$.

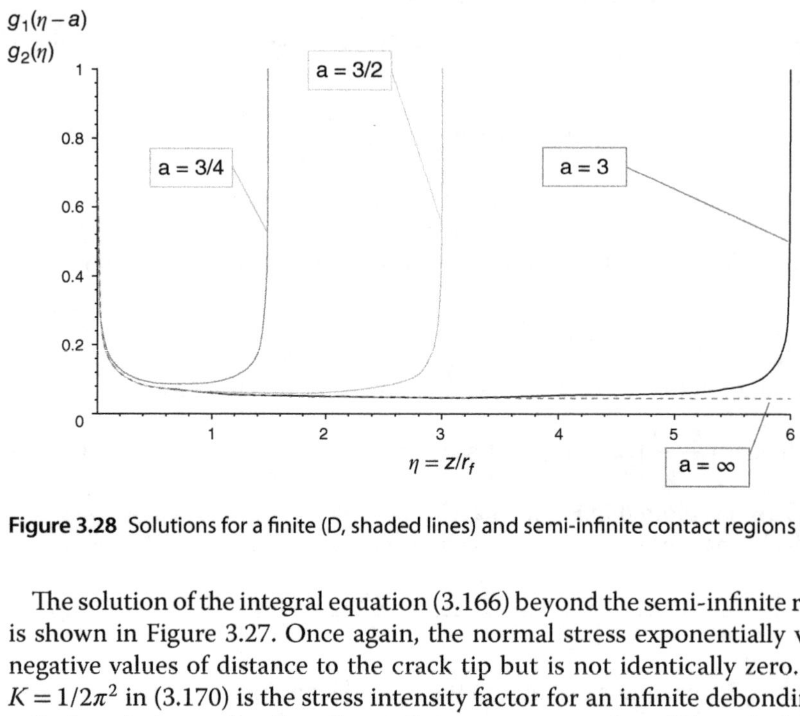

Figure 3.28 Solutions for a finite (D, shaded lines) and semi-infinite contact regions (F, dashed line).

The solution of the integral equation (3.166) beyond the semi-infinite region $0 \leq \eta < \infty$ is shown in Figure 3.27. Once again, the normal stress exponentially vanishes for the negative values of distance to the crack tip but is not identically zero. The coefficient $K = 1/2\pi^2$ in (3.170) is the stress intensity factor for an infinite debonding region.

Both solutions, the first for a finite contact region (D) and the second for a semi-infinite contact region (F), are demonstrated in Figure 3.28. The coefficient of singularity near the tip is equal for both. For calculations, the values $a = 3/4,\ 3/2$ and 3 were used. The difference between the semi-infinite solution and the finite-length solution is practically negligible if the contact length exceeds several fiber diameters.

Summarizing the results of the previous sections, it is now possible to present the solutions of the partially debonding problems.

The Eq. (3.141) expresses the stresses on the interface between matrix and fiber in the case of ideal contact:

$$\sigma_{(f)r}|_{r=r_f} = \sigma_{(m)r}|_{r=r_f} = \frac{\varepsilon_z E_f E_m}{\Delta_d}(\nu_m - \nu_f)({r_m}^2 - {r_f}^2). \tag{3.171}$$

Obviously, stresses (3.171) do not depend on the coordinate z.

If the contact is lost on the part of boundary, the stress on the interface of the partial contact reads:

$$\sigma_{(f)r}|_{r=r_f} = \sigma_{(m)r}|_{r=r_f} = \frac{2\pi^2 \varepsilon_z E_f E_m}{\Delta_d}(\nu_m - \nu_f)({r_m}^2 - {r_f}^2) g_i \left(\frac{z}{r_f}\right). \tag{3.172}$$

For the case of a central contact zone, Eq. (3.159) must be assigned, and for a semi-infinite region, Eq. (3.167).

The provided closed form solution is used later for the study of the semi-infinite contact problem.

3.3.6 Debonding of Fibers from a Matrix Under Cyclic Deformation

An overview of the development of methods for the prediction of fatigue driven delamination growth was given in Pascoe et al. (2013). Four categories of methods were demonstrated: stress/strain-based models, fracture mechanics-based models, cohesive-zone models and models using the numerical methods based on extended finite elements. Delamination resistance records from diverse carbon-fiber reinforced polymer-matrix composites were compared in Stelzer et al. (2014) for different loading modes. The delamination resistance of composites with implications for composite structural design and testing include, for example, the determination of threshold values in cyclic fatigue, the question of conservative mode, approaches for data analysis and possible analogies in short crack cyclic fatigue between fracture behavior of structural metal alloys and composites were analyzed. The cited paper demonstrates the dispersion of curves in Paris-type law data analysis of cyclic fatigue tests and the resulting apparent threshold behavior that has implications for composite structural design.

Testing data of carbon-fiber-reinforced epoxy laminates were analyzed later in Brunner et al. (2016) with different approaches that reduce scatter and to identify parameters for composite structural design. Selected test data comparing load and displacement control for the cyclic fatigue tests were also demonstrated. The threshold values from Paris-law data fitting were compared with values from fitting with a modified Hartman–Schijve approach (Hartman and Schijve 1970).

The comprehensive review of the principal studies on stress ratio effect in delamination growth was given by Khan et al. (2015). The effects of stress ratio effect that investigate the delamination from a mechanistic of physical science perspective were discussed.

The initial flaws around the fibers over the separation surface persist in different concentrations in each fabricated composite material. Harris (1977, 2003) has confirmed that the damage in glass/epoxy composites is accompanied by firer/resin debonding and the composite will often fail by splitting parallel to the fibers.

Horst and Spoormaker (1996) observed that the cyclic axial deformation of a fiber with cylindrical flaws leads to progressive debonding of fibers from matrix, such that the debonding crack extends just outside over the separation surface between matrix and fibers. For some material with poor strength, the debonding crack spreads directly over the separation surface. Horst et al. (1998) demonstrated that a radial tensile stress in short glass fiber composites is generated at the interface as a result of fatigue loading.

In this section, from the viewpoint of the above developed static debonding theory, the spreading of crack over to separation surface between the fiber and matrix will be evaluated for cyclic loading.

The physical application of the elaborated solution is as follows. Fatigue failure occurs when a material is subjected to repetitive loading and unloading cycles. The level of stresses present to cause fatigue failure is usually below values considered as safe for a single static load application. The critical fatigue initiation is usually at a position limited to a small area. There are three stages of fatigue fracture commonly distinguished: initiation, propagation and final rupture (Totten and Tarpani 2008; Fleck et al. 1994). On a microscopic scale, failure occurs along slip planes in the crystalline structure of the materials. Most metals with a body centered cubic crystal structure have a characteristic response to cyclic stresses. These materials have a threshold stress limit below which fatigue cracks will not initiate. This threshold stress value is often referred to as the endurance limit. In steels, the life associated with this behavior is usually accepted to be two million cycles.

The first stage (stage I) of fatigue is referred to as initiation. Initiation is probably the most complex stage of fatigue fracture. The most significant factor about the initiation stage of fatigue fracture is that the irreversible alterations in the metal are caused by cyclic shear stresses. The accumulation of microscopic faults over a large number of load applications leads to cumulative damage. At the location of a severe stress concentration, the number depends on the geometry of the part as well as on environmental, stress, metallurgical and strength conditions, as will become apparent. During the stage I, the spread of fatigue crack per unit cycle decelerates smoothly with number of cycles and the process approaches its second stage.

The second stage of fatigue is known as crack propagation (stage II). The propagation stage of fatigue causes the microscopic crack to change direction and grow perpendicular to the tensile stress. The second or propagation stage of fatigue is typically the most readily recognizable area of a fatigue fracture. The Paris law describes the stage II propagation law (Paris and Erdogan 1963). The traditional form of Paris law pronounces the spread of fatigue crack per unit cycle as a power function of the range of stress intensity factor. Toward the end of the stage II, the spread of fatigue crack per unit cycle begins to accelerate smoothly with number of cycles and the process turns to the third stage.

The final, third stage of fatigue is the final rupture (stage III). As the propagation of the fatigue crack endures, progressively sinking the cross-sectional area of the test specimen, it eventually deteriorates the part so significantly that final, broad fracture occurs with a couple of load cycles. The fracture mode may be either ductile (with a dimpled fracture surface) or brittle (with a cleavage, or intergranular, fracture surface). The combinations of these modes occasionally occur, depending upon the metal concerned, the stress level and the environment. In the course of the stage III, the spread of fatigue crack per unit cycle progressively accelerates with each cycle.

The debonding crack propagation over the separation surface is quantified as a function of the range of the stress intensity factor:

$$K = K_{\max} - K_{\min}. \tag{3.173}$$

The factors in this equation are:

$K_{\max}$ the maximum stress intensity factor;
$K_{\min}$ the minimum stress intensity factor per cycle.

The common form of Paris' law quantifies the fatigue life of a specimen for a given particular debonding crack size a. The range of stress intensity factor reads:

$$K = Y(a)\sqrt{r_f}\Delta\sigma_r. \tag{3.174}$$

The range of stress intensity factor depends on the following factors:
$Y(a)$ is the dimensionless parameter that reflects the geometry (3.165);

$$\Delta\sigma_r = \sigma_{r\max} - \sigma_{r\min} \tag{3.175}$$

is the radial stress range (3.171);

$$\varepsilon_z = (\varepsilon_{z.\max} + \varepsilon_{z.\min})/2 + (\varepsilon_{z.\max} - \varepsilon_{z.\min})\cos\omega t/2. \tag{3.176}$$

cyclic axial strain;

the maximum radial stress per cycle:

$$\sigma_{r.\max} = \frac{\varepsilon_{z.\max}E_f E_m}{\Delta_d}(\nu_m - \nu_f)({r_m}^2 - {r_f}^2); \tag{3.177}$$

and the minimum radial stress per cycle:

$$\sigma_{r.\min} = \frac{\varepsilon_{z.\min}E_f E_m}{\Delta_d}(\nu_m - \nu_f)({r_m}^2 - {r_f}^2). \tag{3.178}$$

For a considerably long debonding length (3.167):

$$Y(a) \to_{a\to\infty} \frac{1}{2\pi^2}, K_\infty = \frac{\Delta\sigma_r\sqrt{r_f}}{2\pi^2}. \tag{3.179}$$

Equation (3.181) demonstrates the remarkable feature. Namely, the stress concentration factor K_∞ does not depend on the length of deboning region, but on the diameter of the fiber if the debonding region is longer than several fiber diameters.

The common form for propagation law is used:

$$c_f\frac{dn}{da} = p_c(K) \quad \text{or} \quad c_f\frac{dn}{da} = p_c(\sigma\sqrt{\pi a}). \tag{3.180}$$

The value $c_f/p_c(K)$ determines the expansion of the fatigue crack per load cycle. The coefficients in Eq. (3.2) are:

$c_f = c_f(R_\sigma)$ the material constant of matrix or resin for a given stress ratio R_σ;

$R_\sigma = \dfrac{K_{\min}}{K_{\max}} < 1$ the stress ratio of cyclic load;

$K_m \equiv \dfrac{K_{\max} + K_{\min}}{2} = \dfrac{1 + R_\sigma}{1 - R_\sigma}\dfrac{K}{2}$. the mean value of stress intensity factor.

The propagation function $p_c(K)$ describes the infinitesimal crack length growths per increasing number of load cycles n, supposing that the load ratio R_σ remains constant over the load history. Thus, the range K and the mean value K_m of stress intensity factor remain constant. The variations of range K and the mean value K_m of stress intensity factor over the load history could be accounted by applications of damage accumulation theories. The crack growth rate function due to Paris–Erdogan is:

$$p_c(K) = K^{-p} \ . \tag{3.181}$$

It is assumed that the cracks with the initial length δ exist in the resin material in the vicinity of the matrix-fiber separation surface. The cycle count starts with the beginning of crack propagation from its initial length. If the initial length δ is small, the stress intensity factor is proportional to the square root of the crack length and is, consequently, low. Thus, the crack spreads slowly at the beginning of cycle history. However, with progressive crack elongation the stress intensity factor initially raises continuously proportional to the square root of the reached crack length. With further elongation, the stress intensity factor increases more and more slowly than the square root of the reached crack length. Thus, the increase of the stress intensity factor gradually decelerates with the elongation of crack. For long cracks, the stress intensity factor reaches its asymptotical limit

$$K_\infty = \Delta\sigma_r\sqrt{r_f}/2\pi^2. \tag{3.182}$$

The stress intensity factor stabilizes below the limit (3.182). As a result, for sufficient long debonding cracks the range of stress intensity factor does not depend on the length of debonding region. Accordingly, the elongation per cycle for the sufficiently long cracks, of several fiber diameters, remains constant. Under cycle loading, the debonding region grows with constant elongation rate (Figure 3.17). The stress intensity range depends linearly on the strain amplitude (3.177) and (3.178) and crack growth rate follows the certain crack propagation law, for example, the Paris Law (3.179).

As a consequence, the elongation of the debonding crack per the load cycle remains constant. The debonding length is directly proportional to the cycle count. This effect leads to the steady degradation of the material over the cycle number. For example, the longitudinal module reduces roughly proportionally to the duration of cyclic load.

3.4 Conclusions

The mechanisms of damage of fiber-reinforced composites on the micromechanical level were discussed. Two main effects were displayed: crack bridging across fiber direction and the debonding of fibers in their direction. These two mechanisms of fracture are frequent for unidirectional fiber bundles. The parametric dependences of crack toughness were derived in closed form. These formulas are necessary for optimization tasks.

References

Achenbach, J.D. (1975). *A Theory of Elasticity with Microstructure for Directionally Reinforced Composites*. CISM International Centre for Mechanical Sciences, Springer-Verlag Wien.

Alexandrov, V.M. and Pozharskii, D.A. (2001). *Three-Dimensional Contact Problems.* Dordrecht/Boston/London: Kluwer Academic Publishers.

Argon, A.S. (2000). Fracture: strength and toughness mechanisms. In: *Comprehensive Composite Materials*, vol. 1 (ed. A. Kelly), 763–802. ISBN: 0-080437192.

Aveston J., Copper A.G., Kelly A. (1971) The properties of fibre composites. In Conference Proceedings, National Physical Laboratory, Guildford, pp. 15–26.

Barbero, E. (1999). *Introduction to Composite Materials Design.* New York: Taylor and Francis.

Beaumont, P.W. (1974). A fracture mechanics approach to failure in fibrous composites. *J. Adhes.* 6: 107–119.

Blackman, B.R.K. and Williams, J.G. (1999). Impact and high rate testing of composites. In: *Mechanics of Composite Materials and Structures* (ed. M. Soares (ed.)), 215–224. Kluwer Academic Publishers. ISBN: 9780792358701.

Brunner, A.J., Stelzer, S., Pinter, G., and Terrasi, G.P. (2016). Cyclic fatigue delamination of carbon fiber-reinforced polymer-matrix composites: data analysis and design considerations. *Int. J. Fatigue* 83: 293–299.

Budianski, B., Hutchinson, J.W., and Evans, A.G. (1986). Matrix fracture in fibre reinforced ceramics. *J. Mech. Phys. Solids* 34: 167–189.

Chamis, C.C. (1969). *Failure Criteria for Filamentary Composites.* Washington: NASA TN D-5367.

Chamis, C.C., Hanson, M.P., and Serafini, T.T. (1971). *Designing for Impact Resistance with Unidirectional Fiber Composites.* Washington: NASA TN D-6463.

Corigliano, A. (2007). Damage and fracture mechanics techniques for composite structures. *Compre. Struct. Integr.* 3: 459–539. https://doi.org/10.1016/B0-08-043749-4/03041-X.

Cox, B.N. and Marshall, D.B. (1994). Concepts for bridged cracks in fracture and fatigue. *Acta Metall. Mater.* 42: 341–363.

Davies, P., Blackman, B.R.K., and Brunner, A.J. (1998). Test methods for delamination resistance of composite materials: current status. *Appl. Compos. Mater.* 5 (6): 345–364.

Dharani, L.R., Jones, W.F., and Goree, J.G. (1983). Mathematical modelling of damage in unidirectional composites. *Eng. Fract. Mech.* 17: 555–574.

Dow, N.F. and Rosen, B.W. (1965). *Evaluation of Filament-Reinforced Composites for Aerospace Structural Applications.* Washington, D.C: NASA CR-207.

Harris, B. (ed.) (2003). *Fatigue in Composites, Science and Technology of the Fatigue Response of Fiber-Reinforced Plastics.* Boca Raton Boston, New York, Washington, D.C: CRC Press.

Fichter, W.B. (1969). *Stress Concentration around Broken Filaments in a Filament-Stiffened Sheet.* Washington: NASA TN D-5453.

Filon, L.N.G. (1901). On the elastic equilibrium of circular cylinders under certain practical systems of load. *Proc. R. Soc. London* A68: 353–358.

Filon, L.N.G. (1902). On the elastic equilibrium of circular cylinders under certain practical systems of load. *Phil. Trans. R. Soc. London* A198: 147–233.

Fleck, N.A., Kang, K.J., and Ashby, M.F. (1994). The cyclic properties of engineering materials, overview no 112,. *Acta Metall. Mater.* 42 (2): 365–381.

Foye R.L.(1966) Compression strength of unidirectional composites. AIAA Paper 66–143, AIAA 3rd Aerospace Science Meeting, New York.

Gohberg, I.C. and Krein, M.G. (1967). *The Theory of Volterra Operators in a Hilbert Space and its Applications [in Russian].* Moscow: Nauka.

Goree, J.G. and Gross, R.S. (1979). Analysis of a unidirectional composite containing broken fibres and matrix damage. *Eng. Fract. Mech.* 13: 563–578.

Greszczuk, L.B. (1972). *Microbuckling of Unidirectional Composites.* AFML TR-71-231. Ohio: Wright Patterson AFB.

Greszczuk, L.B. (1973). *Failure Mechanics of Composites Subjected to Compressive Loading.* AFML TR-72-107. Ohio: Wright Patterson AFB.

Griffith, A.A. (1920). The phenomenon of rupture and flow in solids. *Phil. Trans. R. Soc.* A237: 161–198.

Halpin, J.C. and Meinecke, E.A. (1969). *Delayed Instability (Buckling) of Viscoelastic Materials.* AFML TR-68-331. Ohio: Wright Patterson AFB.

Hancock, J.R. (1970). *Fatigue fracture of metal matrix composites.* AFFDL TR-70-144. Ohio: Wright Patterson AFB.

Hannant, D.J., Hudges, D.C., and Kelly, A. (1983). Toughening of cement and other brittle solids with fibres. *Proc. R. London Soc.* A310: 175–190.

Hartman, A. and Schijve, J. (1970). The effects of environment and load frequency on the crack propagation law for macro fatigue crack growth in aluminum alloys. *Eng. Fract. Mech.* 1 (4): 615–631.

Hart-Smith, L.J. (2000). Strength of Unnotched laminates under multiaxial stress. In: *Comprehensive Composite Materials*, vol. 1 (ed. A. Kelly), 667–681. ISBN: 0-080437192.

Hedgepeth, J.M. and Van Dyke, P. (1967). Local stress concentrations in imperfect filamentary composite materials. *J. Compos. Mater.* 1: 294–309.

Herrmann, K.P. and Ferber, F. (1995). Numerical modelling of crack growth in material models of fibrous composites. In: *Microstructure – Property Interactions in Composite Materials* (ed. R. Pyrz), 129–140. Dordrecht: Kluwer Academic Publishers.

Hollen, J.W.S., Grossberg, P., and Backer, S. (1969). *Structural Mechanics of Fibers, Yarns and Fabrics.* New York: Wiley & Sons.

Horst, J.J. and Spoormaker, J.L. (1996). Mechanisms of fatigue in short glass fiber reinforced polyamide 6. *Polym. Eng. Sci.* 36 (22): 2718–2726.

Horst, J.J., Salienko, N.V., and Spoormaker, J.L. (1998). Fiber-matrix debonding stress analysis for short fiber-reinforced materials with matrix plasticity, finite element modelling and experimental verification. *Composites* 29A: 525–531.

Hull, D. and Clyne, T.W. (1996). *An Introduction to Composite Materials, Second Edition.* Cambridge University Press.

Hutchinson, J.W. and Jensen, H.M. (1990). Models of Fiber Debonding and Pullout in brittle composites with friction. *Mech. Mater.* 9: 139–163.

Jones, R.M. (1975). *Mechanics of Composite Materials.* New York: McGraw-Hill.

Kelly, A. and Zweben, C. (eds.) (2000). *Comprehensive Composite Materials.* Pergamon, Elsevier Ltd.

Khan, R., Alderliesten, R., Badshah, S., and Benedictus, R. (2015). Effect of stress ratio or mean stress on fatigue delamination growth in composites: critical review. *Compos. Struct.* 124: 214–227.

Kobelev, V. (2001). *Failure Criteria for Fiber Composites and Optimization of Their Fracture Behavior (Bruchmechanische Kriterien für Faserverbundwerkstoffe und ihre Optimierung hinsichtlich eines lokalen Bruchverhaltens)*, Fortschritt-Berichte VDI, Reihe 18: Mechanik/Bruchmechanik; 262. Düsseldorf: VDI Verl.

Koeneman, J.B. (1969). *Interfiber Failure of Composite Materials.* Cleveland: Case Western Reserve University.

Korczynskij, Y., Harris, S.J., and Morley, J.G. (1981). The influence of reinforcing fibres on the growth of cracks in brittle matrix composites. *J. Mater. Sci.* 16: 1533–1547.
Kuksenko, V. and Tamuzs, V. (1981). *Fracture Micromechanics of Polymer Materials.* Dordrecht: Springer Science+Business Media.
Landis, C.M., Beyerlein, I.J., and McMeeking, R.M. (2000). Micromechaincal simulation of the failure of fiber reinforced composites. *J. Mech. Phys. Solids* 48: 621–648.
Lekhnitskii, S.G. (1981). *Theory of Elasticity of an Anisotropic Body.* Moscow: Mir.
Lifshitz, J.M. and Rotem, A. (1970). *Longitudinal Strength of Unidirectional Fibrous Composites.* AFML TR-70-194. Ohio: Wright Patterson AFB.
Liu S., Nairn J.A. (1990) Fracture Mechanics Analysis of Composite Microcracking: Experimental Results in Fatigue, *Proceedings of the 5th Technical Conference on Composite Materials, American Society of Composites,* East Lansing, Michigan, June 11–14, 1990
Liu, H.-Y., Zhou, L.-M., and Mai, Y.-W. (1994). On fibre pullout with a rough interface. *Philos. Mag. A* 70: 359–372.
Lockett, F.J. (1970). Fiber stress-concentrations in an imperfect two-dimensional elastic composite. *Nat. Phys. Lab. Rep.* 88: 363.
Lur'e, A.I. (1964). *Three-Dimensional Problems of the Theory of Elasticity.* English translation (ed. J.R.M. Radok). New York: Wiley.
Meda, G. and Steif, P.S. (1994). A detailed analysis of crack bridged by fibres. I: limiting cases of short and long cracks. *J. Mech. Phys. Solids* 42: 1293–1321.
Meleshko, V. (2003). Equilibrium of an elastic finite cylinder: Filon's problem revisited. *J. Eng. Math.* 46: 355. https://doi.org/10.1023/A:1025066408575.
Meng, Q. and Wang, Z. (2015). Theoretical analysis of interfacial debonding and fiber pull-out in fiber-reinforced polymer-matrix composites. *Arch. Appl. Mech.* 85: 745–759. https://doi.org/10.1007/s00419-015-0987-6.
Mikhailov, A.M. (1973). Fracture of a unidirectional glass-plastic. *Mech. Solids* 33: 131–139.
Newaz, G.M. and Yung, J.-Y. (1988). Modelling split crack growth behavior in a unidirectional composite under fatigue load. *Eng. Fract. Mech.* 29: 483–495.
Ochoa, O.O. and Reddy, J.N. (1992). *Finite Element Analysis of Composite Laminates.* Springer Netherlands, Springer Science+Business Media, Dordrecht.
Owen, M.J. and Bishop, P.T. (1974). Crack-growth relationship for glass-reinforced plastics and their application to design. *J. Phys.* 7: 156–161.
Owen M.J., Morris S. (1971) Carbon fibres, their composites and applications. *Int. Conf. Paper 26,* Plastics Institute, London.
Pagano, N.J. (1978). Stress fields in composite laminates. *Int. J. Solids Struct.* 14: 385–400.
Paris, P.C. and Erdogan, F. (1963). A critical analysis of crack propagation laws. *J. Basic Eng.* 85: 528–533. http://dx.doi.org/10.1115/1.3656900.
Pascoe, J.A., Alderliesten, R.C., and Benedictus, R. (2013). Methods for the prediction of fatigue delamination growth in composites and adhesive bonds – a critical review. *Eng. Fract. Mech.* 112–113: 72–96.
Piggott, M.R. (1980). *Load Bearing Fibre Composites.* Oxford: Pergamon Press.
Pipes, R.B. and Pagano, N.J. (1970). Interlaminar stresses in composite laminates under uniform axial extension. *J. Compos. Mater.* 4 (4): 538–548. https://doi.org/10.1177/002199837000400409.
Poe, C.C.J. (1984). *Fracture Toughness of Fibrous Composite Materials.* NASA Technical Paper 2370. Hampton, VI: NASA Scientific and Technical Information Branch.

Polyanin, A.D. and Manzhirov, A.V. (2008). *Handbook of Integral Equations*, 2e ed. CRC 2008 , EQ. 3-6.1, 5. Boca Raton, FL: Chapman & Hall/CRC, Taylor & Francis Group.

Robinson, P., Greenhalgh, E., and Pinho, S. (eds.) (2012). *Failure Mechanisms in Polymer Matrix Composites, Criteria, Testing and Industrial Applications*, Woodhead Publishing Series in Composites Science and Engineering. Woodhead Publishing.

Schapery, R.A. (1972). *Viscoelastic Behaviour and Analysis of Composite Materials*. Mechanical Mater. Research Center Rep. Vol. MM 72-3. Texas A&M University.

Schürmann, H. (2007). *Konstruieren mit Faser-Kunststoff-Verbunden*. Springer-Verlag. ISBN: 3-540-40283-7.

Schürmann H. (2015) Zum werkstofflichen und konstruktiven Potential von Faserverbunden beim Einsatz in Federn, *Deutscher Verband für Materialforschung und -prüfung, DVM-Tag 2015, Federn im Fahrzeugbau*, S. 119–128.

Selvadurai, A.P.S. (1983). Concentrated body force loading of an elastically bridged penny-shaped flaw in a unidirectional fibre reinforced composite. *Int. J. Fract.* 21: 149–159.

Sih, G.C. (1991). *Mechanics of Fracture Initiation and Propagation*. Dordrecht: Kluwer Academic Publishers.

Sih, G.C. and Chen, E.P. (1981). *Cracks in Composite Materials*. The Hague/Boston/London: Martinus Nijhoff Publishers.

Sih, G.C. and Liebowitz, H. (1968). Mathematical theories of brittle fracture, Section IV, Rectilinearly Anisotropic Bodies with Cracks. In: *Fracture: An Advanced Treatise*, Chapter 2, vol. II (ed. H. Liebowitz). Academic Press.

Sih, G.C., Chen, P.E., and Hang, S.L. (1965). On cracks in rectilinearly anisotropic bodies. *Int. J. Fract. Mech.* 1: 189–199.

Sims, D.F. (1972). *Viscoelastic Creep and Relaxation Behavior of Laminated Composite Plates*. Department of Mechanical Engineering, Southern Methodist University.

Slaughter, W.S. and Sanders, J.L. (1991). A model for load-transfer from an embedded Fiber to an elastic matrix. *Int. J. Solids Struct.* 28 (8): 1041–1052.

Sneddon, I.N. and Lowengrub, M. (1969). *Crack Problems in the Classical Theory of Elasticity. (the SIAM Series in Applied Mathematics)*. New York/London/Sydney/Toronto: John Wiley & Sons.

Stang, H.A. (1987). Double inclusion model for microcrack arrest in fibre reinforced brittle materials. *J. Mech. Phys. Solids* 35: 325–342.

Stelzer S., Pinter G., Brunner A.J. (2014) Comparison of Quasi-static and Cyclic Fatigue Delamination Resistance of Carbon Fiber Reinforced Polymer-Matrix Laminates under Different Mode Loading, *20th European Conference on Fracture (ECF20) , Proc. Mat. Sci.*, 3, 1087–1092, doi: https://doi.org/10.1016/j.mspro.2014.06.177.

Sun, C.T. (2000). Strength analysis of unidirectional composites and laminates. In: *Comprehensive Composite Materials*, vol. 1 (ed. A. Kelly), 641–666. ISBN: 0-080437192.

Tirosh, J. (1973). The effect of plasticity and crack blunting on the stress distribution in orthotropic composite materials. *J. Appl. Mech.* 40: 785–790.

Totten, G. and Tarpani, J. (2008). Fatigue crack propagation. *Adv. Mat. Proc.* 166: 39–41.

Tsai, S.W., Adams, D.F., and Doner, D.R. (1966). *Analysis of Composite Structures*. Houston, TX: NASA CR-620.

Vannucci, P. (2018). *Anisotropic Elasticity*. Springer Nature, Singapore Pte Ltd.

Wang, S.S. and Choi, I. (1982). The interface crack between dissimilar anisotropic composite materials. *J. Appl. Mech.* 49: 541–548.

Wu, E.M. and Reuter, R.C. (1965). *Crack Extension in Fiberglass Reinforced Plastics.* TAM Rep. 275. Urbana, IL: University of Illinois.

Xia, Z.C., Hutchinson, J.W., Evans, A.G., and Budiansky, B. (1994). On large scale sliding in fibre reinforced composites. *J. Mech. Phys. Solids* 42: 1139–1158.

Xu, G., Bower, A., and Ortiz, M. (1998). The influence of crack trapping on the toughness of fiber reinforced composites. *J. Mech. Phys. Solids* 46: 1815–1833.

Yao, L., Alderliesten, R.C., and Benedictus, R. (2016). *Compos. Struct.* 140: 125–135. https://doi.org/10.1016/j.compstruct.2015.12.027.

Yin (1998). Delamination: laminate analysis and fracture mechanics. *FFEMS* 21 (4): 509–520. https://doi.org/10.1046/j.1460-2695.1998.00059.x.

Zhang, X., Liu, H.-Y., Mai, Y.-W., and Diao, X.-X. (1999). On steady-state fiber pull-out. Part I: stress field. *Compos. Sci. Technol.* 59: 2179–2189.

Zhou, L.-M., Kim, J.-K., and Mai, Y.-W. (1992). On the single fibre pullout problem: effect of loading methods. *Compos. Sci. Technol.* 45: 153–160.

4

Optimization Principles for Structural Elements Made of Composites

4.1 Stiffness Optimization of Anisotropic Structural Elements

4.1.1 Optimization Problem

The estimation of specific anisotropic stiffness is based on the following technique. The density of the elastic energy of an anisotropic material depends on the orientation of material and the appropriate stress $\widehat{\boldsymbol{\sigma}}$:

$$W_e(\Phi) = \widehat{\boldsymbol{\sigma}}^T \cdot \widehat{\mathbf{S}}(\Phi) \cdot \widehat{\boldsymbol{\sigma}}/2 = \sigma_{ij} S_{ijkl}(\Phi)\sigma_{kl}/2. \tag{4.1}$$

The elastic energy pro mass unit (specific elastic energy) reads:

$$W_e^*(\Phi) = W_e(\Phi)/\rho = \widehat{\boldsymbol{\sigma}}^T \cdot \widehat{\mathbf{S}}(\Phi) \cdot \widehat{\boldsymbol{\sigma}}/2\rho. \tag{4.2}$$

The angle Φ designates the rotation angle in the xy plane and is the angle between axis "1" of material-fixed and the x-axis of element-fixed coordinate systems.

For the optimization, the relations between the components of the tensor $\mathbf{S}(\Phi) = [S_{ijkl}]$ in element-fixed coordinate system and tensor $\mathbf{S}_0 = [s_{ijkl}]$ in the material-fixed coordinate system with the components $s_{ijkl} = S_{ijkl}(\Phi = 0)$, are required.

At first, the rotation is performed in tensor notation of the fourth order. The tensor of the fourth rank $\mathbf{S}(\Phi)$ coordinate system possesses the components:

$$S_{ijkl}(\Phi) = t_{ip}t_{jq}t_{kn}t_{ls}s_{pqns}, \, i,j,k,l = 1,2,3, \tag{4.3}$$

where

$$\mathbf{t}(\varphi) = [t_{ij}] = \begin{bmatrix} c & s & 0 \\ -s & c & 0 \\ 0 & 0 & 1 \end{bmatrix}. \tag{4.4}$$

is the orthogonal rotation matrix. In this chapter, the abbreviations are used: $c = \cos\Phi$, $s = \sin\Phi$.

Similarly, the tensor of the second rank in Kelvin's notation $\widehat{\mathbf{S}}(\Phi)$ depends on the inclination angle Φ as follows:

$$\widehat{\mathbf{S}}(\Phi) = \widehat{\mathbf{T}}(\Phi) \cdot \widehat{\mathbf{S}}^{(0)} \cdot \widehat{\mathbf{T}}^T(\Phi), \tag{4.5}$$

Design and Analysis of Composite Structures for Automotive Applications: Chassis and Drivetrain, First Edition. Vladimir Kobelev.

Companion website: www.wiley.com/go/kobelev/automotive_suspensions

$$\widehat{\mathbf{T}}(\varphi)=\begin{bmatrix} c^2 & s^2 & 0 & 0 & 0 & \sqrt{2}sc \\ s^2 & c^2 & 0 & 0 & 0 & -\sqrt{2}sc \\ 0 & 0 & 1 & 0 & 0 & 0 \\ 0 & 0 & 0 & c & -s & 0 \\ 0 & 0 & 0 & s & c & 0 \\ -\sqrt{2}sc & \sqrt{2}sc & 0 & 0 & 0 & c^2-s^2 \end{bmatrix}, \tag{4.6}$$

$$\widehat{\mathbf{S}}^{(0)}=\begin{bmatrix} E_1^{-1} & -\nu_{12}E_1^{-1} & -\nu_{31}E_3^{-1} & 0 & 0 & 0 \\ -\nu_{12}E_1^{-1} & E_2^{-1} & -\nu_{32}E_3^{-1} & 0 & 0 & 0 \\ -\nu_{31}E_3^{-1} & -\nu_{32}E_3^{-1} & E_3^{-1} & 0 & 0 & 0 \\ 0 & 0 & 0 & \frac{1}{2}G_{23}^{-1} & 0 & 0 \\ 0 & 0 & 0 & 0 & \frac{1}{2}G_{31}^{-1} & 0 \\ 0 & 0 & 0 & 0 & 0 & \frac{1}{2}G_{12}^{-1} \end{bmatrix}. \tag{4.7}$$

The first optimization problem consists of a search of the orientation of the angle Φ that provides the minimum or maximum of the elastic energy, assuming the stress $\widehat{\boldsymbol{\sigma}}$ is given:

$$W_e(\Phi) \to \min_{\Phi} \tag{4.8}$$

or

$$W_e(\Phi) \to \max_{\Phi}. \tag{4.9}$$

The minimum of the elastic energy (4.8) is attained for the stiffest structures, if the slightest distortion of the structure is questioned. The maximum of the elastic energy (4.9) leads to the structure that stores the highest amount of the elastic energy under the action of the given stresses.

The optimization problem (4.8) has been the subject of several studies in the context of orthotropic solids. In Seregin and Troitskii (1981) the coaxiality condition was stated in the cited work. According to the coaxiality condition, a minimum or maximum of strain energy is attained, if the stress and strain tensors share common principal axes. Rovati and Taliercio (2005) stated the coaxiality condition three-dimensional elastic materials with orthotropic and cubic symmetries. The coaxiality condition was also considered by Cowin (1994), using method of Kelvin's notation.

4.1.2 Optimality Conditions

One way to derive the optimality conditions uses the variations of tensor components of the fourth rank in three-dimensional space (Banichuk and Kobelev 1987). At each point of the body let the direction of the anisotropy axes be given by the unit vectors $\mathbf{j}_1, \mathbf{j}_2, \mathbf{j}_3$. The unit vectors that define the directions of a global Cartesian coordinate system are $\mathbf{i}_1, \mathbf{i}_2, \mathbf{i}_3$.

Orientation of the anisotropy axes relative to the axes of the global coordinate system is given by the magnitudes of the cosines of the angles between the directions $\mathbf{j}_l$, and $\mathbf{i}_k$, by the quantities $t_{lk} = (\mathbf{j}_l, \mathbf{i}_k)$. The rotation transformation matrix t_{lk} is orthogonal:

$$t_{lk}t_{kn} = \delta_{ln},$$

where δ_{ln} is the Kronecker delta. Since the orientation varies from point to point, the magnitudes of the direction cosines are functions of the coordinates. The optimization problem consists of seeking the orientation of the anisotropy axes, that is, the functions $t_{lk}(x_1, x_2, x_3)$ from the condition for the maximum or minimum of the elastic energy.

The expressions for the variations δS_{ijkl} due to the variation δt_{lk} of the cosines t_{lk} read:

$$\delta S_{ijkl} = 4S_{ijks}t_{sp}\delta t_{pl}. \tag{4.10}$$

The variation of the orthogonal rotation matrix t_{lk} is:

$$\delta \mathbf{t}(\Phi) = [\delta t_{ij}] = -\begin{bmatrix} s & -c & 0 \\ c & s & 0 \\ 0 & 0 & 0 \end{bmatrix} \delta\Phi. \tag{4.11}$$

Adding six conditions (4.11) with six Lagrange multipliers η_{sl} to Eqs. (4.8) or (4.9), we get the augmented elastic energy:

$$W_e(\Phi) = \sigma_{ij}S_{ijkl}\sigma_{kl}/2 + \eta_{sl}(t_{sp}t_{pl} - \delta_{sl}). \tag{4.12}$$

Using the relations (4.10), the variation of (4.12) reads:

$$\begin{aligned} \delta W_e(\Phi) &= 4\sigma_{ij}S_{ijks}\sigma_{kl}t_{sp}\delta t_{pl}/2 + 2\eta_{sl}t_{sp}\delta t_{pl} \\ &= 2[\sigma_{ij}S_{ijks}\sigma_{kl} + \eta_{sl}]t_{sp}\delta t_{pl}. \end{aligned} \tag{4.13}$$

Consequently, the variation $\delta W(\Phi)$ disappears, if the following condition is satisfied:

$$\sigma_{ij}S_{ijks}\sigma_{kl} = -\eta_{sl}. \tag{4.14}$$

The augmented elastic energy (4.12) behaves as a scalar for an arbitrary rotation transformation. Consequently, the Lagrange multipliers form symmetric tensor $\boldsymbol{\eta} = \boldsymbol{\eta}^T$, or, in component notation, $\eta_{sl} = \eta_{ls}$. From this condition and (4.14) follows the important identity:

$$\sigma_{ij}S_{ijks}\sigma_{kl} = \sigma_{ij}S_{ijkl}\sigma_{ks}, \quad \boldsymbol{\sigma} \cdot \cdot \mathbf{S} \cdot \boldsymbol{\sigma} = \boldsymbol{\sigma} \cdot \mathbf{S} \cdot \cdot \boldsymbol{\sigma}. \tag{4.15}$$

The one and two dots between the symbols denote the simple and double products. Because $\mathbf{S} \cdot \cdot \boldsymbol{\sigma} = \boldsymbol{\varepsilon}$, the identity (4.15) yields to the commutation relation:

$$\boldsymbol{\varepsilon} \cdot \boldsymbol{\sigma} = \boldsymbol{\sigma} \cdot \boldsymbol{\varepsilon}. \tag{4.16}$$

Two symmetric second-rank tensors are said to be coaxial if their principal axes coincide. Two symmetric, second-rank tensors are coaxial if and only they commute (Ogden 1997, problem 1.3.8 in chapter 1). The tensors $\boldsymbol{\sigma}$, $\boldsymbol{\varepsilon}$ commute and, consequently, are coaxial.

Another elegant way to derive the coaxiality conditions in 6×6 tensor notation was proposed by Norris (2005). According to (Norris 2005, Eq. 31), the variation of the elastic energy (4.1) reads:

$$\delta W_e(\Phi) = \frac{\delta\Phi}{2}\hat{\mathbf{\sigma}}^T \cdot [\hat{\mathbf{P}}(\mathbf{n}) \cdot \hat{\mathbf{Q}}(\mathbf{n},\Phi) \cdot \hat{\mathbf{S}}^{(0)} \cdot \hat{\mathbf{Q}}^T(\mathbf{n},\Phi) + \hat{\mathbf{Q}}(\mathbf{n},\Phi) \cdot \hat{\mathbf{S}}^{(0)T} \cdot \hat{\mathbf{Q}}^T(\mathbf{n},\Phi) \cdot \hat{\mathbf{P}}^T(\mathbf{n})] \cdot \hat{\mathbf{\sigma}}. \tag{4.17}$$

By virtue of commutation of the tensors $\hat{\mathbf{P}}(\mathbf{n}), \hat{\mathbf{Q}}(\mathbf{n},\Phi)$, the expression (4.17) reduces to

$$\begin{aligned}\delta W_e(\Phi) &= \hat{\mathbf{\sigma}}^T \cdot \hat{\mathbf{P}}(\mathbf{n}) \cdot \hat{\mathbf{Q}}(\mathbf{n},\Phi) \cdot \hat{\mathbf{S}}^{(0)} \cdot \hat{\mathbf{Q}}^T(\mathbf{n},\varphi)\hat{\mathbf{\sigma}}, \\ \delta\Phi &= \hat{\mathbf{\sigma}}^T \cdot \hat{\mathbf{P}}(\mathbf{n}) \cdot \hat{\mathbf{S}} \cdot \hat{\mathbf{\sigma}}\delta\Phi.\end{aligned} \tag{4.18}$$

In the stationary case, the first variation vanishes: $\delta W(\Phi) = 0$. If the stress tensor is aligned with the fixed axis, the first two terms condense to the vector:

$$\hat{\mathbf{\sigma}}^T \cdot \hat{\mathbf{P}}(\mathbf{n}) = \left[0, 0, 0, \sqrt{2}(\sigma_{III} - \sigma_{II})n_1, \sqrt{2}(\sigma_I - \sigma_{III})n_2, \sqrt{2}(\sigma_{II} - \sigma_I)n_3\right]. \tag{4.19}$$

Consequently,

$$\delta W_e(\Phi) = \hat{\mathbf{\sigma}}^T \cdot \hat{\mathbf{P}}(\mathbf{n}) \cdot \hat{\mathbf{S}}(\Phi) \cdot \hat{\mathbf{\sigma}}\delta\Phi = = 2\begin{bmatrix}(\sigma_{III} - \sigma_{II})n_1 \\ (\sigma_I - \sigma_{III})n_2 \\ (\sigma_{II} - \sigma_I)n_3\end{bmatrix}^T \begin{bmatrix}S_{14} & S_{24} & S_{34} \\ S_{15} & S_{25} & S_{35} \\ S_{16} & S_{26} & S_{36}\end{bmatrix}\begin{bmatrix}\sigma_I \\ \sigma_{II} \\ \sigma_{III}\end{bmatrix}. \tag{4.20}$$

The variation (4.20) must disappear for all directions of the rotation axis $\mathbf{n}$. Accordingly, the condition of the stationarity of the elastic energy $W_e(\varphi)$ reads:

$$\begin{bmatrix}\sigma_{III} - \sigma_{II} & 0 & 0 \\ 0 & \sigma_I - \sigma_{III} & 0 \\ 0 & 0 & \sigma_{II} - \sigma_I\end{bmatrix}\begin{bmatrix}S_{14} & S_{24} & S_{34} \\ S_{15} & S_{25} & S_{35} \\ S_{16} & S_{26} & S_{36}\end{bmatrix}\begin{bmatrix}\sigma_I \\ \sigma_{II} \\ \sigma_{III}\end{bmatrix} = \begin{bmatrix}0 \\ 0 \\ 0\end{bmatrix}. \tag{4.21}$$

If all principal stresses are different, the matrix on the left side is diagonal with non-zero eigenvalues and (4.21) brings us a simpler form:

$$\begin{bmatrix}S_{14} & S_{24} & S_{34} \\ S_{15} & S_{25} & S_{35} \\ S_{16} & S_{26} & S_{36}\end{bmatrix}\begin{bmatrix}\sigma_I \\ \sigma_{II} \\ \sigma_{III}\end{bmatrix} = \begin{bmatrix}0 \\ 0 \\ 0\end{bmatrix}. \tag{4.22}$$

The identity (4.22) results, and the off-diagonal components of the strain tensor in stationary case disappear: $\varepsilon_{23} = \varepsilon_{31} = \varepsilon_{12} = 0$. In other words, the strain tensor and stress tensor contain only the diagonal components in the selected coordinate system:

$$\hat{\mathbf{\varepsilon}} = [\varepsilon_1,\ \varepsilon_2,\ \varepsilon_3,\ 0,\ 0,\ 0]^T, \hat{\mathbf{\sigma}} = \left[\sigma_1,\ \sigma_2,\ \sigma_3,\ 0,\ 0,\ 0\right]^T. \tag{4.23}$$

The stress and strain tensors have also proved to be coaxial for the stationary case of elastic energy. Both presented variational methods can be applied for the following optimization problems.

4.1.3 Optimal Solutions in Anti-Plane Elasticity

The simplest solution of the optimization problem delivers the anti-plane stress state. In this state, all components of the stress tensor except σ_{13} and σ_{23} vanish. The density of the elastic energy reads:

$$W_e(\varphi) = \left(\frac{c^2}{G_{23}} + \frac{s^2}{G_{13}}\right)\frac{\sigma_{23}^2}{2} + sc\left(\frac{1}{G_{23}} - \frac{1}{G_{13}}\right)\sigma_{23}\sigma_{13} + \left(\frac{c^2}{G_{13}} + \frac{s^2}{G_{23}}\right)\frac{\sigma_{13}^2}{2}. \tag{4.24}$$

The exact bounds for the density of the elastic energy are:

$$\frac{\sigma_{23}^2 + \sigma_{13}^2}{2\max[G_{23}, G_{13}]} \leq W(\Phi) \leq \frac{\sigma_{23}^2 + \sigma_{13}^2}{2\min[G_{23}, G_{13}]}. \tag{4.25}$$

These bounds correspond to the inclination angles

$$\Phi = \pm\arctan(\sigma_{13}/\sigma_{23}). \tag{4.26}$$

4.1.4 Optimal Solutions in Plane Elasticity

The case of plane elastic problem could be also investigated in closed form. For the uniaxial tension stress the stress vector reads: $\hat{\boldsymbol{\sigma}}_{(T)} = [\sigma, 0, 0, 0, 0, 0]$. The vector of principal stress is directed parallel to the x-axis. The magnitude of the principal stress is: σ. The plot of the elastic energy density $W_e(\Phi)$ as the function of angle Φ is shown in Figure 4.1. The materials with higher values of elastic energy density for a given stress accumulate more energy in the volume unit in comparison to the materials with lower values the elastic energy. The plot of the specific elastic energy $W_e(\Phi)/\rho$ as the function of angle Φ is shown in Figure 4.2. For calculation the stress magnitude is set to $\sigma = 1MPa$.

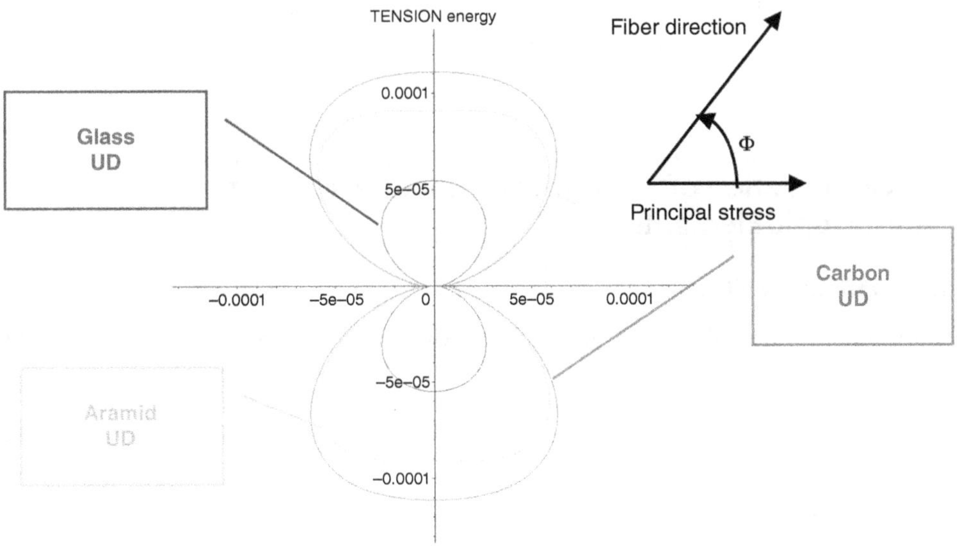

Figure 4.1 Elastic energy for uniaxial tension as a function of angle Φ.

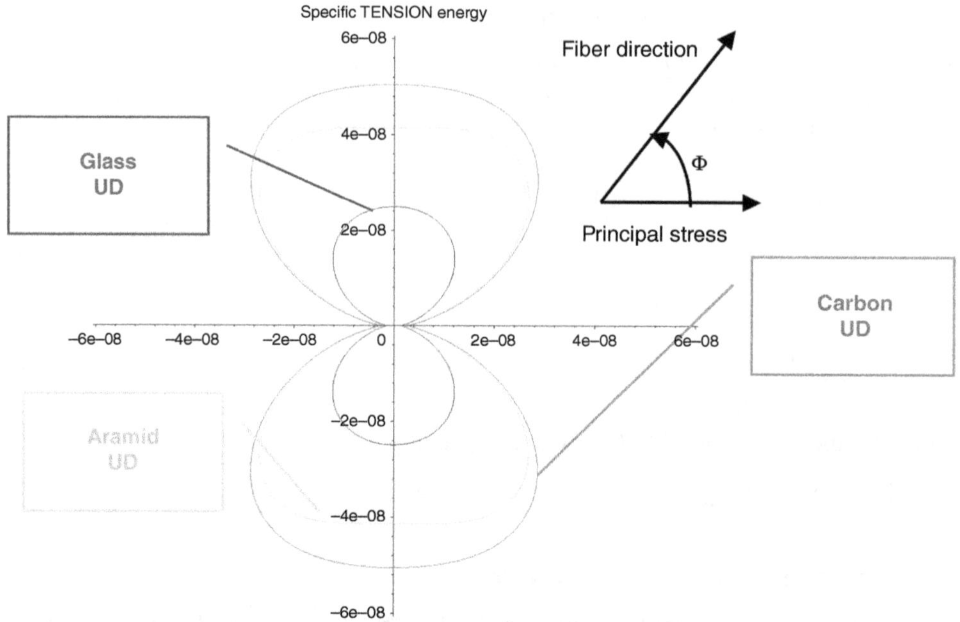

Figure 4.2 Specific elastic energy for uniaxial tension as a function of angle Φ.

If the shear stress in the xy plane is the sole stress in material, the stress vector reads: $\widehat{\boldsymbol{\sigma}}_{(S)} = \sigma[0, 0, 0, 0, 0, 1]$. The plot of the elastic energy $W_e(\Phi)$ as the function of angle Φ is shown in Figure 4.3. The plot of the specific elastic energy $W_e(\Phi)/\rho$ as the function of angle Φ is shown in Figure 4.4.

Note that the situation is similar if the strain is given instead of the predefined stress. The materials with the higher values of the elastic energy density for a given strain accumulate more energy in comparison to the materials with the lower values of the elastic energy density.

4.2 Optimization of Strength and Loading Capacity of Anisotropic Elements

4.2.1 Optimization Problem

We consider a structural element made of composite material that occupies the region Ω with the boundary Γ. For evaluation of failure of the structural element the phenomenological criterion is applied. For definiteness, we use the tensor-polynomial of the second order in Tsai–Wu form. The failure takes place at a certain point, if the stress exceeds the failure criterion:

$$g(\widehat{\boldsymbol{\sigma}}) = \widehat{\boldsymbol{\sigma}}^T \bullet \widehat{\mathbf{F}}^{(4)}(\Phi) \bullet \widehat{\boldsymbol{\sigma}} + \widehat{\mathbf{F}}^{(2)}(\Phi) \bullet \widehat{\boldsymbol{\sigma}} \leq 1, \tag{4.27}$$

$$g(\sigma_{ij}) = \sigma_{ij} F^{(4)}_{ijkl}(\varphi) \sigma_{kl} + F^{(2)}_{ij} \sigma_{ij} \leq 1. \tag{4.28}$$

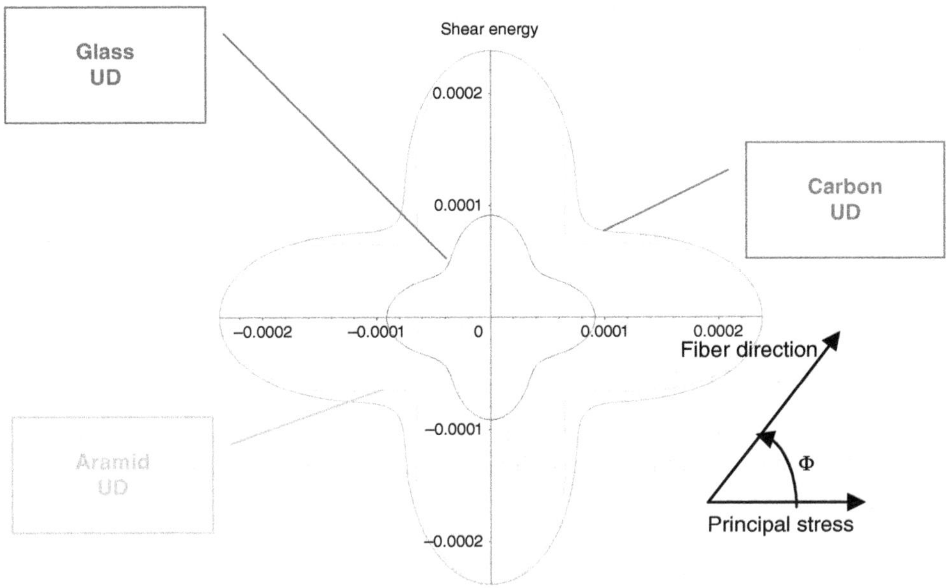

Figure 4.3 Elastic energy for pure shear as a function of angle Φ.

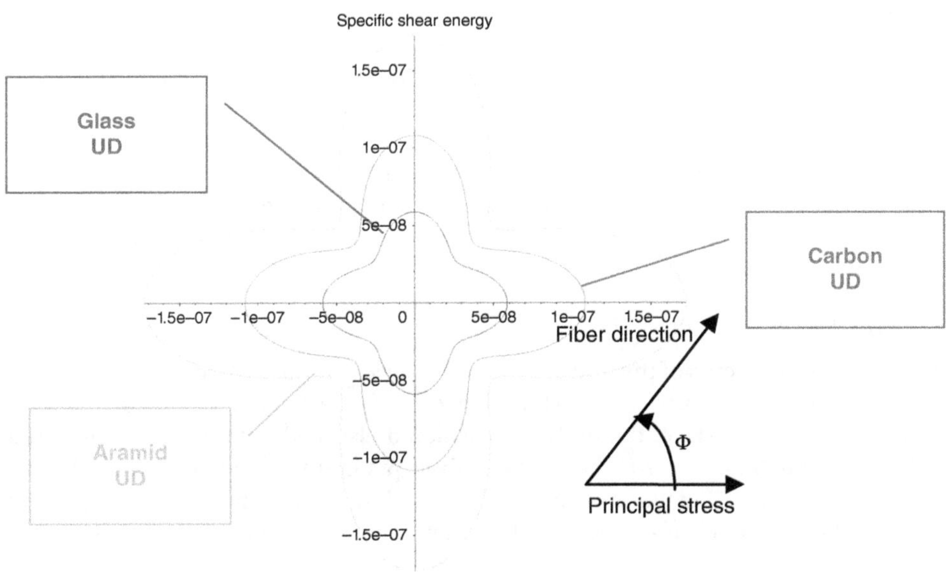

Figure 4.4 Specific elastic energy for pure shear as a function of angle Φ.

The components of tensors $\mathbf{F}^{(4)}$ and $\mathbf{F}^{(2)}$ depend on the orientation Φ of the anisotropy in each point of the structural element:

$$F_{ijkl}^{(4)} = F_{ijkl}^{(4)}(\Phi), F_{ij}^{(2)} = F_{ij}^{(2)}(\Phi)$$

The derivation of the optimality conditions is based on the relations between the components of the tensors $\mathbf{F}^{(4)}$, $\mathbf{F}^{(2)}$ and their components in the material-fixed axes $f_{ijkl}^{(4)} = F_{ijkl}^{(4)}(\Phi = 0)$, $f_{ij}^{(2)} = F_{ij}^{(2)}(\Phi = 0)$. The Equation (2.33) in Chapter 2 provides the transformation laws:

$$F_{ijkl}^{(4)}(\Phi) = t_{ip}t_{jq}t_{kn}t_{ls}f_{pqns}^{(4)}, \quad F_{ij}^{(2)}(\Phi) = t_{ip}t_{jq}f_{pq}^{(2)}. \tag{4.29}$$

Similarly, the pressure-dependent Mises–Hill strength criterion reduces to:

$$g(\hat{\boldsymbol{\sigma}}) = \hat{\boldsymbol{\sigma}}^T \bullet \hat{\mathbf{F}}(\varphi) \bullet \hat{\boldsymbol{\sigma}} \leq 1, \quad g(\sigma_{ij}) = \sigma_{ij}F_{ijkl}\sigma_{kl} \leq 1. \tag{4.30}$$

The second rank tensor depends on the inclination angle Φ:

$$\hat{\mathbf{F}}(\Phi) = \hat{\mathbf{T}}(\Phi) \cdot \hat{\mathbf{F}}_p \cdot \hat{\mathbf{T}}^T(\Phi). \tag{4.31}$$

The second optimization problem consists of a search for the orientation of the angle Φ that provides the minimum of strength criterion, presuming the given stress $\hat{\boldsymbol{\sigma}}$:

$$g(\Phi) \to \min_{\Phi}. \tag{4.32}$$

The optimization problem (4.32) has been intensively studied for load-bearing composite structures. The review is given in the book by Vasiliev and Gurdal (1999). The fulfillment of the extremum in (4.32) assures the maximal load capacity of the structure. The principal applications of the optimization problem (4.32) are, for example, the pressure vessels made of wounded composite fibers or layers. The deformation and stiffness in these applications are insignificant and not accounted for in the optimization problem (4.32). The optimization problem (4.32) leads similarly to the stiffness-optimization problems considered in the previous section, in particular to the co-axiality condition. Equations (4.27) and (4.28) in the six-dimensional stress space yield a family of convex surfaces enclosing the origin. To estimate the carrying capacity, the theorem of ultimate equilibrium is used, according to which the body sustains applied loads if a safe statically possible field of stresses exists, that is, a stress distribution satisfying the equilibrium equations and boundary conditions:

$$\sigma_{ij,j} + q_j = 0, \quad \sigma_{ij}n_j|_{\Gamma_\sigma} = T_i. \tag{4.33}$$

Here n_j are components of the unit external normal vector to the body surface Γ_σ. The Γ_σ is the part of the body surface on which the loads T_i are given, $\Gamma_\sigma \cup \Gamma_u = \Gamma$.

On the rest of the surface, the body is considered clamped; that is, it is considered that the displacements are zero. The Roman subscripts in this section take the values 1, 2, 3 or the values 1 and 2 in plane problems. Summation is over repeated subscripts and the subscript after the comma denotes differentiation with respect to the corresponding coordinate.

If the external loads on the structural element increase proportionally to a load multiplier p:

$$q_i = pq_i^0, T_i = pT_i^0 \tag{4.34}$$

the loading capacity is exhausted; if at any point the failure criterion attains its ultimate admissible value.

The vectors T_i^0, q_i^0 are prescribed.

Accordingly, the problem of maximization of loading capacity is dual to the problem (4.32). The dual optimization problem searches the appropriate orientation of the anisotropy axes at each point of the body that maximizes the admissible load multiplier.

The optimization problem consists of seeking the orientation of the anisotropy axes, that is, the functions $t_{lk}(x_1, x_2, x_3)$ from the condition for the maximum of the admissible load multiplier:

$$p_* = \max_{\Phi} p(t_{lk}), \tag{4.35}$$

upon compliance with the condition (4.27) or (4.30). The failure conditions (4.27) or (4.30) in form of inequalities are equivalent for an arbitrary real value μ to the condition in the equality form:

$$g - 1 + \mu^2 = 0. \tag{4.36}$$

4.2.2 Optimality Conditions

The expressions for the variations $\delta F^{(2)}_{ij}$, $\delta F^{(4)}_{ijkl}$ due to the variation δm_{lk} of the cosines m_{lk} read:

$$\delta F^{(4)}_{ijkl} = 4t_{sp}\delta t_{pl}F^{(4)}_{ijks}, \qquad \delta F^{(2)}_{ij} = 2t_{sp}\delta t_{pj}F^{(2)}_{is}. \tag{4.37}$$

For application of the variation method, the functional being optimized must be represented in the form of a spatial integral:

$$p = \left(\int_\Omega d\Omega\right)^{-1} \int_\Omega p(x_1, x_2, x_3)d\Omega, \qquad p(x_1, x_2, x_3) = const. \tag{4.38}$$

The augmented Lagrange functional accounts for Eqs. (4.33), (4.36) and (4.38):

$$J = \left(\int_\Omega d\Omega\right)^{-1} \int_\Omega p(x_1, x_2, x_3)d\Omega + \int_\Omega \psi_i(\sigma_{ij,j} + q_i)d\Omega + \int_\Omega \lambda(g - 1 + \mu^2)d\Omega + \int_\Omega \eta_{ij}(t_{ik}t_{kl} - \delta_{ij})d\Omega. \tag{4.39}$$

We define the conjugate variables ψ_i functions satisfying the following differential equations

$$\frac{1}{2}(\psi_{i,j} + \psi_{j,i}) = \lambda\frac{\partial g}{\partial \sigma_{ij}} \equiv 2\lambda\varepsilon^*_{ij}, \qquad \varepsilon^*_{ij} = F^{(4)}_{ijkl}\sigma_{kl} + \frac{1}{2}F^{(2)}_{ij}, \tag{4.40}$$

and the boundary condition

$$\psi_i|_{\Gamma_u} = 0.$$

The Lagrange multipliers λ and μ are subject to additional conditions:

$$\lambda\mu = 0 \quad in \quad \Omega. \tag{4.41}$$

Using (4.40) and (4.41), we obtain the variation of the augmented Lagrange functional in the form:

$$\delta J = \int_\Omega (\sigma_{ki}\varepsilon^*_{is} + \eta_{ks})t_{sp}\delta t_{pk}d\Omega. \tag{4.42}$$

On one hand, the tensor η_{ks} is symmetric: $\eta_{ks} = \eta_{sk}$. On the other hand, the variation δt_{pk} is arbitrary. The requirement $\delta J = 0$ and the arbitrariness of δt_{pk} brings us to the necessary optimality condition:

$$\sigma_{ki}\varepsilon^*_{is} = \sigma_{si}\varepsilon^*_{ik}, \text{ or } \quad \boldsymbol{\sigma} \cdot \boldsymbol{\varepsilon}^* = \boldsymbol{\varepsilon}^* \cdot \boldsymbol{\sigma}. \tag{4.43}$$

The necessary optimality condition manifests the commutation of the tensors $\boldsymbol{\sigma}$ and $\boldsymbol{\varepsilon}^*$. The tensors $\boldsymbol{\sigma}$ and $\boldsymbol{\varepsilon}^*$ possess the same eigenvectors. Thus, the commutation relation corresponds to the necessary optimality condition:

$$\left(\boldsymbol{\sigma} \cdot \cdot \mathbf{F}^{(4)} + {}^1\!/\!_2\mathbf{F}^{(2)}\right) \cdot \boldsymbol{\sigma} = \boldsymbol{\sigma} \cdot \left(\boldsymbol{\sigma} \cdot \cdot \mathbf{F}^{(4)} + {}^1\!/\!_2\mathbf{F}^{(2)}\right). \tag{4.44}$$

From the commutation (4.43) and (4.44) follows, that the tensors $\boldsymbol{\sigma}$ and $\boldsymbol{\varepsilon}^* = \boldsymbol{\sigma} \cdot \cdot \mathbf{F}^{(4)} + {}^1\!/\!_2\mathbf{F}^{(2)}$ must be coaxial in case of the optimal orientation of anisotropy axes.

4.2.3 Optimal Solutions in Anti-Plane Elasticity

In the anti-plane stress state, all components of the stress tensor except for σ_{13} and σ_{23} vanish. The strength criterion reads:

$$g(\Phi) = 2L(\sigma_{23}c + \sigma_{13}s)^2 + 2M(\sigma_{13}c - \sigma_{23}s)^2 \tag{4.45}$$

The constants in Eq. (4.45) according to Table 2.2 in Chapter 2 are:

$$2L = \widetilde{\tau}_4^{-2}, 2M = \widetilde{\tau}_5^{-2}.$$

The exact bounds for the density of the elastic energy are:

$$\frac{\sigma_{23}^2 + \sigma_{13}^2}{\max[\widetilde{\tau}_4, \widetilde{\tau}_5]} \leq \frac{g(\Phi)}{2} \leq \frac{\sigma_{23}^2 + \sigma_{13}^2}{\min[\widetilde{\tau}_4, \widetilde{\tau}_5]}. \tag{4.46}$$

These bounds correspond to the inclination angles:

$$\Phi = \pm \arctan(\sigma_{13}/\sigma_{23}). \tag{4.47}$$

4.2.4 Optimal Solutions in Plane Elasticity

For the uniaxial tension stress, the stress vector reads: $\widehat{\boldsymbol{\sigma}}_{(T)} = [\sigma, 0, 0, 0, 0, 0]$. The plot of the strength criterion $g(\Phi)$ for tension stress as the function of angle Φ is shown in Figure 4.5. The value of the strength criterion is reciprocal to the ultimate directional stress. Accordingly, the materials with higher strength criterion values are weaker than those with the lower values. In other words, a higher value of the strength criterion for a given stress means the material is weaker. The plot of the specific strength criterion $g(\Phi)/\rho$ as the function of angle Φ is shown in Figure 4.6. For calculation the stress magnitude is set to $\sigma = 1MPa$.

For the pure shear load in the xy plane, the stress vector is: $\widehat{\boldsymbol{\sigma}}_{(S)} = [0, 0, 0, 0, 0, 1]$. The plot of the ultimate criterion $g(\Phi)$ for shear stress as the function of angle Φ is shown in Figure 4.7. The plot of the specific ultimate criterion $g(\Phi)/\rho$ as the function of angle Φ is shown in Figure 4.8.

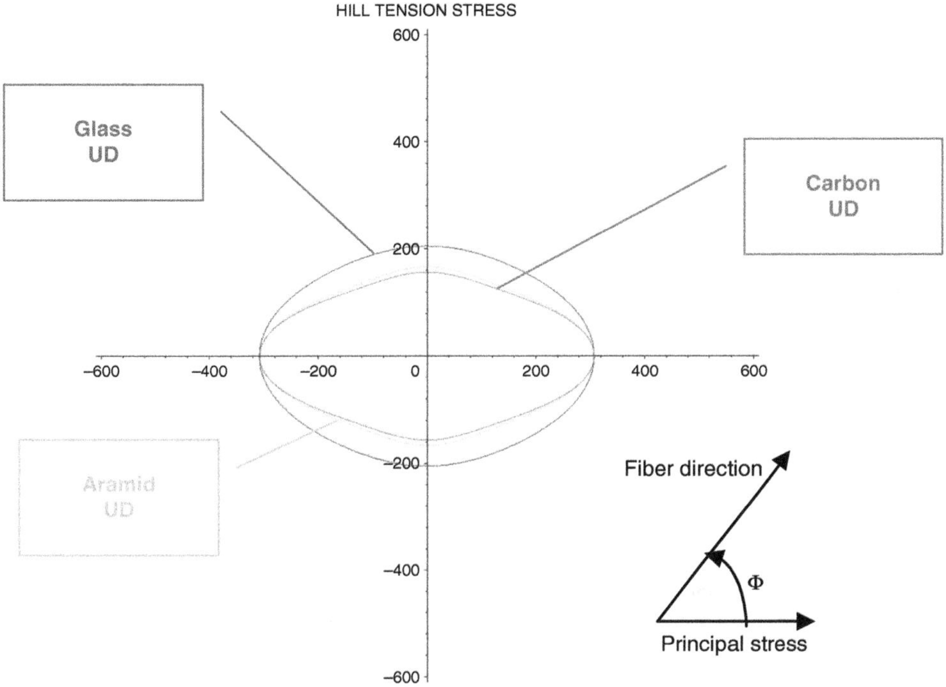

Figure 4.5 Direction-dependent strength criterion $g(\Phi)$ for a uniaxial load.

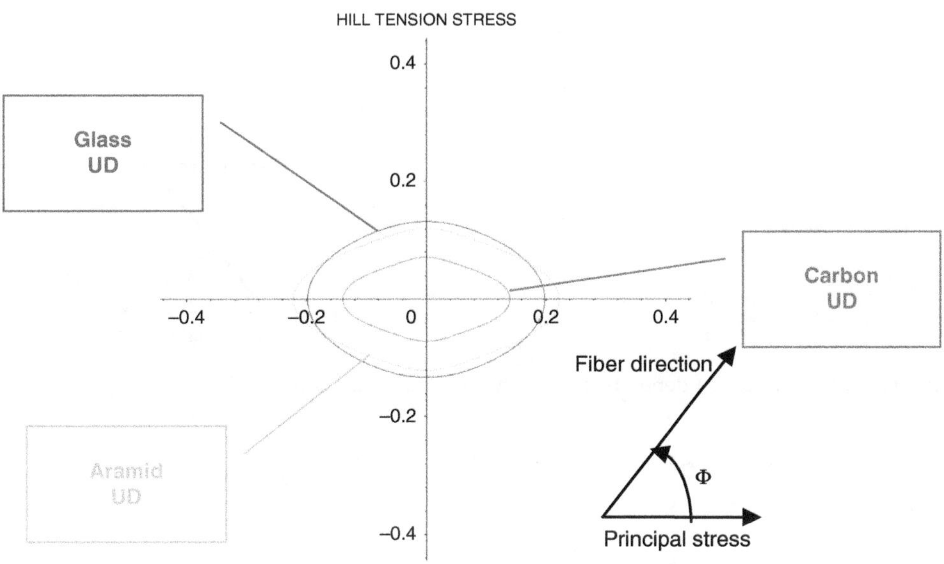

Figure 4.6 Specific, direction-dependent strength criterion $g(\Phi)$ for a uniaxial load.

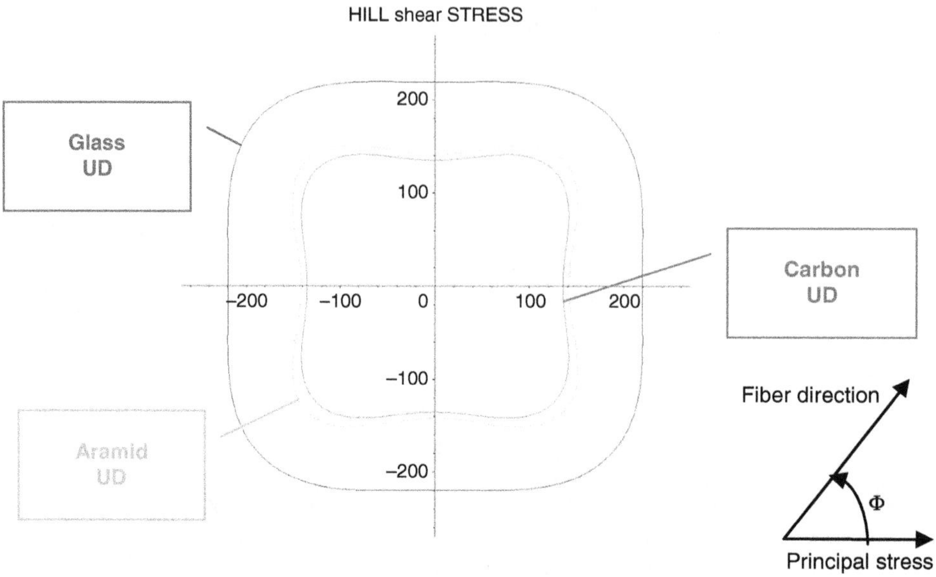

Figure 4.7 Direction-dependent strength criterion $g(\Phi)$ for a pure shear load.

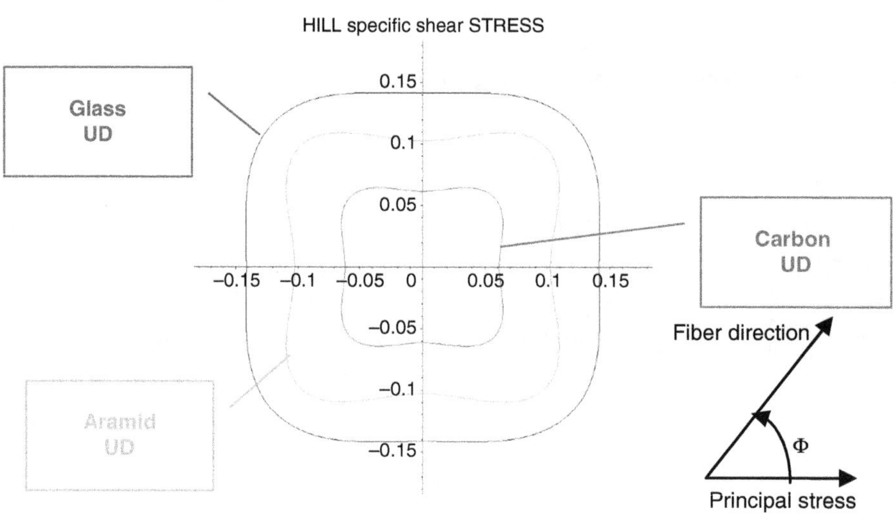

Figure 4.8 Specific, direction-dependent strength $g(\Phi)/\rho$ for a pure shear load.

4.3 Optimization of Accumulated Elastic Energy in Flexible Anisotropic Elements

4.3.1 Optimization Problem

The third optimization problem studies the orientation of the angle Φ that provides the maximum of the stored elastic energy, presuming the given stress $\hat{\boldsymbol{\sigma}}$ fulfills the quadratic strength criterion:

$$W_e(\Phi) = \widehat{\boldsymbol{\sigma}}^T \cdot \widehat{\mathbf{S}}(\Phi) \cdot \widehat{\boldsymbol{\sigma}}/2 \to \max_{\Phi}, \tag{4.48}$$

$$g(\Phi) = \widehat{\boldsymbol{\sigma}}^T \cdot \widehat{\mathbf{F}}^{(4)}(\Phi) \cdot \widehat{\boldsymbol{\sigma}} + \widehat{\mathbf{F}}^{(2)}(\Phi) \cdot \widehat{\boldsymbol{\sigma}} \leq 1. \tag{4.49}$$

In terms of the components of the fourth rank tensors, the optimization problem reads:

$$W_e(\Phi) = \sigma_{ij} S_{ijkl}(\Phi) \sigma_{kl}/2 \to \max_{\Phi}, \tag{4.50}$$

$$g(\Phi) = \sigma_{ij} F^{(4)}_{ijkl} \sigma_{kl} + \sigma_{ij} F^{(2)}_{ij} \leq 1. \tag{4.51}$$

The sense of the optimization problem is the following. The mass of the structure is not affected by the orientation angle and considered as a constant during the optimization. The elastic element of structure of a given mass that stores the maximal elastic energy for a given applied load is pursued. The stored energy is limited by ultimate stresses that follow from the failure criterion. The applications of the optimization problem are the springs and energy harvesters.

The equivalent formulation of the optimization problem regards specific elastic energy:

$$\widetilde{W}_e(\Phi) = \frac{1}{2} \frac{\widehat{\boldsymbol{\sigma}}^T \cdot \widehat{\mathbf{S}} \cdot \widehat{\boldsymbol{\sigma}}}{\widehat{\boldsymbol{\sigma}}^T \cdot \widehat{\mathbf{F}}^{(4)} \cdot \widehat{\boldsymbol{\sigma}} + \widehat{\mathbf{F}}^{(2)} \cdot \widehat{\boldsymbol{\sigma}}} \to \max_{\Phi}, \tag{4.52}$$

$$\widetilde{W}_e(\Phi) = \frac{1}{2} \frac{\sigma_{ij} S_{ijkl} \sigma_{kl}}{\sigma_{ij} F^{(4)}_{ijkl} \sigma_{kl} + \sigma_{ij} F^{(2)}_{ij}} \to \max_{\Phi}. \tag{4.53}$$

Due to anisotropy of the composite material, the specific stored elastic energy depends on the relations between the stress components. The value of the stored elastic energy for a unidirectional load therefore differs from the value in a multi-axial stress state. This circumstance makes the comparison of stored elastic energy to some extent indirect. For comparison of different materials, the stress state must be fixed. Later, we mainly study two principal stress states, namely unidirectional load and pure shear stress, which are typical for common spring elements.

4.3.2 Optimality Conditions

The derivation of the optimality conditions is similar to those of the previously considered optimization problems. The variation of the ultimate elastic energy reads:

$$\begin{aligned}
\delta \widetilde{W}_e = {} & \frac{1}{2} \frac{\sigma_{ij} \delta S_{ijkl} \sigma_{kl}}{\sigma_{ij} F^{(4)}_{ijkl} \sigma_{kl} + \sigma_{ij} F^{(2)}_{ij}} - \\
& \frac{1}{2} \frac{\sigma_{ij} S_{ijkl} \sigma_{kl}}{(\sigma_{ij} F^{(4)}_{ijkl} \sigma_{kl} + \sigma_{ij} F^{(2)}_{ij})^2} (\sigma_{ij} \delta F^{(4)}_{ijkl} \sigma_{kl} + \sigma_{ij} \delta F^{(2)}_{ij}) = \\
& \frac{1}{2} \frac{\sigma_{ij} \delta S_{ijkl} \sigma_{kl} - 2\widetilde{W}_e \sigma_{ij} \delta F^{(4)}_{ijkl} \sigma_{kl} - 2\widetilde{W}_e \delta F^{(2)}_{ij} \sigma_{kl}}{\sigma_{ij} F^{(4)}_{ijkl} \sigma_{kl} + \sigma_{ij} F^{(2)}_{ij}},
\end{aligned} \tag{4.54}$$

Using (4.10) and (4.28), we transform the variation (4.54) to the form:

$$\delta\widetilde{W}_e = \frac{\sigma^{**}_{ks}\sigma_{kl}}{\sigma_{ij}F^{(4)}_{ijkl}\sigma_{kl} + \sigma_{ij}F^{(2)}_{ij}} 2t_{sp}\delta t_{pk}, \tag{4.55}$$

$$\varepsilon^{**}_{ks} = \sigma_{ij}S_{ijks} - 2\left(\sigma_{ij}F^{(4)}_{ijks} + \frac{1}{2}F^{(2)}_{ks}\right)\widetilde{W}_e, \tag{4.56}$$

$$\boldsymbol{\varepsilon}^{**} = \boldsymbol{\varepsilon} - 2\left(\boldsymbol{\sigma}\cdot\cdot\mathbf{F}^{(4)} + 1/2\mathbf{F}^{(2)}\right)\widetilde{W}.$$

The requirement of vanishing of variation (4.55) and the arbitrariness of δt_{pk} delivers the necessary optimality condition:

$$\sigma_{ki}\varepsilon^{**}_{is} = \sigma_{si}\varepsilon^{**}_{ik}, \text{or} \quad \boldsymbol{\sigma}\cdot\boldsymbol{\varepsilon}^{**} = \boldsymbol{\varepsilon}^{**}\cdot\boldsymbol{\sigma}. \tag{4.57}$$

The variation $\delta\widetilde{W}_e$ vanishes, if the tensors $\boldsymbol{\sigma}$ and $\boldsymbol{\varepsilon}^{**}$ commute. From the commutation condition (4.57) follows, that the tensors $\boldsymbol{\sigma}$ and $\boldsymbol{\varepsilon}^{**}$ possess identical eigenvectors and therefore must be coaxial for the optimal orientation of anisotropy axes.

For technical applications and weight evaluation of different materials, the specific ultimate elastic energy is useful:

$$\widetilde{W}^*_e = \frac{1}{2\rho}\frac{\hat{\boldsymbol{\sigma}}^T\cdot\hat{\mathbf{S}}\cdot\hat{\boldsymbol{\sigma}}}{\hat{\boldsymbol{\sigma}}^T\cdot\hat{\mathbf{F}}\cdot\hat{\boldsymbol{\sigma}}} \to \max_{\Phi}, \tag{4.58}$$

$$\widetilde{W}^*_e = \frac{1}{2\rho}\frac{\sigma_{ij}S_{ijkl}\sigma_{kl}}{\sigma_{ij}F^{(4)}_{ijkl}\sigma_{kl} + \sigma_{ij}F^{(2)}_{ij}} \to \max_{\Phi}. \tag{4.59}$$

The expressions (4.58) and (4.59) indicate the ultimate stored elastic energy related to density of the material. The necessary optimality conditions do not alter and are valid also for the expressions (4.58) and (4.59).

4.3.3 Optimal Solutions in Anti-Plane Elasticity

In the anti-plane stress state, all components of the stress tensor except for σ_{13} and σ_{23} vanish. The ultimate elastic energy reads:

$$\widetilde{W}_e(\Phi) = \frac{\left(\frac{c^2}{G_{23}} + \frac{s^2}{G_{13}}\right)\frac{\sigma^2_{23}}{2} + sc\left(\frac{1}{G_{23}} - \frac{1}{G_{13}}\right)\sigma_{23}\sigma_{13} + \left(\frac{c^2}{G_{13}} + \frac{s^2}{G_{23}}\right)\frac{\sigma^2_{13}}{2}}{2L(\sigma_{23}c + \sigma_{13}s)^2 + 2M(\sigma_{13}c - \sigma_{23}s)^2}. \tag{4.60}$$

With Table 2.2 in Chapter 2, the values are: $2L = \widetilde{\tau}_4^{-2}$, $2M = \widetilde{\tau}_5^{-2}$

For the given σ_{13}, σ_{23}, the exact bounds of the density of the elastic energy are:

$$\min_{i=1,..,4}\mathbf{w} \le \widetilde{W}_e(\Phi) \le \max_{i=1,..,4}\mathbf{w}, \qquad \mathbf{w} = \left[\widetilde{W}_e(0), \widetilde{W}_e\left(\frac{\pi}{2}\right), \frac{\widetilde{\tau}^2_4}{4G_{23}}, \frac{\widetilde{\tau}^2_5}{4G_{13}}\right], \tag{4.61}$$

$$\widetilde{W}_e(0) = \frac{\widetilde{\tau}^2_4\widetilde{\tau}^2_5}{4G_{23}G_{13}}\frac{G_{13}\sigma^2_{23} + G_{23}\sigma^2_{13}}{\widetilde{\tau}^2_5\sigma^2_{23} + \widetilde{\tau}^2_4\sigma^2_{13}},$$

$$\widetilde{W}_e\left(\frac{\pi}{2}\right) = \frac{\widetilde{\tau}^2_4\widetilde{\tau}^2_5}{4G_{23}G_{13}}\frac{G_{23}\sigma^2_{23} + G_{13}\sigma^2_{13}}{\widetilde{\tau}^2_4\sigma^2_{23} + \widetilde{\tau}^2_5\sigma^2_{13}}.$$

The bounds (4.61) correspond to the inclination angles:

$$\boldsymbol{\Phi} = [0, \pi/2, \arctan(\sigma_{13}/\sigma_{23}), -\arctan(\sigma_{13}/\sigma_{23})].$$

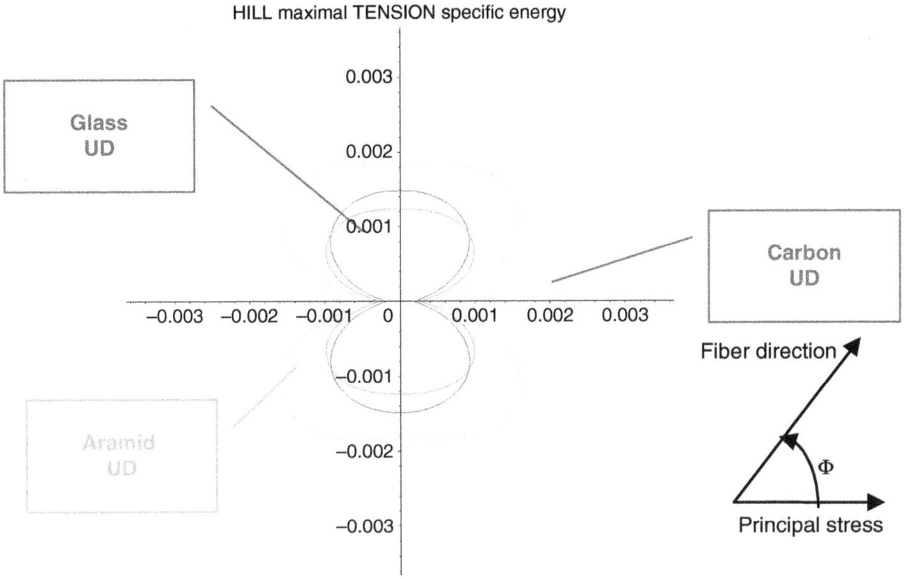

Figure 4.9 Specific direction-dependent ultimate stored energy for uniaxial tension.

4.3.4 Optimal Solutions in Plane Elasticity

The plot of the specific ultimate elastic energy $\widetilde{W}_e^*(\Phi)$ for tension stress as a function of angle Φ is shown in Figure 4.9. The materials with higher values of specific ultimate elastic energy accumulate more elastic energy before breakage than materials with lower values of specific ultimate elastic energy. In other words, the higher the value of specific ultimate elastic energy, the more able the material is to store more elastic energy per mass unit than the lower value of specific ultimate elastic energy. Materials with higher values of specific ultimate elastic energy are preferable for energy storage applications, like springs and energy harvesters.

For the pure shear load in the xy plane, the stress vector reads: $\hat{\boldsymbol{\sigma}}_{(S)} = [0, 0, 0, 0, 0, 1]$. The plot of the specific ultimate elastic energy $\widetilde{W}_e^*(\Phi)$ for shear stress as a function of angle Φ is shown in Figure 4.10.

4.4 Optimal Anisotropy in a Twisted Rod

In the first example of the application of optimization principles examined previously, the simple problem of optimal arrangement of anisotropy in a cross-section of a straight rod in torsion is examined. Let us consider the torsion of a cylindrical rod subjected to moments applied to its ends. We assume that the transverse section of the rod is a simply connected domain Ω. The Ox_3 axis of the $Ox_1x_2x_3$ orthogonal coordinate system is oriented in the direction of the cylinder generator, while the Ox_1, Ox_2 axes are oriented in the plane of the transverse section. The lateral sides of the rod are load-free. It is also assumed that the material is continuously inhomogeneous and has a plane of elastic symmetry normal to the generator (the Ox_3 axis) at each point. The material is orthotropic at each point where the two orthotropy axes are in the plane of the

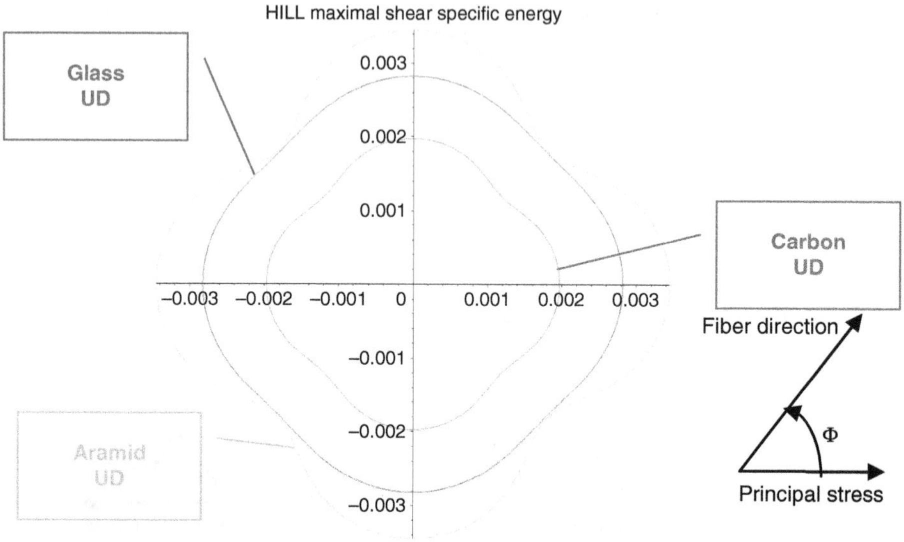

Figure 4.10 Specific direction-dependent ultimate stored energy for a pure shear load.

transverse section and their orientation is independent of the coordinate x_3. Under torsion there are only two stress tensor components: τ_{13}, τ_{23}. The torque is expressed in terms of the stress tensor components as follows:

$$M = \int_{\Omega} (\tau_{23}x_1 - \tau_{13}x_2) d\Omega. \tag{4.62}$$

The equilibrium equation and boundary conditions are the following:

$$\partial\tau_{13}/\partial x_1 + \partial\tau_{23}/\partial x_2 = 0 \quad on \quad \Omega, \tag{4.63}$$

$$\tau_{13}n_1 + \tau_{23}n_2 = 0 \quad on \quad \Gamma. \tag{4.64}$$

Since only two stress tensor components are not zero, criterion (2.4) in Chapter 2 is rewritten in the abbreviated form:

$$g(\tau_{13}, \tau_{23}) = \sum_{i,j=1,2,3} \tau_{i3} F^{(4)}_{i3j3}(\varphi) \tau_{j3} \leq 1. \tag{4.65}$$

The anisotropy parameters are calculated in the global coordinate system $Ox_1x_2x_3$ with directions $\mathbf{i}_1, \mathbf{i}_2, \mathbf{i}_3$. These anisotropy parameters are functions of the coordinates x_1, x_2 and depend on the orientation of the anisotropy axes in the plane of the section. In addition to the $Ox_1x_2x_3$ coordinate system, we introduce an intrinsic coordinate system $oy_1y_2y_3$ whose directions $\mathbf{j}_1$, $\mathbf{j}_2$, $\mathbf{j}_3$ correspond at each point o with the direction of the anisotropy axes of the material. The Ox_3 axis is parallel to the oy_3 axis and the angle between the oy_1 and Ox_1 axes is $\Phi(x_1, x_2)$. The fracture condition in the intrinsic coordinate system $oy_1y_2y_3$ is also a quadratic form of the shear stresses. The parameters $f_{12} \equiv f_{1323}, f_{22} \equiv f_{2323}$, characterizing the anisotropy in the intrinsic coordinate system $oy_1y_2y_3$ are independent of the point coordinates and are constants of material. These quantities are expressed in terms of the shear stress limits with respect to the principal anisotropy axes:

$$\left(\frac{\tau_{13}}{\widetilde{\tau}_{13}}\right)^2+\left(\frac{\tau_{23}}{\widetilde{\tau}_{23}}\right)^2\le 1, \tag{4.66}$$

$$F^{(4)}_{1313}(0)\equiv f^{(4)}_{1313}=1/\widetilde{\tau}^2_{13},\qquad F^{(4)}_{2323}(0)\equiv f^{(4)}_{2323}=1/\widetilde{\tau}^2_{23}. \tag{4.67}$$

The coefficients f_{12}, f_{22} and F_{11}, F_{12}, F_{22} are components of symmetric positive-definite matrices:

$$\mathbf{f}=\begin{bmatrix} f^{(4)}_{1313} & 0\\ 0 & f^{(4)}_{2323}\end{bmatrix},\qquad \mathbf{F}=\begin{bmatrix} F^{(4)}_{1313}(\Phi) & F^{(4)}_{1323}(\Phi)\\ F^{(4)}_{1323}(\Phi) & F^{(4)}_{2323}(\Phi)\end{bmatrix}. \tag{4.68}$$

The relation between the coefficients F_{ij} and f_{ij} can be represented in the form:

$$F_{ij}=t_{ik}t_{jl}f_{kl}. \tag{4.69}$$

The non-vanishing components stress tensor read in terms of the stress function $\Psi(x_1, x_2)$:

$$\tau_1=\partial\Psi/\partial x_2,\qquad \tau_2=-\partial\Psi/\partial x_1\qquad on\qquad \Omega. \tag{4.70}$$

The equilibrium equations are satisfied identically with (4.70). The boundary condition reduces to the function $\Psi(x_1, x_2)=0$ being zero on the boundary Γ.

The condition (4.66) is rewritten in the form:

$$\begin{aligned} &Q_{ij}\Psi_{,i}\Psi_{,j}=\nabla\Psi\cdot\mathbf{Q}\cdot\nabla\Psi\le 1,\\ &\mathbf{Q}=|\mathbf{F}|\mathbf{F}^{-1}=|\mathbf{f}|(\mathbf{t}\cdot\mathbf{f}\cdot\mathbf{t}^T)^{-1},\qquad \mathbf{t}\cdot\mathbf{t}^T=\mathbf{E}\end{aligned} \tag{4.71}$$

The ultimate torque equals the extremal value of the functional:

$$M=2\max_{\Psi}\int_{\Omega}\Psi(x_1,x_2)d\Omega, \tag{4.72}$$

among all continuously differentiable functions Ψ that satisfy:

$$\begin{aligned} &\nabla\Psi\cdot\mathbf{Q}\cdot\nabla\Psi\le 1\qquad on\qquad \Omega,\\ &\Psi=0\qquad\qquad\qquad on\qquad \Gamma.\end{aligned} \tag{4.73}$$

For the functions Ψ, the equilibrium and conditions for ultimate stresses are identically satisfied.

The optimization problem formulated in Section 4.1 for the three-dimensional problem becomes the following for a specific case of the state of stress: determine the optimal orientation of the anisotropy axis from the condition for the maximum of the ultimate torque (4.30):

$$M^*=\max_{\Phi(x_1,x_2)}M. \tag{4.74}$$

According to Eq. (4.43), the necessary optimality conditions are:

$$(F_{11}\tau_1+F_{12}\tau_2)\tau_2=(F_{12}\tau_1+F_{22}\tau_2)\tau_1. \tag{4.75}$$

The necessary optimality conditions (4.75) are satisfied identically if:

$$F_{11}\tau_1+F_{12}\tau_2=\hat{\lambda}\tau_1,\qquad F_{12}\tau_1+F_{22}\tau_2=\hat{\lambda}\tau_2. \tag{4.76}$$

Relying on relationships (4.76), Eq. (4.73) reduces to the form:

$$\mathbf{Q}\cdot\nabla\Psi = \lambda\nabla\Psi, \qquad \lambda = |\mathbf{F}|\hat{\lambda} \equiv |\mathbf{f}|\hat{\lambda} = \max_{\mathbf{u}} \frac{\mathbf{u}\cdot\mathbf{Q}\cdot\mathbf{u}}{\mathbf{u}\cdot\mathbf{u}}. \tag{4.77}$$

The maximum M^* of the integral (4.72) for the torque M is achieved when a regime corresponding to a large eigenvalue $\mathbf{Q}$ is realized in the whole domain Ω:

$$\mathbf{Q}\cdot\nabla\Psi = \lambda_{max}\nabla\Psi, \qquad \lambda_{max} = \max(f_{1313}^{(4)}, f_{2323}^{(4)}). \tag{4.78}$$

The minimum M^{**} of torque M (4.72) is achieved when a regime of lower eigenvalue $\mathbf{Q}$ everywhere in Ω:

$$\mathbf{Q}\cdot\nabla\Psi = \lambda_{min}\nabla\Psi, \qquad \lambda_{min} = \min(f_{1313}^{(4)}, f_{2323}^{(4)}). \tag{4.79}$$

It therefore follows that the ultimate moment reaches a maximum when the directions with the highest yield at each point agree with the direction of action of the maximum tangential stress:

$$M^{**} \le M \le M^*. \tag{4.80}$$

4.5 Optimal Anisotropy of Bending Console

In the second example of the application of optimization principles, the problem of optimal arrangement of anisotropy in a cross-section of a straight rod in a state of bending is studied. The state of stress is examined in a prismatic cantilever rod loaded by surface forces distributed over its unclamped end face. We superpose the origin on the stiffly framed left end face of the rod. Let the side surface of the rod be force-free and P the principal vector of the surface forces on the right end $x_3 = l$ directed along the x_1 axis. The surface loads on the end face produce a moment M_B relative to the center of inertia whose vector is directed along x_2. The stress tensor components τ_{13}, τ_{23} and σ_{33} differ from zero at each section of the rod. The normal stress σ_{33} is represented by the formula:

$$\sigma_{33} = -\frac{M_B + P(l - x_3)}{I}x_1. \tag{4.81}$$

Therefore, a transverse force:

$$P = \int_{\Omega} \tau_{13} d\Omega. \tag{4.82}$$

and a bending moment:

$$M_B = -\int_{\Omega} \sigma_{33} x_1 d\Omega \tag{4.83}$$

act on the cantilever. The statics equations in the volume and on the surface are written in the form:

$$\partial\tau_{13}/\partial x_1 + \partial\tau_{23}/\partial x_2 = -I^{-1}Px_1 \quad \partial\tau_{13}/\partial x_3 = 0, \quad \partial\tau_{23}/\partial x_3 = 0, \tau_{13}n_1 + \tau_{23}n_2 = 0. \tag{4.84}$$

In Eq. (4.81) I is the moment of inertia of a section relative to the x_2 axis. It is assumed that the transverse section has an axis of symmetry that agrees with the force line of action.

The shear stresses are expressed in terms of the stress function $\Psi(x_1, x_2)$:

$$\tau_{13} = \Psi_{,2} - \frac{1}{2} I^{-1} P {x_1}^2 + \widetilde{\Psi}(x_2),\ \tau_{23} = -I^{-1} P \Psi_{,1}. \tag{4.85}$$

The auxiliary function $\widetilde{\Psi}(x_2)$ will be defined later from the boundary condition in (4.92). The boundary condition (4.92) is rewritten with the introduced stress function as:

$$\partial\Psi/\partial s = \left[\frac{1}{2} I^{-1} P {x_1}^2 - \widetilde{\Psi}(x_2)\right] \partial y/\partial s, \tag{4.86}$$

the derivative with respect to the contour is $\partial/\partial s$. If the function $\widetilde{\Psi}(x_2)$ is selected in such a way that the condition:

$$\frac{1}{2} I^{-1} P {x_1}^2 = \widetilde{\Psi}(x_2) \tag{4.87}$$

is satisfied on the contour, then the boundary condition takes the form $\Psi(\Gamma) = 0$.

We will consider the plane of the cantilever transverse section to be the plane of symmetry. Under this assumption, the criterion (2.4) in Chapter 2 is written in the form:

$$g(\tau_{13}, \tau_{23}, \sigma_{33}) = \tau_{i3} F^{(4)}_{i3j3}(\Phi) \tau_{j3} + F^{(4)}_{3333} {\sigma_{33}}^2 \leq 1. \tag{4.88}$$

We assume that the bending moment M_B and the transverse force P vary over load history proportionally to a single parameter κ, such that the ratio $M_B/P = \kappa$ remains constant. The parameter κ depends only on the location x_3 of the section on the rod axis. With some stress function $\Psi(x_1, x_2)$, the stresses satisfy inequality (4.88) at each point.

The optimization problem consists of seeking a distribution of the orientation of the anisotropy axes such that the limit load reaches the maximum value.

The optimal orientation of the anisotropy axes is such that the direction of the shear stress vector $\{\tau_{13}, \tau_{23}\}$ acting in the plane of the rod transverse section will agree with the direction of the eigenvector of the matrix **B** corresponding to its smallest eigenvalue. Such an orientation corresponds to the minimum value of the quadratic form on the left-hand side of (4.88).

4.6 Optimization of Plates in Bending

The estimation of specific anisotropic stiffness of plates in a state of pure bending is based on a similar technique. The density of the elastic energy of an anisotropic material depends on the orientation of material and the given moment vector:

$$W_D(\Phi) = \widehat{\mathbf{M}}^T \cdot \widehat{\mathbf{D}}^{-1}(\Phi) \cdot \widehat{\mathbf{M}}/2, \widehat{\mathbf{M}}^T = [M_{11}, M_{22}, \sqrt{2} M_{12}]. \tag{4.89}$$

The specific elastic energy reads:

$$W_D{}^*(\Phi) = \widehat{\mathbf{M}}^T \cdot \widehat{\mathbf{D}}^{-1}(\Phi) \cdot \widehat{\mathbf{M}}/2\rho. \tag{4.90}$$

The tensor of the second rank $\widehat{\mathbf{D}}^{-1}(\Phi)$ depends on the inclination angle Φ:

$$\widehat{\mathbf{D}}^{-1}(\Phi) = \widetilde{\mathbf{T}}(\Phi) \cdot \widehat{\mathbf{D}}^{(0)-1}(\Phi) \cdot \widetilde{\mathbf{T}}^{-1}(\Phi), \tag{4.91}$$

$$\widetilde{\mathbf{T}}(\Phi) = \begin{pmatrix} c^2 & s^2 & \sqrt{2}sc \\ s^2 & c^2 & -\sqrt{2}sc \\ -\sqrt{2}sc & \sqrt{2}sc & c^2 - s^2 \end{pmatrix}. \tag{4.92}$$

The first optimization problem consists of a search for the orientation of angle Φ that provides the minimum or maximum elastic energy, assuming the stress $\widehat{\boldsymbol{\sigma}}$ is given:

$$W_D(\Phi) \to \min_{\Phi} \tag{4.93}$$

or

$$W_D(\Phi) \to \max_{\Phi}. \tag{4.94}$$

The minimum of the elastic energy (4.94) is attained for the stiffest structures if the smallest distortional amount of the structure is questioned. This is similar for strength optimization.

In second and fourth rank tensor notations, the Mises–Hill strength criterion (2.4) in Chapter 2 in terms of bending moments reduces to:

$$g(\widehat{\mathbf{M}}) = \widehat{\mathbf{M}}^T \bullet \widetilde{\mathbf{F}}(\Phi) \bullet \widehat{\mathbf{M}} \le 1, \tag{4.95}$$

$$\widetilde{\mathbf{F}}^{(0)} = \begin{bmatrix} G+H & -H & 0 \\ -H & F+H & 0 \\ 0 & 0 & 2N \end{bmatrix}.$$

The second rank tensor depends on the inclination angle Φ:

$$\widetilde{\mathbf{F}}(\Phi) = \widetilde{\mathbf{T}}(\Phi) \cdot \widetilde{\mathbf{F}}^{(0)} \cdot \widetilde{\mathbf{T}}^T(\Phi). \tag{4.96}$$

The second optimization problem for the pure bending of plates consists of a search for the orientation of angle Φ that provides the minimum of the strength criterion (4.95), presuming the given stress $\widehat{\boldsymbol{\sigma}}$:

$$g(\Phi) \to \min_{\varphi}.$$

The third optimization problem for the pure bending of plates studies the orientation of angle Φ that provides the maximum of the stored elastic energy (4.94), presuming the given stress $\widehat{\boldsymbol{\sigma}}$ fulfills the quadratic strength criterion (4.95):

$$W_D(\Phi) = \widehat{\mathbf{M}}^T \cdot \widehat{\mathbf{D}}^{-1}(\Phi) \cdot \widehat{\mathbf{M}}/2 \to \max_{\Phi}, \tag{4.97}$$

$$g(\Phi) = \widehat{\mathbf{M}}^T \cdot \widetilde{\mathbf{F}}(\Phi) \cdot \widehat{\mathbf{M}} \le 1. \tag{4.98}$$

The logic of the optimization problem is the following. The mass of the structure is not affected by the orientation angle and considered to be a constant during optimization. The elastic element of structure of a given mass that stores the maximal elastic energy for a given applied bending moment is examined. The stored energy is limited by ultimate stresses that follow from the failure criterion. The applications of the optimization problem are the springs and energy harvesters.

The equivalent formulation of the optimization problem regards the specific elastic energy:

$$\widetilde{W}_D(\Phi) = \frac{1}{2}\frac{\widehat{\mathbf{M}}^T \cdot \widehat{\mathbf{D}}^{-1}(\varphi) \cdot \widehat{\mathbf{M}}}{\widehat{\mathbf{M}}^T \cdot \widetilde{\mathbf{F}} \cdot \widehat{\mathbf{M}}} \to \max_{\Phi}. \tag{4.99}$$

The optimization conditions in this section lead to requirements analogous to those in Sections 4.1, 4.2 and 4.3 of this chapter.

4.7 Conclusions

Three main problems of the optimization for anisotropic materials were investigated. The loads on the structural element were considered to be known and not alter during the deformation of the structural member. The first problem studied the optimal placement of the anisotropy axes, which maximizes or minimizes the elastic energy. The maximization of an elastic energy for the given loads leads to the most resilient design of the structural element, which allows maximal flexibility. The minimization of an elastic energy for the given loads leads to the stiffest design of the structural element, with the lowest flexibility.

The second problem searched for the orientation of anisotropic properties to minimize solely the failure number of the structural member adversely affecting stiffness. This optimization aim is interesting for load bearing structural elements.

The third problem examined the orientation of anisotropic properties to maximize the stored elastic energy in an element but limited the failure number of the structural member. This optimization aim is interesting for energy harvesters or spring elements.

Ashby (2010) overviews a material selection method that is based on the dimensionless material selection numbers. This method allows comparison of different materials for a given application. The method does not account for the anisotropy of the elastic and strength characteristics of the considered materials. Accordingly, the method is suited to principal material selection. The book by Miravete (1996) oversees the application of composite materials in automotive applications, primarily low-stressed structural elements. The studies by Banichuk (1996), Mehrabadi and Cowin (1990), Cowin (1997), Pedersen (1989), Pedersen (1990), Rovati and Taliercio (2005), Sgarra and Vianello (1997a,b), Vianello (1996a,b), Koiter (1960), Chen and Wei (2005), Lipton and Chen (2004), Chen and Lipton (2007), Lipton (1998), Blasques et al. (2011), Sonmez (2017), Ganguli (2013), Qiao et al. (1998) and Del Piero (1979) explore the various aspects of the optimization of accumulated elastic for anisotropic elastic bodies. The closed form solutions are derived for some elements of bodies with basic geometry.

References

Ashby, M.F. (2010). *Materials Selection in Mechanical Design*, 4e. Burlington: Elsevier.

Banichuk, N.V. (1996). Optimization of anisotropic properties for continuum bodies and structural elements using spectral methods of tensor analysis. *Mech. Struct. Mach.* 24: 71–87.

Banichuk, N.V. and Kobelev, V.V. (1987). On optimal plastic anisotropy. *J. Appl. Math. Mech.* 51 (3): 381–385.

Blasques J.P.A.A., Stolpe M., Berggreen C., Branner K. (2011) Optimal Design of Laminated Composite Beams. Kgs. Lyngby, Denmark: Technical University of Denmark (DTU), *DCAMM Special Report*, No. S134.

Chen, T. and Lipton, R. (2007). Bounds for the torsional rigidity of shafts with arbitrary cross-sections containing cylindrically orthotropic fibres or coated fibres. *Proc. R. Soc. A* 463: 3291–3309.

Chen, T. and Wei, C.-J. (2005). Saint-Venant torsion of anisotropic shafts: theoretical frameworks, extremal bounds and affine transformations. *Q. J. Mech. Appl. Math.* 58 (2): 269–287.

Cowin, S.C. (1994). Optimization of the strain energy density in linear anisotropic elasticity. *J. Elast.* 34: 45–68.

Cowin, S.C. (1997). Remarks on coaxiality of strain and stress in anisotropic elasticity. *J. Elast.* 47: 83–84.

Del Piero, G. (1979). Some properties of the set of fourth-order tensors, with application to elasticity. *J. Elast.* 9: 245–261.

Ganguli, R. (2013). Optimal design of composite structures: a historical review. *J. Indian Inst. Sci.* 93: 557–570.

Koiter, W.T. (1960). General theorems for elastic-plastic solids. In: *Progress in Solid Mechanics*, vol. 1 (ed. N. Sneddon and R. Hill), 167–221. Amsterdam: North-Holland.

Lipton, R. (1998). Optimal fiber configurations for maximum torsional rigidity. *Arch. Ration. Mech. Anal.* 144: 79–106.

Lipton, R. and Chen, T. (2004). Bounds and extremal configurations for the torsional rigidity of coated fiber reinforced shafts. *SIAM J. Appl. Math.*, 65 (1): 299–315.

Mehrabadi, M.M. and Cowin, S.C. (1990). Eigentensors of linear anisotropic elastic materials. *Q. J. Mech. Appl. Math.* 43: 15–41.

Miravete, A. (1996). *Optimisation of Design of Composite Structures*. UK: Woodhead Publishing Limited, Cambridge.

Norris, A.N. (2005). Optimal orientation of anisotropic solids. *Q. J. Mech. Appl. Math.* 59 (1): 29–53.

Ogden, R.W. (1997). *Non-linear Elastic Deformations*. Mineola: Dover Publications Inc.

Pedersen, P. (1989). On optimal orientation of orthotropic materials. *Struct. Optim.* 1: 101–106.

Pedersen, P. (1990). Bounds on elastic energy in solids of orthotropic materials. *Struct. Optim.* 2: 55–62.

Qiao, P., Davalos, J.F., and Barbero, E.J. (1998). Design optimization of fiber reinforced plastic composite shapes. *J. Compos. Mater.* 32 (2): 177–196.

Rovati, M. and Taliercio, A. (2005). On stationarity of strain energy density for some classes of anisotropic solids. *Int. J. Solids Struct.* 40: 6043–6075.

Seregin, G.A. and Troitskii, V.A. (1981). On the best position of elastic symmetry planes in an orthotropic body. *J. Appl. Math. Mech.* 45: 139–142.

Sgarra, C. and Vianello, M. (1997a). Directions of coaxiality between pure strain and stress in linear elasticity. *J. Elast.* 46: 263–265.

Sgarra, C. and Vianello, M. (1997b). Rotations which make strain and stress coaxial. *J. Elast.* 47: 217–224.

Sonmez, F.O. (2017). Optimum design of composite structures: a literature survey (1969–2009). *J. Reinf. Plast. Compos.* 36 (1): 3–39.
Vasiliev, V.V. and Gurdal, Z. (1999). *Optimal Design: Theory and Applications to Materials and Structures*, 1e. CRC Press.
Vianello, M. (1996a). Coaxiality of strain and stress in anisotropic linear elasticity. *J. Elast.* 42: 283–289.
Vianello, M. (1996b). Optimization of the stored energy and coaxiality of strain and stress in finite elasticity. *J. Elast.* 44: 193–202.

5

Optimization of Composite Driveshaft

5.1 Torsion of Anisotropic Shafts With Solid Cross-Sections

The function and design of the common driveshafts made of steel are presented in Chapter 6.7 of Stone and Ball (2004). The composite drive shaft has advantages of substantial weight reduction, the dynamic balance of increasing operating speed (Badie et al. 2006, 2011). The material is electrically nonconductive and the proper design allows adapting the end fitting. The advantages of the composite design assure perfect performance from the viewpoint of vibrations and harshness, long fatigue life and reduction in bearing wear. The materials usually have a lower modulus of elasticity, meaning that when torque peaks occur, the drive shaft may work as a shock absorber. A drive shaft commonly operates through constantly changing angles between the transmission and rare axle driveshafts (Figure 5.1).

Composite drive shafts are widely used in automotive and aeronautical applications due to their weight advantage combined with exceptional strength and stiffness (Beardmore and Johnson 1986; Kaviprakash et al. 2014; Cherniaev and Komarov 2015). One of the first composite applications was the driveshaft (propshaft) in the Renault Espace Quadra built in 1988 (Pollard 1989) (Figure 5.2).

The hybrid driveshafts made of composite materials such as E-Glass, carbon fibers and hybrid composite layers of E-Glass, carbon and S-Glass with different fiber orientations were investigated in Sivakandhan and Prabhu (2014). Strength of hybrid composite driveshafts was studied by conducting tensile, bending and torsion tests in the cited paper.

An interesting and well-documented study of the design and production of a composite driveshaft for a historical vehicle was performed by students at the Texas A&M University in Kingsville (Figure 5.3) (Keys et al. 2018).

Consider torsion of a shaft with a solid circular cross-section. One end of the shaft is fixed. The other end acts as the definite traction load. The traction load results in a pure torque over the end surface (Figure 5.4). Referring to cylindrical coordinates (r, θ, z),

Design and Analysis of Composite Structures for Automotive Applications: Chassis and Drivetrain, First Edition. Vladimir Kobelev.

Companion website: www.wiley.com/go/kobelev/automotive_suspensions

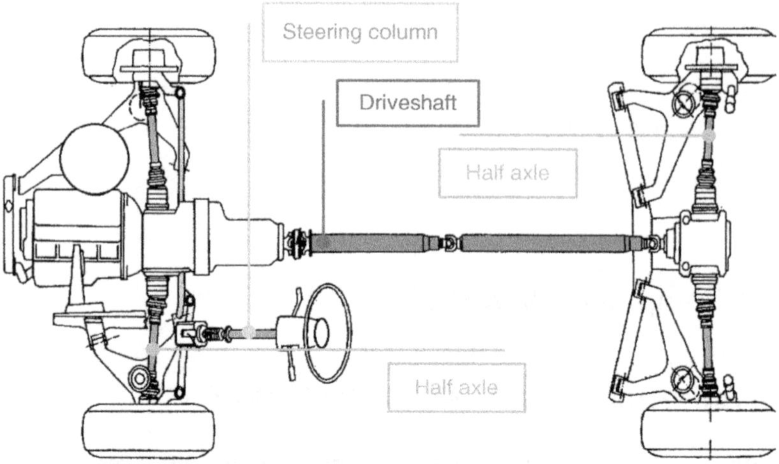

Figure 5.1 Driveshafts of a common passenger car.

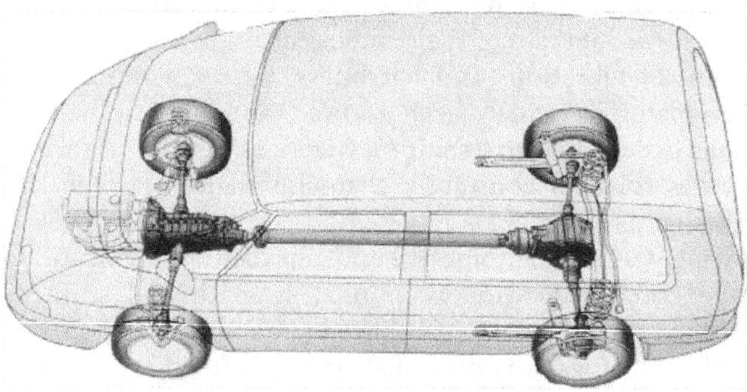

Figure 5.2 Composite propshaft in a Renault Espace Quadra (Pollard, 1989).

Figure 5.3 Composite driveshaft (Keys et al., 2018).

Figure 5.4 Torsion of a composite driveshaft by the terminal torque M_T.

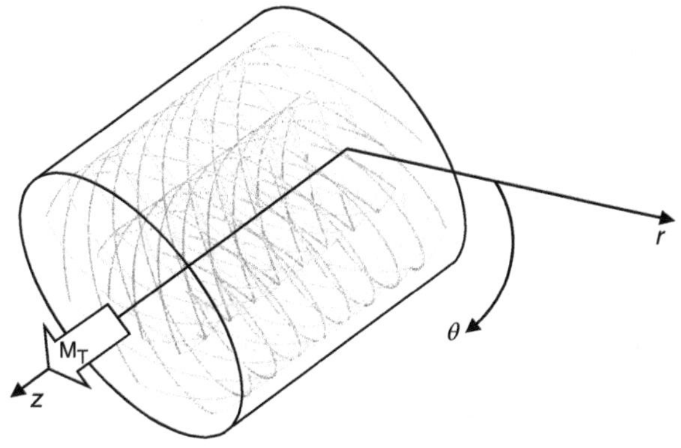

the medium is cylindrically orthotropic and homogeneous. The constitutive equations of the material are (Lekhnitskii 1981):

$$\begin{bmatrix} \sigma_r = 0 \\ \sigma_\theta = 0 \\ \sigma_z = 0 \\ \tau_{\theta z} \\ \tau_{rz} = 0 \\ \tau_{r\theta} = 0 \end{bmatrix} = \begin{bmatrix} C_{11} & C_{12} & C_{13} & 0 & 0 & 0 \\ C_{12} & C_{22} & C_{23} & 0 & 0 & 0 \\ C_{13} & C_{23} & C_{33} & 0 & 0 & 0 \\ 0 & 0 & 0 & C_{44} & C_{45} & 0 \\ 0 & 0 & 0 & C_{45} & C_{55} & 0 \\ 0 & 0 & 0 & 0 & 0 & C_{66} \end{bmatrix} \begin{bmatrix} \varepsilon_r \\ \varepsilon_\theta \\ \varepsilon_z \\ 2\varepsilon_{\theta z} \\ 2\varepsilon_{rz} \\ 2\varepsilon_{r\theta} \end{bmatrix}. \tag{5.1}$$

where $\varepsilon_r, \ldots$ and $\sigma_r, \ldots$ are the strain and stress components, $C_{11}, \ldots$ and $S_{11}, \ldots$ are the independent elastic and compliance constants of the cylindrically orthotropic material.

The non-vanishing strain–displacement relations for twist deformations are:

$$\begin{aligned} &\frac{\partial u_\theta}{\partial \theta} = 0, \frac{\partial w}{\partial z} = 0, \frac{\partial u_\theta}{\partial \theta} - \frac{u_\theta}{r} = S_{46}\tau_{\theta z}, \\ &\frac{1}{r}\frac{\partial w}{\partial \theta} + \frac{\partial u_\theta}{\partial z} = S_{44}\tau_{\theta z}, \frac{\partial w}{\partial r} = S_{45}\tau_{\theta z}, u_r = 0. \end{aligned} \tag{5.2}$$

The solution of Eq. (5.2) is:

$$\tau_{\theta z} = \Theta r / S_{44}, \tag{5.3}$$

$$u_\theta = \Theta r \left(z + \int_{r_i}^{r_a} \frac{S_{46}}{S_{44}} dr \right) + \omega_3 r, \tag{5.4}$$

$$w = \Theta \int_{r_i}^{r_a} \frac{S_{45}}{S_{44}} dr + w_0. \tag{5.5}$$

The values ω_3 and w_0 characterize the shift and elongation of the driveshaft.

The displacements on the fixed end vanish. For Saint-Venant's torsion of circular bars, the traction components over the cinematically unconstrained end sections perform a pure torque. The statically equivalent end conditions are:

$$\begin{aligned} &\int_0^{2\pi} \int_{r_i}^{r_a} [\sigma_z r \sin\theta, \sigma_z r \cos\theta, \tau_{\theta z} r] r dr d\theta = [0, 0, M_T], \\ &\int_0^{2\pi} \int_{r_i}^{r_a} [\tau_{rz}\cos\theta - \tau_{\theta z}\sin\theta, \tau_{rz}\sin\theta + \tau_{\theta z}\cos\theta, \sigma_z] r dr d\theta = [0, 0, 0]. \end{aligned} \tag{5.6}$$

The solution, which is independent of the axial coordinate, with statically equivalent end conditions reads as (Lekhnitskii 1981):

$$u_\theta = \frac{M_T}{C_T} rz, \quad \tau_{\theta z} = C_{44}\frac{M_T}{C_T} r. \tag{5.7}$$

The torsional rigidity of radially inhomogeneous bars is:

$$C_T = \frac{2\pi}{\Theta}\int_{r_i}^{r_a} \tau_{\theta z} r^2 dr = 2\pi \int_{r_i}^{r_a} C_{44}(r) r^3 dr. \tag{5.8}$$

The torsional rigidity in of homogeneous bars is:

$$C_T = \pi C_{44}(r_a^4 - r_i^4)/2. \tag{5.9}$$

The elastic energy of the driveshaft is equal to

$$U_T = M_T^2/2C_T. \tag{5.10}$$

5.2 Thin-Walled Anisotropic Driveshaft with Closed Profile

5.2.1 Geometry of Cross-Section

We will examine a thin-walled driveshaft as a cylindrical anisotropic shell. The middle surface of the shell is the cylindrical surface formed by motion of a straight line. This straight line moves in the same direction and intersects with the given curve. This curve is called a directrix. Straight lines, corresponding to different positions of the straight line at its motion, are called generatrices of a cylindrical surface. Consequently, the middle surface of the shell is part, generally speaking, of the surface of a noncircular cylinder formed by the motion of a closed plane curve (the generatrix, which is assumed to be smooth) along a rectilinear directrix (Figure 5.5).

The form of the shell is such that its length along the axis L (i.e. the length of the generatrix) is much greater than the arc length of the directrix. Hence, the shell is highly elongated in one direction and can be examined as a thin-walled driveshaft. By thin-walled, we mean that the thickness of the walls, h, is considerably less than the length of the arc, while the rate of change in thickness over the arc is very small ($dh/ds << 1$, where s is a

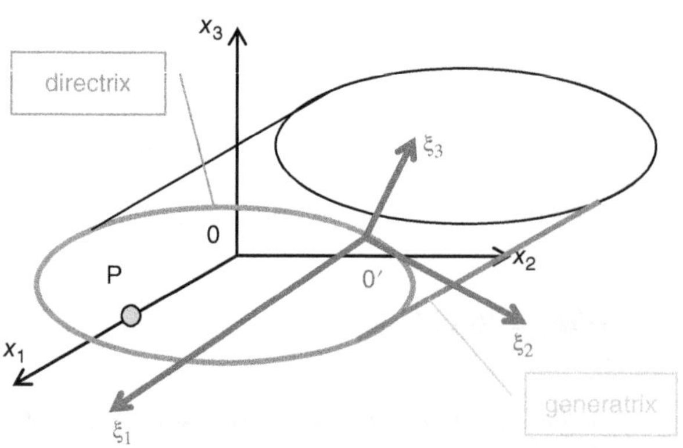

Figure 5.5 Geometry of the cross-section of the rod.

parameter of the arc length of directrix, $0 \leq s \leq l$). Thus, the main assumptions regarding the character of the geometry of the thin-walled driveshaft-shell can be expressed in the form of the inequalities:

$$h << l << L.$$

These inequalities will subsequently allow us to change over from the two-dimensional equations of shell theory to the one-dimensional theory of anisotropic driveshafts. The inequalities just given are needed for proper use of the shell theory equations. The requirement of smoothness of the profile of the contour is also important; if the profile contains points of discontinuity, additional studies will be necessary.

The position of a point on the middle surface of the rod and in space will be determined by assigning its position in one of two coordinate systems. The first system $Ox_1x_2x_3$ (henceforth referred to as the global system) is Cartesian and rectangular. The direction of the Ox_1 axis of this system coincides with the orientation of the generatrix of the rod, while the directrix of the cylindrical surface lies in the plane Ox_2x_3.

The second coordinate system $\widetilde{O}\xi_1\xi_2\xi_3$ (henceforth referred to as the local system) is curvilinear and orthonormal. The axis $\widetilde{O}\xi_1$ of this system is parallel to the axis Ox_1 of the global system ($x_1 = \xi_1$), while the cylindrical surface corresponds to the equation $\xi_3 = 0$. The coordinate ξ_2 locally directed tangential to directrix. The angle between the axes ξ_3 and x_3 is:

$$\theta = \theta(s) = \theta(\xi_2).$$

The transformation matrix from the coordinate system $\widetilde{O}\xi_1\xi_2\xi_3$ to the system $Ox_1x_2x_3$ is $\mathbf{t}$. The components t_{ij} of the transformation matrix $\mathbf{t}$ are:

$$t_{ij} = \partial x_i / \partial \xi_j, \quad i, j = 1, 2, 3.$$

On the surface $\xi_3 = 0$ the components of the matrix are:

$$\mathbf{t} = \begin{bmatrix} 1 & 0 & 0 \\ 0 & \cos\theta & \sin\theta \\ 0 & -\sin\theta & \cos\theta \end{bmatrix}.$$

5.2.2 Main Kinematic Hypothesis

The main kinematic hypothesis assumes that the profile of the rod in each normal section with the plane $x_1 = 0$ is not distorted, but only rotated and displaced in 3D space. Consider the membrane bounded by the directrix that is stretched over the hollow cross-section of the shaft. The kinematic hypothesis postulates that this membrane is not deformed but only displaces as a rigid body.

The point P of intersection of the membrane with the axis x_1 has the coordinates: $x_1 = x_1^{(0)}, x_2 = 0, x_3 = 0$. The components of the displacement at the point P are equal to U_1, U_2, U_3. The components of the vector of the infinitesimal rotation at the point P are equal to $\Omega_1, \Omega_2, \Omega_3$. All quantities are small enough so that the equations of the linear theory of elasticity are valid. The quantities U_1, U_2, U_3 and $\Omega_1, \Omega_2, \Omega_3$ are functions of the coordinate x_1 only.

The kinematic hypothesis determines the components of the displacement and rotation vectors of all points on the driveshaft profile lying in the normal section $x_1 = x_1^{(0)}$. In the global system the displacements are equal to:

$$\begin{aligned} u_1(x_1, x_2, x_3) &= U_1(x_1) + x_2\Omega_3(x_1) - x_3\Omega_2(x_1), \\ u_2(x_1, x_2, x_3) &= U_2(x_1) + x_3\Omega_1(x_1), \\ u_3(x_1, x_2, x_3) &= U_3(x_1) - x_2\Omega_1(x_1). \end{aligned} \tag{5.11}$$

To determine the deformation of the thin shell, the components of deformation should be transformed to the local coordinates. The displacement vector in the local coordinate system reads $\hat{\mathbf{u}} = \mathbf{t} \cdot \mathbf{u}$ with the transformation matrix $\mathbf{t}$. Its components are:

$$\begin{aligned} \hat{u}_1(\xi_1, \xi_2, \xi_3) &= U_1(x_1) + x_2\Omega_3(x_1) - x_3\Omega_2(x_1), \\ \hat{u}_2(\xi_1, \xi_2, \xi_3) &= U_2(x_1)\cos\theta - U_3(x_1)\sin\theta + \omega(\xi_2)\Omega_1(x_1), \\ \hat{u}_3(\xi_1, \xi_2, \xi_3) &= U_2(x_1)\sin\theta + U_3(x_1)\cos\theta + \chi(\xi_2)\Omega_1(x_1), \end{aligned} \tag{5.12}$$

where:

$$\omega(\xi_2) = x_3\cos\theta + x_2\sin\theta, \quad \chi(\xi_2) = -x_2\cos\theta + x_3\sin\theta.$$

The quantities x_2 and x_3 in the Eq. (5.12) are coordinates of the middle surface of the cross-section $\xi_3 = 0$. These coordinates expressed through coordinate of a point on the cross-section line ξ_2, because the profile shape is given: $x_2 = x_2(\xi_2)$, $x_2 = x_2(\xi_2)$, $x_3 = x_3(\xi_2)$.

Hypotheses (5.11) and (5.12) describe the kinematics of deformation of the closed thin-walled rod. Hypotheses of the type in (5.11) were used in Salim and Davalos (2005a,b) and Harursampath et al. (2017a,b) to construct a beam model with allowance for rotation and displacements of the sections. The main difference here is that the proposed model presumes the wall of the rod-shell is thin enough to satisfy the Kirchhoff–Love hypotheses of a straight normal for deformation of the rod. In a thin-walled rod with a closed profile, shear stresses prevent the off-plane distortion (warping) of the cross-section.

On the contrary, in a rod with an open profile the section undergoes off-plane displacement. Conditions with an absence of shear stresses on the contour of open thin-walled beams is used to derive the law of sectoral areas: see Vlasov (1947) and Librescu and Song (2006). The Vlasov hypotheses were used in Bauld et al. (1984) to study a model of an open thin-walled rod made of anisotropic materials.

Once the kinematic of the deformation is defined, the governing equations of the rod can then be obtained by using the theory of thin shells. For open thin-walled rods, the common method for deriving the governing equations was proposed in Vlasov (1947). The method uses the semi-momentless theory of elastic shells. The semi-momentless theory of thin anisotropic shells is slightly different from the common theory of anisotropic thin shells (Mikeladze 1966). The basis of this theory is a set of simplifying assumptions of a static nature. Namely, the twisting moments as well as the longitudinal bending moment and the corresponding transverse shearing force are neglected (Axelrad 1987). Elastic thin shell theories, such as semimoment-less, Donnell shallow shell (Donnell 1977), Sanders theory (Sanders 1959) and Koiter theory (Koiter 1960) are based on the following assumptions; that is, Love's "first approximation":

1) The thickness of the shell is small compared with the radius and the shell is sufficiently thin.
2) The strains and the displacements are sufficiently small enough that the quantities of the second and higher order magnitude in the strain–displacement relations may be neglected in comparison with the first order terms.
3) The transverse normal stress is small compared with the other normal stress components and may be neglected.
4) The normal to the undeformed middle surface remains straight and normal to the deformed middle surface, and it suffers no extension.
5) The rotary inertia and transverse shear deformation effects may be neglected.

5.3 Deformation of a Composite Thin-Walled Rod

5.3.1 Equations of Deformation of a Anisotropic Thin-Walled Rod

The equations of deformation for a thin-walled rod are obtained by variation methods.

The variation method is based on substitution of the expressions for displacements (5.12) into the potential energy functions of an anisotropic thin-walled shell (Ventsel and Krauthammer 2001). The potential energy is expressed in terms of the displacements of points of the shell surface in the local coordinate system. The potential energy functional is expressed through the functions U_1, U_2, U_3 and Ω_1, Ω_2, Ω_3. The Eulerian equations of the potential energy functional provide the equilibrium equations. The components of the strains ε_1, ε_2, γ and curvatures κ_1, κ_2, τ of the middle surface of the shell read (Ventsel and Krauthammer 2001):

$$\varepsilon_1 = \frac{\partial \hat{u}_1}{\partial \xi_1} = \sum_{i=1}^{3} \frac{\partial \hat{u}_1}{\partial x_i} a_{i1} = U_1'(x_1) + x_2\Omega_3'(x_1) - x_3\Omega_2'(x_1), \tag{5.13}$$

$$\varepsilon_2 = \frac{\partial \hat{u}_2}{\partial \xi_2} + \frac{\hat{u}_3}{R} = \sum_{i=1}^{3} \frac{\partial \hat{u}_2}{\partial x_2} a_{i2} + \frac{\hat{u}_3}{R} = 0,$$

$$\gamma = \frac{\partial \hat{u}_1}{\partial \xi_2} + \frac{\partial \hat{u}_2}{\partial \xi_1} = \sum_{i=1}^{3} \left(\frac{\partial \hat{u}_1}{\partial x_i} a_{i2} + \frac{\partial \hat{u}_2}{\partial x_i} a_{i1} \right)$$
$$= (U_2' + \Omega_3)\cos\theta + (-U_3' + \Omega_2)\sin\theta + (x_1) + \Omega_1'\omega(\theta),$$

$$\kappa_1 = -\frac{\partial^2 \hat{u}_3}{\partial {\xi_1}^2} = -\sum_{j}^{3} a_{j1} \frac{\partial}{\partial \xi_j} \left[\sum_{i=1}^{3} \frac{\partial \hat{u}_3}{\partial x_i} a_{i1} \right] = -U_2'' \sin\theta - U_3'' \cos\theta - \Omega_3'' \chi(\theta),$$

$$\kappa_2 = -\frac{\partial^2 \hat{u}_3}{\partial {\xi_2}^2} - \frac{\hat{u}_3}{R^2} = -\sum_{j}^{3} a_{j2} \frac{\partial}{\partial \xi_j} \left[\sum_{i=1}^{3} \frac{\partial \hat{u}_3}{\partial x_i} a_{i2} \right] - \frac{\hat{u}_3}{R^2} = 0,$$

$$\tau = -2\frac{\partial^2 \hat{u}_3}{\partial \xi_1 \partial \xi_2} - \frac{1}{R}\left(\frac{\partial \hat{u}_1}{\partial \xi_2} - \frac{\partial \hat{u}_2}{\partial \xi_1} \right) = -2\sum_{j=1}^{3} a_{i1} \frac{\partial}{\partial \xi_j} \sum_{i=1}^{3} a_{i2} \frac{\partial \hat{u}_3}{\partial x_i} - \frac{1}{R}\left(\frac{\partial \hat{u}_1}{\partial \xi_2} - \frac{\partial \hat{u}_2}{\partial \xi_1} \right)$$
$$= -(U_2' \cos\theta - U_3' \sin\theta + \Omega_1'\omega(\theta) + \Omega_2 \sin\theta + \Omega_3 \cos\theta)/R + 2\Omega_1'.$$

A prime denotes a derivative with respect to the coordinate $f' = \partial f / \partial x_1 = \partial f / \partial \xi_1$. In all equations the coordinates along the axis of the rod coincide: $x_1 = \xi_1$.

The rule of summation over repeating indices in the range of 1–3 is used. In deriving Eq. (5.13), the following expressions are valid:

$$\partial \sin\theta/\partial x_2 = \cos^2\theta/R, \quad \partial \cos\theta/\partial x_2 = -\sin\theta\cos\theta/R.$$

Here, R is the radius of curvature of the rod profile, $R^{-1} = \partial\theta/\partial\xi_2$.

The equilibrium equations follow on from the principle of the minimum of potential energy. The potential strain energy in the theory of thin anisotropic shells is given by Ventsel and Krauthammer (2001):

$$\begin{aligned}
&U_e = U_A + U_D,\\
&U_A = \frac{1}{2}\int_0^L\int_0^l W_A d\xi_1 d\xi_2,\ W_A = A_{11}{\varepsilon_1}^2 + A_{66}\gamma^2 + 2A_{16}\varepsilon_1\gamma,\\
&U_D = \frac{1}{2}\int_0^L\int_0^l W_D d\xi_1 d\xi_2,\ W_D = D_{11}{\kappa_1}^2 + D_{66}\tau^2 + 2D_{16}\kappa_1\tau.
\end{aligned} \tag{5.14}$$

The components of stiffness matrices of the shell in a plane stress state and, correspondingly, in bending are A_{ij}, D_{ij}, $i, j = 1$ *or* 6, see Chapter 1 Eq. (1.37). The coupling quadrant is assumed to be absent. In (5.14), it was considered that, in accordance with (5.13), $\varepsilon_2 = 0$, $\kappa_2 = 0$ and the corresponding terms were omitted. The total potential energy of the rod is equal to:

$$\Pi = U_e - U_f,\ U_f = \int_0^L \left[\sum_{i=1}^{3} p_i U_i + m_i\Omega_i\right] d\xi_1.$$

The distributed forces and moments p_i, m_i, $i = 1, 2, 3$ are known functions of the coordinate ξ_1 and are constant in the course of deformation. Calculating the variation of total potential

$$\delta\Pi = \delta U_e - \delta U_f,$$

and equating it to zero, we obtain

$$\delta U_e = \frac{1}{2}\int_0^L\int_0^l (\delta W_A + \delta W_D) d\xi_1 d\xi_2,\quad \delta U_f = \int_0^L \left[\sum_{i=1}^{3} p_i \delta U_i + m_i \delta\Omega_i\right] d\xi_1,$$

$$\begin{aligned}
&\delta W_A = 2A_{11}\varepsilon_1\delta\varepsilon_1 + 2A_{66}\gamma\delta\gamma + 2A_{16}(\varepsilon_1\delta\gamma + \gamma\delta\varepsilon_1),\\
&\delta W_D = 2D_{11}\kappa_1\delta\kappa_1 + 2D_{66}\tau\delta\tau + 2D_{16}(\kappa_1\delta\tau + \tau\delta\kappa_1).
\end{aligned}$$

Replacing $\partial\varepsilon_1$, $\partial\gamma$, $\partial\tau$, $\partial\kappa_1$ with their expressions (5.13) and transforming one series by parts, we obtain the equations of equilibrium:

$$\begin{aligned}
&-\int_0^l \sigma_1' d\xi_2 = p_1,\\
&-\int_0^l (\sigma_{12}'\cos\theta + \Sigma_1''\sin\theta) d\xi_2 = p_2,\\
&\int_0^l (\sigma_{12}'\sin\theta - \Sigma_1''\cos\theta) d\xi_2 = p_3,\\
&-\int_0^l (\sigma_{12}'\omega + \Sigma_1''\chi + 2\Sigma_{12}'\chi) d\xi_2 = m_1,
\end{aligned}$$

$$\int_0^l (\sigma_1' x_3 + \sigma_{12} \sin\theta) d\xi_2 = m_2,$$
$$-\int_0^l (\sigma_1' x_2 - \sigma_{12} \cos\theta) d\xi_2 = m_3. \tag{5.15}$$

Here are:

$\sigma_1 = A_{11}\varepsilon_1 + A_{16}\gamma$, membrane tensile stress in the axis direction;
$\sigma_{12} = A_{16}\varepsilon_1 + A_{66}\gamma - (D_{16}\kappa_1 + D_{66}\tau)/R$, membrane shear stress;
$\Sigma_1 = D_{11}\kappa_1 + D_{16}\tau$, tensile stress due to bending in the axis direction;
$\Sigma_{12} = D_{16}\kappa_1 + D_{66}\tau$, shear stress due to torsion.

Substitution of (5.13) into (5.15) delivers the equilibrium equations in terms of displacements. In matrix form, the governing system of equations is written as follows:

$$\left\{ [M]\frac{d^4}{dx_1^4} + [R]\frac{d^3}{dx_1^3} + [N]\frac{d^2}{dx_1^2} + [Q]\frac{d}{dx_1} + [L] \right\} [V] = [P]. \tag{5.16}$$

Here, $[M]$, $[N]$, $[L]$ are symmetric matrices; $[R]$, $[Q]$ are skew-symmetric matrices of the dimensions 6×6. The six-dimensional vector of generalized displacements reads:

$$[V] = [U_1, U_2, U_3, \Omega_1, \Omega_2, \Omega_3]^T. \tag{5.17}$$

The six-dimensional vector of external loads is:

$$[P] = [p_1, p_2, p_3, m_1, m_2, m_3]^T. \tag{5.18}$$

The non-vanishing components of the matrices $[M]$, $[N]$, $[L]$, $[R]$, $[Q]$ are:

$$[M] = \int [m] d\xi_2, \tag{5.19}$$

$$[N] = \int D_{11}[n] d\xi_2, \tag{5.20}$$

$$[L] = \int D_{11}[l] d\xi_2, \tag{5.21}$$

$$[Q] = \int D_{11}[q] d\xi_2, \tag{5.22}$$

$$[R] = \int D_{16}[r] d\xi_2. \tag{5.23}$$

The matrices of the dimensions 6×6 in (5.19) and (5.23) are:

$$[m] = \begin{bmatrix} -A_{11} & -A_{16}c & A_{16}s & -A_{16}\omega & A_{11}x_3 & -A_{11}x_2 \\ & -A_{66}c^2 & A_{66}cs & -A_{66}c\omega & A_{16}cx_3 & -A_{16}cx_2 \\ & & -A_{66}s^2 & A_{66}s\omega & -A_{16}sx_3 & A_{16}sx_2 \\ & & & -A_{66}\omega^2 & A_{16}\omega x_3 & -A_{16}\omega x_2 \\ & & & & -A_{11}{x_3}^2 & A_{11}x_3x_2 \\ & \text{symm} & & & & -A_{11}{x_2}^2 \end{bmatrix},$$

$$[n] = \begin{bmatrix} 0 & 0 & 0 & 0 & 0 & 0 \\ 0 & s^2 & sc & \chi c & 0 & 0 \\ 0 & sc & c^2 & \chi s & 0 & 0 \\ 0 & \chi c & \chi s & \chi^2 & 0 & 0 \\ 0 & 0 & 0 & 0 & 0 & 0 \\ 0 & 0 & 0 & 0 & 0 & 0 \end{bmatrix}, \quad [l] = \begin{bmatrix} 0 & 0 & 0 & 0 & 0 & 0 \\ 0 & 0 & 0 & 0 & 0 & 0 \\ 0 & 0 & 0 & 0 & 0 & 0 \\ 0 & 0 & 0 & 0 & 0 & 0 \\ 0 & 0 & 0 & 0 & -c^2 & cs \\ 0 & 0 & 0 & 0 & cs & -s^2 \end{bmatrix},$$

$$[q] = \begin{bmatrix} 0 & 0 & 0 & 0 & 0 & 0 \\ 0 & 0 & 0 & 0 & sc & s^2 \\ 0 & 0 & 0 & 0 & -c^2 & -sc \\ 0 & 0 & 0 & 0 & -\chi s & \chi c \\ 0 & -sc & c^2 & \chi s & 0 & 0 \\ 0 & -s^2 & sc & -\chi c & 0 & 0 \end{bmatrix}, \quad [r] = \begin{bmatrix} 0 & 0 & 0 & 0 & 0 & 0 \\ 0 & 0 & -R^{-1} & -2s & 0 & 0 \\ 0 & R^{-1} & 0 & -2c & 0 & 0 \\ 0 & 2s & 2c & 0 & 0 & 0 \\ 0 & 0 & 0 & 0 & 0 & 0 \\ 0 & 0 & 0 & 0 & 0 & 0 \end{bmatrix}.$$

Integration is performed over the arc of the profile of the cross-section of a thin-walled beam. To abbreviate the notation, we employ the symbols: $c = \cos\theta$, $s = \sin\theta$.

5.3.2 Boundary Conditions

The boundary conditions on free and fastened ends can be obtained as natural and principal boundary conditions of the functional of total potential energy (5.14). We will examine the following main types of fastening.

5.3.2.1 Ideal Fixing

In this case, the rod end ($x_1 = 0$) cannot rotate or undergo plane displacement or shifting, since it is welded as a whole to a rigid base (Figure 5.6a). Here, the nine conditions of the main type of variational problem for functional (5.14) reduce to the equalities

$$U_1 = U_2 = U_3 = 0, \tag{5.24}$$

$$\Omega_1 = \Omega_2 = \Omega_3 = 0, \Omega_1' = \Omega_2' = \Omega_3' = 0. \tag{5.25}$$

5.3.2.2 Ideally Free End

In the case of an ideally free end (Figure 5.6b), all the variable functions U_i, Ω_i, $i = 1$, 2, 3 and the first derivatives U_2', U_3', Ω_1' are assumed to be independent variables. The corresponding natural boundary conditions of the variational problem will be the nine equations written here in matrix form

$$\left\{ [M^*]\frac{d^3}{dx_1^3} + [R^*]\frac{d^2}{dx_1^2} + [N^*]\frac{d}{dx_1} + [Q^*] \right\} [V] = [P^*], \tag{5.26}$$

where $[M^*]$, $[R^*]$, $[N^*]$, $[Q^*]$ are matrices containing six rows and nine columns; $[P^*]$ is a five-dimensional vector, all the components of which are equal to zero on the unloaded end. The non-vanishing components of the matrices $[M^*]$, $[N^*]$ are:

$$M_{ij}^* = M_{ij}, \quad 2 \le i,j \le 4; N_{ij}^* = N_{ij}, \quad 1 \le i,j \le 6; \tag{5.27}$$

$$N_{27}^* = -N_{38}^* = \int D_{16} csR^{-1} d\xi_2, N_{28}^* = \int D_{16} c^2 R^{-1} d\xi_2, \tag{5.28}$$

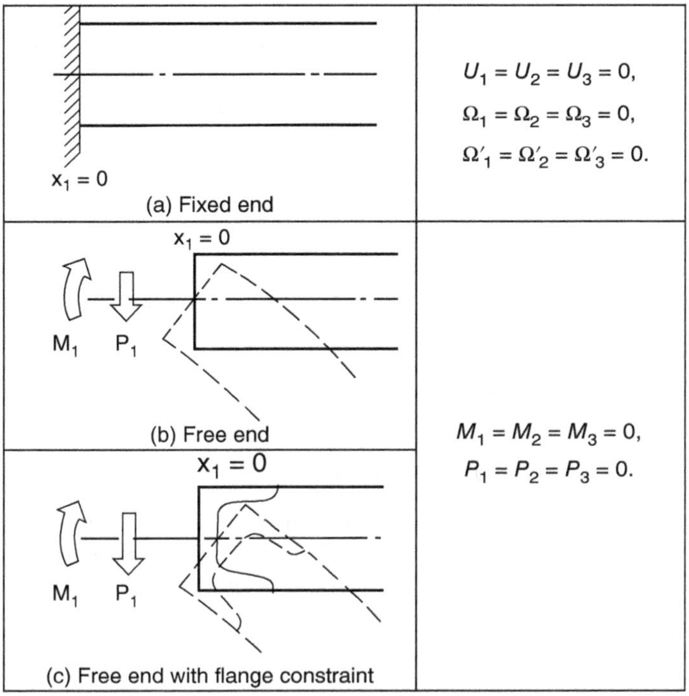

Figure 5.6 Types of fastening of the ends of the rod.

$$N_{37}^{*} = -\int D_{16}s^2R^{-1}d\xi_2, N_{29}^{*} = \int D_{16}c\chi R^{-1}d\xi_2, \tag{5.29}$$

$$N_{39}^{*} = -\int D_{16}s\chi R^{-1}d\xi_2, N_{49}^{*} = -3\int D_{16}\chi d\xi_2, \tag{5.30}$$

$$N_{47}^{*} = R_{24}, N_{48}^{*} = R_{34}. \tag{5.31}$$

The components of the matrix $[R^*]$ follow from the rules:

$$R_{ij}^{*} = R_{ij} \quad for \quad 2 \le i,j \le 3, \tag{5.32}$$

$$R_{44}^{*} = N_{49}^{*}, R_{34}^{*} = R_{34}, R_{27}^{*} = M_{22}, R_{28}^{*} = R_{37}^{*} = M_{23}, \tag{5.33}$$

$$R_{29}^{*} = R_{47}^{*} = M_{24}, R_{38}^{*} = M_{33}, R_{39}^{*} = R_{48}^{*} = M_{34}, R_{49}^{*} = M_{44}. \tag{5.34}$$

The components of the matrix $[Q^*]$ are:

$$Q_{12}^{*} = -N_{13}, Q_{13}^{*} = N_{12}, Q_{22}^{*} = -N_{23}, Q_{23}^{*} = N_{22}, Q_{32}^{*} = -N_{33}, \tag{5.35}$$

$$Q_{33}^{*} = N_{23}, Q_{42}^{*} = -N_{34}, Q_{43}^{*} = N_{24}, Q_{52}^{*} = -N_{36}, \tag{5.36}$$

$$Q_{53}^{*} = N_{26}, Q_{62}^{*} = -N_{35}, Q_{63}^{*} = N_{25}, Q_{72}^{*} = N_{37}^{*}, Q_{73}^{*} = N_{27}^{*}, \tag{5.37}$$

$$Q_{82}^{*} = N_{38}^{*}, Q_{83}^{*} = N_{28}^{*}, Q_{92}^{*} = N_{39}^{*}, Q_{93}^{*} = N_{29}^{*}. \tag{5.38}$$

5.3.2.3 Boundary Conditions of the Intermediate Type

Boundary conditions of the intermediate type are also possible. For example, if plane displacement of the free end of the rod is constrained by a blank flange (Figure 5.6c) that prevents mutual displacement of the surfaces of the shell, then the quantities U_2', U_3' will no longer be independent variables of the functional. Thus, the seventh and eighth equations of (5.26) are replaced by the following conditions on the end $x_1 = 0$:

$$U_2' = \Omega_2, U_3' = -\Omega_3. \tag{5.39}$$

If forces P_1, P_2, P_3 and moments M_1, M_2, M_3 act on the end $x_1 = 0$, then the components of the five-dimensional load vector [P*] from (5.26) are calculated in accordance with the rule

$$[P^*] = [P_1, P_2, P_3, M_1, M_2, M_3, 0, 0, 0]^T. \tag{5.40}$$

5.3.3 Governing Equations in Special Cases of Symmetry

Governing Eq. (5.26) and boundary conditions are written in the general case. In the general case, the orientation of the principal symmetry axis of the material does not coincide with the generatrix, the contour of the rod (directrix) is arbitrary and the position of the coordinate system $Ox_1x_2x_3$ in the cross-section is also chosen freely. Generally speaking, the origin of $Ox_1x_2x_3$ differs from the center of inertia of the cross-section. Equation (5.16) and the corresponding boundary conditions are simplified considerably in the special cases of symmetry of elastic properties and geometry of a cross-section.

5.3.3.1 Orthotropic Material

The first substantial simplification is achieved for an orthotropic material. The orthotropy axes of material correspond to the directions of the axes $\tilde{O}\xi_1\xi_2\xi_3$ at each point of the shell surface. In this case the tension-shear coupling coefficients vanish:

$$A_{16} = D_{16} = 0.$$

Consequently, the coefficients:

$$N_{12}, N_{13}, N_{14}, N_{25}, N_{26}, N_{35}, N_{36}, N_{45}, N_{46},$$
$$Q_{15}, Q_{16}, Q_{56}, R_{23}, R_{24}, R_{34}$$

also disappear.

5.3.3.2 Constant Elastic Properties Along the Arc of a Cross-Section

The simplifications occur if the thickness and elastic constants remain constant along the arc of the directrix. Then all quantities $A_{11}, A_{16}, A_{66}, D_{11}, D_{16}, D_{66}$ also remain constant along the arc ξ_2. The constant factors within an integrand that contain these quantities can be separated from the integrand.

5.3.4 Symmetry of Section

By choosing the coordinate origin to be at the center of inertia of the cross-section, we can make the coefficients N_{15}, N_{16}, N_{56} equal to zero.

First, if the cross-section is symmetric relative to the axis Ox_3, the functions $\sin\theta$, χ, x_2 are the odd functions of the angle $\theta = \theta(s) = \theta(\xi_2)$. Similarly, the functions $\cos\theta$, R, x_3, ω are the even functions of the angle. In this case, all the coefficients containing integrals of the odd functions of the angle θ vanish. Also, the coefficients that contain products of the even functions with the odd functions of the angle θ vanish.

The following coefficients disappear:

$$N_{16}N_{13}, N_{23}, N_{26}, N_{34}, N_{35}, N_{46}, N_{56}, M_{23}, M_{34},$$
$$L_{56,}Q_{15}, Q_{25}, Q_{36}, Q_{45}, R_{21}.$$

Second, if the cross-section is symmetric relative to the axis Ox_2, then $\cos\theta$, χ, x_3 are even functions of the angle θ; $\sin\theta$, ω, x_2 are the odd functions of the angle. The coefficients, which contain $\cos\theta$, χ, x_3 and the products of these functions with $\sin\theta$, ω, x_2, turn all to zero:

$$N_{15}N_{12}, N_{23}, N_{24}, N_{26}, N_{35}, N_{45}, N_{56}, M_{23}, M_{24},$$
$$L_{56,}Q_{16}, Q_{25}, Q_{36}, Q_{46}, R_{34}.$$

Finally, if the cross-section has two symmetry axes, then the system acquires a particularly simple form.

The model proposed for thin-walled closed rods can be used to calculate the stiffness, strength and stability of different driveshafts made of composite materials.

5.4 Buckling of Composite Driveshafts Under a Twist Moment

5.4.1 Greenhill's Buckling of Driveshafts

The minor counterpart for the celebrated Euler buckling problem is the Greenhill problem, which studies the forming of a loop in an elastic bar under torsion (Greenhill 1883). This problem was studied for several conditions on the ends of the driveshaft by Ziegler (1977) and Godoy (1999). The simplest form of supports is studied here, namely the periodic supports for an infinite shaft. The other boundary conditions lead to different values of critical moment and the corresponding solutions are observed in Ziegler (1977). Our purpose is to examine the cross-sectional properties for the maximal stability of the driveshaft for the applied torque.

We revise the buckling of the straight composite member made of anisotropic material, twisted by couples M_T applied at its ends alone (Greenhill's problem for twisted composite driveshaft or spring) (Ziegler 1977). The z-axis of the driveshaft is also the orthotropy axis of the material. No bending or extension accompanies the torsion of the driveshaft. Consider a straight composite driveshaft of an infinite length, twisted by couples applied at its ends alone (Greenhill's problem for an equivalent column). The driveshaft is holding on the periodically spaced supports. The distance between the supports is Δ_z. The ends of the driveshaft are assumed to be attached to the supports by ideal spherical hinges and are free to rotate in any direction.

The twisting couple M_T retains its initial direction during buckling. First, consider a driveshaft with constant bending stiffness along the length of the spring (Figures 5.7 and 5.8).

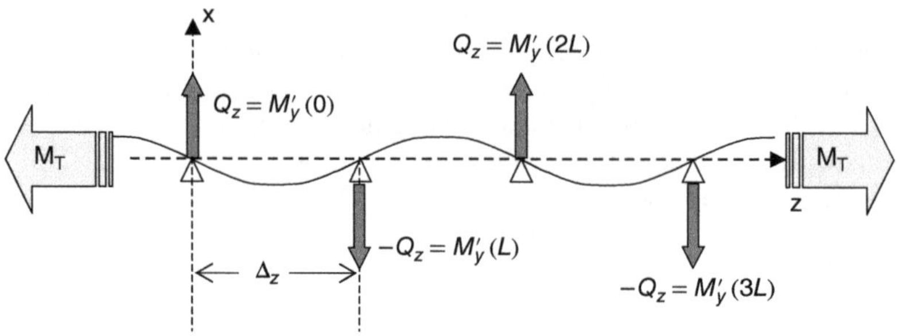

Figure 5.7 Driveshaft on the periodically spaced momentless supports.

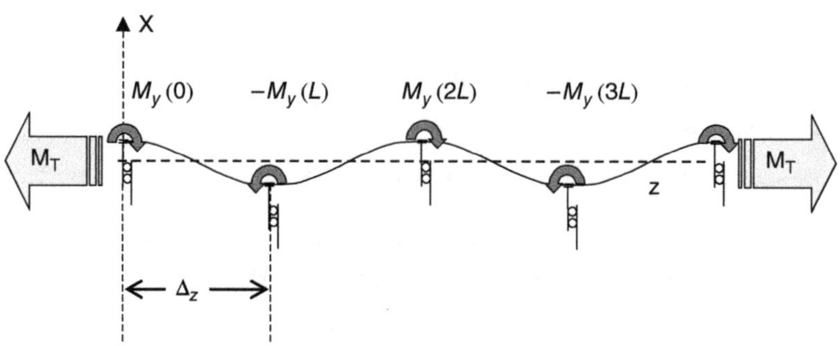

Figure 5.8 Driveshaft on the periodically spaced rotation-free supports.

To begin, we consider all possible convex, simply connected shapes of the cross-section. Let I_{b1} and I_{b2} denote the moments of inertia of Ω about the principal axes of inertia through the center of gravity. The bending moments in terms of curvatures of the driveshaft are

$$M_z = C_{33}I_{b2}u'' \quad M_y = C_{33}I_{b1}v''. \tag{5.41}$$

The equations for an infinite periodic uniform beam resting on an arbitrary number of simple supports in terms of lateral displacements $u(z)$, $v(z)$ read

$$M_z'' = -\Lambda v''', \quad M_y'' = \Lambda u'''. \tag{5.42}$$

Assume the initial conditions for $k = 0$:

$$M_y(0) = a, \quad M_z(0) = b. \tag{5.43}$$

The solution of the Eqs. (5.41) and (5.42) with the initial conditions (5.43) with arbitrary constants a, b. Is:

$$\begin{aligned} M_y(z) &= a\sin G(z) + b\sqrt{I_{b1}/I_{b2}}\cos G(z), \\ M_z(z) &= -a\sqrt{I_{b2}/I_{b1}}\cos G(x) + b\sin G(z), \end{aligned} \tag{5.44}$$

where:

$$G(z) = \frac{\Lambda}{E\sqrt{I_{b1}I_{b2}}}z. \tag{5.45}$$

The periodicity conditions for the functions (5.44) require that:

$$M'_y(0) = -M'_y(\Delta_z) \leftrightarrow -b = -a \sin G(\Delta_z) + b \cos G(\Delta_z),$$
$$M'_z(0) = -M'_z(\Delta_z) \leftrightarrow \quad a = -b \sin G(\Delta_z) - a \cos G(\Delta_z). \tag{5.46}$$

The corresponding periodicity conditions (5.46) for the functions (5.44) are satisfied for arbitrary constants a, b, only if:

$$2 + 2\cos\ G(\Delta_z) = 0. \tag{5.47}$$

This equation delivers the stability equation with the first positive solution:

$$G(\Delta_z) = \pi. \tag{5.48}$$

The critical eigenvalue of the driveshaft of the distance between supports Δ_z is:

$$\Lambda^* = \frac{C_{33}\pi\sqrt{I_{b1}I_{b2}}}{\Delta_z}. \tag{5.49}$$

5.4.2 Optimal Shape of the Solid Cross-Section for Driveshaft

The next step is to determine the optimal solid, convex shape of the cross-section. The solution grounds on the isoperimetric equation (Ting 1963). Namely, of all the convex domains the equilateral triangle yields the maximum of I_{b1}, I_{b2}, such that

$$\sqrt{I_{b1}I_{b2}} \leq \frac{\sqrt{3}}{18}A_0^2. \tag{5.50}$$

Thus, the driveshaft with the cross-section in the form of an equilateral triangle delivers the maximum-for-critical eigenvalue for all convex domains of the same cross-sectional area A_0 (Figure 5.9).

The isoperimetric inequality is stated:

$$\Lambda^*[A] \leq \frac{\pi C_{33}\sqrt{I_{b1}I_{b2}}}{\Delta_z} \leq \pi C_{33}\frac{\sqrt{3}}{18}\frac{A_0^2}{\Delta_z} = \Lambda^{**}. \tag{5.51}$$

The driveshaft with a solid circular cross-section correspondingly delivers the minimum-for-critical eigenvalue.

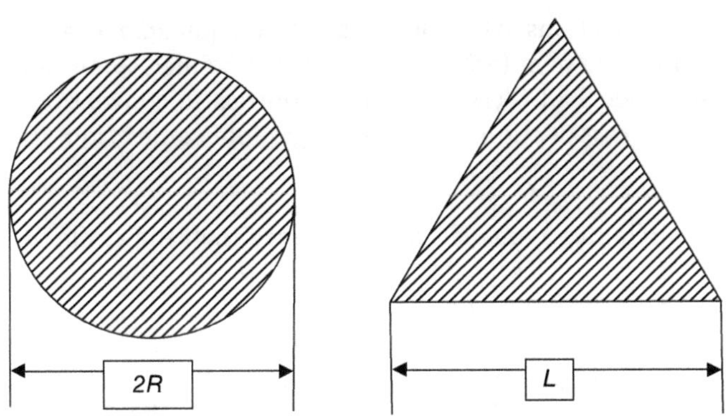

Figure 5.9 Solid triangular and circular cross-sections of the driveshaft of the same cross-sectional area.

Table 5.1 Areas and area moments for solid triangular and circular cross-sections.

	Solid equilateral triangle	Solid circular cross-section
Moments of inertia	$I_\Delta = \frac{\sqrt{3}L^4}{96}$	$I_\oplus = \frac{\pi R^4}{4}$
Cross-sectional area	$A_\Delta = \frac{\sqrt{3}L^2}{4}$	$A_\oplus = \pi R^2$
Size for given cross-sectional area A_0	$L = \sqrt{\frac{4A_0}{\sqrt{3}}}$	$R = \sqrt{\frac{A_0}{\pi}}$
Moments of inertia for given cross-sectional area A_0	$I_\Delta = \frac{\sqrt{3}A_0}{18}$	$I_\oplus = \frac{A_0}{4\pi}$

Table 5.1 compares both cross-sections.

First, the geometrical moments of inertia of the cross-sections determined for both cross-sections are listed for the solid equilateral triangle I_Δ and for the solid circular cross-section $I_\oplus$. The side length of the equilateral triangle is L and the radius of the circle is R. Then the cross-sectional areas of both figures are given: A_Δ and $A_\oplus$, correspondingly. We assume that the area of cross-sections of both driveshafts is the same, A_0. From this condition, the side length of the equilateral triangle and radius of the circle are determined. We substitute the determined sizes in the expression for the moments of inertia for a given cross-sectional area A_0. The relation of both moments of inertia of the solid cross-sections guarantees that the cross-section in the form of a solid equilateral triangle is nearly 21% stiffer to bending in comparison to circular solid cross-section of the same area of cross-section:

$$\frac{I_\Delta}{I_\oplus} = \left(\frac{\sqrt{3}A_0}{18}\right) \Big/ \left(\frac{A_0}{4\pi}\right) = \frac{2\pi\sqrt{3}}{9} \approx 1.2091. \tag{5.52}$$

5.4.3 Hollow Circular and Triangular Cross-Sections

The next step is to compare the cross-sections in the form of a hollow equilateral triangle and hollow circle (Figure 5.10). The document (SWIX Triac 2018) contained some facts about their recently designed ski poles SWIX Triac, which in contrast to conventional (round) ski poles have a triangular cross-section. The following quote comes from the document (SWIX Triac 2018):

> A triangle encircled by a circle has a smaller circumference than the circle. By using a triangle geometry less material is needed and weight is saved. The distance from the center of the shaft, to the outer edge of the shaft is the dimension that gives a pole shaft its stiffness. A triangle has the same distance from the center to a corner as the circle has although only in three directions. By orienting the corners in optimum directions, we can achieve the same geometry stiffness (if not increase) with a reduced amount of material and thus decreased overall weight of the shaft.

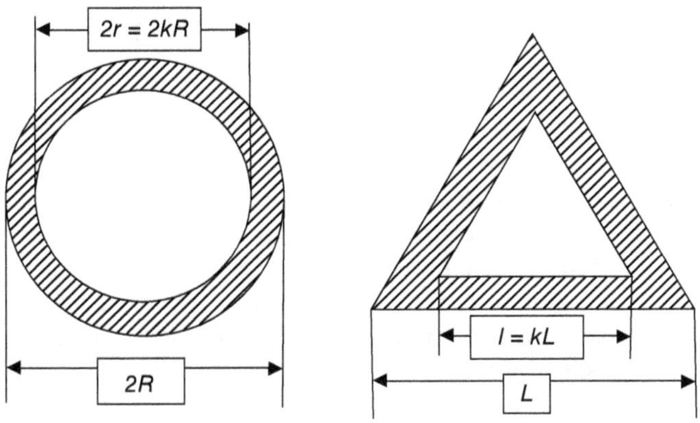

Figure 5.10 Hollow triangular and circular cross-sections of the driveshaft.

The only remark is the following. It is not necessary to orient the corners of the equilateral triangle in any direction, because the bending stiffness of the beam with equilateral triangle cross-section is direction independent (Ting 1963). Exactly this symmetry is critically important for the stability of rotating driveshafts because otherwise the driveshaft will buckle in the direction of the lower stiffness.

We approve the statements in the previous quote. These statements are also immediately applicable for the design of the automotive driveshaft that is subjected to torque instability. For confirmation, we compare the hollow cross-sections with similar outer and inner profiles. The linear sizes of inner profiles are factor k_s smaller than the sizes of outer profiles: $l = k_s L$, $r = k_s R$ (Figure 5.9). The factor k_s is referred to as the similarity factor.

A demonstration of the statement is given in Table 5.2.

Now we use the determined sizes in the expression for the moments of inertia for a given cross-sectional area, A_0. The relation of both moments of inertia of the hollow cross-sections guarantees that the cross-section in the form of a hollow

Table 5.2 Areas and area moments for hollow triangular and circular cross-sections.

	Hollow equilateral triangle	**Hollow circular cross-section**
Moments of inertia	$\widetilde{I}_\Delta = \dfrac{\sqrt{3}(L^4 - l^4)}{96}$	$\widetilde{I}_\oplus = \dfrac{\pi(R^4 - r^4)}{4}$
Cross-sectional area	$\widetilde{A}_\Delta = \dfrac{\sqrt{3}(L^2 - l^2)}{4}$	$\widetilde{A}_\oplus = \pi(R^2 - r^2)$
Similarity factor	$l = k_s L$	$r = k_s R$
Size for a given cross-sectional area A_0	$L = \dfrac{1}{\sqrt{1 - k_s^2}}\sqrt{\dfrac{4A_0}{\sqrt{3}}}$	$R = \dfrac{1}{\sqrt{1 - k_s^2}}\sqrt{\dfrac{A_0}{\pi}}$
Moments of inertia for a given cross-sectional area A_0	$\widetilde{I}_\Delta = \dfrac{\sqrt{3}A_0}{18}\dfrac{1 + k_s^2}{1 - k_s^2}$	$\widetilde{I}_\oplus = \dfrac{A_0}{4\pi}\dfrac{1 + k_s^2}{1 - k_s^2}$

equilateral triangle is once again 21% stiffer to bend in comparison to a circular hollow cross-section of the same area of material:

$$\frac{\widetilde{I}_{\Delta}}{\widetilde{I}_{\oplus}} = \left(\frac{\sqrt{3}A_0}{18}\frac{1+k_s^{\ 2}}{1-k_s^{\ 2}}\right) \Bigg/ \left(\frac{A_0}{4\pi}\frac{1+k_s^{\ 2}}{1-k_s^{\ 2}}\right) = \frac{2\pi\sqrt{3}}{9} \equiv \frac{I_{\Delta}}{I_{\oplus}} \approx 1.2091. \tag{5.53}$$

The estimation of twist buckling effects is necessary for design evaluation of the driveshafts and axles. The strength of the composite driveshafts is limited by the shear strength of the material and the preferable orientation of fibers is 45° to the axis of the driveshaft. However, the preferable orientation of fibers for the maximal buckling moment is parallel to the axis of the shaft. Consequently, both criteria lead to different desirable orientations of principal anisotropy axes.

5.5 Patents for Composite Driveshafts

This section displays the assortment of the related patents and patent applications for the composite driveshafts in the automotive powertrains.

US3651661 (A) 1972-03-28, https://www.uspto.gov/,

> *Composite shaft with integral end flange:* A composite shaft with an integral end flange fabricated of a plurality of layers of high modulus fibers of selected fiber orientation positioned between and bonded to protective outer and inner layers as required and to one another such that each layer extends the full length of the shaft and the flange. The shaft layers may be of sandwich construction and may be asymmetric and the flange periphery may include doubles of bearing material (Figure 5.11).

US4171626 (A) 1978-03-27, https://www.uspto.gov/,

> *Carbon fiber-reinforced composite drive shaft*: A carbon fiber-reinforced composite tubular drive shaft is provided wherein the fibrous reinforcement is positioned in a structural configuration capable of yielding improved service characteristics. The wall of the drive shaft comprises at least four bonded circumferentially disposed layers composed of appropriate fibrous reinforcement situated within a resin matrix (as described). The carbon fibers are disposed generally longitudinally in a layer at an intermediate position within the tube wall and are capable of efficiently increasing the stiffness of the shaft. The overall requirement for the carbon fiber component of the fibrous reinforcement is minimized. Glass fibers likewise are disposed generally longitudinally in a layer at an intermediate position within the tube wall and are capable of economically contributing resistance to torsion buckling to the overall drive shaft without deleteriously influencing the critical speed of the drive shaft. The drive shafts of the present invention, accordingly, are light weight and surprisingly capable of yielding reliable service in the absence of harmful secondary powertrain resonance and the concomitant noise associated therewith (Figure 5.12).

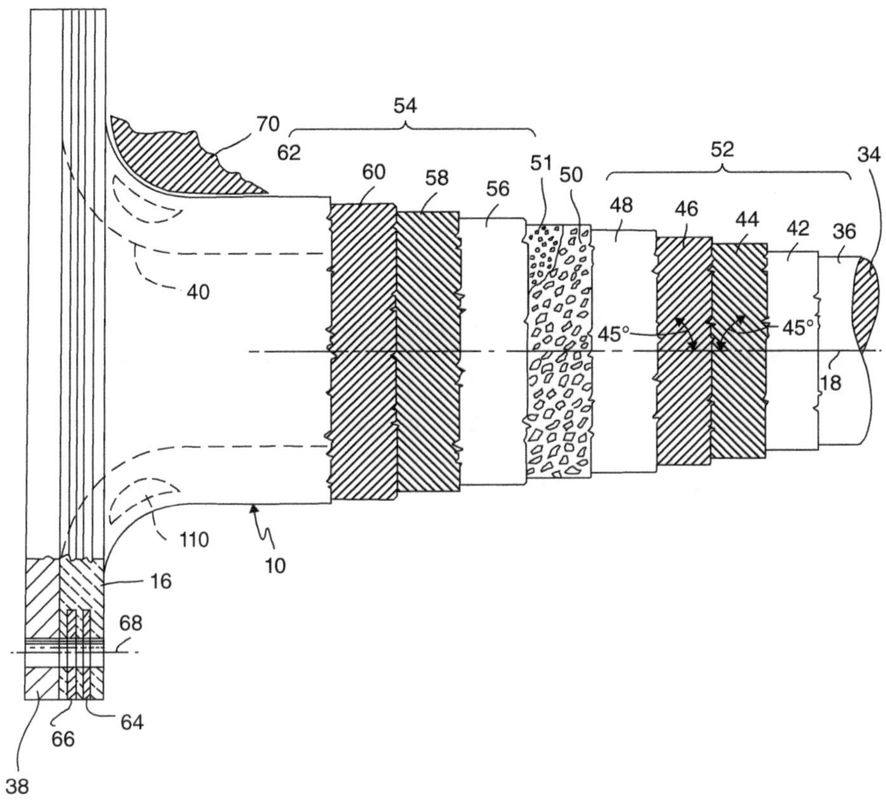

Figure 5.11 US3651661(A), 1972: "Composite shaft with integral end flange."

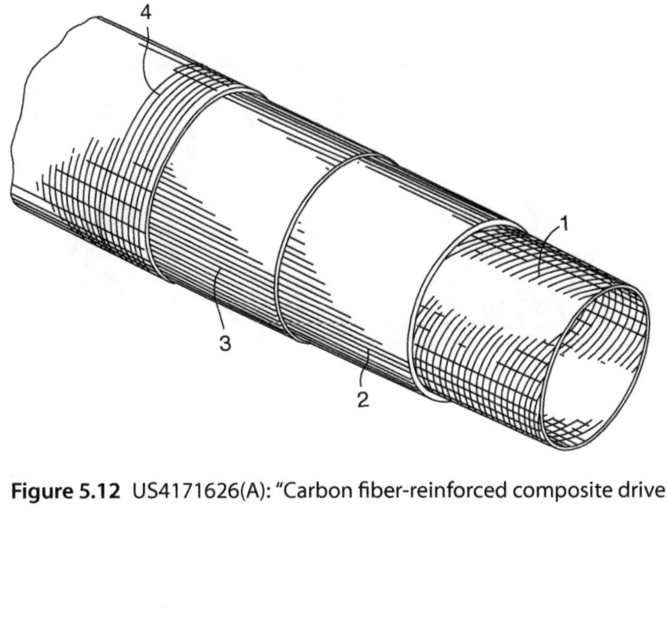

Figure 5.12 US4171626(A): "Carbon fiber-reinforced composite drive shaft."

US4173128 (A), 1979-11-06, https://www.uspto.gov

Composite drive shaft: A laminated drive shaft having integral universal provisions adjacent ends attached to driving and driven members joined by the shaft to transmit torque there between while permitting angular misalignment and displacement of these members (Figure 5.13).

US4551116 (A) (1983-06-09) https://www.uspto.gov

Drive shaft assembly: A drive shaft assembly which consists essentially of two shaft parts both made of fiber-reinforced plastic material at least one having a circumferential radially outwardly extending fold formation provided for enabling articulation of the shaft assembly with an intermediate support bearing being located between the two shaft parts, the support bearing having an inner race positioned by steps at the region of connection between the two shaft parts in such a way that the intermediate bearing forms an integral part of the shaft and is assembled so as to form a permanent connection (Figure 5.14).

US5725434 (A) – 1998-03-10, https://www.uspto.gov,

Shaft of fiber reinforced material: A shaft with integrated angle compensating elements of fiber-reinforced material includes an elastically-deformable coupling

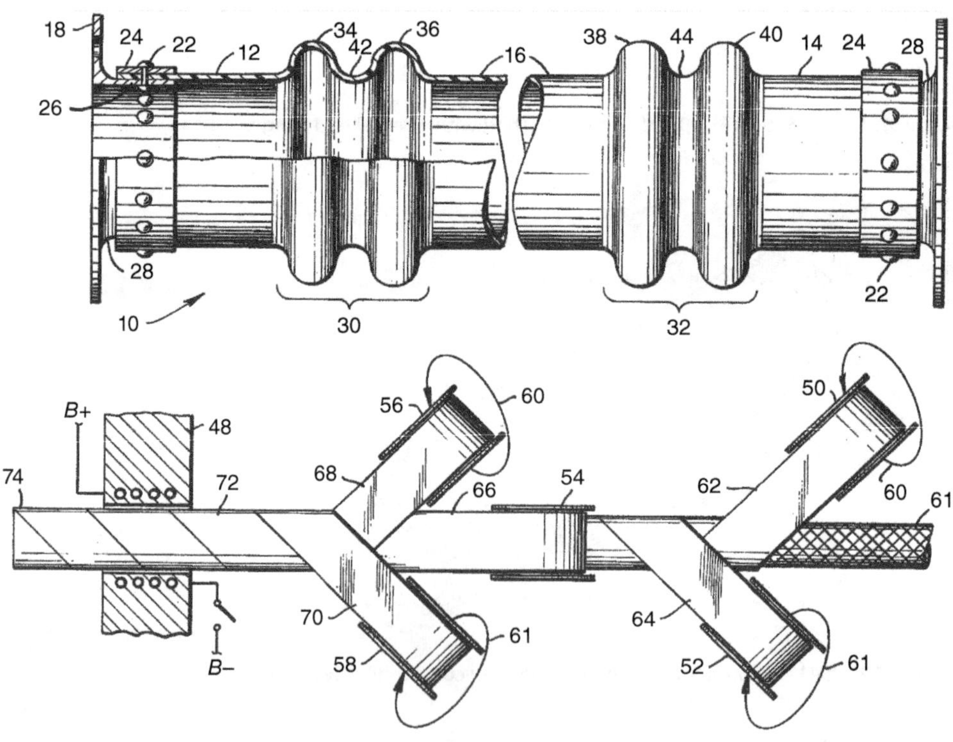

Figure 5.13 US4173128(A): Composite drive shaft.

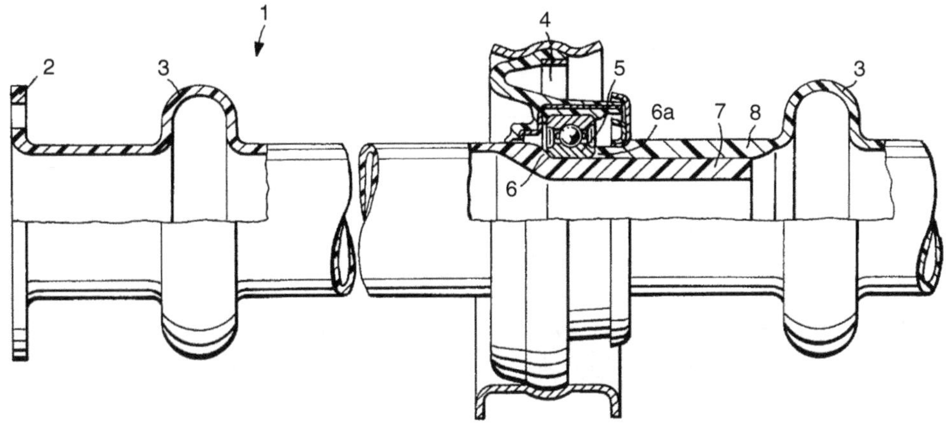

Figure 5.14 US4551116(A): Drive shaft assembly.

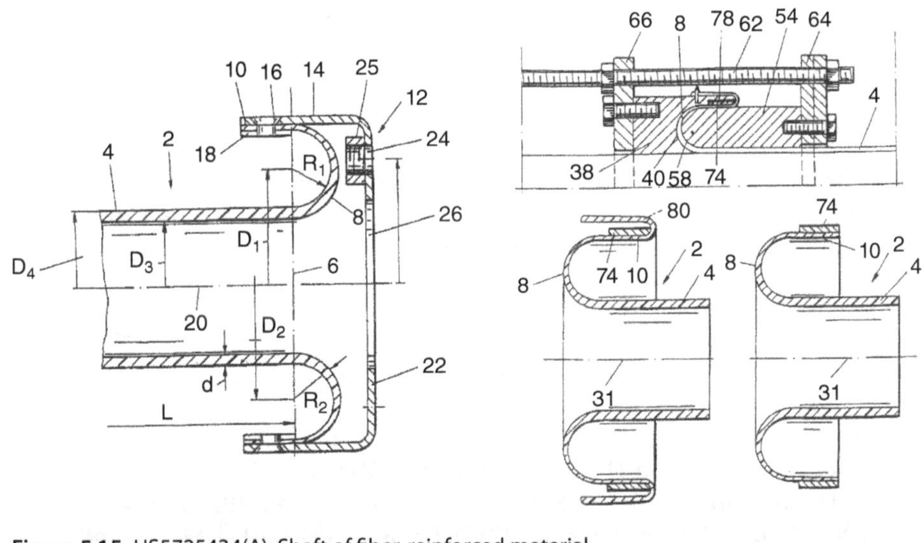

Figure 5.15 US5725434(A): Shaft of fiber-reinforced material.

flange that is curved backward to form an annular portion which connects the flange to the inner wall of a surrounding disk-shaped connection element. The shaft comprises fibers cross-wound at an angle to the axis of the shaft which extend up to the free edge of the flange (Figure 5.15).

US2017082149 (A1) 2017-03-23, https://www.uspto.gov,

Composite driveshaft for a rotary system: A composite driveshaft includes a body having a first end, a second end, and an intermediate portion defining a driveshaft axis (DSA). The first end defines a first coupling region and the second end defines a second coupling region. At least one of the first and second coupling regions defines a virtual hinge configured to accommodate both bending moments and axial changes of the body (Figure 5.16).

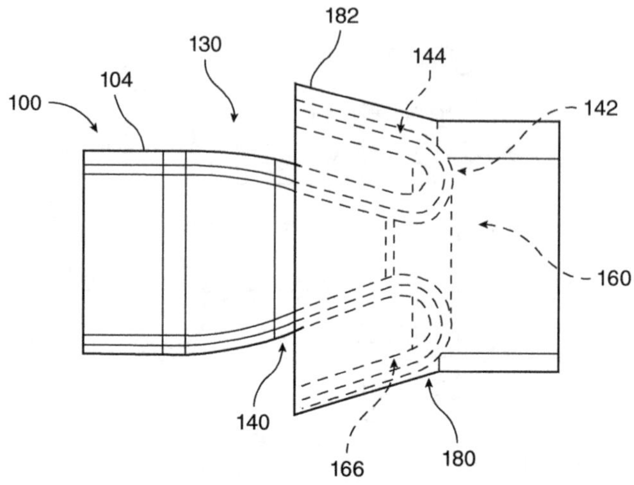

Figure 5.16 US2017082149(A1): Composite driveshaft for a rotary system.

5.6 Conclusions

In this chapter, the deformation of the straight members of closed solid and closed tubular cross-sections was studied. The twist and flexure of the members were investigated for different orientation of anisotropy axes with respect to the DSA. Torsional stability was studied using the Greenhill method. The results were shown to be applicable to the components of the powertrain, as for driveshafts of ground vehicles.

There are important works that study problems similar to those considered in this chapter. Without discussing the methods and results of these studies in detail, we recommend them for further reading: Tarn (2002), Bert and Francis (1974), Libai and Simmonds (1988), Vinson (1993), Whitney and Pagano (1970), Washizu (1982), Timoshenko (1945), Barbero et al. (1993), Rios (2009), Borri et al. (1992), Kroker (2013), Shadmehri et al. (2011), Ahmad and Hoa (2016), Erbaşua et al. (2017), Maceri and Vairo (2009), Berdichevsky et al. (1992), Kosmatka (1992), Whitney and Halphin (1968), Pai (2011), Hodges and Peters (2001), Yu et al. (2005), Kollár and Pluzsik (2012), Klinkel and Govindjee (2003), Tarn and Chang (2008), Budan and Manjunatha (2010), Chiu et al. (2007), Hendry and Pobert (1986), Grigor'ev et al. (2012), Katsikadelis and Tsiatas (2016), Gubran (2005), Lee et al. (2004), Rastogi (2004), Viroliya and Viroliya (2018), Rangaswamy and Vijayarangan (2005), Dharmadhikari et al. (2014), Gibson (2016), Whitney (1968) and Halpin (1992).

References

Axelrad, E.L. (1987). *Theory of Flexible Shells*. North-Holland: Elsevier.

Badie, M., Mahdi, A., Abutalib, A. et al. (2006). Automotive composite driveshafts: investigation of the design variables effects. *Int. J. Eng. Technol.* 3: 227–237.

Badie, M.A., Mahdi, E., and Hamouda, A.M.S. (2011). An investigation into hybrid carbon/glass fiber reinforced epoxy composite automotive drive shaft. *Mater. Des.* 32: 1485–1500.

Barbero, E.J., Lopez-Anido, R., and Davalos, J.F. (1993). On the mechanics of thin-walled laminated composite beams. *J. Compos. Mater.* 27 (8): 806–809.

Bauld, Je, N.R., T., and Tzeng, L.-S. (1984). A Vlasov theory for fiber reinforced beams with thin-walled open cross sections. *Int. J. Solids Struct.* 20 (3): 277–297.

Beardmore, P. and Johnson, C.F. (1986). The potential for composites in structural automotive applications. *J. Compos. Sci. Technol.* 26: 251–281.

Berdichevsky, V., Armanios, E., and Badir, A. (1992). Theory of anisotropic thin-walled closed-cross-section beams. *Compos. Eng.* 2 (5–7): 411–432.

Bert, C.W. and Francis, P.H. (1974). Composite material mechanics: structural mechanics. *AIAA J.* 12 (9): 1173.

Borri, M., Ghiringhelli, G.L., and Merlini, T. (1992). Linear analysis of naturally curved and twisted anisotropic beams. *Compos. Eng.* 2 (5–7): 433–456.

Budan, A.D. and Manjunatha, T.S. (2010). Investigation on the feasibility of composite coil spring for automotive applications. *World Acad. Sci. Eng. Technol.* 70: 1035–1039.

Cherniaev, A. and Komarov, V. (2015). Multistep optimization of composite drive shaft subject to strength, buckling, vibration and manufacturing constraints. *Appl. Compos. Mater.* 22: 475–487.

Chiu, C.-H., Hwan, C.-L., Tsai, H.-S., and Lee, W.-P. (2007). An experimental investigation into the mechanical behaviors of Helical composite springs. *Compos. Struct.* 77: 331–340.

Dharmadhikari, S.R., Mahakalkar, S.G., Giri, J.P., and Khutafale, N.D. (2014). "Design and analysis of composite drive shaft using ANSYS and genetic algorithm" A critical review. *Int. J. Mod. Eng. Res.* 3 (1, Jan-Feb. 2013): 490–496.

Donnell, L.H. (1977). *Beams, Plates and Shells.* McGraw-Hill Book Company.

Erbaşua, R., Ţăpuşia, D., and Voiculescua, D. (2017). Bending of straight bars made of anisotropic materials. *Energy Procedia* 112: 621–631.

Geuchy Ahmad, M.I. and Hoa, S.V. (2016). Flexural stiffness of thick-walled composite tubes. *Compos. Struct.* 149: 125–133.

Gibson, R.F. (2016). *Principles of Composite Material Mechanics.* Boca Raton: CRC Press.

Godoy, L. (1999). *Theory of Elastic Stability: Analysis and Sensitivity*, 450. CRC Press.

Greenhill, A.G. (1883). On the strength of shafting when exposed both to torsion and to end thrust. *Proc. Inst. Mech. Eng.* 34 (1): 182–225.

Grigor'ev, S.N., Krasnovskii, A.N., and Khaziev, A.R. (2012). The design of composite anisotropic rods. *Int. Poly. Sci. Technol.* 40 (4): 37–42.

Gubran, H.B.H. (2005). Dynamics of hybrid shafts. *Mech. Res. Commun.* 32 (4): 368–374.

Halpin, J.C. (1992). *Primer on Composite Materials Analysis.* Taylor & Francis Group, LLC.

Harursampath, D., Harish, A.B., and Hodges, D.H. (2017a). Model reduction in thin-walled open-section composite beams using variational asymptotic method. Part I: theory. *Thin-Walled Struct.* 117: 356–366.

Harursampath, D., Harish, A.B., and Hodges, D.H. (2017b). Model reduction in thin-walled open-section composite beams using variational asymptotic method. Part II: applications. *Thin-Walled Struct.* 117: 367–377.

Hendry, J.C. and Pobert, C. (1986). Carbon fiber coil springs. *Mater. Des.* 7 (6): 330–337.

Hodges, D.H. and Peters, D.A. (2001). Lateral-torsional buckling of cantilevered elastically coupled composite strip- and I-beams. *Int. J. Solids Struct.* 38: 1585–1603.

Katsikadelis, J.T. and Tsiatas, G.C. (2016). Saint-Venant torsion of non-homogeneous anisotropic bars. *J. Appl. Comput. Mech.* 2 (1): 42–53. https://doi.org/10.22055/jacm.2016.12270 Flexural stiffness of thick-walled composite tubes.

Kaviprakash, G., Kannan, C.R., Lawrence, I.D., and Regan, A.P. (2014). Design and analysis of composite drive shaft for automotive application. *Int. J. Eng. Res. Technol.* 3 (11): 429–436.

Keys C. Kinkler W., Santiago A. (2018) *Composite Driveshaft*, MEEN 4263/4264, Texas A&M University-Kingsville, http://users.tamuk.edu/kfldp00/MEIE_Peel_website/Courses/MEEN4264/Student%20Projects/driveshaft_senior_design_report.pdf (Retrieved November 03, 2018).

Klinkel, S. and Govindjee, S. (2003). Anisotropic bending-torsion coupling for warping in a non-linear beam. *Comput. Mech.* 31: 78–87.

Koiter, W.T. (e.) (1960). A consistent first approximation in the general theory of thin elastic shells, Proceedings of the Symposium on. In: *Thin Elastic Shells*, 12–33. North-Holland, Amsterdam.

Kollár, L.P. and Pluzsik, A. (2012). Bending and torsion of composite beams (torsional-warping shear deformation theory). *J. Reinf. Plast. Compos.* 31 (7): 441–480.

Kosmatka J.B. (1992) The Use of Cross-Section Warping Functions in Composite Rotor Blade Analysis, Report No. Ssrp-92/11, NASA Research Grant Number Nag-1-1151, University of California, San Diego.

Kroker A.M. (2013) *Strukturmechanische Modellierung und Analyse des Tragverhaltens von dünnwandigen hochbelasteten Composite-Biege- und Querkraftträgern*, Forschungsberichte des Instituts für Mechanik der Technischen Universität Darmstadt, Band 28, 2013.

Lee, D.B., Kim, H.S., Kim, J.W., and Kim, J.K. (2004). Design and manufacture of an automotive hybrid aluminium/composite drive shaft. *Compos. Struct.* 63: 87–99.

Lekhnitskii, S.G. (1981). *Theory of Elasticity of an Anisotropic Body*. Moscow: Mir.

Libai, A. and Simmonds, J.G. (1988). *The Nonlinear Theory of Elastic Shells, One Spatial Dimension*. London: Academic Press.

Librescu, L. and Song, O. (2006). *Thin-Walled Composite Beams, Theory and Application, Solid Mechanics and Its Applications*, vol. 131. Dordrecht, The Netherlands: Springer.

Maceri, F. and Vairo, G. (2009). Anisotropic thin-walled beam models: a rational deduction from three-dimensional elasticity. *J. Mech. Mater. Struct.* 4 (2): 371–394.

Mikeladze, M.S. (1966). Semi-momentless theory of thin plastic shells slightly different from cylindrical ones. *J. Mech. Phys. Solids* 14 (2): 89–94.

Pai, F.P. (2011). Geometrically exact beam theory without Euler angles. *Int. J. Solids Struct.* 48: 3075–3090.

Pollard A. (1989) *Polymer Matrix Composites in Driveline Applications*, GKN Tech., UK, http://www.spinning-composites.com/images/CompositesInDriveline.PDF (Retrieved November 3, 2018).

Rangaswamy, T. and Vijayarangan, S. (2005). Optimal sizing and stacking sequence of composite drive shafts. *Mater. Sci.* 11 (2): 133–139.

Rastogi N. (2004) Design of composite drive shafts for automotive applications. *SAE Technical Paper Series*. 2004 SAE World Congress.

Rios G. (2009) *A Unified Analysis of Stiffener Reinforced Composite Beams with Arbitrary Cross-Section*. The University of Texas, Arlington, December 2009.

Salim, H.A. and Davalos, J.F. (2005a). Torsion of open and closed thin-walled laminated composite sections. *J. Compos. Mater.* 39 (6): 497–524.
Salim, H.A. and Davalos, J.F. (2005b). Shear lag of open and closed thin-walled laminated composite beams. *J. Reinf. Plast. Compos.* 24 (7): 673–690.
Sanders J.L., Jr., (1959) An improved first approximation theory for thin shells, *NASA. No. 24.*
Shadmehri, F., Derisi, B., and Hoa, S.V. (2011). On bending stiffness of composite tubes. *Compos. Struct.* 93: 2173–2179.
Sivakandhan, C. and Prabhu, P.S. (2014). Investigation of hybrid composite drive shaft under tensile, bending and torsion testing. *Int. J. Veh. Struct. Syst.* 6 (4): 110–114. https://doi.org/10.4273/ijvss.6.4.05.
Stone, R. and Ball, J.K. (2004). *Automotive Engineering Fundamentals*, 15096–15595. Warrendale, PA: SAE.
SWIX Triac (2018) *Swix Triac booklet*, available online at: www.swixtriac.com/press_kit/swix_triac_booklet.pdf, Retrieved: January 10, 2018.
Tarn, J.Q. (2002). A state space formalism for anisotropic elasticity. Part II: cylindrical anisotropy. *Int. J. Solids Struct.,* 39: 5157–5172.
Tarn, J.-Q. and Chang, H.-H. (2008). Torsion of cylindrically orthotropic elastic circular bars with radial inhomogeneity: some exact solutions and end effects. *Int. J. Solids Struct.* 45: 303–319.
Timoshenko, S. (1945). Theory of bending, torsion, and buckling of thin-walled members of open cross section. *J. Franklin Inst.* 239: 201–219, 249–268, 343–361.
Ting, T.W. (1963). Isoperimetric inequality for moments of inertia of plane convex sets. *Trans. Am. Math. Soc.* 107 (3): 421–431.
Ventsel, E. and Krauthammer, T. (2001). *Thin Plates and Shells, Theory, Analysis, and Applications.* Basel: Marcel Dekker AG.
Vinson, J.R. (1993). *The Behavior of Shells Composed of Isotropic and Composite Materials.* Netherlands: Springer.
Viroliya, C.C. and Viroliya, V.C. (2018). A review on "design and development of composite drive shaft". *Int. J. Adv. Res. Technol.* 7 (1): 1–6.
Vlasov V.Z. (1947) *Some New Problems on Shells and Thin Structures*, Technical Memorandum No. 1204.
Washizu, K. (1982). *Variational Methods in Elasticity and Plasticity.* Oxford: Pergamon Press.
Whitney, J.M. (1968). Engineering constants of laminated composite materials. *J. Compos. Mater.* 2 (2): 261–263.
Whitney, J.M. and Halphin, J.C. (1968). Analysis of laminated anisotropic tubes under combined loading. *J. Compos. Mater.* 2 (3): 360367.
Whitney, J.M. and Pagano, N.J. (1970). Shear deformation in heterogeneous anisotropic plates. *ASME J. Appl. Mech.* 37 (4): 1031–1036.
Yu, W., Hodges, D.H., Volovoi, V.V., and Fuchs, E.D. (2005). A generalized Vlasov theory for composite beams. *Thin-Walled Struct.* 43: 1493–1511.
Ziegler, H. (1977). *Principles of Structural Stability*, 2e. Basel, Stuttgart: Birkhäuser.

6

Dynamics of a Vehicle with Rigid Structural Elements of Chassis

6.1 Classification of Wheel Suspensions

6.1.1 Common Designs of Suspensions

A chassis consists of several systems that perform different tasks in the entire chassis design (Pauwelussen 2015; Reimpell et al. 2001; Reimpell and Betzler 1986; Leiter et al. 2008). The structure of a chassis and thus the selection of individual systems are strongly dependent on the vehicle class in which the chassis is to be used and the drive concept of the motor vehicle used. The requirements placed on a chassis are always very similar; only the weighting of the criteria differs depending on the vehicle class under consideration. The most important requirements for a chassis can be summarized in the upper grips, driving dynamics, safety and comfort and are to be fulfilled considering the boundary conditions of low weight and low cost. For common axles there are three different suspension concepts, which can be seen in Figure 6.1.

Independent wheel suspensions represent a technically complex and expensive concept of a wheel suspension and are therefore mainly used in vehicles of the middle and upper class or in sports cars. Semi-rigid axle suspensions are designed to be less complex, consist of fewer components and are therefore less expensive to manufacture and install than independent wheel suspensions. Semi-rigid axle suspensions are usually used in mini, small and compact cars. Commercial vehicles are subject to different requirements than vehicles of the upper and middle class, which mean that in this vehicle class the use of a rigid axle is in most cases satisfactory. To withstand the centrifugal forces and correct the lateral guide of the solid axles, the Watt linkage or Panhard rod are used.

Besides several characteristics of the wheel position, still to be presented, the coupling of the two wheels is characteristic of the properties of the three concepts of wheel suspension. The coupling of the wheels increases from independent suspensions to solid axles. Namely, independent wheel suspensions (C) allow the left and right wheel to move independently; that is, there is no coupling between the two wheels. This is not the case with the rigid axle (A), where there is a direct coupling between the left and right wheels, which results in a change in the position and angle of one wheel being forced on the other. The semi-rigid axle (B) suspension should be seen as an intermediate type between the independent wheel suspension and rigid axle suspension. The coupling of the two wheels here depends on the position of the cross beam, which also leads to a division of the semi-rigid axles into different groups.

Design and Analysis of Composite Structures for Automotive Applications: Chassis and Drivetrain, First Edition. Vladimir Kobelev.

Companion website: www.wiley.com/go/kobelev/automotive_suspensions

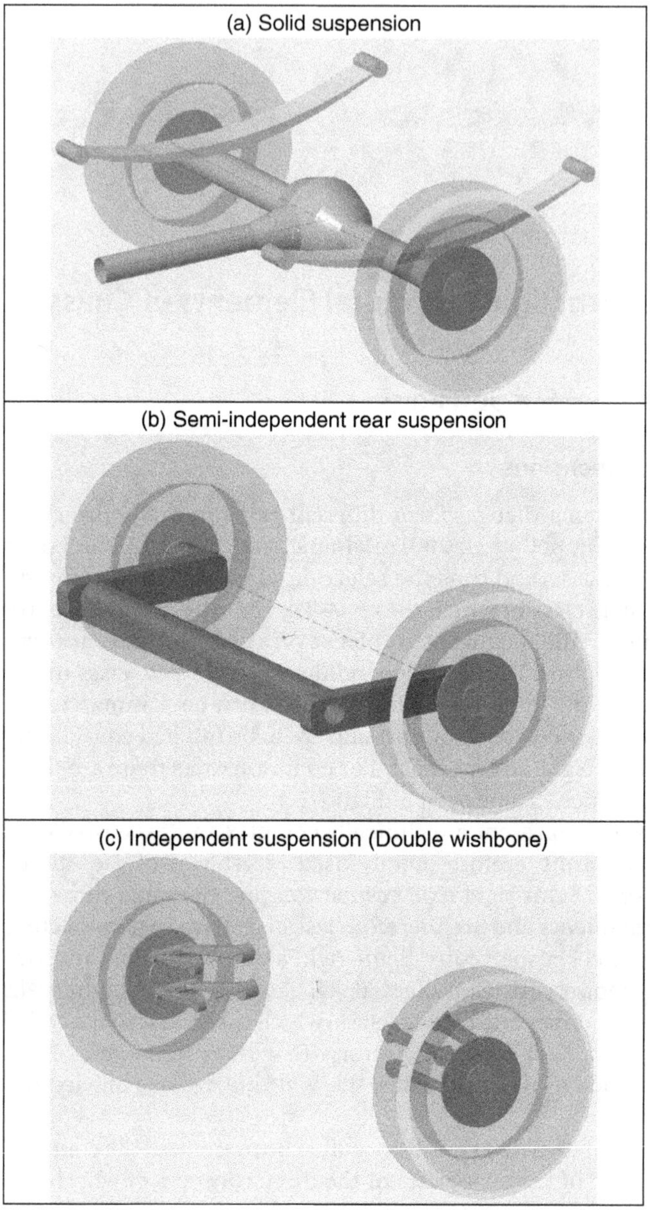

Figure 6.1 Basic types of suspension.

6.1.2 Types of Twist-Beam Axles

The twist-beam axle is a very economical suspension for non-driven rear axles. In addition to its good kinematic properties, the twist-beam axle is characterized by a small space requirement, which permits a favorable partitioning of space in the rear of the vehicle (Figure 6.2).

Figure 6.2 Semi-independent axles with Watt linkage wheel guide correction.

The twist-beam axle offers the following kinematic advantages

- Insignificant variation of toe-in and track gauge during equal-sided and reciprocal vertical movement of wheels;
- Low camber change;
- Minor brake pitch.

On the other hand, there is a strong tendency toward oversteer lateral forces, which cannot be remedied by track-correcting rubber bearings either. In addition, the twist-beam axle is not particularly well suited for use with driven rear axles and could only be used for this purpose with complex design changes to the cross beam. To bear up the centrifugal forces and to improve the lateral guide of the twist-beam axles, the Watt linkage could be used.

The main structural components of a twist-beam axle, or generally the semi-rigid axle, are the two trailing arms that are firmly connected to each other by the cross-member (twist-beam). As for the spring damper concept for twist-beam axles, spatially separated monotube dampers and coil springs are usually used.

Systems that characterize the entire chassis structure include, for example, the spring and damping concept, the steering concept and the concept of a brake system. The front and rear axle concept, that is, the actual wheel suspension, is also part of the chassis systems. The required driving characteristics of the vehicle can be achieved through optimal coordination of the concepts.

6.1.3 Kinematics of Wheel Suspensions

The stiffness and compliance of the trailing arms and wishbones, also known as elasto-kinematics, are of decisive importance for the handling of semi-rigid axles in this suspension concept. The trailing arms are rigid in terms of bending and torsion and have the task of absorbing the longitudinal forces. However, lateral forces and camber moments must be absorbed by the cross-member, which must be rigid for this reason, but also torsional soft to allow the rear wheels to compress independently. At the same time, the cross beam serves as a stabilizer. For cross beams, mainly semi-open C- or V-shaped cross-sections are used, which are produced by a forming process and must be designed in such a way that they meet the requirements. The stiffness of

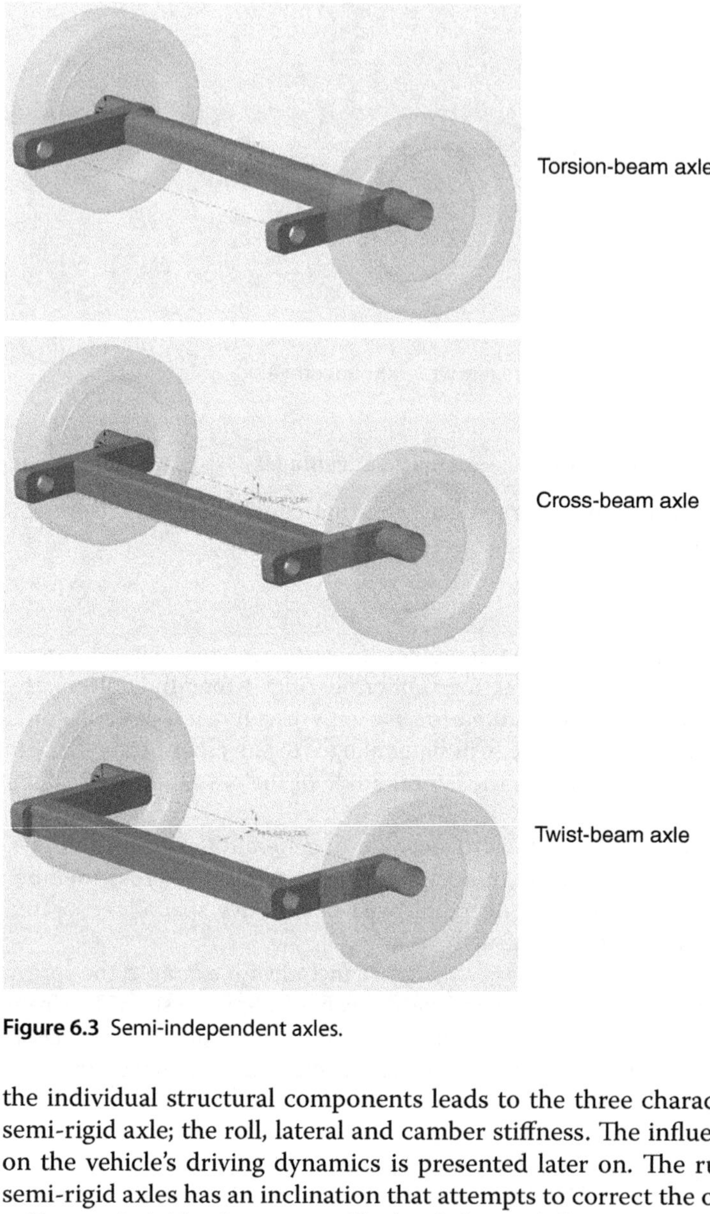

Figure 6.3 Semi-independent axles.

the individual structural components leads to the three characteristic stiffnesses of a semi-rigid axle; the roll, lateral and camber stiffness. The influence of these stiffnesses on the vehicle's driving dynamics is presented later on. The rubber mounting of the semi-rigid axles has an inclination that attempts to correct the overdrive tendency.

The semi-rigid axles are usually divided into different groups (Figure 6.3). The transition of the three groups depends on the position of the twist-beam. The "twist-beam axle" is when the cross beam is close to the rubber bearing and thus has similar properties to a trailing arm axle. The "cross beam axle" of a coupling arm axle is located in the middle of the trailing arms. In contrast to the cross beam axle, the connecting member of the "torsion beam axle" is very close to the wheel center. The torsion beam axle has similar properties to the rigid axle. The Panhard rods are frequently used for the correction of the lateral guide of the torsion beam axle, similar to the solid axles. Watt links are also occasionally used for the same purpose.

6.2 Fundamental Models in Vehicle Dynamics

6.2.1 Basic Variables of Vehicle Dynamics

The standard ISO 8855:2013 "Road vehicles – Vehicle dynamics and road-holding ability" defines coordinate systems and variables relevant to driving dynamics (Mitschke and Wallentowitz 2014). These are briefly introduced next and are also used in the other chapters. Since vehicle models are usually multi-mass models, each mass must be assigned its own coordinate system.

The kinematics of the wheel suspension is largely responsible for the requirements placed on a chassis, as it determines the position of the wheel in relation to the road surface. The points listed on the subject of kinematics of the wheel suspension are also summarized under the generic term wheel position parameters and are clearly defined in ISO 8855. The requirements placed on the wheel suspensions can be seen in Figure 6.4.

The definition of the wheelbase of a motor vehicle can be described as the distance between the wheel contact points of the front and rear wheels in the longitudinal transverse plane of the vehicle. A long wheelbase increases driving safety, while a short wheelbase improves the handiness of the vehicle. The distance of the wheel contact points of an axis in the projection of the longitudinal-vertical plane is called track width B. A wide track width ensures better handling and less rolling. The semi-rigid axles are characterized by a small change in track width.

The chamber angle between the wheel center plane and a plane perpendicular to the road (which runs parallel to the intersection line of the wheel center plane with the road surface plane) influences, for example, the lateral dynamics, steering return and chassis tuning. The size of the camber change of a semi-rigid axle is also small and strongly

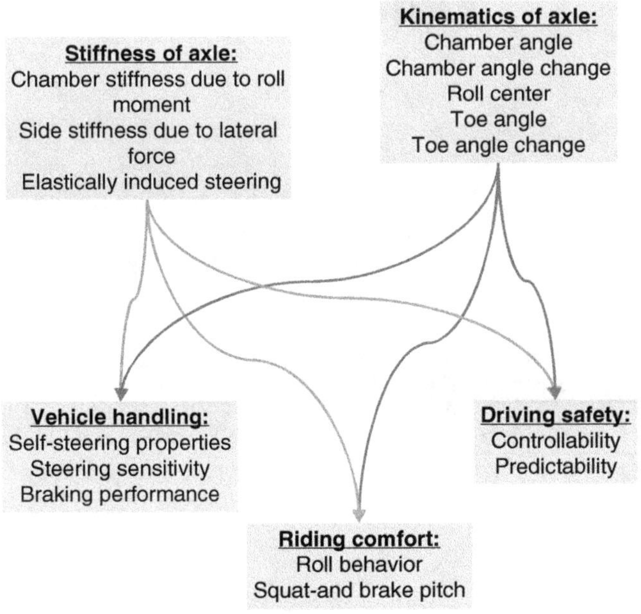

Figure 6.4 Vehicle dynamics effects.

dependent on the position of the cross beam and is the smallest for torsion crank axles, since it is purely due to elasticity. During jounce and rebound, the camber angle changes the sensitivity of the camber angle influencing position of the wheel on the ground and consequently the driving performance. From this viewpoint it is desirable to minimize or at least control the variation of the chamber angle. When driving along a straight line, a slightly positive camber angle suppresses wheel wobble, however, a slightly negative camber angle increases the car's ability to cope with lateral forces present during cornering.

If the distance between the rim horns in front of the wheel center is smaller than the distance between the rim horns behind the wheel center, we speak of toe-in. The total toe-in angle results from the toe-in angle on both wheels. A small toe angle is used to control handling, because toe angle effects straight line driving stability, cornering behavior and suspension tuning. Sometimes, the toe angle is used to compensate for elastic deformation caused by longitudinal driving forces.

The roll center is the current pivot point of the vehicle body in an axle plane around which the vehicle body tilts laterally during rolling movements (Figure 6.5). A high roll center, W_P, causes low rolling movements of the vehicle body, because the lever arm roll center W_P to the center of gravity of the vehicle is small. A deep roll center W_P ensures low track gauge and camber changes. To determine the roll center, W_P, of a twist-beam

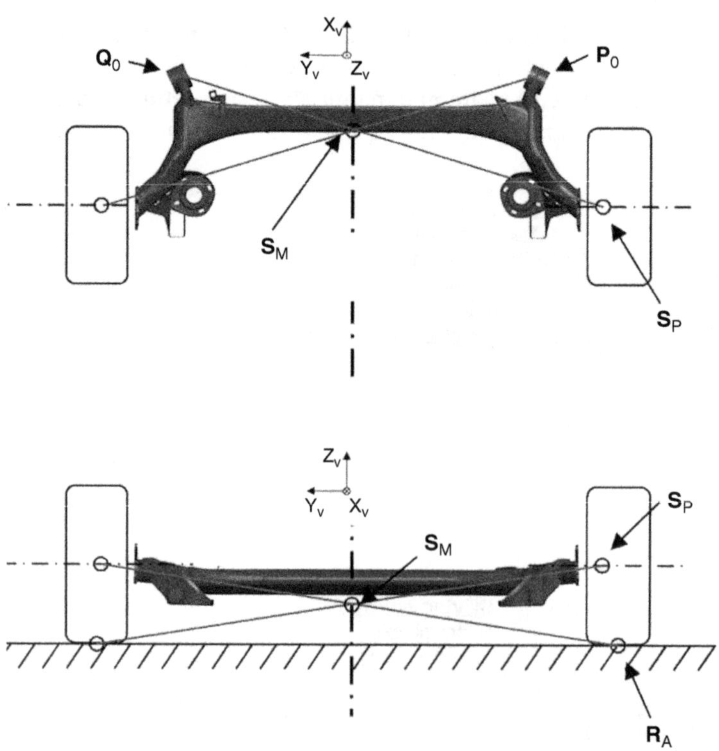

Figure 6.5 Roll center of semi-independent axles.

axle, the shear center S_M of the cross beam must be known. First, the intersection S_P of the wheel center with the imaginary line between the bearing point P_0 and the shear center S_M or the bearing point Q_0 and the shear center S_M is determined. The intersection S_P is then projected into the rear view and the roll center W_P results as the intersection of the straight lines running through the intersection S_P and the wheel contact point R_A. The shear center S_M of the cross-member therefore directly influences the driving behavior of the vehicle. The roll axis of the motor vehicle is obtained by connecting the roll centers of the front and rear axle by a straight line.

6.2.2 Coordinate Systems of Vehicle and Local Coordinate Systems

The coordinate systems used are described next. Figure 6.6 shows the four coordinate systems, the angular position of the systems to each other is visible. The earth-fixed and vehicle-fixed coordinate system as well as a wheel coordinate system on a vehicle are shown.

Besides the mentioned coordinate systems of the whole vehicle, other local coordinate systems use location. These local systems are linked to a structural element of the chassis or powertrain. Usually, the axis z is oriented along the axis of an elongated structural element, and two other axes x, y are in the cross-section of the element. The axes of the local coordinate systems are marked with lowercase letters. The only exception is the local coordinate system of the wheel.

6.2.2.1 Earth-Fixed Coordinate System

The earth-fixed coordinate system is a coordinate system bound to a location. The movement of the vehicle in space can be described relative to it. The X_E and Y_E axes of this

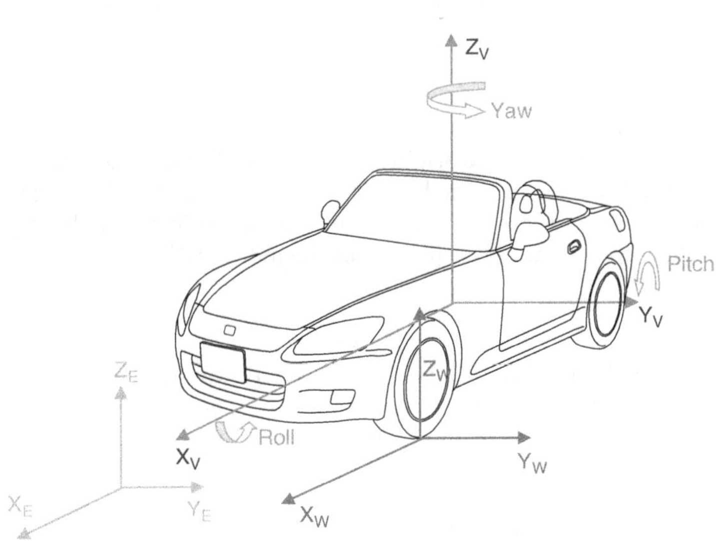

Figure 6.6 Coordinate systems of vehicle (DIN ISO 8855).

right system lie in the road surface plane, the Z_E axis points upward from the road surface. This coordinate system shows the movement of a vehicle from the point of view of a non-moving observer. It receives the index "E."

6.2.2.2 Vehicle-Fixed Coordinate System

The origin of the vehicle-fixed coordinate system is (usually) in the center of gravity of the vehicle. Its X_V axis is horizontal and forward-facing and is located on the median longitudinal plane of the vehicle. The Y_V axis is perpendicular to the vehicle longitudinal median plane and points to the left, the Z_V axis points upward. The vehicle-fixed coordinate system is also called the superstructure coordinate system. It is assigned the index "V."

6.2.2.3 Horizontal Coordinate System

The horizontal coordinate system is also a right-angled coordinate system whose XY plane lies in the $X_E Y_E$ plane of the fixed coordinate system (and thus in the roadway plane). The X axis is the projection of the X_V axis onto the $X_E Y_E$ plane, the z-axis points upward. The horizontal coordinate system moves with the vehicle; its origin is usually below the vehicle's center of gravity. The SAE vehicle axis system uses the z-axis downwards (Gillespie 1992, ch. 1).

Rotational motions around these three axes are referred to as roll, pitch and yaw. Rotations around the x-axis are referred to as roll; rotations around the y-axis as pitch and rotations around z as yaw. Pitching is commonly divided into dive, squat and lift. They occur during breaking or acceleration. Dive ensues when the nose of the car dips during breaking. Simultaneously, at the back of the car lift occurs. Squat happens during acceleration when the back of the car moves toward the earth and the front of the car lifts up at the same time.

Roll takes place mainly during cornering but is present in all situations where a side load or torque along the drive axis is applied. Roll happens around the roll axis of the car. The roll axis is an imaginary line between the roll centers of the front and rear suspension.

6.2.2.4 Wheel Coordinate System

Each wheel has its own coordinate system. The origin of a fixed wheel coordinate system lies in the wheel contact point, the $X_W Y_W$ plane lies in the $X_E Y_E$ plane. The X_W axis is the intersection line of the wheel plane with the $X_E Y_E$ plane and points forward, the Z_W axis points upward. The wheel coordinate system receives the index "W."

6.2.3 Angle Definitions

The angular position of the vehicle-fixed coordinate system relative to the fixed coordinate system results from the sequence of rotations around the three coordinate axes. The angles are expressed in radians throughout the book. If the value in degrees is required, take the value in radians and multiply it by factor of $180^\circ/\pi = 57.3^\circ$. The angle between Y and Y_V axis is called the roll angle φ, the angle between X and X_V axis the pitch angle θ and the angle between X_E and X axis is called the yaw angle ψ (Tables 6.1 and 6.2).

Different views of the geometry of the wheel suspension are shown in Figures 6.7–6.10. The angles, which are necessary for evaluation of the lateral dynamic of the vehicle, are shown in Figure 6.11.

Table 6.1 Angle definitions in vehicle dynamics.

Earth fixed coordinate system	Right-angled coordinate system linked to the place. X_E and Y_E axes lie in a horizontal plane. Z_E axis is directed upward.
Vehicle fixed coordinate system	Rectangular right system with any coordinate origin in the vehicle such that: X_V axis is always horizontal and directed forward and is located in the vehicle longitudinal center plane; Y_V axis is perpendicular to the vehicle's median longitudinal plane and points to the left; Z_V axis points upward.
Vehicle velocity **v**	Vector quantity describing the speed of the origin of the vehicle fixed coordinate system in the fixed coordinate system.
Longitudinal speed v_x	Component of the vehicle speed in the direction of the X axis.
Transversal speed v_y	Component of the vehicle speed in the direction of the Y axis.
Yaw angle ψ	Angle resulting from a rotation around the Z_E axis.
Pitch angle θ	Angle resulting from rotation about the Y axis.
Roll angle φ	Angle resulting from rotation about the X_V axis.
Sideslip angle β of vehicle	Angle between X axis and the direction of the horizontal speed resulting from rotation about the Z axis.
Trajectory, path	Course of the origin of the vehicle fixed coordinate system projected onto the $X_E Y_E$ plane.
Course angle $\varsigma = \psi - \beta$	Angle between X_E axis and horizontal speed resulting from rotation about the Z axis.
Path radius	Distance between a point on the trajectory (path curve) and the corresponding instantaneous center.

Table 6.2 Angle definitions in geometry of chassis.

Static toe angle	The angle bet the longitudinal axis of the vehicle and the line of intersection of the wheel plane and the road surface
Camber angle	The inclination of the wheel plane to the vertical Z_V axis
Caster angle	The angle in side elevation between the steering axis and the vertical Z_V axis

6.2.4 Components of Force and Moments in Car Dynamics

6.2.5 Degrees of Freedom of a Vehicle

Depending on the level of detail, a different number of degrees of freedom is necessary for a vehicle modeling (Table 6.3).

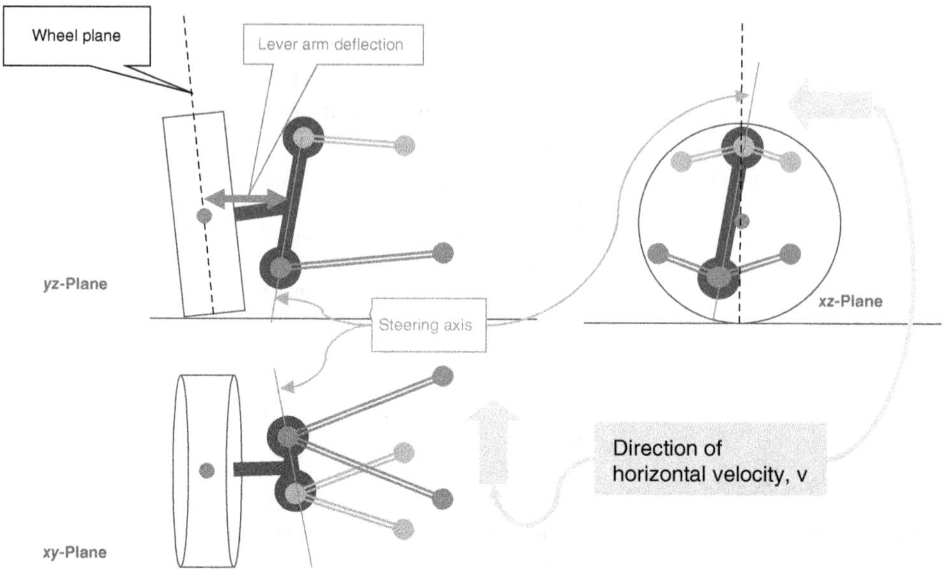

Figure 6.7 Suspension geometry.

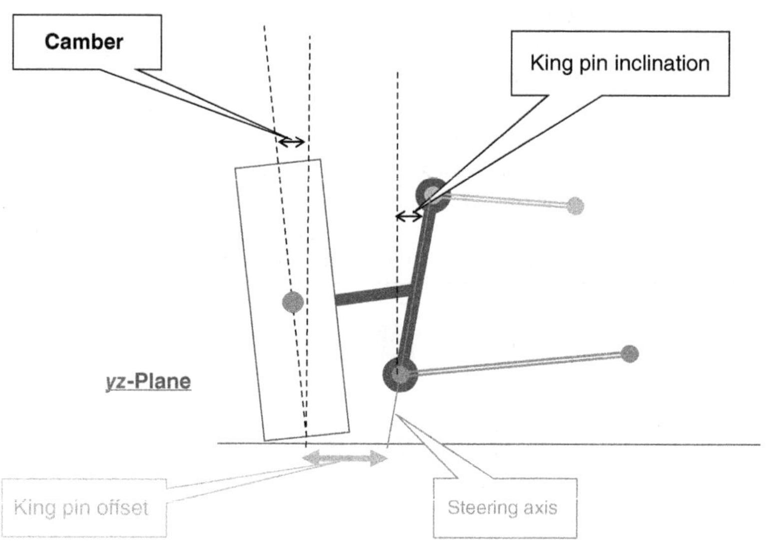

Figure 6.8 Suspension geometry, projection on the *yz*-plane.

A rigid vehicle has a total of six degrees of freedom, which can be divided into three translational and three rotational degrees of freedom. The translational degrees of freedom include the movements of the vehicle in the direction of the three main axes of the fixed coordinate system of the vehicle; the rotatory comprise the rotations of the vehicle about the longitudinal, transverse and vertical axis.

For dependence of required model accuracy and the effects in question, the degrees of freedom necessary must be decided on. For example, it may be sufficient to model

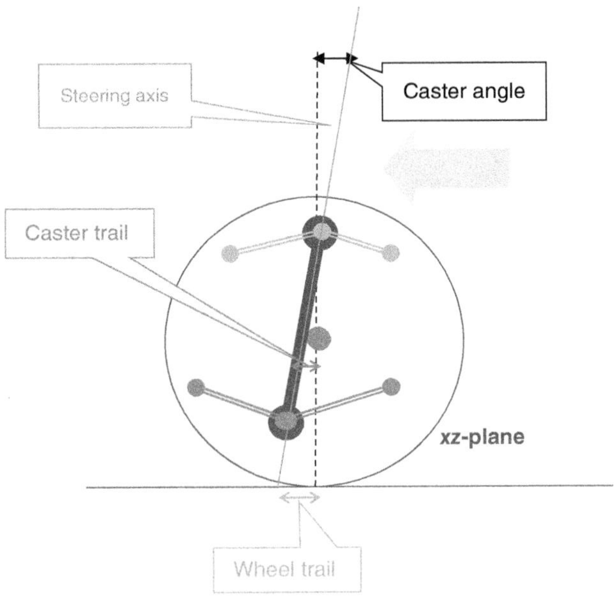

Figure 6.9 Suspension geometry, projection on the *xz*-plane.

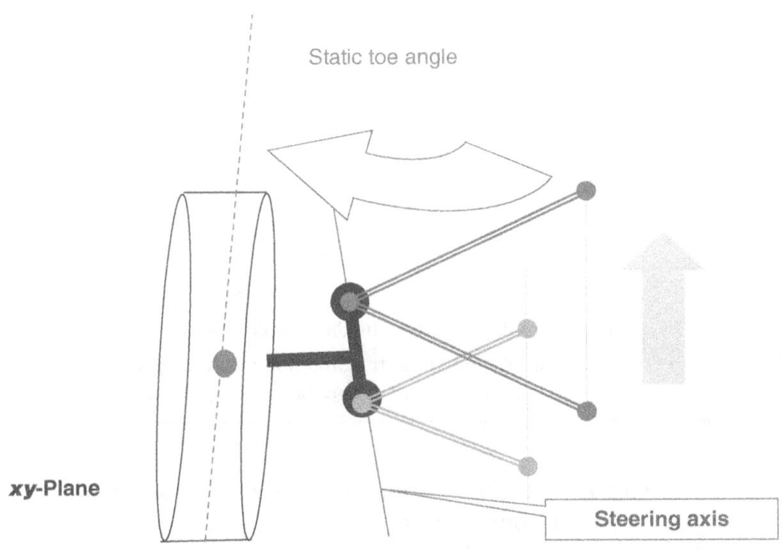

Figure 6.10 Suspension geometry, projection on the *xy*-plane.

only the degree of freedom of longitudinal movement of the vehicle. For a transverse dynamic model, on the other hand, at least two degrees of freedom must be modeled, namely the side slip motion and yawing movement of the vehicle.

There are different models that are based on the degrees of freedom of the model to be designed. Model types are mutually exclusive due to the choice of degrees of freedom acquaintances. Model types are the single and two-track models as well as multi-body

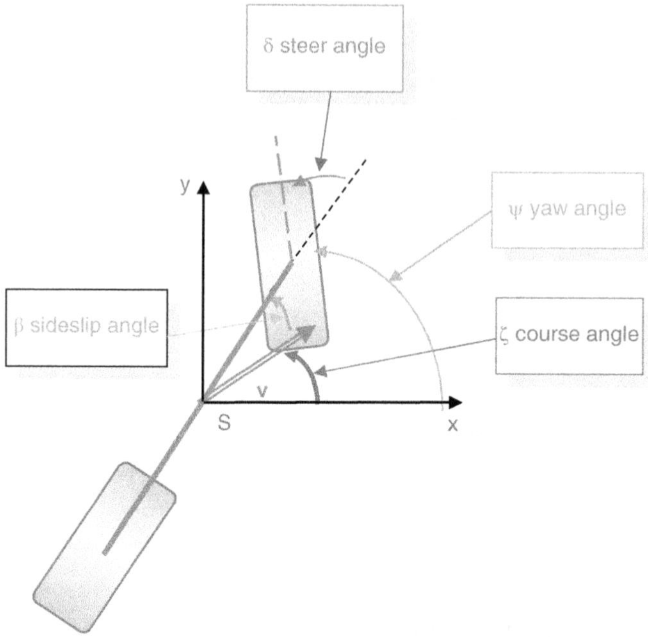

Figure 6.11 Bicycle model of the vehicle in an earth-fixed coordinate system.

Table 6.3 Components of forces and moments.

Components of force	
Longitudinal force F_x	Component of force in the direction of the$X$ axis
Lateral force F_y	Component of force in the direction of the Y axis
Vertical force F_z	Component of force in the direction of the Z axis
Components of the moment	
yawing moment M_z	Component of moment in the direction of the Z axis
pitching moment M_y	Component of moment in the direction of the Y axis
rolling moment M_x	Component of moment in the direction of the X axis

models. Hybrid models are also possible in which, for example, a two-track model can be coupled with a multi-body model for one chassis.

Furthermore, for the vehicle components to be modeled, such as the powertrain, brake system or steering, models can be developed. Therefore, assistance systems, such as the ABS anti-lock braking system, can be integrated into a vehicle model. The last step is the selection of a tire model. Depending on the area of application, decide on which model to use. The different tire models will be discussed later. Generally speaking, hybrid models are those in which an attempt is made to compensate for the disadvantages of individual models by combining several sub-models from different types of modeling.

The term "complete vehicle model" refers to a model that contains the components vehicle dynamics, chassis, powertrain, brake and steering. Input variables are the brake pedal position specified by the driver, the accelerator pedal position, the steering wheel angle and the position of the automatic transmission selector lever as well as ambient conditions such as side and headwind, adhesion coefficient and road unevenness.

Longitudinal, transverse and vertical dynamics are directly coupled. Modern vehicle dynamics control systems and driver assistance systems intervene in vehicle dynamics via various actuators, such as brakes, powertrain, steering, active stabilizers or active chassis. Apart from wind and air forces, forces between vehicle and environment can only be transmitted via the tires.

In addition, the use of composite materials requires dedicated models of car dynamics that describe some distinct effects. On one hand, the composite materials possess lower elastic moduli and a comparative strength to the common steel alloys that are traditionally used in automotive suspensions. On the other hand, the anisotropy of composite materials allows for correction of vehicle dynamics. The implementation of composite materials requires detailed study of the effects of elastic deformation in suspensions.

The entire handling behavior of a car is characterized using several maneuvers that try to evaluate the handling characteristics of the vehicle, see chapter 6 of Heißing and Ersoy (2013). These maneuvers are generally included in the most common International Standards (ISO). The common tests to characterize the handling performance of a vehicle are:

- Braking in turn (ISO 7975:2006),
- Transient response test with one period of sinusoidal input (ISO/TR 8725),
- Straight-ahead braking (ISO 14512:1999)
- Steady-state circular driving behavior (ISO 4138:2012).

Steering parameters of passenger vehicles depict the cornering controllability, straight-ahead controllability and first impression behavior. The acquisition of these parameters encompasses low to high lateral acceleration maneuvers as random input (ISO 7401:2011) or continuous sinusoidal input (ISO 7401:2011).

In order to model the definite effects of dynamics of passenger cars, specific vehicle dynamics models are developed. Theoretical modeling summarizes the basic equations and includes significant elements that lead to ordinary and/or partial differential equations for the vehicle. If the equations cannot be solved explicitly, further simplifying assumptions are made. Simplification can be achieved by linearization, reduction of the model order or approximation.

6.3 Forces Between Tires and Road

6.3.1 Tire Slip

Forces between the tires and the road are transmitted via the so-called tire slipper. The force that can be transmitted by each tire depends on the tire slip. The circumferential slip is defined according to DIN ISO 8855:

$$S_{X;V} = \omega/\omega_0 - 1. \tag{6.1}$$

Here, ω is the rotational speed of the wheel, ω_0 is the rotational speed of a straight and freely rolling wheel. The slip is normalized to values between 0 and 1. The slip angle α

of the wheel is a measure for the lateral slip of the tire and is defined according to DIN ISO 8855:

$$\tan\alpha = V_{Y,W}/V_{X,W}. \tag{6.2}$$

Coefficients of circumferential and lateral force are defined according to DIN ISO 8855 as follows:

$$\mu_{X,W} = F_{X,W}/F_{Z,W}, \mu_{Y,W} = F_{Y,W}/F_{Z,W}. \tag{6.3}$$

To calculate the rotational speed of a wheel $\dot{\omega}$, the moment balance must be formed. The balanced equation is as follows:

$$J_R\dot{\omega} = M_A - M_B - M_R - r_{dyn}F_{X,W}, \tag{6.4}$$

where are:

M_A the driving torque,
M_B the braking torque,
M_R the torque caused by rolling friction and
$F_{X,W}$ the wheel circumferential force.

The constants J_R and r_{dyn} are the moment of inertia and the dynamic radius of the wheel. The torque M_R results from the rolling friction force.

Since tire forces on the rolling wheel can only be transmitted if slip occurs, the following equations give the slip definition. In the case of braked wheels, the following applies for longitudinal and transverse slip, respectively:

$$S_{X,W} = 1 - v^{-1}r_{dyn}\omega\cos\alpha, \quad S_{Y,W} = v^{-1}r_{dyn}\omega\ \sin\alpha. \tag{6.5}$$

These formulas in the case of driven wheels transform to:

$$S_{X,W} = \cos\alpha - vr_{dyn}{}^{-1}\omega^{-1}, \quad S_{Y,W} = \sin\alpha. \tag{6.6}$$

The term $r_{dyn}\omega$ corresponds to the wheel circumference speed, v is the speed at which the wheel contact point moves. It must be calculated from the translational and rotational speeds of the vehicle for each wheel contact point. The resulting total slip can be specified as a geometric sum of longitudinal and transverse slip:

$$S_W = \sqrt{S_{X,W}{}^2 + S_{Y,W}{}^2}. \tag{6.7}$$

6.3.2 Side Slip Curve and Lateral Force Properties

Determining the transmission of force between the tire and the road is one of the most important challenges, because the determination of wheel forces depends largely on the statistically distributed condition of the road surface. The coefficient of adhesion μ between road and tire tread is anisotropic. The common tire models use the slip, the slip angle and the camber at the wheel as input variables in order to then determine tire forces. With the help of the μ-slip curve the resulting slip can be transformed into the resulting tire force. Figure 6.12 displays a qualitative example of a side slip curve. The lateral force dependence on slip angle changes depending on speed, wheel load and character of contact surface between the tire and the road. For example, if all other factors

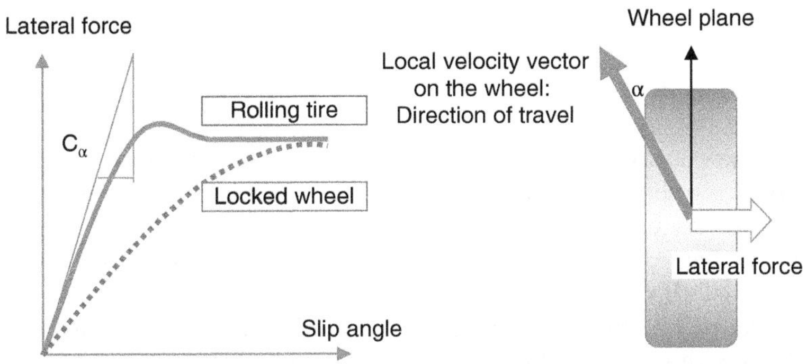

Figure 6.12 Side slip curve and lateral force properties.

are the fixed, the resulting adhesion coefficient μ_W can thus be determined as a function of the resulting slip on the wheel S_W, and $\mu_W = \mu_W(S_W)$.

An appropriate tire model is applied for the numerical evaluation of the lateral and longitudinal reaction forces and moments of the tire. The frequently used model is the approach (Pacejka and Bakker 1992), which is based on the combination of trigonometric functions. In this model, the reaction forces or moments on the tire arising from the tire–road interaction are presented by the "magic formula." This formula is an empirical equation based on four fitting coefficients, which allows the qualitatively good fitting of tire reaction curves. Tire properties such as compliance and inertia are customarily ignored. For the simple case of pure side or pure longitudinal slip the magic formula can be employed (Pacejka 2012). The formula expresses the sideforce, aligning torque and longitudinal force as a function of the slip angle and the longitudinal slip, respectively. By selecting an input value s and specifying a form factor, it can be determined whether lateral force $F_{Y,W}$, longitudinal force $F_{X,W}$ or restoring torque M_Z are to be calculated. The function has the following form (magic formula) (Pacejka and Bakker 1992; Clover and Bernard 1998)

$$f = \hat{D}\sin(\hat{C} \bullet \arctan(\hat{B}x - \hat{E} \bullet (\hat{B}x - \arctan(\hat{B}x)))). \tag{6.8}$$

Here, the following apply:

$\hat{B}$ factor for stiffness;
$\hat{C}$ form factor for selecting the force or torque to be calculated;
$\hat{D}$ maximum value of the curve;
$\hat{E}$ bending factor.

The role the input parameter plays is in the resulting slip on the wheel S_W or the slip angle α. Needless to say, the coefficients differ depending upon the actual input parameter and output value. The Taylor series of this general expression over the definite input variable reads:

$$f = \hat{D}\hat{C}\hat{B}x - \hat{D}\hat{C}\hat{B}^3(2 + 2\hat{E} + \hat{C}^2)x^3/6 + O(x^5). \tag{6.9}$$

A linear model is commonly used to calculate the lateral tire force for small slip angles α. The model is based on the assumption that the lateral force $F_{Y,W}$ in the linear range of the lateral force slip angle α can be calculated using the linear relationship between these values. In the linear model, the lateral force $F_{Y,\,W}$ is the first term of the Taylor series:

$$F_{Y,W} = c_\alpha \alpha, \tag{6.10}$$

where $c_\alpha = \widehat{D}\widehat{C}\widehat{B}$ is the skew stiffness of tire and α the slip angle on the wheel. It is the simplest relationship between slip angle of wheel and lateral wheel force. The linear relationship, provided by the first term of the magic formula, is sufficient for the rough estimations of the linear single-track model. The application possibilities of this model are certainly limited to small slip angles.

Influencing factors are mainly the condition of the ground and the properties and condition of the tire (air pressure, ambient temperature, tire compound, wheel load and rotation speed). With the help of the resulting adhesion coefficient μ_W and the wheel upright force $F_{Z,\,W}$, the resulting tire force can be determined:

$$F_W = \mu_W F_{Z,W}, F_W = \sqrt{F_{X,W}{}^2 + F_{Y,W}{}^2}. \tag{6.11}$$

The force F_W can be divided into wheel circumferential $F_{X,\,W}$ and wheel lateral force $F_{Y,\,W}$:

$$F_{X,W} = S_{X,W}\mu_W F_{Z,W}/S_W, F_{\dot{Y},W} = S_{Y,W}\mu_W F_{Z,W}/S_W. \tag{6.12}$$

6.4 Dynamic Equations of a Single-Track Model

6.4.1 Hypotheses of a Single-Track Model

A stiff body of a vehicle is described in its environment by six degrees of freedom in space; three translatory and three rotational degrees of freedom of movement. In the linear single-track (or bicycle) models, the vehicle is traced back to a flat problem, that is, two degrees of translational freedom and one degree of rotational freedom are considered. The vehicle model has only one lane, that is, only one front and rear tire. Both translational movements in the roadway plane are controlled by the dynamic equations of forces for the corresponding coordinate directions.

Evaluation of lateral dynamic vehicle behavior is performed using simplified models. Using a linearized model based on the hypotheses of Riekert and Schunck, the evaluation of modest lateral accelerations is done. This model is based on the following hypotheses:

- Constant wheel loads, no lifting and pitching motion,
- Small movements,
- Linear tire forces,
- Lots of wheels,
- Constant travel speed, no longitudinal acceleration,
- Vanishing vehicle width, no rolling movement.

This simple vehicle model presented is suitable for explaining the basic steering characteristics of a two-axle vehicle. The center of gravity of the model lies on the road surface, so that acceleration forces in the longitudinal and transverse direction have no lever arm relative to the road surface. Therefore, in the bicycle model the longitudinal

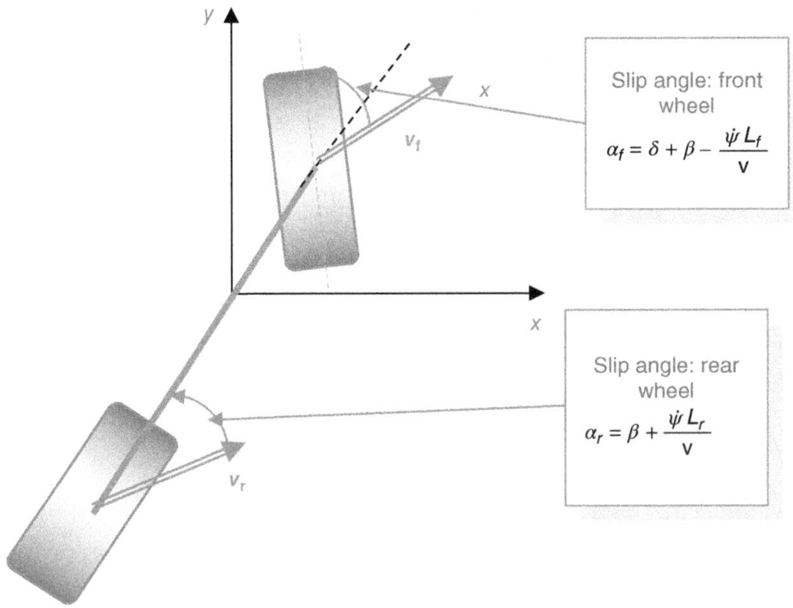

Figure 6.13 Wheel slip angles in a single-track model.

acceleration does not lead to pitching and lateral acceleration does not lead to rolling (rotations around the transverse or vertical axis).

In the linear single-track model, the wheels of an axis are formally aligned and can be combined into one wheel element each. This procedure halves the number of unknown tire forces. The sum of the tire forces is centered in the vehicle's longitudinal axis. The tire overtravel at the front or the rear is neglected in this model for the time being. The tire overrun takes into account the effect that the wheel forces are not exactly centered in the contact points of the tires.

Figure 6.13 shows the single-track model with kinematic relationships and dependencies between the various variables. The movement of a vehicle during cornering is characterized such that the longitudinal axis of the vehicle is twisted by a certain angle with respect to the speed vector of the center of gravity or the tangent of the path curve.

6.4.2 Moments and Forces in a Single-Track Model

In Figures 6.13–6.15, L_f is the horizontal distance from the front axle to the center of mass, L_r is the horizontal distance from the rear axle to center of mass. The movement of a vehicle on a circular movement can be described by a rotation around the instantaneous center O (Figures 6.14 and 6.15). The current pole O represents the coordinate origin for the fixed coordinate system. The distance between the instantaneous pole and the center of gravity S is called the swivel radius. The speed of the vehicle v is a product of the angular speed of the vehicle around the current center of rotation of the vehicle $\dot{\varsigma}$ with the current swivel radius r:

$$\mathrm{v} = \dot{\varsigma}\, r. \tag{6.13}$$

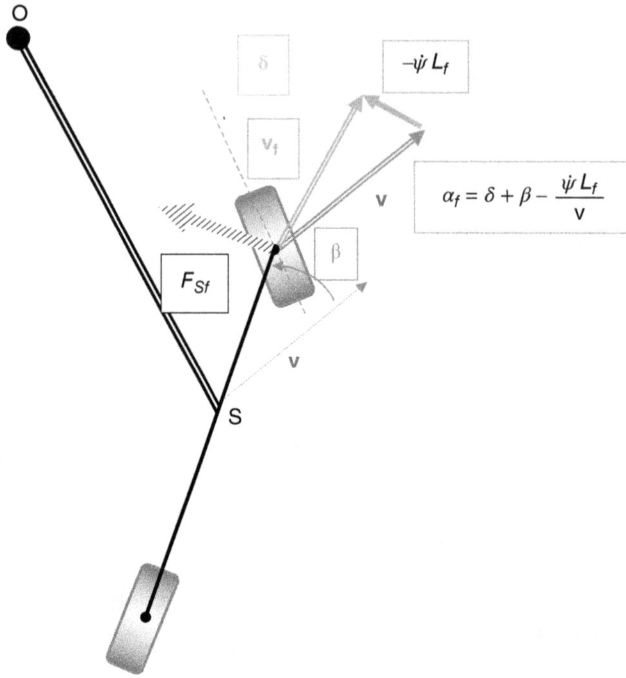

Figure 6.14 Slip angle and lateral force on the front wheel.

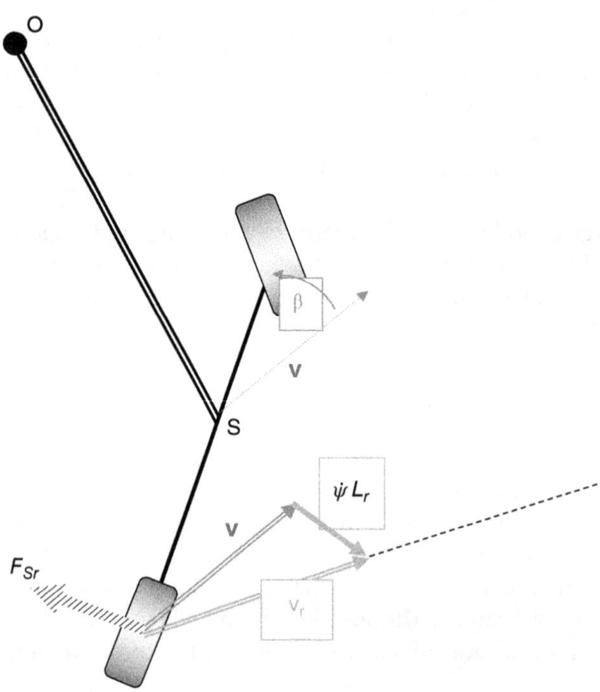

Figure 6.15 Slip angle and lateral force on the rear wheel.

We assume that the speed remains constant while driving. On the contrary, the angular speed $\dot{\varsigma}$ and radius r change during cornering. The angular speed of the vehicle around the current center of rotation of the vehicle $\dot{\varsigma}$, is the difference between the yaw speed $\dot{\varsigma}$ and the speed of the change in the sideslip angle of vehicle, $\dot{\beta}$:

$$\dot{\varsigma} = \dot{\psi} - \dot{\beta}. \tag{6.14}$$

The swivel radius r and the yaw angle speed $\dot{\varsigma}$ are used to indicate the velocity of center of gravity speed of the vehicle. Inserting the angular velocity over yaw velocity and floating angle velocity results in:

$$\mathrm{v} = (\dot{\psi} - \dot{\beta})\, r. \tag{6.15}$$

The lateral acceleration (centrifugal acceleration) of a vehicle as a function of speed and turning radius is

$$a_y = v^2/r \equiv \dot{\varsigma}\, r \equiv v\dot{\varsigma}.$$

Using the expression (6.13) for angular velocity and (6.15), lateral acceleration results in:

$$a_y = \mathrm{v}\,(\dot{\psi} - \dot{\beta}). \tag{6.16}$$

In stable driving situations, the sideslip angle of a vehicle is small. Therefore, linearization of the tangent function is also possible here without significant loss of accuracy.

6.4.3 Balance of Forces and Moments in a Single-Track Model

The rotational freedom is orthogonal to the road surface and describes the yaw of the vehicle around the vertical axis. The equation of motion is represented by the balance of moments around the center of gravity. The calculations start with the balance of moments around the vertical axis. For small steer angles:

$$I_z\ddot{\psi} = F_{Sf}L_f - F_{Sr}L_r, \tag{6.17}$$

I_z is the mass moment of inertia of the vehicle around the vertical axis.

The Newton theorem describes the lateral movements of the body. The second start equation is the equilibrium of forces at the center of gravity:

$$ma_y = F_{Sf} + F_{Sr}. \tag{6.18}$$

The side forces on the front and rear wheels depend on the corresponding slip angles. The slip angles α_v and $\alpha_h$are determined by geometric relationships on the front wheel or at the rear wheel of the linear single-track model (Figures 6.14 and 6.15). The front and rear side forces are sums of the side forces acting on both wheels of each corresponding axis. The vector of local velocity on the wheel is inclined to the wheel plane. As already mentioned, the slip angle of wheel is defined by the vehicle fixed speed variables in the direction of the wheel plane and normal to the wheel plane α. Centripetal acceleration results in a certain relationship between float angle, yaw rate, lateral acceleration and velocity. Similarly, dynamic wheel load shifts (changes in wheel contact force due to longitudinal or lateral acceleration) do not occur.

Two component equations in the x- and y-directions can be used and the speed of the front or rear wheel must be specified. By eliminating the wheel speeds, the required relationships are for the slip angles on the wheels. As an example, using this procedure on the rear wheel with the aid of Figures 6.14 and 6.15, two speed balance equations are obtained.

The equation for the rear slip angle follows directly by quotient formation and, analogously for the front axle, taking into account the front wheel steer angle. Linearization of the slip angles is justified if the slip angle is <5°. From the free body image in the tire plane, simple relationships for calculating the tire forces can be determined if the coefficient of cohesion μ, defined as the ratio of the normal force to the horizontal resistance force, is taken into account.

A starting point for further considerations concerns the differential equations of the linear single-track model. The linear tire model is used in these equations. The lateral tire forces (wheel guiding forces) are assumed to be proportional to the slip angle, the proportionality factor is the slip stiffness. The approximation for the slip angles is valid for small steer angles and small yaw speeds:

$$\begin{cases} F_{Sf} = c_{\alpha f}\alpha_f = c_{\alpha f}(\delta + \beta - \dot{\psi}\, L_f/\mathrm{v}), \\ F_{Sr} = c_{\alpha r}\alpha_r = c_{\alpha r}(\beta + \dot{\psi}\, L_r/\mathrm{v}). \end{cases} \tag{6.19}$$

The side force on steering axis (6.19) additionally contains the influence of the steer angle. The values $c_{\alpha f}$ $c_{\alpha r}$ are the skew stiffnesses for the front and rear axis, correspondingly. If there are additional effects of vehicle roll, bending and elasticity of the axle, induced steering of the wheel has to be added to the ideal steer angle:

$$\begin{cases} F_{Sf} = c_{\alpha f}(\delta + \delta_V + \beta - \dot{\psi}\, L_v/\mathrm{v}), & \delta_f = \delta_f(F_{Sf,}\theta, M_x, \ldots), \\ F_{Sr} = c_{\alpha r}(\delta_R + \beta + \dot{\psi}\, L_h/\mathrm{v}), & \delta_r = \delta_r(F_{Sr,}\theta, M_x, \ldots). \end{cases} \tag{6.20}$$

The additional effects in the dynamical equations are considered in detail in Chapter 7.

6.4.4 Steady Cornering

6.4.4.1 Necessary Steer Angle for Steady Cornering

For a stationary round trip, that is, in the event that

$$\ddot{\psi} = 0 \text{ and } \dot{\beta} = 0 \tag{6.21}$$

the moment force vanishes and the forces on the front and rear axes are in balance:

$$F_{Sv}L_v - F_{Sh}L_h = 0. \tag{6.22}$$

Denoting the wheel base as $L_w = L_r + L_f$, the resolution of these equations according to wheel guiding forces during stationary circular travel provides:

$$\begin{cases} F_{Sf} = \frac{L_r}{L_w} m\, a_y, \\ F_{Sr} = \frac{L_f}{L_w} m\, a_y. \end{cases} \tag{6.23}$$

Use the lateral forces in Eq. (6.23):

$$\begin{cases} c_{\alpha f}\left(\delta + \beta - \dot{\psi}\, \frac{L_f}{\mathrm{v}}\right) = \frac{L_r}{L_w} m\, a_y, \\ c_{\alpha r}\left(\ \beta + \dot{\psi}\, \frac{L_r}{\mathrm{v}}\right) = \frac{L_f}{L_w} m\, a_y, \end{cases} \tag{6.24}$$

From Eq. (6.24) we obtain the relation for sideslip angle of a vehicle and its yaw velocity:

$$\begin{cases} \delta + \beta - \dot{\psi}\,\frac{L_f}{\mathrm{v}} = \frac{L_r}{c_{\alpha f} L_w} m\, a_y, \\ \beta + \dot{\psi}\,\frac{L_r}{\mathrm{v}} = \frac{L_f}{c_{\alpha r} L_w} m\, a_y. \end{cases}$$

Subtraction of the second equation from the first results in the following for the required steer angle on front wheels:

$$\delta - \frac{\dot{\psi}\, L_f}{\mathrm{v}} - \frac{\dot{\psi}\, L_r}{\mathrm{v}} = \left[\frac{L_r}{c_{\alpha f}} - \frac{L_f}{c_{\alpha r}}\right] \frac{m}{L}\, a_y \tag{6.25}$$

In the steady rotational trip of the vehicle, its velocity reads:

$$\mathrm{v} = \dot{\psi} r. \tag{6.26}$$

Substitution of (6.26) for (6.25) leads to the required steer angle, δ, as:

$$\delta = \frac{L_w}{r} + \left[\frac{L_r}{c_{\alpha f}} - \frac{L_f}{c_{\alpha r}}\right] \frac{m}{L_w} a_y = \delta_{\mathrm{a}} + \left[\frac{L_r}{c_{\alpha f}} - \frac{L_f}{c_{\alpha r}}\right] \frac{m}{L_w} a_y. \tag{6.27}$$

This is the required steer angle. The angle in (6.27) differs from the steer angle according to Ackermann:

$$\delta_{\mathrm{a}} = \frac{L_w}{r} \tag{6.28}$$

The higher the lateral acceleration of the vehicle a_y, the greater the difference between the required steer angle (6.27) and the Ackermann steer angle (6.28).

6.4.4.2 Yaw Gain Factor and Steer Angle Gradient

Now, let's set the expressions for lateral acceleration a_y and the velocity of the vehicle:

$$\delta = \frac{\dot{\psi}\, L_w}{\mathrm{v}} + \frac{1}{L_w}\left[\frac{L_r}{c_{\alpha f}} - \frac{L_f}{c_{\alpha r}}\right] m\, \frac{\mathrm{v}^2}{r} = \frac{\dot{\psi}\, L_w}{\mathrm{v}} + \frac{1}{L_w}\left[\frac{L_r}{c_{\alpha f}} - \frac{L_f}{c_{\alpha r}}\right] m\, \dot{\psi}\, \mathrm{v}.$$

The ratio of yaw rate to steering wheel angle is called the yaw gain factor:

$$\frac{\dot{\psi}}{\delta} = \frac{1}{\frac{L_w}{\mathrm{v}} + \frac{m}{L_w}\left[\frac{L_r}{c_{\alpha f}} - \frac{L_f}{c_{\alpha r}}\right] \mathrm{v}} = \frac{1}{\frac{L_w}{\mathrm{v}} + \mathrm{v}\cdot\frac{\partial \delta}{\partial a_y}}, \tag{6.29}$$

whereby

$$\frac{\partial \delta}{\partial a_y} = \left[\frac{L_r}{c_{\alpha f}} - \frac{L_f}{c_{\alpha r}}\right] \frac{m}{L_w} \tag{6.30}$$

is the steer angle gradient. This factor pronounces how fast the vehicle rotates around the vertical axis at a certain steer angle.

The steer angle gradient indicates whether the vehicle is under-, over- or neutral steering. Along with the steer angle gradient, is the required steer angle:

$$\delta = \frac{L_w}{r} + \frac{\partial \delta}{\partial a_y}\cdot a_y = \delta_{\mathrm{a}} + \frac{\partial \delta}{\partial a_y}\cdot a_y.$$

A steer angle gradient of zero means that the yaw gain increases exactly linearly with the speed (vehicle "neutrally controlling"). Such a vehicle would behave when stationary as if it were "on rails," that is, no self-steering effect would be noticeable regardless of the speed in addition to the steer angle specified by the driver.

6.4.4.3 Classification of Self-Steering Behavior

If the lateral acceleration of the vehicle is insignificant, pure geometrical steering take place. This steering is referred to as Ackermann steering (Figure 6.15). The radius of turn in this case will be referred to as the "Ackermann path radius." The Ackermann path radius depends only on the wheelbase and steer angle, see Eq. (6.28). However, if the vehicle velocity is high and the radius of turn is low, the lateral acceleration cannot be neglected any more. In this case, the actual radius of turn will differ from the Ackermann path radius.

Classically, the driving dynamics of motor vehicles are considered separately according to the three different degrees of translational freedom of movement of the vehicle body. The term longitudinal dynamics is used when movements in the longitudinal direction of the vehicle, that is, starting and braking processes, are investigated.

Motion processes transverse to the longitudinal axis of the vehicle are summarized under the term transverse ceramic. Here the driving stability, cornering behavior and lane and course posture are considered. The vibration behavior of the body in the direction of the vertical axis of a vehicle is called vertical dynamics. The focus here is on tuning suspension and damping behavior; for example, to increase driving comfort for the occupants.

For the actual contest, only the lateral dynamics of the vehicle is of interest and is now considered in more detail for this reason. The change in course of a motor vehicle without the intervention of the driver is referred to as self-steering behavior. For cornering, self-steering behavior is divided into three areas; over- and understeering and neutral behavior, which are shown in Figures 6.16–6.18.

The term "self-steering" refers to a change in the course of a vehicle without a steering command from the driver. In this context, the terms understeering, neutral and over-controlling driving behavior are often used. In the real vehicle, four identical tires are usually used so that the lateral stiffness of the front and rear tires is equal. However, the wheels can change their track (i.e. the steering lock angle) under lateral force. This is achieved through the targeted use of elasticities in the steering and axles as well as through the geometry of the wheel suspensions, which can lead to track changes when the vehicle tilts sideways. These two effects are called elastokinematics and kinematics (Figure 6.19).

There are various criteria according to which one can classify the self-steering behavior. If the radius of turn is smaller than the Ackermann path radius, we speak of oversteer; that is, the rear of the vehicle breaks or "steps" out.

If the two radii are equally large, one speaks of a neutral self-steering behavior. Understeer occurs when the dynamic radius of turn becomes larger than the Ackermann path radius and the vehicle thereby pushes over the front wheels to the outer edge of the curve.

From the mathematical viewpoint, the track changes can be included in the skew stiffness according to their effect. The so-called effective skew stiffness must then be used in the equations. If the following condition is satisfied:

$$\frac{\partial \delta}{\partial a_y} = \left[\frac{L_r}{c_{Sf}} - \frac{L_f}{c_{Sr}}\right]\frac{m}{L_w} > 0, \tag{6.31}$$

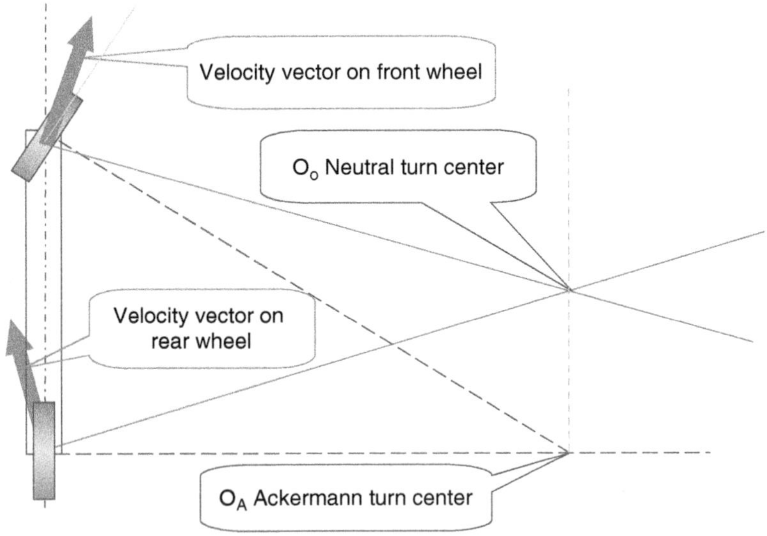

Figure 6.16 Neutral steer in the bicycle model.

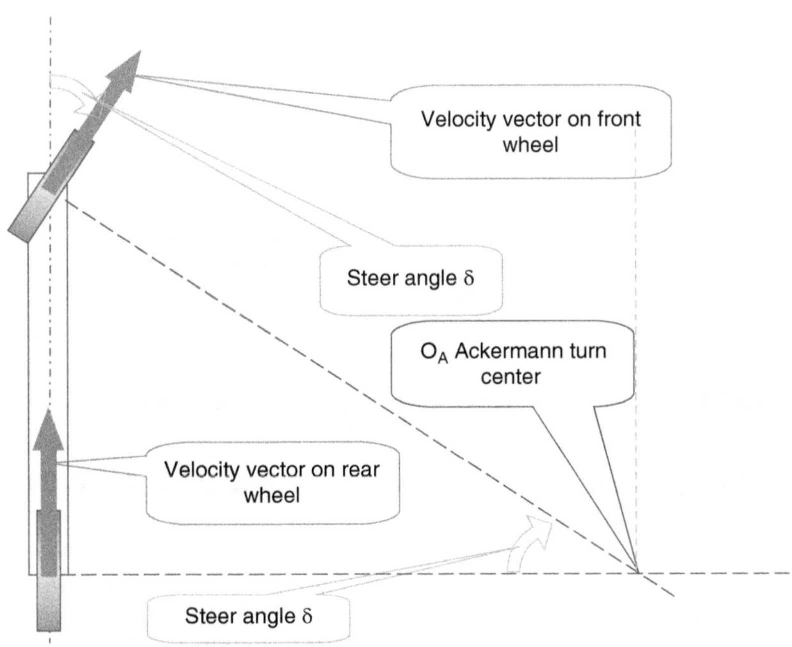

Figure 6.17 Steer angle for low speed turning. "Ackermann" cornering in the bicycle model.

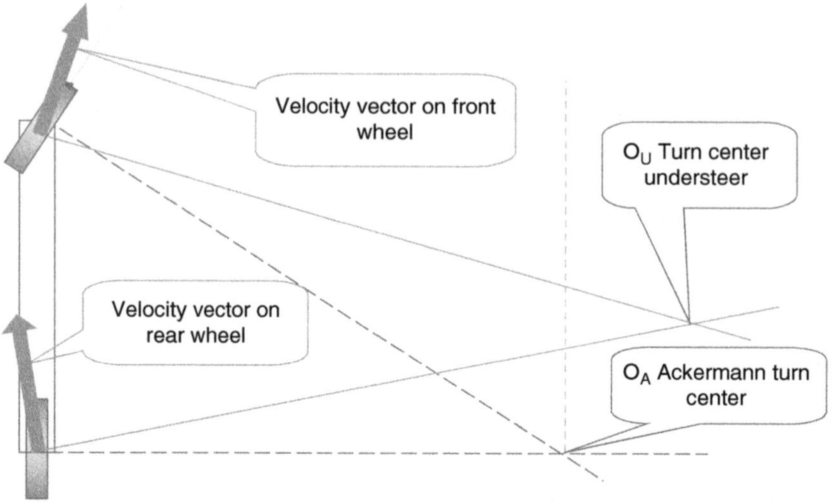

Figure 6.18 Explanation of the understeer effect from the viewpoint of the bicycle model.

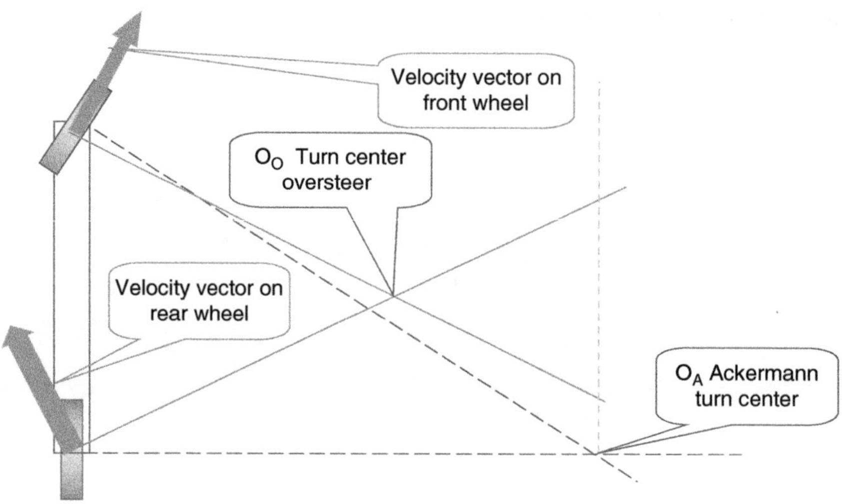

Figure 6.19 Explanation of the oversteer effect from the viewpoint of the bicycle model.

a transverse force at the front leads to a greater angle of skew than at the rear; that is, the vehicle is understeered.

The opposite effect is also present. The response to steering commands increases and the vehicle overdrives. This effect is self-reinforcing and increases disproportionately with increasing driving speed: lateral forces reduce the radius of curvature and thus increase themselves. In general, the steering elasticity of the front axle leads to a yielding of the steering impact under lateral force so that the effective lateral running stiffness is lower than the lateral running stiffness of the tire. Conversely, most rear axles have (elasto-)kinematics that additionally tilts the wheel under lateral force and raises the

Table 6.4 Under-, over- or neutral-driving vehicle.

Oversteer	Neutral	Understeer
$\frac{L_r}{c_{\alpha f}} < \frac{L_f}{c_{\alpha r}}$	$\frac{L_r}{c_{\alpha f}} = \frac{L_f}{c_{\alpha r}}$	$\frac{L_r}{c_{\alpha f}} > \frac{L_f}{c_{\alpha r}}$
$\Rightarrow \frac{\partial \delta}{\partial a_y} < 0$	$\Rightarrow \frac{\partial \delta}{\partial a_y} = 0$	$\Rightarrow \frac{\partial \delta}{\partial a_y} > 0$
$\delta < \delta_a$	$\delta = \delta_a$	$\delta > \delta_a$

effective skew stiffness above the skew stiffness of the tire. This guarantees the understeering design of vehicles common today (Table 6.4).

6.4.5 Non-Steady Cornering

6.4.5.1 Equations of Non-Stationary Cornering

In the non-stationary circular movement, the insertion of the wheel guiding forces into the principle of angular momentum and Newton's equation leads to the following equations of motion:

$$m\,\mathrm{v}\,(\dot{\psi} - \dot{\beta}) = c_{\alpha f}\left(\delta + \beta - \frac{L_f}{\mathrm{v}}\dot{\psi}\right) + c_{\alpha r}\left(\beta + \frac{L_r}{\mathrm{v}}\dot{\psi}\right),$$

$$I_z\,\ddot{\psi} = c_{\alpha f}L_f\left(\delta + \beta - \frac{L_f}{\mathrm{v}}\dot{\psi}\right) + c_{\alpha r}L_r\left(\beta + \frac{L_r}{\mathrm{v}}\dot{\psi}\right). \quad (6.32)$$

The first equation of motion (6.32) is the linear equation with respect to yaw velocity. The resulting expression for yaw velocity is the linear function of the steer angle and the sideslip angle.

The derivation of yaw velocity by time provides the yaw acceleration:

$$\ddot{\psi} = \frac{m\,v\,\ddot{\beta} + c_{\alpha f}(\dot{\delta} + \dot{\beta}) + c_{\alpha r}\dot{\beta}}{m\,v - c_{\alpha r}\mathrm{v}^{-1}L_r + c_{\alpha f}\mathrm{v}^{-1}L_f}.$$

Yaw angular acceleration $\ddot{\psi}$ and velocity $\dot{\psi}$ is then used in the second equation of motion (6.32). After substitution, derivatives of sideslip angle from the second equation of motion could be grouped in the form:

$$\ddot{\beta} + \mathrm{A}\,\dot{\beta} + \mathrm{B}\beta = \Delta_V,$$

$$\Delta_V = p\delta - \frac{c_{\alpha f}}{m\mathrm{v}}\dot{\delta},$$

$$\mathrm{A} = \frac{c_{\alpha f} + c_{\alpha r}}{m\mathrm{v}} + \frac{c_{\alpha f} + c_{\alpha r}}{I_z\mathrm{v}}L_w^{\,2},$$

$$\mathrm{B} = \frac{c_{\alpha r}L_r - c_{\alpha f}L_f}{I_z} + \frac{c_{\alpha f}c_{\alpha r}}{I_z m\mathrm{v}^2}L_w^{\,2},$$

$$p = \frac{c_{\alpha f}L_f}{I_z} - \frac{c_{\alpha f}c_{\alpha r}}{I_z m\mathrm{v}^2}(L_rL_f + L_r^{\,2}). \quad (6.33)$$

The Eq. (6.33) is an inhomogeneous linear ordinary second order differential equation. The similar equation describes the linear oscillator. Forced oscillations are performed by an oscillator that is stimulated to oscillate by time-varying external excitation $\Delta_V(t)$. In (6.33) external excitation depends on time-dependent steering angle of wheel δ and its first time derivative $\dot{\delta}$. Of practical importance are, above all, impulse and stepwise excitation functions. Besides them, the periodic excitations are important for analysis. Among them, the harmonic, sinusoidal excitation is of interest for stability investigation of lateral vehicle dynamics. In the case of the periodic excitation Δ_V, its frequency is called the excitation frequency.

First, we investigate the already considered stationary case of the circular route. From the equations

$$\mathrm{B}\beta = p\delta. \tag{6.34}$$

Here p/B is magnification factor for the sideslip angle of vehicle:

$$\beta = \frac{p}{\mathrm{B}}\delta,$$
$$\frac{p}{\mathrm{B}} = \frac{c_{\alpha f}L_f m\mathrm{v}^2 - c_{\alpha f}c_{\alpha r}(L_r L_f + {L_r}^2)}{(c_{\alpha r}L_r - c_{\alpha f}L_f)m\mathrm{v}^2 + c_{\alpha f}c_{\alpha r}{L_w}^2}. \tag{6.35}$$

6.4.5.2 Oscillatory Behavior of Vehicle During Non-Steady Cornering

In the general, non-stationary case we get the representation of the equations of motion in the form of a linear oscillator:

$$\ddot{\beta} + \mathrm{A}\,\dot{\beta} + \mathrm{B}\beta = \Delta_V,$$
$$\mathrm{A} = 2D_V\omega_V,$$
$$\mathrm{B} = \omega_V^2. \tag{6.36}$$

With corresponding natural frequency and damping factor:

$$\omega_V = \sqrt{\mathrm{B}}, \qquad D_V = \frac{\mathrm{A}}{2\sqrt{\mathrm{B}}}. \tag{6.37}$$

Inserting the formulas (6.33) for A and B returns:

$$D_V = \frac{1}{2\omega_e}\left[\frac{c_{\alpha f} + c_{\alpha r}}{m\mathrm{v}} + \frac{c_{\alpha f} + c_{\alpha r}}{I_z\mathrm{v}}{L_w}^2\right],$$
$$\omega_V = \sqrt{\frac{c_{\alpha r}L_r - c_{\alpha f}L_f}{I_z} + \frac{c_{\alpha f}c_{\alpha r}}{I_z m\mathrm{v}^2}{L_w}^2},$$
$$\omega_{VD} = \sqrt{1 - {D_V}^2}\,\omega_V. \tag{6.38}$$

An oscillatory stability of motion is assured if natural frequency of the vibrations, which are excited by a disturbing force, is a positive real value:

$${\omega_V}^2 = \frac{c_{\alpha r}L_r - c_{\alpha f}L_f}{I_z} + \frac{c_{\alpha f}c_{\alpha r}}{I_z m\mathrm{v}^2}{L_w}^2 > 0. \tag{6.39}$$

The occurring vibration amplitudes should be as small as possible and should subside after the excitation is finished. A non-stationary driving maneuver is therefore stable if a stable stationary motion is established after the excitation has disappeared. If the inequality (6.39) is violated, an oscillatory instability of motion occurs.

6.4.6 Anti-Roll Bars Made of Composite Materials

A stabilizer pushes down at the outer side and weights the outer wheel more, thus relieving the spring of that wheel. At the same time, it relieves the wheel inside the curve and loads its spring. Therefore, the stiffer stabilizer increases the wheel load difference on the axle. The lateral forces of the tires degressively increase above the wheel load. If a higher lateral force with a higher wheel load is to be produced on the outer tire, a larger slip angle is required. Consequently, a stronger suspension (and a stronger stabilizer) on the front axle increases the inclination to understeer. On the contrary, a stronger suspension (and a stronger stabilizer) on the rear axle increases the tendency to oversteer.

With the same weight distribution between the front and rear axle and the same tires, the axle with the (stronger) stabilizer thus reaches the limit of the transferable lateral guiding force more quickly.

The anti-roll bar made of composite material was patented (US2002000703 2002, https://www.uspto.gov). The composite beam (Figure 6.20, Pos. 16) is composed of a fiber-reinforced thermoset or thermoplastic polymer laminate material. Preferably, the polymer used in the composite laminate layers is an epoxy resin.

The patent (EP2423012 2012, https://www.epo.org) reveals a stabilizer bar that consists of a tubular rear part and two bending legs. The legs are made of a fiber composite material or a metallic material. The stabilizer rear part made of a metal material is wound around by the composite material. Coupling points are form-fit of force-locking coupled between the stabilizer rear part and legs. The legs in the coupling point are in part enfolded with the fiber composite material. Reinforcement fibers could be selected from glass fibers, carbon fibers or basalt fibers (Figure 6.21).

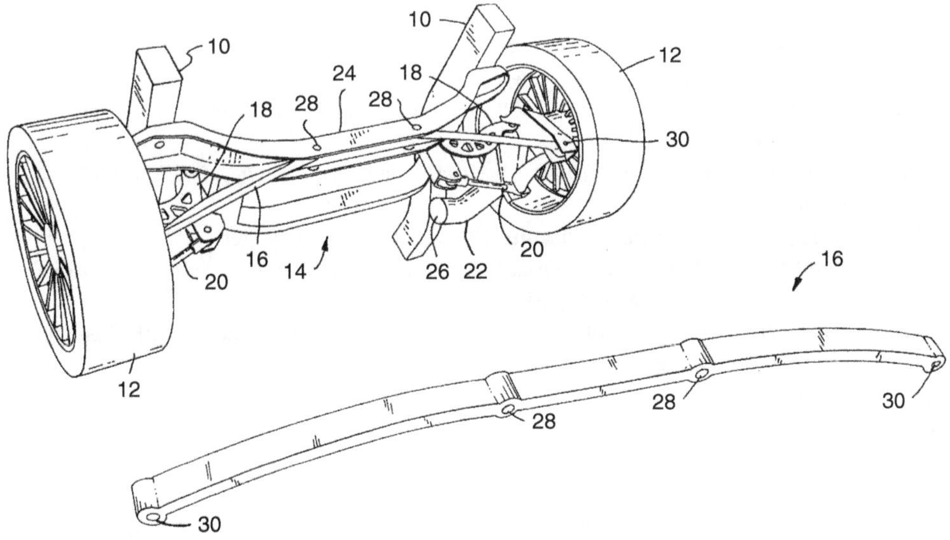

Figure 6.20 US2002000703 (A1) Wheel suspension system with an integrated link, spring and anti-roll bar, https://www.uspto.gov.

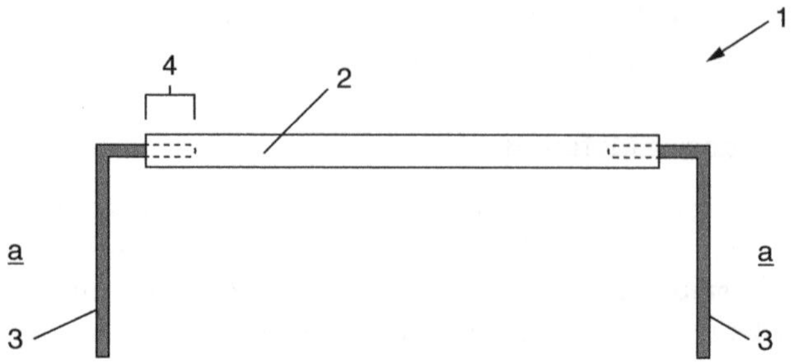

Figure 6.21 EP2423012 (A2) Fiber composite anti-roll bar, https://www.epo.org.

6.5 Conclusions

The models of dynamics for vehicles with solid, stiff elements of suspensions were presented. Parameters were usually assumed to be known or could be estimated. The aim of modeling is to obtain a mathematical-analytical description of the dynamic behavior of a vehicle with coupling between its principal degrees of freedom, but neglecting the elasticity of axle components.

References

Clover, C.L. and Bernard, J.E. (1998). Longitudinal tire dynamics. *Veh. Syst. Dyn.* 29 (4): 231–260. https://doi.org/10.1080/00423119808969374.

DIN ISO 8855:2013-11, Road Vehicles – Vehicle Dynamics and Road-Holding Ability – Vocabulary.

EP2423012A2 (2012) Fiber Composite Anti-Roll Bar.

Gillespie, T. (1992). *Fundamentals of Vehicle Dynamics, R-114*. Warrendale, PA: SAE.

Heißing, B. and Ersoy, M. (2013). *Grundlagen · Fahrdynamik, · Komponenten, · Systeme, · Mechatronik, Perspektiven*. Springer Vieweg.

Leiter, R., Mißbach, S., and Walden, M. (2008). *Fahrwerk, Lenkung, Reifen und Räder*. Würzburg: Vogel Buchverlag.

Mitschke, M. and Wallentowitz, H. (2014). *Dynamik der Kraftfahrzeuge*, 5e. Wiesbaden: Springer Fachmedien.

Pacejka, H. (2012). *Tire and Vehicle Dynamics*, 3e. Amsterdam: Elsevier.

Pacejka, H.B. and Bakker, E. (1992). The magic formula tire model. *Veh. Syst. Dyn.* 21 (S1): 1–18. https://doi.org/10.1080/00423119208969994.

Pauwelussen, J.P. (2015). *Essentials of Vehicle Dynamics*. Kidlington, Oxford, UK: Elsevier Ltd., Butterworth-Heinemann.

Reimpell, J. and Betzler, J. (2001, 1986). *Fahrwerktechnik: Radaufhängungen*. Würzburg: Vogel Buchverlag.

Reimpell, J., Stoll, H., and Betzler, J.W. (2001). *The Automotive Chassis: Engineering Principles*, 2e. Oxford, UK: Butterworth-Heinemann.

US2002000703 A1 (2002) Wheel suspension system having an integrated link, spring, and anti-roll bar.

7

Dynamics of a Vehicle With Flexible, Anisotropic Structural Elements of Chassis

7.1 Effects of Body and Chassis Elasticity on Vehicle Dynamics

7.1.1 Influence of Body Stiffness on Vehicle Dynamics

The influence of elasticity on the dynamic of vehicles is explained in detail in Section 32.8 of Mitschke and Wallentowitz (2014). The twist stiffness of the car body is a significant parameter of the vehicle design. It expresses resistance to torque that a body has when an axial twist moment is applied, see Chapter 6 in Heißing et al. (2011). This value is estimated for typical passenger car bodies between 15 and 40 kNm/°.

Bending stiffness refers to a difference in pitch angle between front and rear part of the body. A vehicle body bends when it is accelerated and load transfer occurs. This situation appears during accelerations and decelerations of the car. These effects are the so-called dive and squat behavior. The recovering of the pitch angle after striking a speed bump is another situation in which bending stiffness plays a role. While designing vehicle bodies it is believed that a high value of twist stiffness will also yield to an adequate bending stiffness value of the body.

Local stiffness refers to the stiffness of the suspension attachment points on the body. For evaluation of local stiffness values and its influence on car dynamics, the following full-car tests are applied: camber and toe, kinematics and compliance. A force or a torque is applied to the center of the wheel; accordingly, the resulting displacements and rotations at the wheel center must be acquired. An estimation of local stiffness of the car body is 0.2° per kN.

Lateral stiffness denotes the load case if a force is applied in the lateral direction at the front and rear axle. Centrifugal forces acting through the wheels affect cornering. The lateral stiffness distribution between front and rear axle affects the yaw performance of the car. Consequently, yaw behavior of the vehicle is possibly influenced. The typical lateral stiffness is about 1 kN mm^{-1}.

These mechanisms represent the effect of body stiffness on vehicle dynamics. Another frequently significant effect on the vehicle dynamics results from the stiffness of the suspensions. Two common types of suspensions repeatedly possess an insufficient lateral and longitudinal stiffness.

The first is the Hotchkiss rear suspension with leaf springs. If the lateral bending stiffness of the leaf springs is not high enough, the side deflection of the axis influences the

Design and Analysis of Composite Structures for Automotive Applications: Chassis and Drivetrain, First Edition. Vladimir Kobelev.

Companion website: www.wiley.com/go/kobelev/automotive_suspensions

steering behavior of the vehicle. Additionally, if the longitudinal bending stiffness of the leaf spring is low, drift of the axis in the driving direction occurs during braking. The brake drift of the axis in the driving direction due to brake forces also affects cornering behavior. These cornering errors could be significant for small and mid-size commercial motor and utility vehicles.

The second is the twist-beam axle, which frequently possesses minor lateral and chamber stiffness. The cornering errors due to flexibility of twist-beam axles are frequent because the twist-beam axis is the prevalent suspension in light passenger cars. The use of composite material could additionally provoke cornering errors because the perspective twist-beam axles made of composite materials will possess low values of chamber, toy and roll stiffness. These stiffness values differ from the known values for typical, steel-made twist-beam axles. The explanation of these effects for composite-made twist-beam axles is given in the next sections.

7.1.2 Lateral Dynamics of Vehicles With Stiff Rear Axles

Twist-beam axles for motor vehicles are made in various forms. Its sheet steel twist-beams, for example, are state of the art in shell construction. In addition, the practice has assembled cross beams from tubular profiles with various cross-sections and trailing arms to form twist-beam axles. When designing such a composite suspension, a general distinction is made between "torsion crank axle" and "twist-beam axle."

The control arms of the torsion crank axle are firmly connected to the axle bridge, a U-profile open at the bottom or in the rear direction. The axle bridge is referred to as twist-beam or cross-member. Occasionally, Watt linkage is added to withstand the centrifugal forces and correct the lateral guide. The Panhard rod leads to an asymmetry of the wheel guide and is not suited for passenger vehicles. With parallel spring movement, the axle swivels around the connecting line of the control arm bearings on the vehicle and, with rolling suspension, the open profile twists in the first approximation around the neutral torsion axis specified by its shear center.

When cornering at higher speeds, the wheels have a sideslip angle α, which results in the lateral tire forces required to support centrifugal forces. The overdriving self-steering behavior of motor vehicles with twist-beam axles is due to the flexibility of the trailing arms and the transverse beam. Figure 7.1 shows the relationship between the compliance of trailing arm and wishbone, the lateral forces of the tires and the overriding self-steering behavior.

For the beginning the slip motion of the vehicle due to later acceleration is neglected. In this case, the motor vehicle with an absolutely stiff, rigid rear axle drives through the curve without self-steering of the rear wheels and the track radius r_a (Ackermann radius) can be calculated by neglecting the sideslip angle as follows:

$$r_a = \frac{L_w}{\delta_f}, L_w = L_f + L_r. \tag{7.1}$$

The values L_f, L_r are the distances in longitudinal direction from the mass center to the front and rear wheelbases. The angle δ_f is the front wheel steer angle, which is arbitrary and adjustable by the car driver.

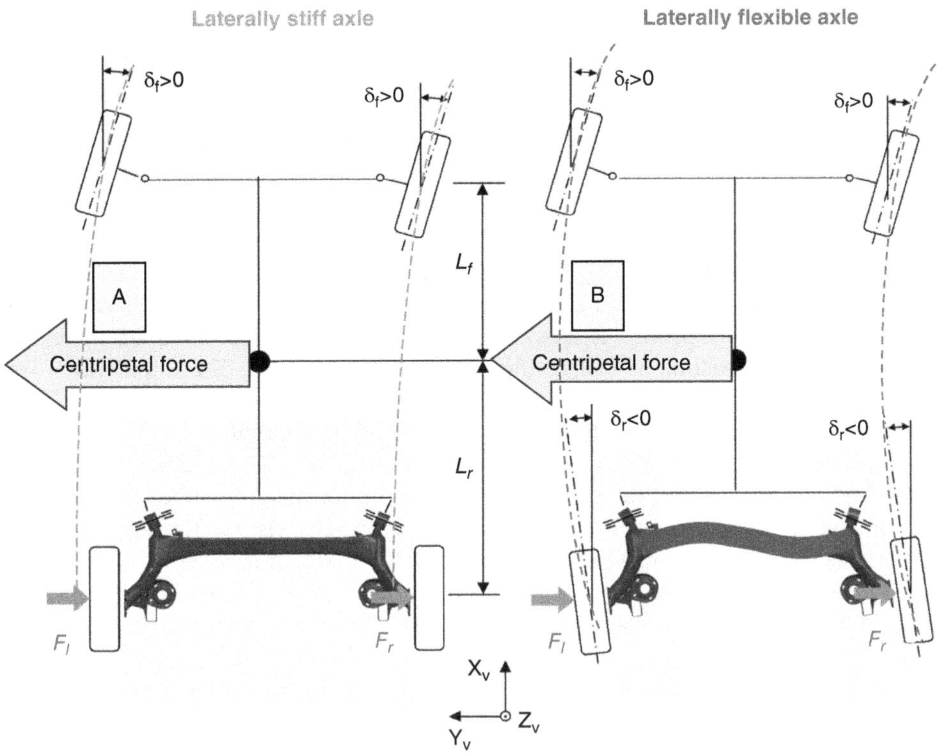

Figure 7.1 Cornering of a car with stiff and with flexible rear axles.

In the case of absolutely stiff axles and the absence of sideslip of the wheels, the front wheel steer angle is referred to as the Ackermann angle: $\delta_f \equiv \delta_f^A$.

7.1.3 Induced Effects on Wheel Orientation and Positioning of Vehicles with Flexible Rear Axle

The absolutely rigid twist-beam axle demonstrates neutral self-steering behavior for low vehicle velocity. With a flexible twist-beam axle, the lateral forces acting on the rear wheels deform the twist-beam axle in such a way that the trailing arms bend toward the momentary turning point, resulting in compliance steering of the rear and front wheels. In this section, we describe only the compliance steer of the rear axis. The compliance steer of the front axis is usually less significant and could be studied in the same way.

With a flexible twist-beam axle, the path radius r_e, of the vehicle with a flexible rear axis is expressed as:

$$r_e = \frac{L_W}{\delta_f + \delta_r}. \tag{7.2}$$

The compliance steer angle at the rear axle in (7.2) is δ_r. This angle depends upon the external lateral forces on the vehicle (Figure 7.1). Particularly, for a twist-beam axle, the lateral centrifugal forces cause bending of the axle and shift the car body outside from the turning point. Simultaneously, the centrifugal force moves the wheels, respectively, from car body toward the cornering center. This side wheel movement induces the compliance steer angle. The compliance steer angle is equal to the side movement of the rear wheel with respect to the car body divided by the length of trail arm of the rear axle. This calculation gives the compliance steer angle in radians, for the value in degrees this angle must be multiplied by $180/\pi$. The bending moment in the twist beam is equal to the length of the trail arm multiplied by the lateral force on the wheel. With bending stiffness of the twist beam, the compliance steer angle can be easily calculated, as shown later in this chapter. The maximal bending stiffness delivers cross-section in form of the double-T profile. Remarkably, the twist stiffness of this open profile is not too high, so the reasonable roll spring rate could be realized. The calculations are discussed next.

The compliance steer angle must be accounted for if the lateral forces on the rear axle cause significant lateral elastic deformation of links of the rear suspension. A comparison of Eqs. (7.1) and (7.2) shows that the radius r_e is smaller than the radius r_a, resulting in the known oversteering behavior of vehicles with a flexible twist-beam axle. Figure 7.1 demonstrates the oversteering behavior of a car with a flexible rear axle.

Another effect is camber alternation due to the side forces on the wheels. The lateral forces lead to the reduction of camber angle on the curve outer wheel and to increase of camber angle on the curve inner wheel. The outer wheel is overloaded due to centrifugal forces. The reduction of camber angle on the outer wheel increases the overdriving effect as well. If the wheel angle at the rear wheel due to elastic deformation is negligible, $\delta_r \approx 0$, track radius is approximately equal to the Ackermann radius r_a.

The principle of optimal adjustment of a twist beam could be summarized as:

- the motor vehicle swings around the roll axis when cornering due to centrifugal force,
- the centrifugal force generates a lateral force at the wheel,
- the lateral force on the wheel causes a bending moment on the trailing arm,
- bending of the trailing arm leads to compliance steer of rear wheels and causes oversteering,
- the rolling of the motor vehicle affects the drill of the cross-member that connects the right and left trailing arms,
- the coupling between twisting and bending of the torsion profile around the z-axis results in the coupling bending moment around the upwardly directed z-axis,
- the coupling bending moment compensates for the bending moment due to centrifugal force and increase understeer,
- the coupling between twisting and bending of the torsion profile about the x-axis reduces the variation of the camber angle.

The common vehicle model is based on the simplified single-track (bicycle) model from Riekert and Schunck (1940). This model is based on the following assumptions: Small steering and sideslip angles; aerodynamic forces are not considered; longitudinal forces are not considered and the relation between cornering stiffness and tire cornering force is considered to be linear. The simplified single-track model is, however, not suitable for analyzing the coupling of bending and torsion of the rear axle, as the vehicle width disappears and no rolling movements are therefore possible.

However, when the axle deforms, the wheel forces and angles are altered leading to a changed vehicle behavior dependent on axle stiffness values. To get a fundamental understanding of the relative influence from these global stiffness values on vehicle characteristics, the understeer gradient for varying axle stiffness values during steady state cornering is studied, based on the dedicated bicycle model of the vehicle. The model accounts for the influence of body stiffness on roll steer, the effect from varying lateral load transfer from varying torsion stiffness on axle cornering stiffness and the influence from lateral compliance on front and rear wheel angles.

The dedicated bicycle vehicle model is based on the following assumptions,

- no lifting and pitching movements,
- small movements and linear tire forces,
- massless wheels,
- constant travel speed and no longitudinal acceleration,
- roll angle is a linear function of the roll torque,
- vertical forces on the right and left wheel are linear functions of the rolling moment,
- rolling moment is linearly dependent on lateral acceleration,
- elastic wheel lock on the front wheel is a linear function of the lateral force on the front wheel,
- elastic wheel lock at the rear wheel is a linear function of the lateral force at the rear wheel,
- compliance steer angle of rear wheels depends linearly on the angle of torsion of the cross-member.

The twist-beam axles are frequently characterized by undesired self-steering behavior that tends to oversteer. A simple explanation of this behavior is the following. In order to drive through a certain curve, lateral forces on wheels are necessary to withstand the mass force of the vehicle (centrifugal force). With an absolutely rigid suspension (Figure 7.2a), the rear wheels are not self-steering. Neglecting the sideslip angle, the path radius (half diameter of curvature) is given by Eq. (7.1). In this case, the wheel steer angle δ_f is identical to the Ackermann angle, whose tangent is the wheelbase $L_f + L_r$ through the path radius of the rear axle center.

The lateral forces F_r at the rear wheels (Figure 7.2b) deform the wheel suspension in such a way that the longitudinally oriented links of the axle bend toward the momentary turning point. The reaction forces cause the suspension to bend, so that with a flexible suspension the rear wheels are self-steering. Neglecting the sideslip angle, the path radius (half diameter of curvature) is given by Eq. (7.2). The angle $\delta_r < 0$ is the induced wheel steer angle at the rear axis. Since the track radius on self-steering rear wheels r_e is smaller than the track radius r_r on rigid rear wheels:

$$r_r > r_e. \tag{7.3}$$

In this case, the undesired oversteer occurs.

The influence of laterally directed centrifugal forces on the camber angle could be studied analogously (Figure 7.3). For visibility, the roll motion of the vehicle is not displayed. The change of camber angle leads to the variation of the skew compliance of the tire, and consequently varies the side slip force of the tire. This effect could be accounted for with a proper tire model.

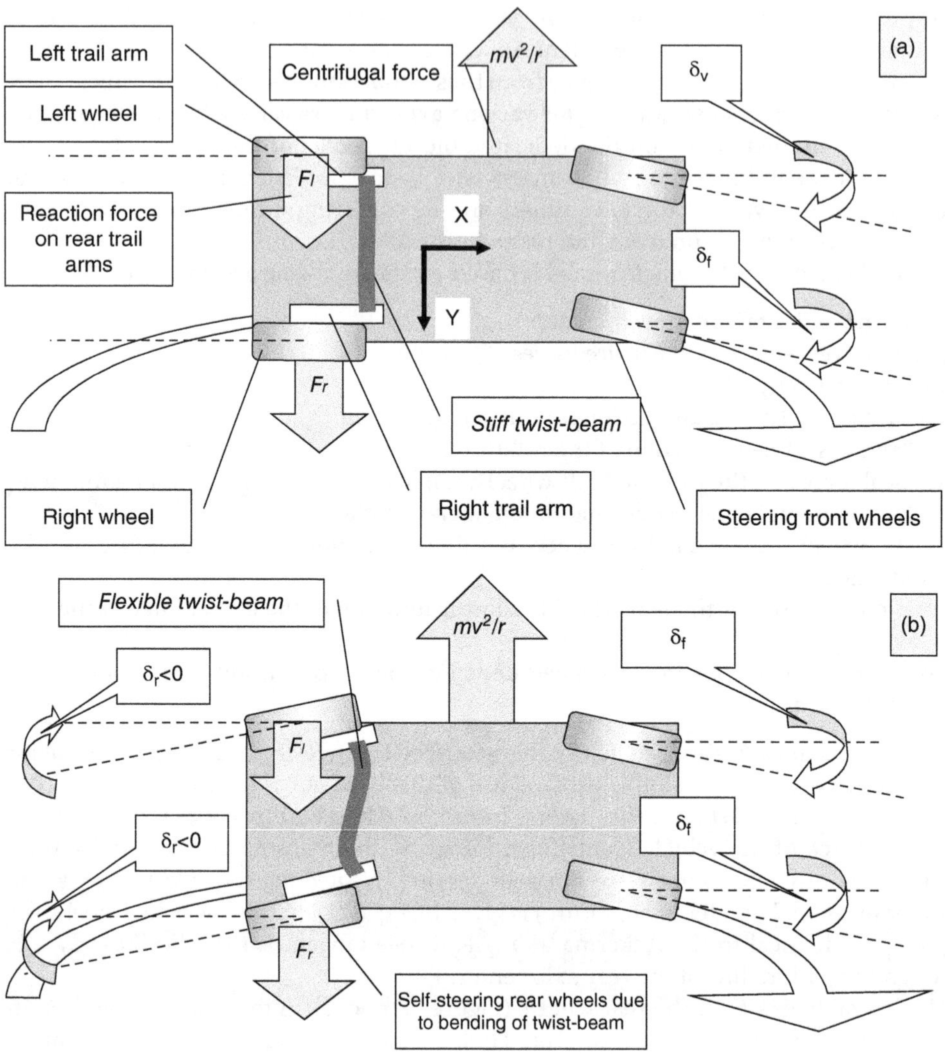

Figure 7.2 Steering of vehicle with (a) absolutely rigid axle and (b) induced steer angle of vehicle with flexible rear axle.

7.2 Self-Steering Behavior of a Vehicle With Coupling of Bending and Torsion

7.2.1 Countersteering for Vehicles with Twist-Beam Axles

7.2.1.1 Countersteering Mechanisms

The lateral stiffness of the twist-beam axle is occasionally low and causes oversteering of the vehicle. The solutions to this problem known from the state of the art are usually based on the modification of the handlebar geometry by generating a steering axle on the rear wheel or by using an open U-profile or V-profile with an offset torsion axle. There are a number of patents and patent applications that correct the steering behavior of

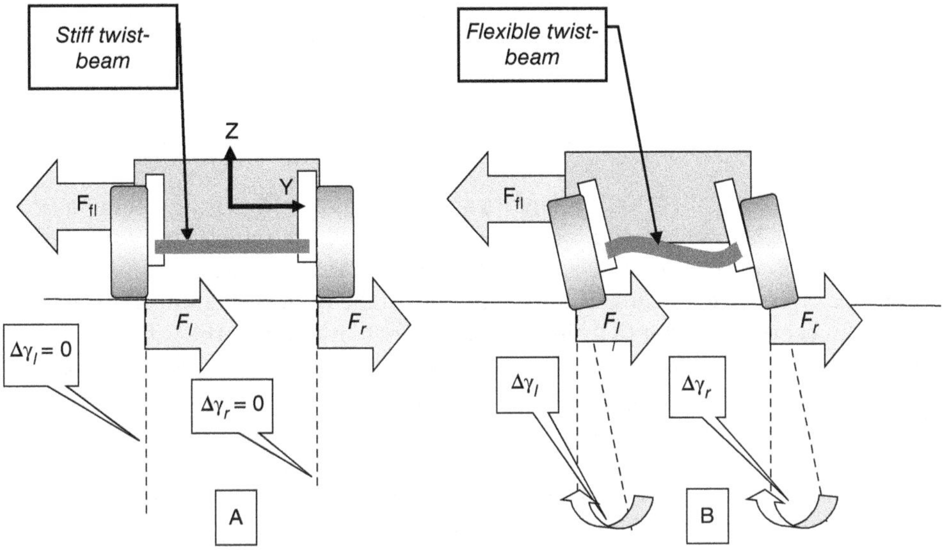

Figure 7.3 Induced camber angle for a flexible rear axle (roll motion is not displayed).

a vehicle with twist-beam axles. The inventions exploit the effect that is referred to as "countersteering."

The patent US8220810B2 (2008) introduces a countersteering vehicle rear axle that induces a rotational motion of a wheel carrier about a virtual spread axis in the toe-in direction under the influence of a lateral force on the outside cornering wheel mounted on the wheel carrier.

The invention (DE10321877 2004) relates to a single-wheel suspension of an inclined, longitudinal or composite strut variety comprising a wheel-guiding inclined, longitudinal or composite strut consisting of a wheel carrier mounted in several elastic pivoting bearings. The elastic pivoting bearings at least partially consist of two concentric rings or a ring and a bolt between which an elastomer body is non-detachably integrated in an at least partial manner. The wheel carrier is supported by means of an oscillating plate both in front of the axis of rotation of the wheel and to the rear of the said axis of rotation. Four elastic pivoting bearings are integrated into the oscillating plate. Half of the fixing elements thereof are secured to the strut and the other half are secured to the wheel carrier. According to the invention, a single-wheel suspension with a separately elastically mounted wheel carrier is developed. This wheel carrier enables a high amount of driving comfort in the presence of lateral, longitudinal and vertical forces.

The patent (EP1159147 2001) reveals a composite pivot pin for a motor vehicle, comprising two longitudinal side members, extending in a substantially longitudinal direction of said vehicle and a torsional cross-member that is joined to each longitudinal side member.

The invention (EP2019023 2009) describes the steering rear axle for automobile, comprising a deformable wheel pivot. The system has a steering actuator for controlling steering of rear wheels and fixed to a frame of a motor vehicle. The steering control causes distortion of a wheel pivot, which drives rotation of the steering axles following a vertical axis for steering of the rear wheels.

The patent (FR2860751 2005) proposes the steering axle, comprising two longitudinal arms, connected by a cross-member, each arm having a front end connected to the body by a hinge means adapted to deform elastically, to introduce a displacement of the wheel relative to the vehicle body during longitudinal impacts on the wheel. The upper hinge means being able to deform elastically according to a substantially longitudinal direction.

The invention (US6173977 2001) relates to a motor vehicle comprising steerable front wheels supporting the vehicle and rear wheels supporting the vehicle, at least one rear wheel of which is coupled by means of an at least substantially horizontal axle journal extending transversely to the longitudinal direction of the vehicle, to a wheel suspension arm extending at least substantially in the longitudinal direction of the vehicle.

The invention (WO2009124658 2009) improves a countersteering automotive rear axle that affects, under the impact of a lateral force onto the wheel on the outside of the bend, which wheel is fastened to a wheel carrier, a rotational movement of wheel carrier by a virtual axis of spread toward positive toe. Two supporting elements are provided above and below the wheel center, being flexible to torsion but resistant to bending and are tilted with respect to the vertical direction to the same extent as the virtual axis of spread when seen in a lateral projection.

The patent (WO2010000363 2010) introduces a countersteering automotive rear axle. The axis induces a rotational motion of a wheel carrier about a virtual axis of spread in the direction of positive toe-in under the effect of a side force on the wheel on the outside of a curve and mounted on wheel carrier, wherein the support comprises a segment substantially parallel to the plane of the wheel carrier and a segment extending substantially in the transverse vehicle direction and abutting said segment behind the wheel center point in the direction of travel.

7.2.1.2 Countersteering by Anisotropic Coupling of Bending and Torsion

The anisotropic elastic behavior of fiber composite materials can be used to couple bending and torsion deformation of the elastic connection profile. This principle is known as "anisotropic tailoring." This has been known in aviation for a while, but is still not widely used. The idea to influence the deformation behavior of airfoils with the help of fiber composites dates back to 1980. At this time, the articles by Krone (1975), Weisshaar (1980) and Piening (1984) appeared. These works dealt with the problem of divergence of forward arrowed wings. The torsion of the elastic axis by laminate rotation was regarded as one way of avoiding divergence. By pivoting the laminates in the top layers (upper and lower shell) of the wing box, the bending and torsion deformations are coupled. This type of tailoring is referred to as Circumferentially Asymmetric Stiffness (CAS) Configuration (Librescu and Song 2006). There is another Circumferentially Uniform Stiffness (CUS) Configuration, which leads to extension-torsion coupling of the box.

In addition to the use in airfoils, a number of theories have also been developed that aim at the use of coupled cross-sections (tension-torsion coupling) in helicopter rotor blades. For this purpose, analytical methods are also shown that are opposite for demonstrating the deformation behavior of coupled cross-sections (primarily using box girders as an example).

Orthotropic laminate families and classic (symmetric) anisotropic laminates are particularly appropriate for use in load-bearing structures. In principle, efforts are being made to keep the coupling between disk and plate quadrants as small as possible in order to minimize additional stresses. The anisotropic laminates offer the possibility to directly influence deformation behavior. In the case of classically anisotropic laminates, the coupling between tensile and shear deformation is decisive. When anisotropic laminates are used in the upper and lower surfaces of the box structure (symmetric twisting of the laminates), this coupling effect leads to a torsion bending coupling of the deformations (Kobelev, 2018).

Applying the principle of the CAS configuration to the automotive discipline, it states that, for an appropriate tuning, profile torsion in the transverse direction of the vehicle coupled to the bending of twist beam reduces the oversteering of the vehicle. The proposed countersteering design of the twist-beam axle advances dynamic driving characteristics and improves comfort properties. The central point here is the following. The coupling of bending and torsion in the cross-beam counteracts oversteer. The bending-torsion coupling can be achieved by anisotropy of the material of the cross-beam and distinctive orientation of the fiber composite.

On the other hand, the CUS configuration leads to extension-torsion coupling. The extension of torsion profile changes the angle between the trailing arms of the twist-beam axle. Consequently, the angles between the wheel planes, that is, the toe angle, change. This effect influences as well the cornering characteristics of the vehicle.

The qualitative statements about the dynamic of vehicles can usually be obtained with the greatly simplified "bicycle model" (Riekert and Schunck 1940). This model is based on the following statements:

- Constant wheel loads, no lifting and pitching motion
- Small movements, linear tire forces
- Wheels' planes normal to ground plane (lot position of wheels)
- Constant travel speed, no longitudinal acceleration
- Vanishing vehicle width, no rolling movement.

The aforementioned bicycle model is not suitable for analysis of countersteering with the coupling of bending and torsion of the rear axle. For optimum tuning of a twist-beam axle, the dedicated bicycle vehicle model is used. The dedicated model is based on the following hypotheses:

- No lifting and pitching motion.
- Small movements, linear tire forces.
- Constant travel speed, no longitudinal acceleration.
- The roll angle is a linear function of the rolling moment.
- The vertical forces on the right and left wheel are linear functions of the rolling moment.
- The rolling moment depends linearly on lateral acceleration.
- Elastic wheel inclination (camber angle change) on the front wheel is a linear function of the lateral force on the front wheel.
- Elastic wheel inclination (camber angle change) at the rear wheel is a linear function of the lateral force at the rear wheel.
- Compliance steer angle of the rear axle depends linearly on the torsion angle of the twist-beam (due to coupling of bending and torsion of the twist-beam axle).

7.2.2 Bending-Twist Coupling of a Countersteering Twist-Beam Axle

During cornering, vertical forces act on the wheels:

- Z_{rL} on the left rear wheel,
- Z_{rR} on the right rear wheel,
- Z_{fL} on the left front wheel,
- Z_{fR} on the right front wheel.

The sum of all vertical forces corresponds to the weight of the vehicle

$$Z_{fR} + Z_{rR} + Z_{fL} + Z_{rL} = m\,g. \tag{7.4}$$

The moments of the vertical forces around the front axle and the rear axle are compensated for by the moment of the vehicle weight

$$\begin{cases} (Z_{rR} + Z_{rL})L_w = m\,g\,L_f, \\ (Z_{fR} + Z_{fL})L_w = m\,g\,L_r. \end{cases} \tag{7.5}$$

The rolling moment around the longitudinal axis of the vehicle M_x is the product of the lateral force due to centripetal acceleration of the vehicle F_y and h is the vertical distance between the center of mass of the vehicle and its roll axis is:

$$M_x = h\,F_y, \tag{7.6}$$

where:

- a_y the centrifugal (lateral) acceleration of the vehicle;
- h the distance between the center of mass of the vehicle and its roll axis. For rough estimations, this value is approximately equal to the height of the mass center point of the vehicle with respect to the road surface.

The inertia force acting on the center of gravity of the vehicle $F_y = m\,a_y$ is the product of vehicle mass and centrifugal acceleration resulting from the momentary curvature of the path:

$$a_y = \frac{\mathrm{v}^2}{r} = \left(\frac{d\psi}{dt} - \frac{d\beta}{dt}\right)\mathrm{v}. \tag{7.7}$$

Here:

- v travel speed;
- r actual path radius (cornering radius);
- $d\mathrm{v}/dt$ path angular speed;
- $d\psi/dt$ yaw angular speed;
- $d\beta/dt$ slip angular speed.

The rolling moment is balanced with the moment of the vertical forces:

$$(-Z_{fR} - Z_{rR} + Z_{fL} + Z_{rL})B/2 = M_x. \tag{7.8}$$

Here, B is the rear track width (thread).

From Eqs. (7.4), (7.5), (7.6) and (7.8) the vertical forces result in:

$$Z_{fR} = \frac{L_r}{L_w}\left(\frac{m\,g}{2} - \frac{m\,a_y h}{B}\right), Z_{fL} = \frac{L_r}{L_w}\left(\frac{m\,g}{2} + \frac{m\,a_y h}{B}\right), \tag{7.9}$$

$$Z_{rR} = \frac{L_f}{L_w}\left(\frac{m\,g}{2} - \frac{m\,a_y h}{B}\right), Z_{rL} = \frac{L_f}{L_w}\left(\frac{m\,g}{2} + \frac{m\,a_y h}{B}\right). \tag{7.10}$$

If the lateral acceleration disappears ($a_y = 0$), the vertical forces are:

$$Z_{fR}^{(0)} = Z_{fL}^{(0)} = \frac{L_r}{L_w}\frac{m\,g}{2}, \tag{7.11}$$

$$Z_{rR}^{(0)} = Z_{rL}^{(0)} = \frac{L_f}{L_w}\frac{m\,g}{2}. \tag{7.12}$$

7.2.3 Roll Angle of Vehicle

7.2.3.1 Relationship Between Roll Angle and Centrifugal Force

There is a linear relationship between the roll angle φ and rolling moment of the vehicle $M_R = F_y\,h$ due to the centrifugal force $F_y = m\,a_y$:

$$\varphi = \frac{M_R}{R_r} = \frac{m\,a_y\,h}{R_r}. \tag{7.13}$$

The roll spring rate of the vehicle is R_r.

For simplicity, we consider only one elastic axle per vehicle, such that the rolling moment of the vehicle is identically equal to the roll moment of an elastic axle. Evidently, for calculation, all axles of the vehicle must be accounted for so that the roll spring rate of the vehicle is the sum of the roll spring rates per axle.

7.2.3.2 Lateral Reaction Forces on Wheels

The elastic wheel is the most important element for keeping vehicles on course. The cause of the side forces is the sideslip angle and the transverse or lateral slip connected to it. For small sideslip angles and a constant vertical force, the lateral force at the wheel depends on the sideslip angle of the corresponding wheel α_{fR}, α_{rR}, α_{fL}, α_{rL}. The lateral forces on the wheels are the linear functions of vertical forces and sideslip angles of the wheels:

$$F_{fL} = \frac{Z_{rL}}{Z_{rL}^{(0)}}\frac{\alpha_{fL}}{C_{\alpha f}^{(i)}},\ F_{fR} = \frac{Z_{rR}}{Z_{rR}^{(0)}}\frac{\alpha_{fR}}{C_{\alpha f}^{(i)}}, \tag{7.14}$$

$$F_{rL} = \frac{Z_{rL}}{Z_{rL}^{(0)}}\frac{\alpha_{rL}}{C_{\alpha r}^{(i)}},\ F_{rR} = \frac{Z_{rR}}{Z_{rR}^{(0)}}\frac{\alpha_{rR}}{C_{\alpha r}^{(i)}}. \tag{7.15}$$

The coefficients in Eqs. (7.14) and (7.15) are:

$C_{\alpha r}^{(i)} = c_r^{-1}$, $C_{\alpha f}^{(i)} = c_f^{-1}$ the skew compliance of the rear and front tires,
c_r, c_f lateral skew tire stiffness of the rear and front wheels, respectively.

The slip angles of the wheels are the angles between the wheel circumference direction and speed vector in the wheel position points. Considering geometric dependencies, the slip angles read:

$$\alpha_{fL} = \delta_{fL} + \beta - \frac{L_f}{\mathrm{v}}\frac{d\psi}{dt},\ \alpha_{fR} = \delta_{fR} + \beta - \frac{L_f}{\mathrm{v}}\frac{d\psi}{dt}, \tag{7.16}$$

$$\alpha_{rL} = \delta_{rL} + \beta + \frac{L_r}{\mathrm{v}}\frac{d\psi}{dt},\ \alpha_{rR} = \delta_{rR} + \beta + \frac{L_r}{\mathrm{v}}\frac{d\psi}{dt}. \tag{7.17}$$

In Eqs. (7.16) and (7.17), yaw angle is ψ, sideslip angle is β and δ_{VR}, δ_{HR}, δ_{VL}, δ_{HL} are the total steer angles on wheels. These angles are the sums of controllable and compliance-dependent steer angles.

7.2.3.3 Steer Angles on Front Wheels

The steer angles on the front wheels account for the elasticity of the steering and wheel guides due to vehicle reactions. If the rigidity of wheel guidance and steering is infinitely high, the steer angle at the front wheel corresponds to the ideal or controllable steer angle $\delta_{fL}^{(i)} = \delta_{fR}^{(i)}$.

For simplification, we consider large steering radii so that ideal steer angles on the left and right wheels are equal $\delta_{fL}^{(i)} = \delta_{fR}^{(i)} = \delta_f^{(i)}$, so that the difference of geometric steer angle between both wheels needs not to be accounted for.

However, there is the difference between the ideal steer angle at the front wheel $\delta_f^{(i)}$ and the total steer angle at the left δ_{fL} and right front wheel δ_{fR}. This difference is caused by the elastic flexibility of the wheel suspension. Namely, the difference $\delta_{fL} - \delta_f^{(i)}$ between the ideal steer angle at the front wheel $\delta_V^{(i)}$ and the actual turning angle at the left wheel δ_{fL} is a linear function of the lateral force at the left wheel F_{fL}:

$$\frac{\delta_{fL} - \delta_f^{(i)}}{S_f} = F_{fL}. \tag{7.18}$$

Here, S_f is the side compliance of the front wheel guidance ($S_f = s_f^{-1}$, where s_f side stiffness constant of the front wheel guidance).

Similarly, for the right front wheel, the following applies:

$$\frac{\delta_{fR} - \delta_f^{(i)}}{S_f} = F_{fR}. \tag{7.19}$$

7.2.3.4 Steer Angles on Rear Wheels

The steer angles of rear wheels are affected by the elasticity of the axles in the case of centripetal forces on the one hand, and the coupling between roll angle φ and the corresponding additional steer angle on the rear wheel on the other.

If the rigidity of the wheel suspension and steering is infinitely high and the coupling vanishes, the steer angles of the rear axle disappear:

$$\delta_{rR}^{(i)} = \delta_{rL}^{(i)} = 0. \tag{7.20}$$

Otherwise, the torsion of the twist-beam could lead to the inclination of the wheel plane relative to the midplane of the vehicle. In this case, the steer angle at the rear wheel $\delta_H^{(i)}$ will be:

$$\delta_r^{(i)} = k_\vartheta \vartheta. \tag{7.21}$$

We make no difference between the steer angles of both wheels of the rear axle: $\delta_{rR}^{(i)} = \delta_{rL}^{(i)} \equiv \delta_r^{(i)}$. In (7.21), k_ϑ is the constant of the coupling between the torsion angle ϑ of the twist-beam axis and the steer angle.

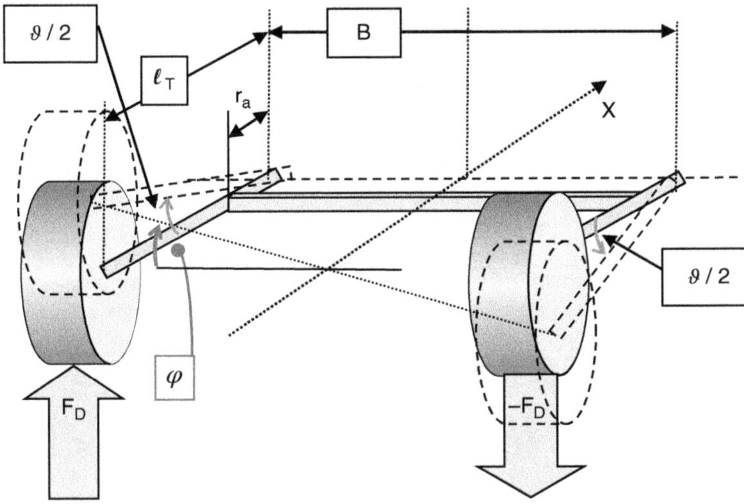

Figure 7.4 Relationship between the rolling motion of the vehicle and twisting of the cross-member.

In turn, the torsion angle ϑ depends on the roll angle φ:

$$\vartheta = \frac{\varphi B}{2l_T}. \tag{7.22}$$

Here, l_T is the length of the trailing arm (Figure 7.4). This relationship provides the dependence of the coupling steer angle at the rear wheel $\delta_H^{(i)}$ on the roll angle φ:

$$\delta_H^{(i)} = k_\varphi \varphi. \tag{7.23}$$

With (7.18) and (7.19), the following relation results:

$$k_\varphi = \frac{B}{2l_T} k_\vartheta. \tag{7.24}$$

The difference between the ideal turning angle at the rear wheel $\delta_r^{(i)}$ and the actual turning angle at the left δ_{rL} and right rear wheel δ_{rR} is caused by the elastic flexibility of the wheel suspension.

The difference $\delta_{rL} - \delta_{rL}^{(i)}$ between the ideal turning angle at the rear wheel $\delta_{rL}^{(i)}$ and the actual turning angle at the left wheel δ_{rL} is a linear function of the lateral force at the left wheel F_{rL}:

$$\frac{\delta_{rL} - \delta_r^{(i)}}{S_r} = F_{rL}. \tag{7.25}$$

Here, S_r is the side flexibility of the rear axle ($S_r = s_r^{-1}$, s_r stiffness constant of the rear wheel guidance).

For the right front wheel, the following applies accordingly:

$$\frac{\delta_{rR} - \delta_r^{(i)}}{S_r} = F_{rR}. \tag{7.26}$$

7.3 Steady Cornering of a Flexible Vehicle

7.3.1 Stationary Cornering of a Car With a Flexible Chassis

For stationary circular motion of a vehicle, the travel speed, yaw speed and floating angle are all constant (Mitschke and Wallentowitz 2014). The condition for equilibrium of moments around the center of the front axle is

$$(F_{fR} + F_{fL})L_w = m\,a_y\,L_r. \tag{7.27}$$

The condition for equilibrium of moments around the center of the rear axle is

$$(F_{rR} + F_{rL})L_w = m\,a_y\,L_f. \tag{7.28}$$

For the stationary round trip, the following applies

$$d\psi/dt = \text{v}/r \tag{7.29}$$

for yaw rate and

$$d\beta/dt = 0 \tag{7.30}$$

for sideslip angle rate.

Equations (7.27)–(7.30) provide the linear system of equations for the unknown steer angles at the front wheel $\delta_V^{(i)}$ and for the sideslip angle β. The speed of the vehicle v and the cornering radius r are kept constant.

The equations are simplified for a small lateral acceleration a_y. The Taylor series of the equations according to a_y, leads to the following equations:

$$\delta_f^{(i)} + \beta + \frac{m\,a_y L_r (S_f - C_{\alpha f})}{2L_w} - \frac{L_f}{r} = 0, \tag{7.31}$$

$$\beta + \frac{m\,a_y L_f (S_r - C_{\alpha r})}{2L_w} + \delta_r + \frac{L_r}{r} = 0, \text{ with } \delta_r = k_\varphi \varphi. \tag{7.32}$$

The solution of the Eq. (7.32) with (7.13) provides the formula for the sideslip angle:

$$\beta = -\frac{m\,a_y L_f (S_f - C_{\alpha r})}{2L_w} - \frac{m\,a_y h}{R_r} k_\varphi - \frac{L_r}{r}. \tag{7.33}$$

The substitution of (7.33) into (7.31) provides finally the steer angle to the cornering on the curve with the given radius r:

$$\delta_f^{(i)} = \frac{m\,a_y L_f (S_r - C_{\alpha r})}{2L_w} - \frac{m\,a_y L_r (S_f - C_{\alpha f})}{2L_w} + \frac{m\,a_y h}{R_r} k_\varphi + \frac{L_w}{r}. \tag{7.34}$$

7.3.2 Necessary Steer Angles for Coupling and Flexibility of Chassis

7.3.2.1 Limit Case: Lateral Acceleration Vanishes

The lateral acceleration a_y in this case (A) disappears, $a_y = 0$. Equations (7.26) and (7.27) are reduced to:

$$\delta_f^{(i)} + \beta - \frac{L_f}{r} = 0, \tag{7.35}$$

$$\beta + \frac{L_r}{r} = 0. \tag{7.36}$$

The solution (7.28) provides:

$$\beta = -\frac{L_r}{r}, \tag{7.37}$$

$$\delta_f^{(i)} = -\beta + \frac{L_f}{r} = \frac{L_w}{r}. \tag{7.38}$$

The steer angle required to drive on a curve corresponds to the angle that depends solely on the geometric data (Ackermann angle):

$$\delta_f^A = \frac{L_w}{r}. \tag{7.39}$$

7.3.2.2 Absolutely Rigid Front and Rear Wheel Suspensions

In this case, the front and rear wheel suspensions are absolutely rigid:

$$S_f - C_{\alpha f} = 0, S_r - C_{\alpha r} = 0. \tag{7.40}$$

Equations (7.26) and (7.27) reduce to

$$\delta_f^{(i)} + \beta - \frac{L_f}{r} = 0, \tag{7.41}$$

$$\beta + \frac{m\, a_y h}{R_r} k_\varphi + \frac{L_r}{r} = 0. \tag{7.42}$$

The solution (3.2) provides

$$\beta = -\frac{L_r}{r} - \frac{m\, a_y h}{R_r} k_\varphi, \tag{7.43}$$

$$\delta_f^{(i)} = -\beta + \frac{L_f}{r} = \frac{L_w}{r} + \frac{m\, a_y h}{R_r} k_\varphi. \tag{7.44}$$

The steer angle required to drive on a curve δ_f^B is the sum of the Ackermann angle δ_f^A and the steer angle Δ_k that characterizes the coupling between bending and torsion:

$$\delta_f^B = \delta_f^A + \Delta_k., \tag{7.45}$$

$$\Delta_k = m\, a_y h k_\varphi / R_r. \tag{7.46}$$

The steer angle required to drive on a curve δ_f^B can be larger or smaller, such as the Ackermann angle δ_f^A. If the angle is $\Delta_k > 0$ the understeering behavior of the vehicle is increased.

7.3.2.3 Bending and Torsion of a Twist Member Completely Decoupled

The steer angle required to drive on the curve δ_f^C is the sum of the Ackermann angle δ_f^A and the angle Δ_s, which characterizes the flexibility of the wheel suspension:

$$\delta_f^C = \delta_f^A + \Delta_s, \tag{7.47}$$

$$\Delta_s = \Delta_f + \Delta_{r,} \tag{7.48}$$

$$\Delta_r = \frac{L_f}{2L_w}(S_f - C_{\alpha f}) m\, a_y, \tag{7.49}$$

$$\Delta_f = -\frac{L_r}{2L_w}(S_f - C_{\alpha f}) m\, a_y. \tag{7.50}$$

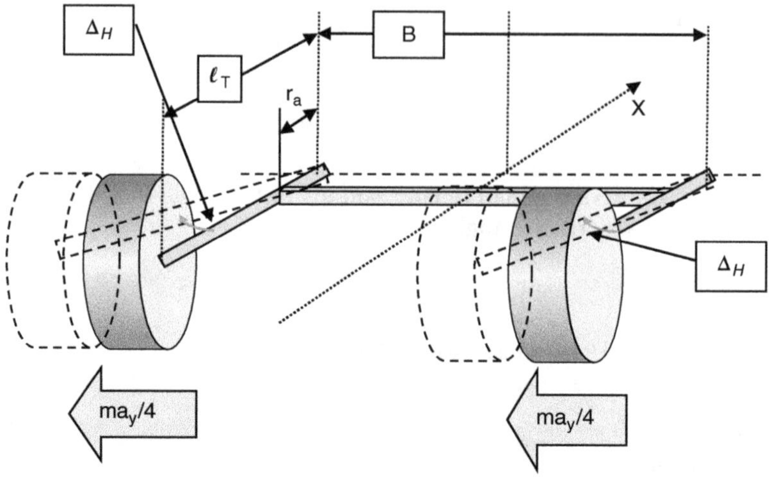

Figure 7.5 Lateral shift of the wheel and additional steer angle due to centrifugal force.

The angle Δ_r, characterizes the flexibility of the rear suspension (Figure 7.5). A negative sign of Δ_r describes the fact that the elastic rear axle requires lower steer angles compared to the Ackermann angle and vehicle overdriven. The angle Δ_f represents the influence of the elasticity of the front suspension. A positive sign of Δ_f describes the fact that the elastic steering requires larger steer angles compared to the Ackermann angle, that is, the vehicle is understeered. Both angles depend on the elasticity of the axles and for calculations are assumed to be prescribed by suspension design.

7.3.2.4 General Case of Coupling Between Bending and Torsion of a Twist Member

The steer angle required to drive on the curve δ_f^D is composed of the Ackermann angle and the "angle" Δ_s, which characterizes the flexibility of the wheel guides and the "angle" Δ_k which characterizes the coupling between bending and torsion:

$$\delta_f^D = \delta_f^A + \Delta_s + \Delta_k. \tag{7.51}$$

7.3.2.5 Neutral Steering Caused by Coupling Between Bending and Torsion of a Twist Member

The full compensation is achieved, if the condition satisfies:

$$\Delta_s + \Delta_k = 0, \text{or } \Delta_f + \Delta_r + \Delta_k = 0. \tag{7.52}$$

As the angles Δ_r, Δ_f, are given and cannot be changed, the only constant to fit the suspension stiffness is the coupling constant Δ_k from (7.46). Substitution of (7.46), (7.49) and (7.50) into (7.52) gives:

$$\frac{L_f}{2L_w}(S_r - C_{\alpha r})ma_y - \frac{L_r}{2L_w}(S_f - C_{\alpha f})ma_y + \frac{ma_y hk_\varphi}{R_r} = 0. \tag{7.53}$$

The factor $m\,a_y$ shortens in Eq. (7.53). The resolution of Eq. (7.53) with respect to coupling constant k_φ expresses the coupling constant, which leads to neutral steering behavior of vehicle:

$$k_{\varphi.neutral} = \frac{R_r}{2h}\frac{L_r(S_f - C_{\alpha f}) - L_f(S_r - C_{\alpha r})}{L_w}. \tag{7.54}$$

The proper coupling between bending and torsion of twist member leads to a full compensation of oversteer. In this case of full compensation, neutral steering is caused by coupling between bending and torsion of twist member. For full compensation, the coupling coefficient k_φ between steer angle and roll angle is to be equal to $k_{\varphi.\,neutral}$.

If the condition (7.52) will be satisfied, the vehicle always steers neutrally for all speeds and lateral accelerations. The cornering character of such a vehicle corresponds to Ackermann cornering. The steer angle on the wheel is equal to the Ackermann steer angle δ_f^A. For slightly different coupling constant a small desirable understeer could be achieved.

7.4 Estimation of Coupling Constant for a Twist Member

7.4.1 Coupling Between Vehicle Roll Angle and Twist of Cross-Member

The following calculation models for the individual stiffnesses of the twist-beam axle are based on assumptions for simplifying the mechanical replacement model of the twist-beam axle. The width of the cross-beam corresponds to the width of the twist-beam axle, the center line of the cross-beam cross-section runs through the connecting line of the rubber bearing and the cross-section is constant over the entire width. The elastic properties of the rubber mounting are neglected and the trailing arms are assumed to be rigid. Figure 7.6 displays the effect of the rolling motion of the vehicle on the twisting of the cross-beam. Namely, the rolling motion of the vehicle during cornering causes the outer wheel of the curve to spring in and the inner wheel of the curve to come out. The flexural trailing arms of the twist-beam axle thus generate twisting in the cross-beam. The roll movement is described by the roll angle of the vehicle φ and the twisting by the torsion angle ϑ. The proportional relationship between

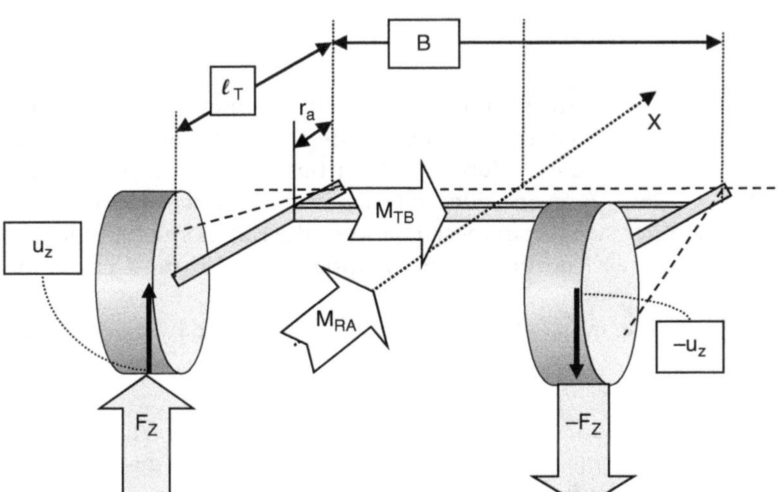

Figure 7.6 Relation between the torque M_{TB} in the twist-beam and the roll moment M_{RA} of the axle. B = rear track width = thread (is approximately the twist-beam length) l_T = trailing arm length.

the steer angle δ_r of the rear wheels and the torsion angle of the cross-member ϑ is:

$$\delta_r = k_\varphi \cdot \vartheta, \tag{7.55}$$

where k_φ is coupling constant between twist angle of cross-member and the rear wheel steer angle axis and ϑ is the twist angle of the cross-member.

Furthermore, there is a linear relationship between the torsion angle ϑ and the roll angle of the vehicle φ:

$$\vartheta = {}^{B\varphi}\!/_{2l_T}. \tag{7.56}$$

The coupling constant k_φ links the steer angle at the rear wheel $\delta_r^{(i)}$ to the roll angle of the vehicle. In its turn, this constant depends upon the coupling constant between twist angle of cross-member and the induced rear wheel steer angle:

$$\delta_r = k_\varphi \cdot \varphi, \quad k_\varphi = {}^{k_\vartheta \vartheta}\!/_{\varphi} = {}^{Bk_\vartheta}\!/_{2l_T}. \tag{7.57}$$

7.4.2 Stiffness Parameters of a Twist-Beam Axle

7.4.2.1 Roll Spring Rate

The overall flexibility of the twist-beam axle is made up of the roll, side and camber stiffness. The roll stiffness of the twist-beam axle allows the rear wheels to deflect in the opposite direction when cornering.

$$Z_{\mathrm{rR}} = \frac{L_f}{L_w}\left(\frac{m\,g}{2} - \frac{m\,a_y h}{B}\right) \equiv Z_{rR}^{(0)} - F_z, \tag{7.58}$$

$$Z_{\mathrm{rL}} = \frac{L_f}{L_w}\left(\frac{m\,g}{2} + \frac{m\,a_y h}{B}\right) \equiv Z_{rL}^{(0)} + F_z. \tag{7.59}$$

From (7.59), the dynamical vertical forces on both wheels are:

$$F_z = \frac{m a_y h L_f}{B L_w}. \tag{7.60}$$

The roll moment of the vehicle is the sum of the partial roll moments of all axles of a vehicle. To calculate the roll moment of the twist-beam axle, the change of wheel vertical force F_Z is considered (Figure 7.6). The two forces (7.60) on both wheels contribute to the roll moment of the axis $M_{RA} = F_Z B$ on one side, and cause the torque in the twist-beam $M_{TB} = F_Z(l_T - r_a)$ on the other side. From these relations, the roll moment of the axis expresses over the torque in twist-beam: $M_{RA} = M_{TB}B/(l_T - r_a)$. For most cases, the value of r_a could be neglected (Figure 7.7).

With (7.58)–(7.60), the rolling moment of the axis M_{RA} is calculated by the equation:

$$M_{RA} = F_z \cdot B. \tag{7.61}$$

The opposite vertical forces F_z are caused by the rolling motion of the vehicle during cornering and cause the cross-member to twist (Figure 7.6). The moment due to vertical forces (7.61) winds round the twist-beam axle during cornering. The roll angle φ can be calculated for moderate roll according to equation:

$$\varphi = 2u_z/B. \tag{7.62}$$

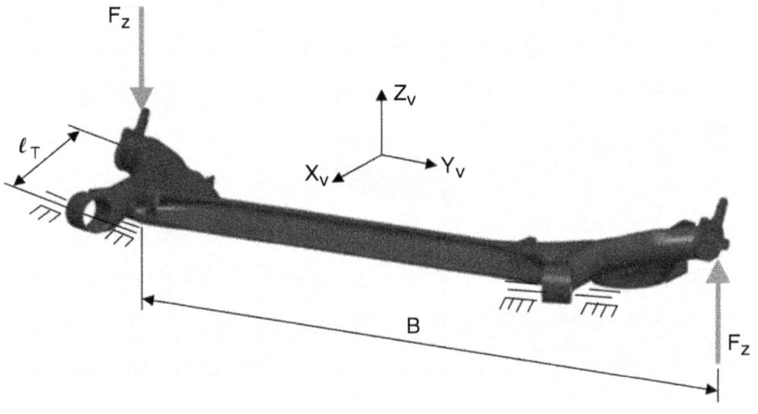

Figure 7.7 Suspension roll stiffness of a twist-beam axle.

In (7.62) u_z is the wheel stroke (Figure 7.6). The roll spring rate R_r of the twist-beam axle is consequently:

$$R_r = \frac{M_{RA}}{\varphi} = \frac{F_z \cdot B^2}{2 \cdot u_z}. \tag{7.63}$$

The roll spring rate R_r of the twist-beam axle is linearly dependent on the torsion stiffness $C_{44}I_t$ of the cross-beam. The value C_{44} is the shear modulus in the local coordinate system xyz, which is associated with the axis z of the twist beam. The torsion moment in the cross-member can be determined as:

$$M_t = F_z \cdot l_T. \tag{7.64}$$

The torsion angle ϑ, which describes the twisting of the cross-beam due to the introduced torsion moment M_t, can be determined with Eq. (7.5) or equation:

$$\vartheta = u_z / l_T. \tag{7.65}$$

The torsion stiffness $C_{44}I_t$ of the twist-beam is:

$$C_{44}I_t = \frac{M_t}{\vartheta}. \tag{7.66}$$

The roll spring rate R_r is a function of the torsion stiffness $C_{44}I_t$ of the cross-beam R_r:

$$R_r = C_{44}I_t \cdot \frac{{l_T}^2}{2 \cdot B^2}. \tag{7.67}$$

The torsion stiffness $C_{44}I_t$ composed of the moment of inertia I_t and the corresponding elasticity constant.

7.4.2.2 Lateral Stiffness

When cornering, the trailing arms are loaded by a bending moment M_L due to the lateral or lateral force F_L·. The deformation of the twist-beam axle is determined by the lateral rigidity. The lateral forces F_L·, acting on the twist-beam axle in this case, can be seen in Figure 7.8.

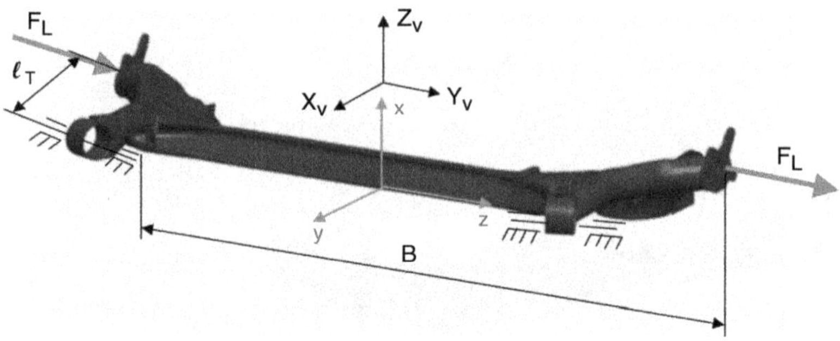

Figure 7.8 Lateral rigidity of a twist-beam axle.

The bending moment is determined with the equation:

$$M_L = F_L \cdot l_T. \tag{7.68}$$

The lateral displacement of the trailing arm reads (Φ_L is the bending angle):

$$u_L = \Phi_L \cdot l_T. \tag{7.69}$$

The lateral spring rate of the twist-beam axle is:

$$R_L = F_L / u_L. \tag{7.70}$$

The bending moment can also be expressed as:

$$M_L = C_{33} I_z \cdot \Phi_L. \tag{7.71}$$

with $C_{33} I_z$ as the bending stiffness (in the component coordinate system). The value C_{33} is the longitudinal elasticity modulus in the local coordinate system *xyz*, which is associated with the *z*-axis of the twist beam.

The lateral spring rate of the twist-beam axle reads (Figure 7.8):

$$R_L = C_{33} I_z / l_T^{\,2}. \tag{7.72}$$

Equation (7.72) demonstrates that the lateral stiffness of the semi-trail axis is directly dependent on the corresponding bending stiffness of the cross-beam.

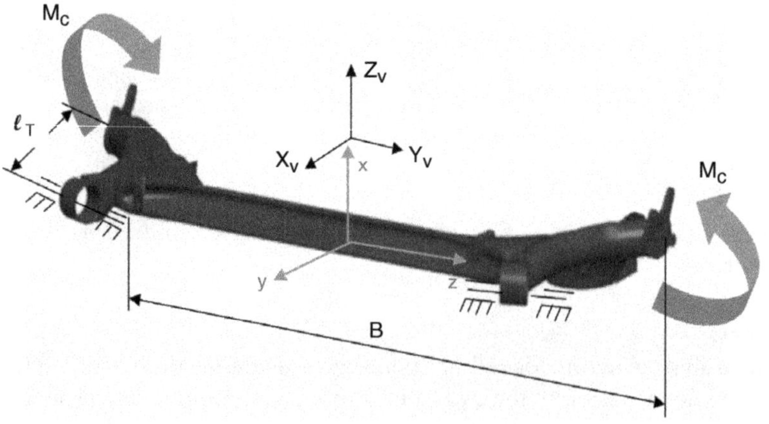

Figure 7.9 Camber stiffness of a twist-beam axle.

7.4.2.3 Camber Stiffness

The camber stiffness R_C is equal to the bending stiffness EI_y of the cross-beam and is calculated by the equation (Figure 7.9):

$$R_C = M_C/\varepsilon_c. \tag{7.73}$$

In Eq. (7.73) M_C is the overturning moment and ε_c camber angle.

7.5 Design of the Countersteering Twist-Beam Axle

7.5.1 Requirements for a Countersteering Twist-Beam Axle

This section describes how the cross-beam of the countersteering twist-beam axle is designed so that it reduces the tendency to oversteer according to the principle of optimal adjustment of a twist-beam axle. The requirements for the countersteering twist-beam axle are formulated in a specification sheet. Then the cross-beam profile is selected and calculated and a prototype of the twist-beam axle is constructed. Particular attention is paid to the connection between the two trailing arms and the cross-member; that is, the application of force into the cross-member.

The structure, which is mirror-symmetric to the symmetry plane of the vehicle, leads to the desired tension-torsion coupling. The mirror-symmetric coating is produced by an oblique wrapping of the structure box (Figure 7.10). The cross-beam is to be preferably considered in the local coordinate system of the structural component. The z-axis is directed along the length axis of the twist-beam and x- and y-axes are in the cross-section plane of the beam. This choice of the local coordinate system is conforming to the alignment of the coordinate axes from Chapter 4.

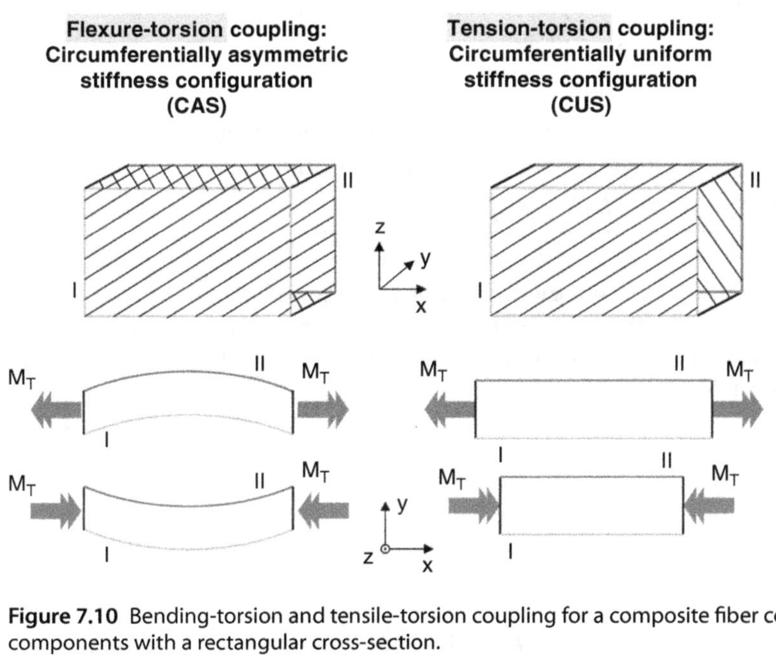

Figure 7.10 Bending-torsion and tensile-torsion coupling for a composite fiber component for components with a rectangular cross-section.

The connection in the two mutually oriented symmetric laminates generates the bending moments in the z-axis on the handlebars and compensates for the overdrive of the vehicle. For the optimal dynamical cornering of the vehicle, the bending-torsion coupling provides the necessary compensation of side forces. The compensation leads to the neutral self-steering behavior, improving driving dynamics. The principle of optimal suspension tuning can be illustrated as follows:

- the vehicle swings around the roll axis when cornering due to centrifugal force;
- the centrifugal force generates a lateral force at the wheel;
- the lateral force causes a bending moment on the handlebars that bends the handlebars and can lead to oversteer;
- the rolling of the vehicle causes the torsion profile to twist between the left and right handlebars;
- the coupling between twisting and bending of the torsion profile leads to the coupling bending moment around the upwardly directed z-axis;
- the coupling bending moment compensates for the bending moment due to centrifugal force and compensates for the negative tendency to oversteer.

The axis corresponding to the invention is clearly defined by a coupling parameter. The coupling parameter is the ratio of the steer angle on the wheel carrier to the rotation angle of the twist-beam axis. If the coupling parameter is positive, the head bending moment compensates for the bending moment due to centrifugal force and compensates for the negative tendency to oversteer. If the coupling parameter is positive, the coupling bending moment increases the bending moment due to centrifugal force and increases the undesirable tendency to oversteer.

The use of coupled laminates opens up new possibilities to influence the deformation and self-steering behavior of the wheel suspension. In principle, bending and torsion in an open profile are coupled solely by the geometric conditions. This effect can be further influenced by the bending-torsion coupling of the anisotropic structure. The material coupling stiffness is the additional parameter to control the elastic-kinematics depending on the load.

The system equations for the modified vehicle model lead to one formula, with which one can determine the steer angle necessary to drive on a curve.

$$\delta_f = ma_y\left[\frac{L_f(S_r - C_{\alpha r})}{2L_w} - \frac{L_r(S_f - C_{\alpha f})}{2L_w} + \frac{h}{R_r}k_\vartheta\right] + \frac{L_w}{r}. \tag{7.74}$$

The first term of Eq. (7.3) results in the angle, which characterizes the flexibility of the wheel suspension and the coupling between bending and torsion. The Ackermann angle is described by the last term in Eq. (7.74). If the first term vanishes, the neutral steering of the vehicle will be achieved.

The specification formulates the requirements for the countersteering twist-beam axle, which must be accounted for in the design so that the functional model reduces the tendency to overdriving self-steering behavior according to the principle of optimum adjustment of a twist-beam axle.

With the derived equations, it is possible to determine the cross-section of the cross-member with given material properties via the coupling coefficient in such a way that a neutral self-steering behavior is achieved. The countersteering twist-beam axle should have the identical roll spring rate R_r as the reference axle. Furthermore, the new twist-beam must have at least equal or higher lateral stiffness R_L and camber stiffness R_C as the reference axis.

In the local coordinates of the twist-beam, the prescribed components of stiffness for the twist-beam axle with bending-twist coupling must be achieved:

- the torsional stiffness $C_{44}I_t$ must be equal to the obligatory torsional stiffness of the reference axis in order to guarantee the roll stiffness of the vehicle;
- the bending stiffness $C_{33}I_y$ must be greater than the bending stiffness of the reference axis in order to guarantee the camber stiffness of the axis;
- the bending stiffness $C_{33}I_x$ must also be greater than the bending stiffness of the reference axis in order to guarantee the lateral stiffness of the axis.

For the optimal adjustment of a twist-beam axle, a bending-torsion coupling in the cross-beam is also necessary. This coupling constant is proportional to $C_{34}I_t$. The left and right parts of the beam must possess the CAS Configuration (Librescu and Song 2006; Rehfield and Atilgan 1989) (Figure 7.10).

The structure of the laminate is described as symmetric for the bending-torsion coupling and mirror-symmetric for the tensile-torsion coupling. If the component is loaded with a positive torsion moment $\mathbf{M}_T$ with symmetric laminate for the bending-torsion coupling, it bends around the positive z-axis. A load with a negative torsion moment $\mathbf{M}_T$ leads to a bending around the negative z-axis. A fiber composite component with mirror-symmetric laminate construction is extended by the load with a positive torsion moment M_t and shortens under a load with a negative torsion moment $\mathbf{M}_T$.

Figure 7.11 shows the laminate structure for creating a bending-torsion coupling in a component made of fiber-reinforced material, applied to the principle of optimum adjustment of a composite beam axle. Figure 7.11 demonstrates that the cross-beam (A) consists of two CAS laminates interconnected in the middle of the vehicle. Both CAS laminates are mirror-symmetric with respect to the lateral plane of the vehicle. The structure of two in the midplane of the vehicle interconnected symmetric laminates assures the spatial S-shaped deformation of the cross-beam. The fiber angle α should in the range $\alpha = 10$–$20°$ to the x-axis to achieve optimum stiffness.

Another alternative design (B) of the cross-beam could also improve the camber angle. The cross-beam (A) again consists of two CAS laminates interconnected in the middle plane of the vehicle. To improve the camber angle, the twist-beam is mounted inclined to the YZ plane of the vehicle. If the inclination angle vanishes, $\varepsilon_{YZ} = 0$, the design (B) turns into the design (A), which corrects only the compliance steer angle. If the inclination plane will be parallel to XY plane and inclination angle is $\varepsilon_{YZ} = 90°$, the design (B) only corrects the camber change. For the intermediate inclination angles $0° < \varepsilon_{YZ} < 90°$, the design (B) corrects both the compliance steer angle and the compliance camber angle.

The optimal setting could be determined in the full-scale vehicle test. For a possible installation of the countersteering twist-beam axle in a test vehicle, the outer dimensions of the cross-member should fit into the installation space for the cross-member of the reference axle as far as possible.

7.5.2 Selection and Calculation of the Cross-Section for the Cross-Member

The following figures show a possible constructive design of the axis according to the revealed principle (Köther 2011). The torsion profile is made of at least two curved,

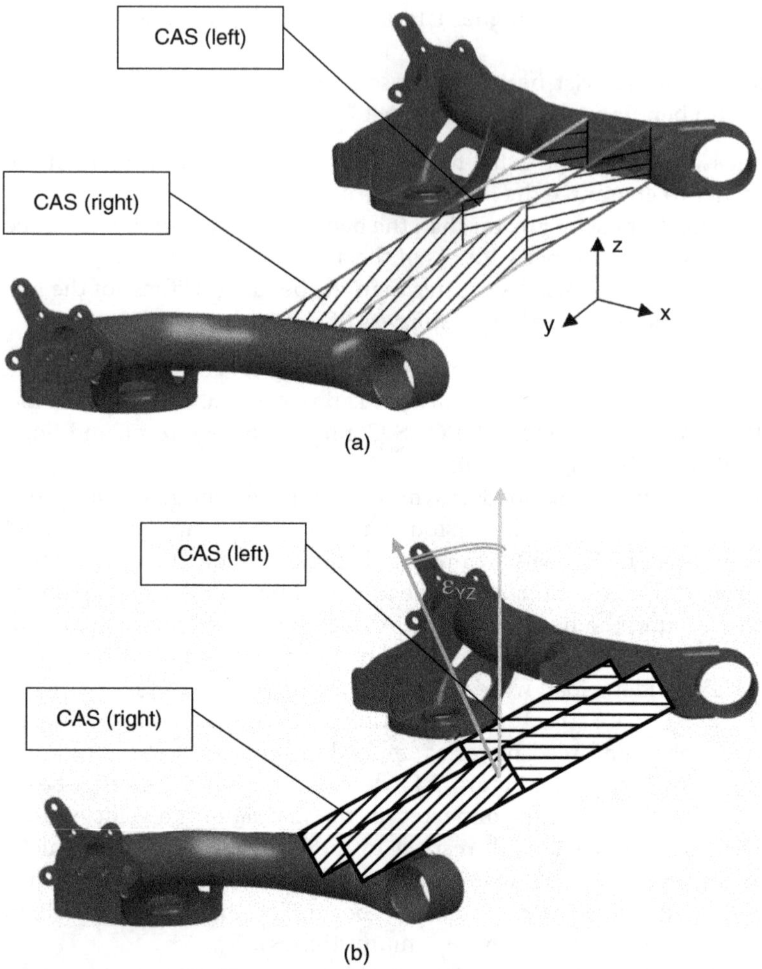

Figure 7.11 Laminate structure in the cross-beam for the countersteering composite link axle. Local coordinate system *xyz* is associated with the twist-beam axis *z*.

interconnected shells. The lower (or, as shown in the picture, middle) shell is arched underneath. When the carrier is rotated about its axis, both wheel carriers move in the same direction and coupled to the rotation. According to the invention, this shift compensates for undesired self-steering behavior. The coupling parameter describes the ratio of the steer angle on the wheel carrier to the angle of rotation of the axis. Reversing the curvature of the elastic shell upwards leads to the negative coupling parameter and increases the undesirable tendency to oversteer due to the self-steering behavior.

The coupling parameter is influenced by increasing the curvature and by the number of layers. At the same time, the bending stiffness and torsional stiffness of the axle are adapted. The torsional stiffness of the axle directly defines the total roll spring rate of the vehicle.

The basic structure of the cross-member by means of a symmetric laminate, as shown in Figure 7.10, determines the basic shape for the cross-section of the cross-member. This means that the cross-section of the cross-beam consists of two flanges that must be connected to each other by one or more webs. If the requirements for the basic properties of a cross-beam are justified, that is, it must be rigid and torsional soft, only open cross-sections could fit for the cross-beam. The use of the hollow, rectangular or round cross-sections is problematic. This problem occurs if the bending stiffness in both bending directions is endorsed to a high value, whereas the resulting torsional stiffness estimates are too high for the prescribed low roll stiffness of the axle. In this case, the use of the open cross-sections is justified. The open cross-sections possess far lower torsion stiffness compared with bending stiffness. Open cross-sections consisting of two vertical flanks and a web are either H-shaped or Π-shaped. The final calculation of the dimensions of the cross-section for the cross-beam must be performed using an optimization algorithm for a prescribed cross-sectional form and dimensions. When selecting the cross-section, the axial moments of inertia in both bending directions and the moment of inertia are determined and compared for both cross-sections. Furthermore, the warp resistance for the H- and Π-cross-section is to be determined, as warp torsion occurs with open cross-sections. This is an advantage for the countersteering twist-beam axle, because the fiber composite material not only provides a coupling bending moment due to its anisotropic material properties, which reduces the tendency of oversteer but allows the strong connection to the trailing arm.

Figure 7.12 shows the H-shaped and Π-shaped cross-sections. Figure 7.11 demonstrates that if the dimensions of both cross-sections are selected identically, the same surface area results for both cross-sections. Thus, one can directly evaluate which cross-section is stiffer to bend or softer to torsion. The warp deformation of the cross beams due to torsion favors bending-torsion coupling in the cross-beam. A higher warp resistance together with the higher modulus of elasticity C_{33} leads to a higher warp torsional stiffness. Hence, a further criterion in the selection of the cross-section should be the warp resistance.

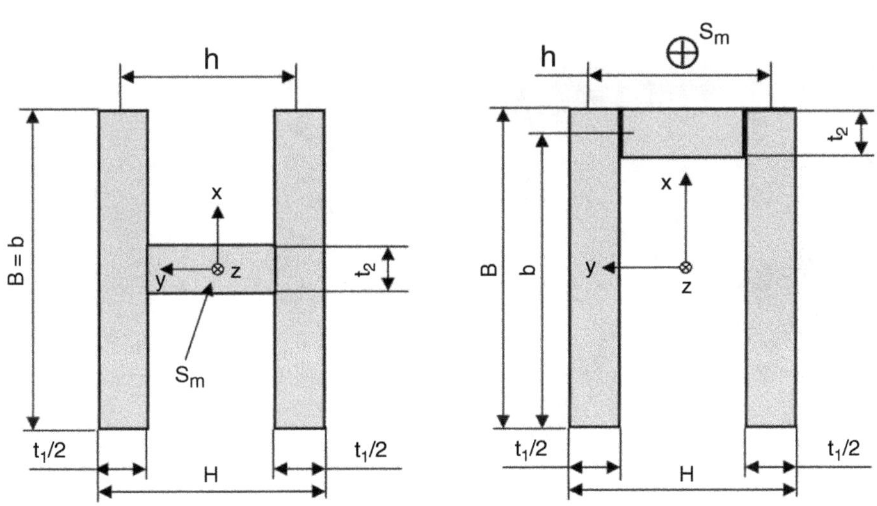

Figure 7.12 H-shaped and Π-shaped cross-sections.

7.5.3 Elements of a Countersteering Twist-Beam Axle

The so-called bucket design has established itself as the introduction of force for the functional model (Köther 2011). The design principle of the bucket construction can be seen schematically in Figure 7.13 in a sectional view through the XZ plane of the cross-member.

With this design principle, the trailing arms are enclosed by the cross-member and the so-called buckets and glued together. Due to the slants at the ends of the cross-section, the cross-member has the shape of a bathtub. The bevel in combination with the buckets ensures an even stiffness transition from the wishbone to the trailing arms. The disadvantage is the small adhesive surface between the wishbone, bucket and trailing arm, which limits the maximum forces and torques that can be transmitted.

Figure 7.14 shows the cross-beam designed with the help of the 3D CAD software in two views. View 1 in Figure 7.13 shows the cross-member in a position similar to the position it is installed in the vehicle. The slants at the end of the cross-section are clearly visible and the semicircles to accommodate the trailing arms. View 2 shows the

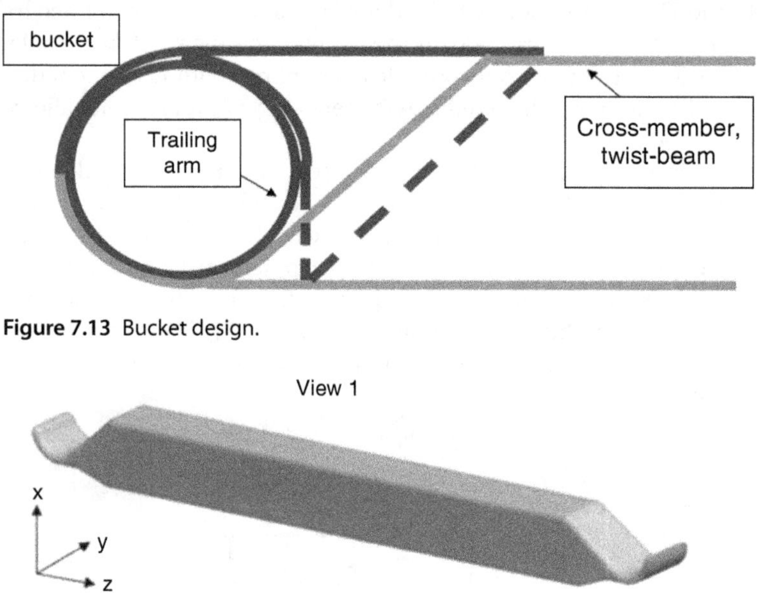

Figure 7.13 Bucket design.

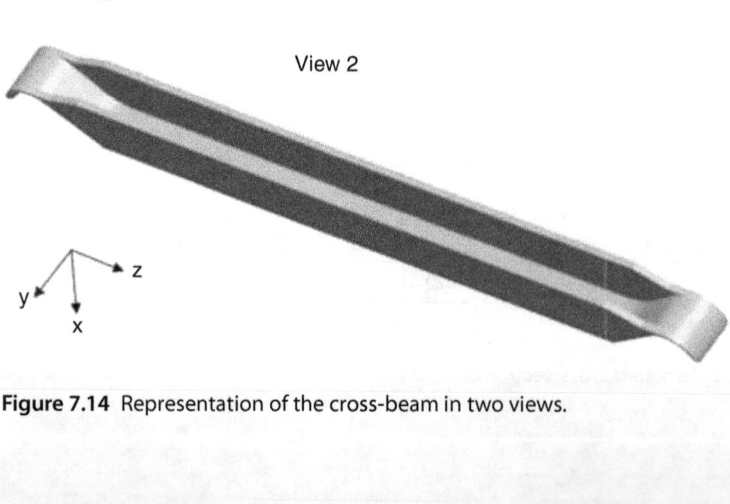

Figure 7.14 Representation of the cross-beam in two views.

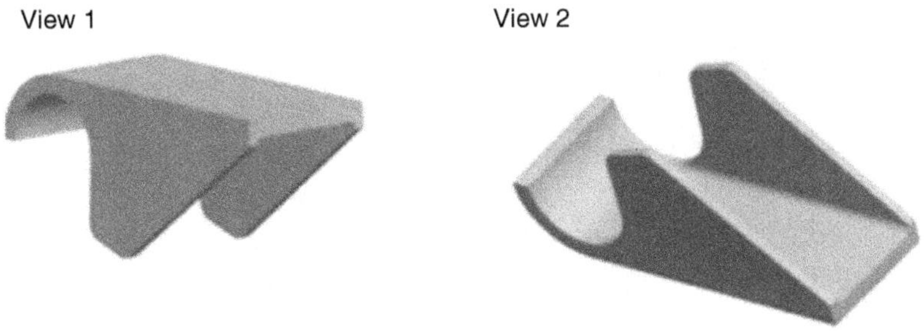

Figure 7.15 Display of the bucket design in two views.

cross-member in a rotated position so that you can look into it "from below," a radius must be applied to the edge where the flanges and the web meet so that the mold can be milled.

The buckets must now be designed in such a way that they fit both on the wishbone and on the trailing arms. Figure 7.15 shows the bucket in two views. The bucket can be seen in view 1 in Figure 7.15 in the position in the vehicle and in view 2 in a rotated view "from below." The dimensions on the inside of the bucket are chosen so that they fit exactly on the cross-member. In addition, the radius must be smaller than that on the cross-member and the bucket must also have demolding. The buckets are mirror-images, that is, they can connect the wishbone to the right trailing arm and the wishbone to the left trailing arm.

To check whether the designed components will fit together, they are collected in an assembly file in the 3D CAD software. Figure 7.16 displays the step-by-step assembly

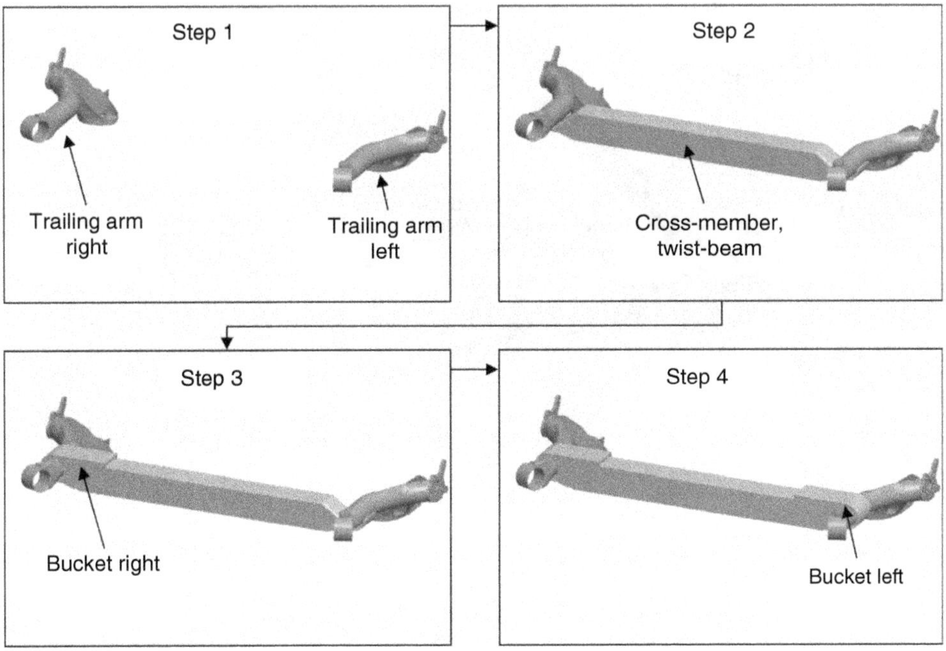

Figure 7.16 Step-by-step assembly of the countersteering twist-beam axle.

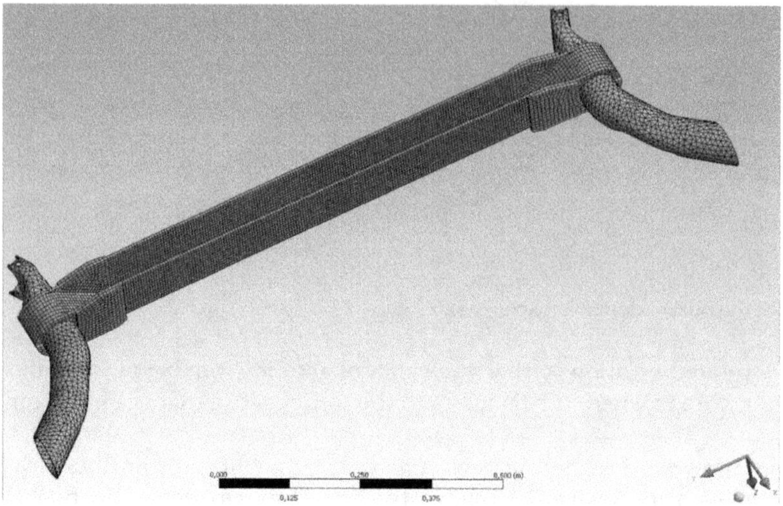

Figure 7.17 Finite-element model of the countersteering twist-beam axle.

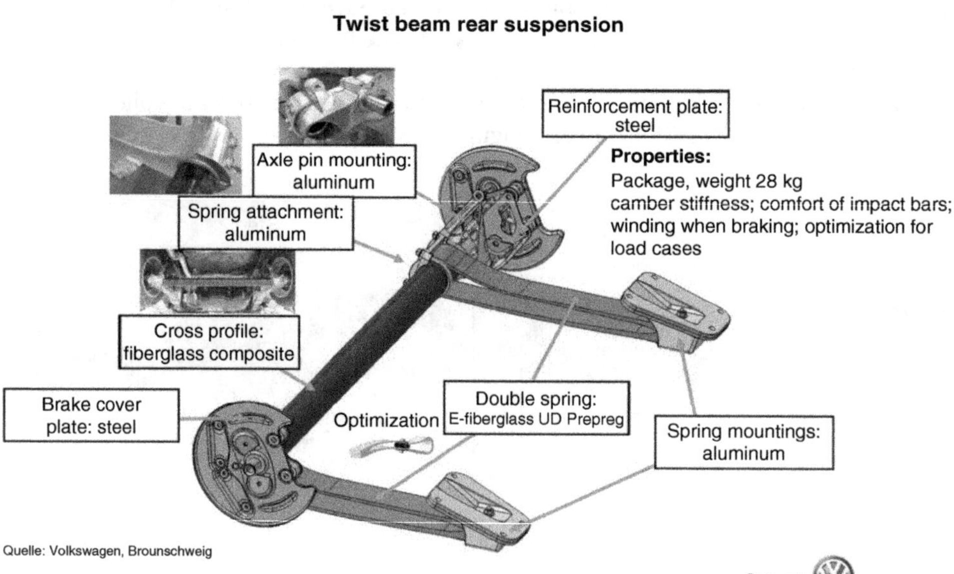

Figure 7.18 All-synthetic plastic concepts: twist-beam rear suspension, Volkswagen (Kersten 2018). Source: Courtesy of Dr. Thomas Kersten/Volkswagen.

of the countersteering twist-beam axle with the trailing arms of the reference axle and the designed components made of fiber composite material. First, the wishbone is positioned between the trailing arms. Then the right and left buckets are aligned so that they fit onto the wishbone and the respective trailing arm. The designed cross-beam and bucket made of fiber composite material is the basis for determining the mass of the respective component.

The finite-element model of the countersteering twist-beam axle is shown on Figure 7.17. The modeling was performed to check the analytical formulas for spatial spring constants on one side, and to evaluate and optimize the local stress distributions on the other. Finite-element models could be transformed to the multibody-system simulation codes and used for numerical modeling of elastic vehicle dynamics.

A modern design of the all-synthetic plastic concept for the rear suspension was presented by Kersten (2018). Figure 7.18 displays this concept design of the twist-beam axle for the rear suspension.

7.6 Patents on Twist-Beam Axles

This section exhibits several interesting patents and patent applications for the composite twist-beam axles.

EP3174744B1 (2014, https://www.epo.org) Motor vehicle suspension with glued composite torsion bar:

> The invention relates more particularly to a motor vehicle suspension, in particular a rear axle, of the type comprising an axially oriented torsion bar constituting the cross-member transversely oriented with respect to the general longitudinal orientation of the vehicle, design generally tubular and at least each of the two axial end portions is tubular in shape. This torsion bar forming a cross is associated, at each of its two axial ends to a hanger which is associated a complementary longitudinal arm each of which supports a rear wheel, left and right, respectively, of the vehicle (Figure 7.19).

DE202013004035U1 (2013, https://www.epo.org): Assembly for a vehicle:

> Assembly (100) for a vehicle, with two longitudinal members (1, 2) and an interposed transverse support (3) extending approximately transversely to the longitudinal supports (1, 2); wherein said cross-member (3) is an elastically twistable torsion beam, and the cross-member (3) mainly made of fiber composite material and the longitudinal beams (1, 2) are mainly made of a metallic material, and the cross-member (3) with the longitudinal beams (1, 2) is connected, wherein is the transverse support (3) at its junction (6) to the longitudinal supports (1, 2) having its respective ends (7) on one of the two longitudinal members (1, 2) is clamped or mounted, and at this joint (6), torsional forces are introduced into the transverse support (3), characterized, in that the connection point (6) between the transverse support (3) and the longitudinal beams (1, 2) comprises an overlapping portion (B) in which a connecting portion (8) of predominantly metal material longitudinal member having (1, 2) with a

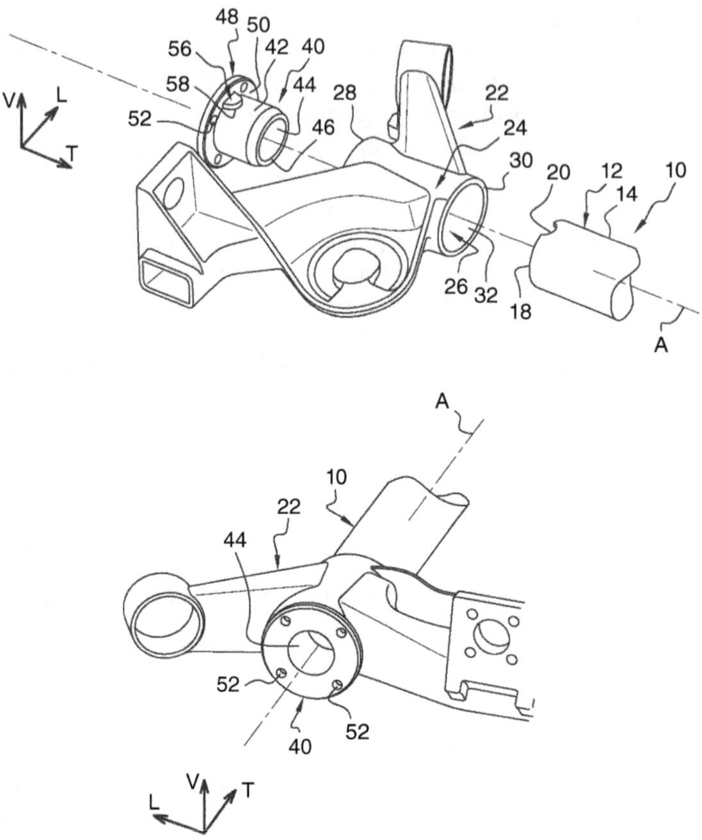

Figure 7.19 EP3174744B1, Motor vehicle suspension with glued composite torsion bar.

> portion of predominantly fiber composite material having cross-member (3) at least partially overlaps (Figure 7.20).

US6382649B1 (1999-07-16, https://www.uspto.gov): Wheel suspension in a motor vehicle:

> A suspension for a wheel in a motor vehicle, such as a passenger car, having a link which is made of fiber composite and is mounted at one end, by means of a mounting, on a chassis of the vehicle and at the other end holds the wheel, and having spring means which are made of fiber composite and support the link on the chassis in a resilient manner. In order to obtain a particularly compact construction for the wheel suspension, the mounting has a bearing in which a bearing part, which is fastened to the chassis, is mounted such that it can be adjusted relative to a bearing part, which is fastened to the link, by pivoting about a pivot axis, and the spring means having a torsion-bar suspension which extends coaxially to the pivot axis and is fastened at one end to the chassis or to the bearing part fixed to the chassis, and is fastened at the other end to the link or to the bearing part fixed to the link (Figure 7.21).

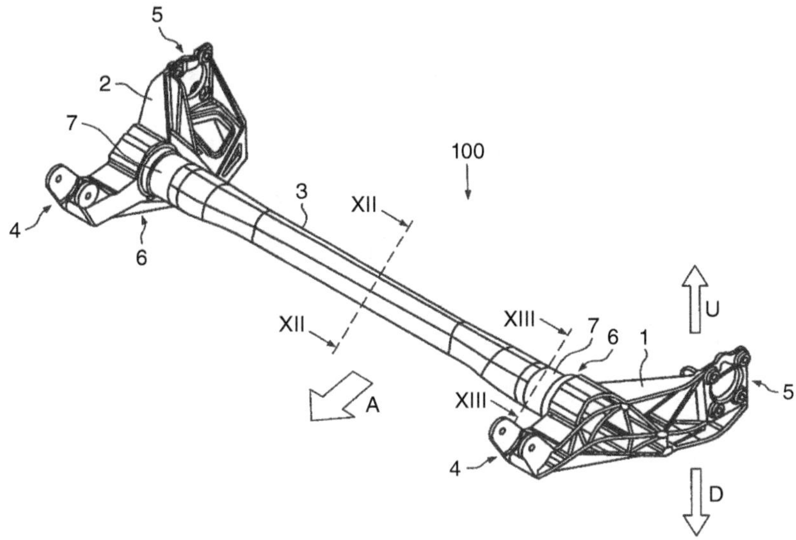

Figure 7.20 DE202013004035U1 (2013): Assembly for a vehicle.

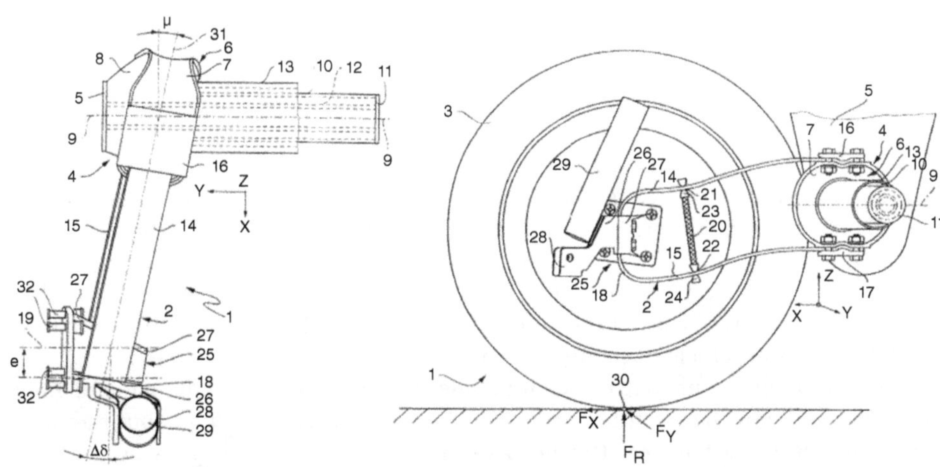

Figure 7.21 US6382649B1 (1999-07-16): Wheel suspension in a motor vehicle.

DE202016105937U1 (2016-10-21, https://www.epo.org) Chassis axle and vehicle:

> Chassis axle (100) for a motor vehicle (102), comprising: two link elements (104) for pivotal connection of wheels (116) of the motor vehicle (102) to a supporting structure of the motor vehicle (102); a connecting member (112), by means of which the link elements (104) are connected to each other to transmit force, wherein the connecting member (112) is at least partially formed from a plastic material (Figure 7.22).

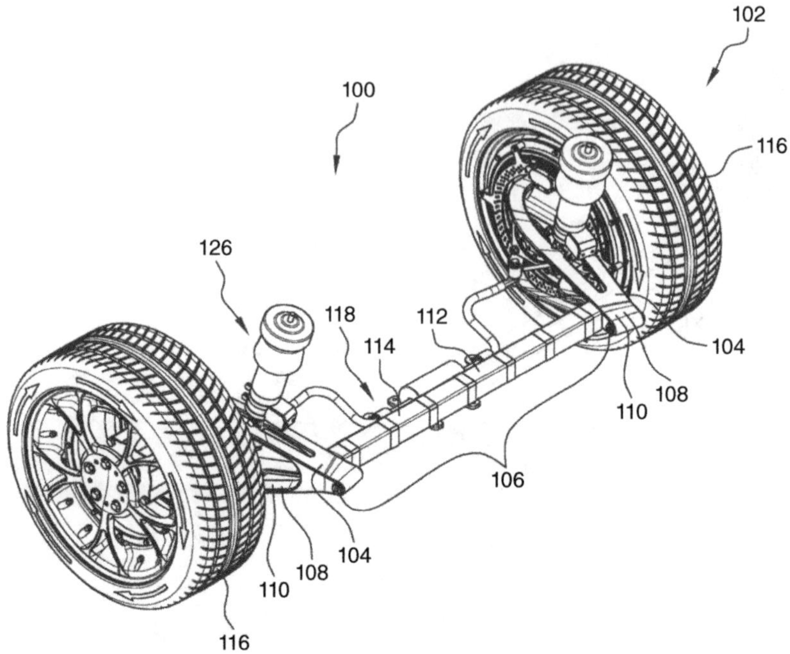

Figure 7.22 DE202016105937U1, "Chassis axle and vehicle."

7.7 Conclusions

Unidirectional layers were used for the laminate structure of the cross-beam. The cross-section for the cross-beam of the countersteering twist-beam axle was selected and calculated later. The first functional pattern of the countersteering twist-beam axle had to use the trailing arms of the reference axle to ensure comparability during testing when measurements are made on the reference axle and the functional pattern. The aim is also to reduce the mass of the countersteering twist-beam axle in order to save fuel and improve the vehicle's driving dynamics by reducing unsprung mass.

It should be pointed out that the countersteering twist-beam axle is made of fiber composite material primarily not in order to achieve a reduction in mass, but rather to improve self-steering behavior with a flexure-torsion coupling in the cross-beam.

The following articles are intended for further reading: Bampton and Craig (1968); Milliken and Milliken (1995); Sampo (2011); Sampo et al. (2010); Helsen et al. (2010); Andrew et al. (2000); Bhaskar et al. (2010); Hari Krishnan et al. (2011); Coox et al. (2012) and Cocana (2015).

References

Bampton, M.C.C. and Craig, R.R. (1968). Coupling of substructures for dynamic analyses. *AIAA J.* 6 (7): 1313–1319.

Bhaskar C., Deepak R., Mugundaram R. (2010) Correlation of Vehicle Dynamics & NVH Performance with Body Static & Dynamic Stiffness through CAE and Experimental

Analysis. SAE Technical Paper. SAE International, April 2010. doi: 10.4271/2010-01-1137. http://dx.doi.org/10.4271/2010-01-1137.

Cocana, O.D.A.G. (2015). *Influence of Body Stiffness on Vehicle Dynamics Characteristics in Passenger Cars*. Göteborg, Sweden: Chalmers University Of Technology.

Coox L., Vivet M., Tamarozzi T., et al. (2012) Numerical assessment of the impact of vehicle body stiffness on handling performance. LMS International, *Proceedings of ISMA2012-USD2012*, pp. 3711–3724

Deakin A., Crolla D., Ramirez J., Hanley R. (2000) The Effect of Chassis Stiffness on Race Car Handling Balance. SAE Technical Paper. SAE International, November 2000. doi: 10.4271/2000-01-3554. http://dx.doi.org/10.4271/2000-01-3554.

DE10321877 (2004) Wheel carrier mount with an oscillating plate, https://www.epo.org.

EP1159147 (2001) Composite pivot pin, https://www.epo.org.

EP2019023 (2009) Steering rear end for automobile, comprising a deformable wheel pivot, https://www.epo.org.

FR2860751 (2005) Essieu directeur d'un vehicule automobile, https://www.epo.org.

Heißing, B., Ersoy, M., and Gies, S. (2011). *Fahrwerkhandbuch*. Vieweg+Teubner Verlag|Springer Fachmedien Wiesbaden GmbH.

Helsen J., Cremers L., Mas P., Sas P. (2010) Global static and dynamic car body stiffness based on a single experimental modal analysis test. Modal testing: Methods and Case Studies. *Proceedings of ISMA 2010 including USD 2010*.

Kersten T. (2018) Leichtbautrends in der Fahrwerkentwicklung am Beispiel der Hinterachse, Volkswagen AG, *2. Siegener Leichtbau Kolloquium*, 16.-17.10.2018, Siegen.

Kobelev, V. (2018). Approximate static aeroelastic analysis of composite wings. *Multidiscipline Modeling in Materials and Structures.* 15 (2): https://doi.org/10.1108/MMMS-02-2018-0019.

Köther S. (2011) *Bemusterung und Erprobung intelligenten Verbundlenkerachse aus Faserverbundwerkstoff, Sampling and testing of intelligent twist beam axle made of fibre composite material*, Diploma thesis, University of Siegen.

Krishnan, M.H., Sreeraj, N., Bhaskar, C. et al. (2011). Establishing correlation between torsional and lateral stiffness parameters of BIW and vehicle handling performance. *SAE Int. J. Passeng. Cars – Mech. Syst.* 4: 22–31. https://doi.org/10.4271/2011-01-0089.

Krone Jr., N.J. (1975): Divergence Elimination with Advanced Composites, *AIAA Paper 75-1009*, Los Angeles, CA.

Librescu, L. and Song, O. (2006). *Thin-Walled Composite Beams, Theory and Application, Solid Mechanics and its Applications*, vol. 131. Dordrecht, The Netherlands: Springer,.

Milliken, W.F. and Milliken, D.L. (1995). *Race Car Vehicle Dynamics*. Society of Automotive Engineers.

Mitschke, M. and Wallentowitz, H. (2014). *Dynamik der Kraftfahrzeuge*, 5e. Springer Fachmedien Wiesbaden.

Piening, M. (1984). Die statische Aeroelastizität des anisotropen Tragflügels: *DGLR-Symposium Entwicklung und Anwendung von CFK-Strukturen, Technische Universität Berlin, 8.-9.11.1984, Jahrestagung der Deutschen Gesellschaft für Luft- und Raumfahrt eV, DGLR; 84–158*, Braunschweig.

Rehfield, L.W. and Atilgan, A.R. (1989). Toward understanding thetailoring mechanisms for thin-walled composite tubular beams. In: *Proceedings of the First USSR–U.S. Symposium on Mechanics of Composite Materials, 23–26 May, Riga* (ed. S.S.R. Latvia, S.W. Tsai, J.M. Whitney, et al.), 187–196. ASME Publishing House.

Riekert, P. and Schunck, T.E. (1940). Zur Fahrdynamik des gummibereiften Kraftfahrzeuges. *Ing.-Arch.* 11: 210–224.

Sampo E.. (2011) *Modelling chassis flexibility in vehicle dynamics simulation*. PhD thesis. University of Surrey, 2011.

Sampo E., Sorniotti A., Crocombe A. (2010) Chassis Torsional Stiffness: Analysis of the Influence on Vehicle Dynamics. *SAE Technical Paper*. SAE International, Apr. 2010. doi: http://dx.doi.org/10.4271/2010-01-0094.

US6173977 (2001) Motor vehicle, https://www.uspto.gov.

US8220810B2 (2008) Countersteering rear axle of a vehicle, https://www.uspto.gov.

Weisshaar, T.A. (1980). Divergence of forward swept composite wings. *J. Aircraft* 17 (6): 442.

WO2009124658 (2009) Countersteering automotive rear axle, https://www.epo.org.

WO2010000363 (2010) Countersteering automotive rear axle, https://www.epo.org.

8

Design and Optimization of Composite Springs

8.1 Design and Optimization of Anisotropic Helical Springs

8.1.1 Forces and Moments in Helical Composite Springs

For conservation of natural resource and economy of energy, automobile manufacturers in recent years have intensively reduced the mass of cars. The potential weight savings, the possibility of reducing noise, vibrations and ride harshness due to their high damping factors, the absence of corrosion problems and lower tooling costs are the declared aims for development of structural elements made of fiber-reinforced polymers. These aims are the focus of actual development of helical composite springs.

Wong et al. (2004) investigated the feasibility of utilizing the nickel–titanium alloy wires in altering the spring rates of a composite circular spring. The study by Kumbhar et al. (2016) presents experimental and computational analysis of a helical spring for automotive applications. The manufacturing of a helical spring made of composite material is explained.

The rectangular wire helical spring (Kotaro et al. 2001) is contrived, which joins the first coil part and the second coil part with a twisting part. If the spring is used for suspension of rally cars, good control stability can be maintained while absorbing impact from the load.

Chiu et al. (2007) studied four different types of helical composite springs. The springs were made of structures including unidirectional laminates, rubber core unidirectional laminates, unidirectional laminates with a braided outer layer and rubber core unidirectional laminates with a braided outer layer. The effects of rubber core and braided outer layer on the mechanical properties were experimentally evaluated. The helical composite spring with a rubber core improved the compression load to failure by about 12%. The braided outer layer increases the compression load to failure by about 18%. It was shown that the uppermost mechanical properties possess a helical spring with a braided outer layer structure.

The methods for evaluation of the optimal ply angles and wire diameters of carbon fiber-epoxy composite coil springs were developed in Choi and Choi (2015). The optimization aim was settled to achieve a prescribed spring rate.

The review of the development and analysis of a helical spring with combination of conventional and composite materials was presented in Jadhav and Gawande (2015). The

Design and Analysis of Composite Structures for Automotive Applications: Chassis and Drivetrain, First Edition. Vladimir Kobelev.

Companion website: www.wiley.com/go/kobelev/automotive_suspensions

main objective of the study was evaluation of the multi-material design, which combines composite materials and metal alloys.

In an effort to reduce weight, fiberglass suspension springs were presented in Ehrich (2015). The lightweight springs constructed from a glass-fiber reinforced polymer composite enabled a weight saving of 2.5 kg per spring and a 40% reduction related with steel springs. The springs were made of long glass fibers twisted together and impregnated with epoxy resin. This structure was then wrapped with more glass fibers at counter-poised 45° angles. As well as providing similar strength to steel springs, fiberglass springs resist corrosion or chemicals and use less energy in manufacture. In the development process of the GRP coil spring, the requirement was to develop a clone spring of the steel spring, which can be installed in the vehicle without further adaptation. The corresponding manufacturing method of the spring is shown in Sardou and Djomseu (2000). It consists of a glass-fiber strand impregnated with epoxy resin, which consists of a core made of twisted glass fibers around which the load-bearing layers are wound in layer angles of ±45°. The springs are shaped by winding them onto a metal core on which the material is cured under heat.

Ekanthappa et al. (2016) fabricated the helical composite spring with the rectangular cross-section by means of coiling the composite over the mandrel, which has the desired shape of spring profile. The mandrel is made of steel alloy and is fixed on a lathe. A silicone gel covers the surface of mold, forming a releasing agent. The fibers pass through the bath, which contains fluid epoxy resin and are wound on the mandrel. After winding of the necessary length of the spring, the mandrel is covered by a rubber tape. The final stage is the polymerization of the spring and releasing of the spring from the mandrel. After the manufacturing process, the mechanical characteristics of the composite springs are acquired.

Petrich et al. (2017b) formalized the material selection procedure for composite springs. Starting from the given row of materials for matrix and fiber, the probable combinations of the components were evaluated. For each one of the possible combinations the mechanical properties were evaluated.

Petrich et al. (2017c) presented the composite volute spring with rectangular cross-section (Figure 8.1a). The manufacturing process consists of preparation of a winding core and carrier film, pre-cutting of fibers or textiles. The next step is lamination. In this step the fibers are layered and the matrix impregnates the fibers. The winding and curing in the core steps follow. Finally, the spring is released from the core and the covered carrier film is removed.

An actual review of the design and analysis methods and fabrication of fiber-reinforces composite helical springs is given by Kara (2017). The resin-transfer molding process is explained in detail in Figure 8.1b.

Helical composite springs, studied in this chapter, are formed by wrapping composite wire of uniform cross-section around a cylinder. We take as a reference frame the cylindrical polar coordinate system (r, θ, z). The axis z of the cylindrical polar coordinate system is aligned with the axis of the cylinder. The analysis of a cylindrical helical composite spring with a circular wire is based on the following formulas. The design variables defining a specific design of a spring are: d, D, n, L_0, where n is a number of active coils, d is the diameter of wire, D is the mean coil diameter and L_0 is a free spring length.

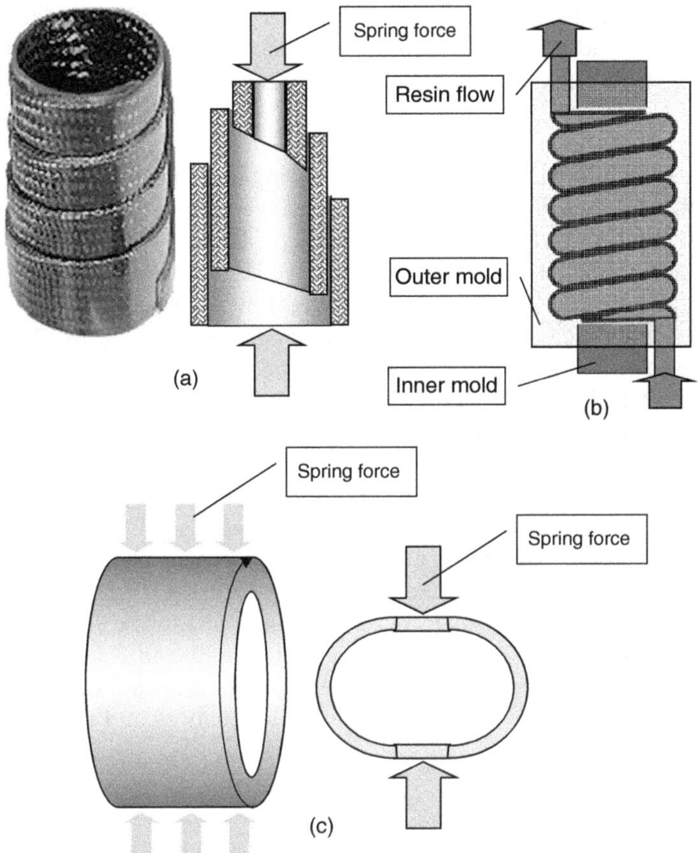

Figure 8.1 Glass fiber reinforced plastic (GFRP) springs: (a) fiberglass volute spring (Petrich et al. 2017c); (b) resin transfer molding (RTM) manufacturing of a coil (Kara 2017) and (c) circular and ring springs.

The outer and inner diameters are, correspondingly:

$$D_e = D + d, D_i = D - d.$$

Ignoring at first other certain complexities of spring technology, we can write simplified basic relations for an analysis of the anisotropic compression spring. This simplified analysis considers spring ends as "plain," so that only active coils are considered. The force acts ideally along the z-axis of the cylindrical coordinate system.

The helical composite springs also store elastic energy by means of bending a wire. A torsion spring is a helical composite spring that works by torsion or twisting. The twist of helically coiled wire occurs about the axis of the coil by sideways forces or terminal moments applied to the ends of springs. The terminal moments twist the coil tighter or looser.

Consider a cylindrical helical composite spring with the mean diameter D. A coil spring can be wound in either a left-hand or right-hand direction. The lower spring end is assumed to be fixed. For now, right-hand helical springs are considered.

Let the angle α be the angle of inclination of the helix with any plane perpendicular to the axis of the coil (pitch angle, lead angle). We take as a reference frame the cylindrical polar coordinate system (r, θ, z). The z-axis of the cylindrical polar coordinate system is aligned with the axis of the cylinder. A fixed distance between the successive coils of a spring is maintained so that the axis of the wire forms a helix. When the distance between coils is small, the spring is called a closed-coiled spring and in the consideration of stress in the wire torsion theory may be applied. Consider a close-coiled helical composite spring, which is subjected to axial loading by the axial force F and total axial torque M_θ. The torque on the upper spring acts clockwise, if we look on the spring from above $M_\theta = F_\theta R$. The circumferential force F_θ pushes toward the wire, applying pressure on the wire cross-section (Figure 8.1a). The line of force coincides with the symmetry axis of the spring and, consequently, the z-axis of the cylindrical coordinate system.

Consider the middle region of the spring where the transition effects due to the end bars disappear and can be neglected. Due to the symmetry of the spring, all sections of the spring are deformed identically. In Figure 8.1b a portion of the spring has been isolated. In the absence of total axial torque, the cut section must have acting upon it a resultant force equal to F, and a couple, $M = FD/2$, for the free body is in equilibrium. The force and couple must lie in the same vertical plane. A view of the spring wire at the cut section has been taken from the right and is shown in Figure 8.3. The couple M, acting in the vertical plane, has been shown as a vector and is resolved into two components, $M \sin\alpha$ and $M \cos\alpha$, which lie in planes that are tangential and normal to the helix, respectively. The couple $M \sin\alpha$ tends to cause bending of the isolated portion of the spring wire, and $M \cos\alpha$ acts so as to cause twisting. The bending and twisting couples are equal, respectively, to:

$$M_B = M \sin\alpha - M_\theta \cos\alpha, M_T = M \cos\alpha + M_\theta \sin\alpha. \tag{8.1}$$

The similar resolution of force F into components, F_N and F_S, causes normal and shearing components of stress, respectively:

$$F_N = F \sin\alpha, F_S = F \cos\alpha. \tag{8.2}$$

For an extreme closed-coiled spring, when the angle α is very small, the couple M_B is insignificant and M_T is roughly equal to M. Correspondingly, the force F_N is negligible so that F_S is approximately equal to F. Any segment of the coil is therefore in equilibrium under two opposite axial forces F with the same magnitude. Non-zero components of the shear stress in cylindrical coordinates are $\tau_{r\theta}$, $\tau_{\theta z}$ (Figure 8.2).

On the other hand, the material of the circular springs or ring-shaped springs (Figures 8.1c and 8.11) is stressed by bending moments. Such springs were analyzed by Reid et al. (2002), Wong et al. (2004) and Reid et al. (2004). The springs, displayed in Figures 8.11–8.15 are also loaded by the flexure moments so that the dominant stress state is bending of the thin-walled shells. The design principles of the corresponding structural elements are analyzed in Chapters 10 and 11 in detail.

8.1.2 Symmetrically Designed Solid Bar With Circular Cross-Section

The mechanical properties of multilayered composites are calculated based on the fiber and matrix properties, and the fiber architectures result from the geometric model. The multilayered composite is usually manufactured as a crimp yarn with the symmetric

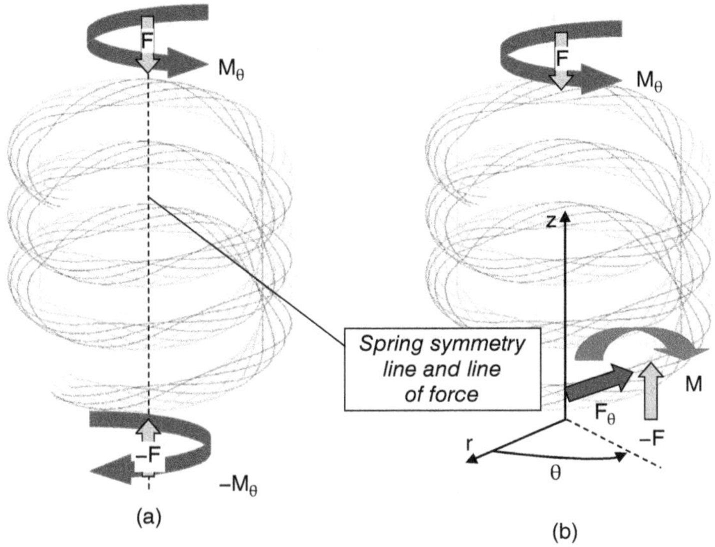

Figure 8.2 Helical composite spring, subjected to axial load and axial torque.

respective to the axis of the rod structure. Since the principal material direction of single layers does not coincide with the principle directions of the cylindrical coordinate system, transformation of the stress–strain relations from the local coordinate system $[\mathbf{C}]_L$, associated with fiber direction, to cylindrical coordinate system $[\mathbf{C}]_C$, reads (Chapter 1, Eq. (3.5)):

$$[\mathbf{C}]_C = [\mathbf{T}_\sigma][\mathbf{C}]_L[\mathbf{T}_\varepsilon]^T, [\mathbf{S}]_C = [\mathbf{T}_\varepsilon][\mathbf{S}]_L[\mathbf{T}_\sigma]^T. \tag{8.3}$$

The local axis 3 coincides with the direction of fibers, so that for the components of compliance matrix $[\mathbf{S}]_L$ of the unidirectional material we have:

$$s_{33} = E_L^{-1}, s_{22} = E_T^{-1}, s_{23} = -\nu_{LT}E_L^{-1}, s_{44} = G_{TL}^{-1}. \tag{8.4}$$

Consider a multilayered material with a symmetric structure with respect to direction "3" of the global coordinate system. The effective elasticity matrix in global coordinates can be obtained by averaging the transformed local compliance matrix of the infinitesimal yarn segment through the crimp angle. For the first direction of roving, the crimp or roving angle is equal to Φ:

$$\begin{cases} S_{44}^+ = 4(s_{33} + s_{22} - 2s_{23} - s_{44})\sin^2\Phi\cos^2\Phi + s_{44}, \\ S_{33}^+ = s_{33}\cos^4\Phi + (2s_{23} + s_{44})\sin^2\Phi\cos^2\Phi + s_{22}\sin^4\Phi. \end{cases} \tag{8.5}$$

For the second direction of roving, the crimp or roving angle is equal to $-\Phi$ we get the same values:

$$\begin{cases} S_{33}^- = s_{33}\cos^4\Phi + (2s_{23} + s_{44})\sin^2\Phi\cos^2\Phi + s_{22}\sin^4\Phi, \\ S_{44}^- = 4(s_{33} + s_{22} - 2s_{23} - s_{44})\sin^2\Phi\cos^2\Phi + s_{44}. \end{cases} \tag{8.6}$$

For unidirectional layers the material properties are locally orthotropic in the local coordinate system and, therefore, $c_{45} = 0$. With the symmetric structure there are equal number of layers with roving angles of Φ and $-\Phi$.

After the averaging the relevant components of the effective compliance matrices of multilayered material in the cylindrical coordinate system are:

$$\begin{cases} S_{44} = \frac{1}{2}(S_{44}^{+} + S_{44}^{-}) = \\ \quad 4(E_L^{-1} + E_T^{-1} + 2\nu_{LT}E_L^{-1} - G_{TL}^{-1})\sin^2\Phi\cos^2\Phi + G_{TL}^{-1}, \\ S_{33} = \frac{1}{2}(S_{33}^{+} + S_{33}^{-}) = \\ \quad E_L^{-1}\cos^4\Phi + (-2\nu_{LT}E_L^{-1} + G_{TL}^{-1})\sin^2\Phi\cos^2\Phi + E_T^{-1}\sin^4\Phi. \end{cases} \tag{8.7}$$

The following question concerns the ultimate stresses of both layers. Consider the local coordinate system, which is associated with the direction of fiber states. The tensor-polynomial criterion of the second rank (Chapter 2) in this coordinate system reads:

$$\boldsymbol{\sigma} \bullet \widetilde{\mathbf{F}}^{(0)} \bullet \boldsymbol{\sigma} = F\sigma_{22}{}^2 + G\sigma_{11}{}^2 + H(\sigma_{11} - \sigma_{22})^2 + 2N\sigma_{12}{}^2 \leq 1, \tag{8.8}$$

$$H = G = \frac{1}{2}\widetilde{\sigma}_L^{-2}, F = \widetilde{\sigma}_T^{-2} - \frac{1}{2}\widetilde{\sigma}_L^{-2}, N = \frac{1}{2}\widetilde{\tau}_{LT}^{-2}, F + G + 4H = \widetilde{\sigma}_T^{-2} + 2\widetilde{\sigma}_L^{-2}.$$

In case of pure shear, $\tau_{\theta z}$, the local stresses in the laminate coordinate system are:

$$\begin{aligned} \sigma_{11} &= 2\tau_{\theta z}\cos\Phi\sin\Phi, \\ \sigma_{22} &= -2\tau_{\theta z}\cos\Phi\sin\Phi, \\ \sigma_{12} &= \tau_{\theta z}(\cos^2\Phi - \sin^2\Phi). \end{aligned} \tag{8.9}$$

The substitution of (8.9) into (8.8) leads to the following form of tensor-polynomial criterion of the second rank failure criteria for each family of fibers in the case of pure shear or torsion:

$$4(F + G + 4H - 2N)\cos^2\Phi\sin^2\Phi + 2N \leq \tau_{\theta z}^{-2}, \tag{8.10}$$

$$4\left(\frac{1}{\widetilde{\sigma}_T^2} + \frac{2}{\widetilde{\sigma}_L^2} - \frac{1}{\widetilde{\tau}_{TL}^2}\right)\cos^2\Phi\sin^2\Phi + \frac{1}{\widetilde{\tau}_{TL}^2} \leq \frac{1}{\tau_{\theta z}^2}.$$

In the case of bending, there is the only component of stress σ_B in the coordinate system, which is associated with the axis of the beam. The transformation to the local stresses in the laminate:

$$\sigma_{11} = \sigma_B\cos^2\Phi, \sigma_{22} = \sigma_B\sin^2\Phi, \sigma_{12} = -\sigma_B\sin\Phi\cos\Phi. \tag{8.11}$$

The substitution of (8.11) into (8.8) leads to the following form of the Tsai–Hill fracture criteria for each of the families of fibers:

$$(F + H)\sin^4\Phi + 2(N - H)\sin^2\Phi\cos^2\Phi + (G + H)\cos^4\Phi \leq \sigma_B{}^{-2} \tag{8.12}$$

or

$$\frac{\sin^4\Phi}{\widetilde{\sigma}_T^2} + \left(\frac{2}{\widetilde{\tau}_{LT}^2} - \frac{1}{\widetilde{\sigma}_L^2}\right)\cos^2\Phi\sin^2\Phi + \frac{\cos^4\Phi}{\widetilde{\sigma}_L^2} \leq \frac{1}{\sigma_B{}^2}. \tag{8.13}$$

The developed formulas are appropriate for the design and analysis of torque-stressed springs, like solid twist bars, exposed in Figures 7.19–7.22 in Chapter 7, as well for a cross-bar with a solid circular cross-section (Figure 8.10). The twist momentum causes the dominant stresses in these elements.

8.1.3 Stiffness and Stored Energy of Helical Composite Springs

Let the stroke of a helical composite spring in the absence of axial torque M_θ be the spring travel from released length L_{rel} to compressed length L_{comp}:

$$s = L_{rel} - L_{comp}.$$

Consider for definiteness a compression spring with a free length L_0. Solid length L_c is the length at which the coils of the compressed spring close up. For the compression spring, the following is valid (Figures 8.3 and 8.4):

$$L_0 > L_{inst} > L_{fin} > L_c > 0.$$

$F_{\max}$, $F_{\min}$ and F_c are the spring loads at lengths L_{comp}, L_{rel} and L_c:

$$F_{\min} = c\,(L_0 - L_{rel}), \quad F_{\max} = c\,(L_0 - L_{comp}), F_c = c\,(L_0 - L_c).$$

The value c denotes the axial compression spring rate, or spring constant. The extension spring is handled in the same way. The stored elastic energy of the linear spring could be expressed either in terms of spring travel or spring force:

$$U_e = \frac{c}{2}\left[(L_{comp} - L_0)^2 - (L_{rel} - L_0)^2\right] = \frac{1}{2c}({F_{\max}}^2 - {F_{\min}}^2). \tag{8.14}$$

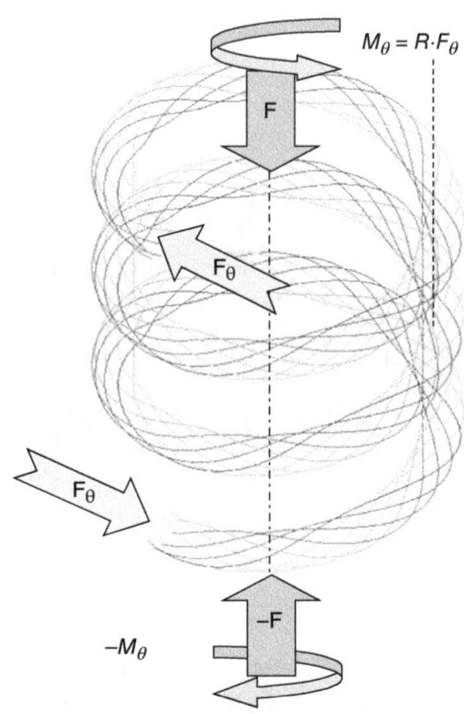

Figure 8.3 Helical spring loaded by torque and axial force.

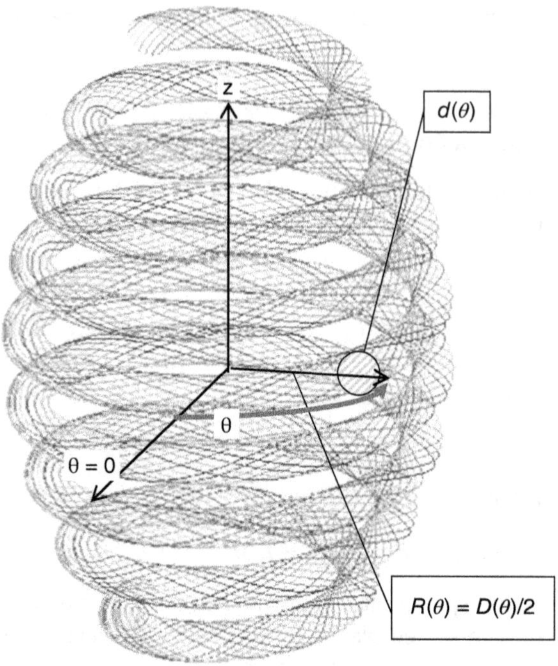

Figure 8.4 Helical spring with variable wire diameter and non-cylindrical form.

The stored elastic energy is equal to the work of applied forces on the total spring travel:

$$U_f = \frac{1}{2}(F_{\max} + F_{\min})s.$$

The energy capacity of the linear spring loaded from its free state with the axial force F and the axial torque M_θ reads:

$$U_e = \frac{F^2}{2c} + \frac{FM_\theta}{c_{\theta F}} + \frac{{M_\theta}^2}{2c_\theta}.$$

The volume and the mass of the spring material of a cylindrical spring with a constant, round cross-section is given by:

$$V = \frac{1}{4}\pi^2 d^2 Dn, \qquad m = \rho V. \tag{8.15}$$

where ρ is the density of the spring material.

The spring stiffness or spring rate is the force required to produce a unit deflection. For close-coiled helical composite springs the force-deflection characteristic is approximately linear and can be calculated from the geometry and effective shear modulus S_{44} of the spring material:

$$c = \frac{d^4}{8\,S_{44} D^3\, n}. \tag{8.16}$$

One must differentiate the basic and corrected shear stress in the spring with an axial load F. The formula for basic stress τ, or uncorrected stress, is obtained by dividing the torsion moment acting on the wire:

$$M_T = \frac{1}{2} F D \tag{8.17}$$

by the section modulus in torsion (Tables 8.3 and 8.4) giving:

$$\tau = \frac{M}{w_T} = \frac{8D F}{\pi d^3}. \tag{8.18}$$

8.1.4 Spring Rates of Helical Composite Springs

The study of the optimization problem requires some generalization of common design formulas that account for the variation in the mean diameter of the spring body as well the wire diameter along its length.

The formulas for stiffness and spring rate for a general non-cylindrical helical composite spring with an arbitrary variable cross-section are derived next following Castigliano's method (Teodorescu 2013). The elastic energy stored in the spring:

$$2U_e = \int_l \left[A^* M_\theta{}^2 + 2B^* M M_\theta \frac{2}{D} + C^* M^2 \left(\frac{2}{D}\right)^2 \right] dl, \tag{8.19}$$

with the following notations for the stiffness of the anisotropic wire:

$$A^* = S_{44} \frac{\sin^2\alpha}{I_T} + S_{33} \frac{\cos^2\alpha}{I_b},$$

$$B^* = \cos\alpha \sin\alpha \left(\frac{S_{44}}{I_T} - \frac{S_{33}}{I_b} \right) \frac{D}{2},$$

$$C^* = \left(S_{44} \frac{\cos^2\alpha}{I_T} + S_{33} \frac{\sin^2\alpha}{I_b} \right) \left(\frac{D}{2}\right)^2.$$

The relevant components of the effective elasticity and compliance matrices of a multilayered material in the cylindrical coordinate system are:

$$S_{44} = 4(s_{33} + s_{22} - 2s_{23} - s_{44})\sin^2\Phi\cos^2\Phi + s_{44},$$

$$S_{33} = s_{33}\cos^4\Phi + (2s_{23} + s_{44})\sin^2\Phi\cos^2\Phi + s_{22}\sin^4\Phi,$$

or, equivalently,

$$S_{44} = 4(E_L^{-1} + E_T^{-1} + 2\nu_{TT}\mathrm{E}_\mathrm{T}^{-1} - G_{TL}^{-1})\sin^2\Phi\cos^2\Phi + G_{TL}^{-1},$$
$$S_{33} = E_L^{-1}\cos^4\Phi + (-2\nu_{TT}\mathrm{E}_\mathrm{T}^{-1} + G_{TL}^{-1})\sin^2\Phi\cos^2\Phi + E_T^{-1}\sin^4\Phi.$$

The angle α is inclination of the helix with any plane perpendicular to the axis of the coil, the pitch angle. For the element of length, the following expression is valid:

$$dl = \frac{D}{2\cos\alpha} d\theta.$$

The total length and the mass of the spring wire, correspondingly, are:

$$l = \int_0^{2\pi n} \frac{D}{2\cos\alpha} d\theta, m = \rho \int_0^{2\pi n} \frac{D a}{2\cos\alpha} d\theta. \tag{8.20}$$

With the substitution for the moment $M(\theta) = FD(\theta)/2$ the formula for the stored energy reduces to:

$$2U_e = M_\theta^{\,2} \int_l A^* dl + 2FM_\theta \int_l B^* dl + F^2 \int_l C^* dl \tag{8.21}$$

The expression for elastic energy delivers the compression spring rate c, the compression-twist springs rate $c_{\theta F}$ and the twist springs rate c_θ of an arbitrary non-cylindrical helical composite spring:

$$\begin{aligned}
\frac{1}{c} &= \frac{\partial^2 U_e}{\partial F^2} = \int_l C^* dl,\\
\frac{1}{c_{\theta F}} &= \frac{\partial^2 U_e}{\partial F\,\partial M_\theta} = \int_l B^* dl,\\
\frac{1}{c_\theta} &= \frac{\partial^2 U_e}{\partial M_\theta^{\,2}} = \int_l A^* dl.
\end{aligned} \tag{8.22}$$

Assuming for small pitch ($\cos\alpha \cong 1$, $\sin\alpha \cong 0$) we get the following representation for compression (or extension) spring rate:

$$\frac{1}{c} = \int_0^{2\pi\, n} \frac{S_{44}}{I_T} \left(\frac{D}{2}\right)^3 d\theta\ . \tag{8.23}$$

Similarly, the compression-twist springs rate $c_{\theta F}$ describes the moment M_θ, which causes the axial travel of the spring in the absence of the axial force. This value is equal to the force F, which causes the axial travel of the spring in the absence of the axial twist moment. For small pitch compression-twist springs rate $c_{\theta F}$ disappears.

Finally, the twist spring rate c_θ describes the moment M_θ, which causes the twist angle of spring of one radiant in the absence of the axial force. The formula for the twist spring rate under the assumption of small pitch reads:

$$\frac{1}{c_\theta} = \int_0^{2\pi\, n} \frac{S_{33}}{I_b} \frac{D}{2} d\theta\ .$$

The volume of wire with variable cross-section is given by the integral of the cross-sectional area $a(\theta)$ over the wire length, so that the mass of the spring is:

$$m = \frac{1}{2}\rho \int_0^{2\pi n} a\, D d\theta.$$

For the springs with a circular cross-section with diameter d expressions for spring rates:

$$c = \pi \left(4 \int_0^{2\pi n} \frac{S_{44} D^3}{d^4} d\theta\right)^{-1},\ c_\theta = \pi \left(32 \int_0^{2\pi\, n} \frac{S_{33} D}{d^4} d\theta\right)^{-1}. \tag{8.24}$$

and for the spring mass:

$$m = \frac{1}{8}\pi\rho \int_0^{2\pi n} d^2\, D\, d\theta. \tag{8.25}$$

For the springs with elliptic and rectangular cross-section shapes, the geometric characteristics depend upon the orientation of axes B, T. The height of the cross-section is measured in the direction of the axis. The equations for spring rates and masses of

Table 8.1 Spring rates and masses of the linear springs with a non-circular wire cross-section.

	Spring rate	Mass
$a = konst$ $D = konst$	$c = \dfrac{4I_T}{\pi\, n\, S_{44} D^3}$	$m = \pi\, n\, a\, \rho\, D$
$a = konst$ $D \neq konst$	$c = \dfrac{8}{I_T^{-1} \int_0^{2\pi n} S_{44} D^3(\varphi) d\varphi}$	$m = \dfrac{1}{2} a\, \rho \displaystyle\int_0^{2\pi n} D(\varphi) d\varphi$
$a \neq konst$ $D = konst$	$c = \dfrac{8}{D^3 \int_0^{2\pi n} S_{44} I_T^{-1}(\varphi) d\varphi}$	$m = \dfrac{1}{2} \rho\, D \displaystyle\int_0^{2\pi n} a(\varphi) d\varphi$
$a \neq konst$ $D \neq konst$	$c = \dfrac{8}{\int_0^{2\pi n} S_{44} D^3(\varphi) I_T^{-1}(\varphi) d\varphi}$	$m = \dfrac{1}{2} \rho \displaystyle\int_0^{2\pi n} a(\varphi)\, D(\varphi) d\varphi$

Table 8.2 Spring rates and masses of the linear springs with a circular wire cross-section.

	Spring rate	Mass
$a = konst$ $D = konst$	$c = \dfrac{d^4}{8n\, D^3 S_{44}}$	$m = \dfrac{1}{4} \pi^2\, n\, d^2\, \rho\, D$
$a = konst$ $D \neq konst$	$c = \dfrac{\pi\, d^4}{4 \int_0^{2\pi n} S_{44} D^3(\varphi) d\varphi}$	$m = \dfrac{1}{8} \pi\, d^2\, \rho \displaystyle\int_0^{2\pi n} D(\varphi) d\varphi$
$a \neq konst$ $D = konst$	$c = \dfrac{\pi}{4D^3 \int_0^{2\pi n} S_{44} d^{-4}(\varphi) d\varphi}$	$m = \dfrac{1}{8} \pi\, \rho\, D \displaystyle\int_0^{2\pi n} d^2(\varphi) d\varphi$
$a \neq konst$ $D \neq konst$	$c = \dfrac{\pi}{4 \int_0^{2\pi n} S_{44} D^3(\varphi) d^{-4}(\varphi) d\varphi}$	$m = \dfrac{1}{8} \pi \rho \displaystyle\int_0^{2\pi n} d^2(\varphi)\, D(\varphi) d\varphi$

Table 8.3 Stiffness of wires with different cross-sections of wire.

Type of wire	Area and moments of inertia	Torsion constant
Cross-section	a, I_b, I_r	I_T
Circular, diameter d	$a = \dfrac{\pi d^2}{4}$, $I_b = I_r = \dfrac{\pi d^4}{64}$	$\dfrac{\pi d^4}{32}$
Elliptic, T-height B-width	$a = \dfrac{\pi}{4} BT$ $I_b = \dfrac{\pi B T^3}{64}$, $I_r = \dfrac{\pi B^3 T}{64}$	$\dfrac{\pi\, B^3 T^3}{16(B^2 + T^2)}$
Rectangular, T-height B-width	$a = BT$, $I_b = \dfrac{BT^3}{12}$, $I_r = \dfrac{B^3 T}{12}$	if $T > B : \xi_1 \left(\dfrac{T}{B}\right) TB^3$ if $T < B : \xi_1 \left(\dfrac{B}{T}\right) BT^3$

$$\xi_1(x) = \frac{1}{3}\left(1 - \frac{192}{\pi^5 x} \sum_{k=1,3,5..}^{\infty} \frac{1}{k^5} \tanh \frac{k\pi x}{2}\right) \approx \frac{1}{3}\left(1 - \frac{0.63..}{x} + \frac{0.052..}{x^5}\right)$$

the springs with circular and non-circular cross-sections of wire are given in Tables 8.1–8.3.

8.1.5 Helical Composite Springs of Minimal Mass

8.1.5.1 Optimization Problem

The designer of the springs deals with the problems that require minimum weight or volume of the spring material because of space limitations or material cost limitations. Formulas are established that express minimum volume and weight in terms of the given requirements. The designer obtains the boundaries for weight and volume and what parameters must be changed for further weight reduction. The mean diameter of the spring and the wire diameter are the functions of the polar angle along the spring wire (Figure 8.3), so that:

$$D = D(\theta), \quad d = d(\theta), \quad \theta = 0..2\pi n.$$

The following optimization problem is investigated: minimize the mass of the spring:

$$m \to \min_{D,d},$$

assuming the spring rate is equal to a given positive constant c^*:

$$c(D,d) = c^*, \tag{8.26}$$

the forces at installed height F_1 and full stroke F_2 are prescribed, and the ideal stresses at full stroke are limited by the value τ_w:

$$\tau = \frac{M_T}{W_T} \le \tau_w.$$

Another practically important optimization aim for the non-cylindrical springs with variable circular cross-section is as follows (Table 8.4). The ideal stress at solid height must be less than a given constant τ_w to protect the spring from inadvertent damage. This restriction, applied on the basic shear stress at solid height, is:

$$\tau \equiv \frac{8F_c D}{\pi d^3} \le \tau_w \equiv \frac{1}{\sqrt{4(F+G+4H-2N)\cos^2\Phi\sin^2\Phi + 2N}},$$
$$F_c = c(L_c - L_0). \tag{8.27}$$

8.1.5.2 Optimal Composite Spring for the Anisotropic Mises–Hill Strength Criterion

Both optimization problems – full stroke and solid height – are solved analogously. For definiteness consider the stress restriction solid height (8.27). The inequality (8.27) could be resolved for wire diameter:

$$d(\theta) \ge d_{opt}(\theta).$$

In this inequality, the optimal diameter of wire is the solution of the algebraic equation $\tau = \tau_w$ with respect to $d(\theta)$:

$$d_{opt}(\theta) \equiv \sqrt[3]{\frac{8\,F_c\,D(\theta)}{\pi\,\tau_w}}. \tag{8.28}$$

Table 8.4 Section modules of wires with different cross-section.

Type of wire	Bending section modulus	Twist section modulus
Cross-section	w_b, w_{br}	w_t
Circular, diameter d	$w_b = w_{br} = \dfrac{\pi d^3}{32}$	$\dfrac{\pi d^3}{16}$
Elliptic T-height B-width	$w_b = \dfrac{\pi BT^2}{32}$ $w_{br} = \dfrac{\pi B^2 T}{32}$	$\dfrac{\pi}{16}\min(B^2T, BT^2)$
Rectangular, T-height B-width	$w_b = \dfrac{BT^2}{6}$ $w_{br} = \dfrac{TB^2}{6}$	if $T > B$: $\xi_2\left(\dfrac{T}{B}\right)TB^2$ if $T < B$: $\xi_2\left(\dfrac{B}{T}\right)BT^2$
$\xi_2(x) = 1 - \dfrac{8}{\pi^2}\sum\limits_{k=1,3,5..}^{\infty}\dfrac{1}{k^2\cosh\dfrac{k\pi x}{2}} \approx \dfrac{1+x^2}{0.35+x^2}\xi_1(x)$		

Rewrite the formula for spring rate (8.25), considering that for all possible cross-sections the stress conditions require $d(\theta) \geq d_1(\theta)$:

$$c(D,d) = \pi\left(4\int_0^{2\pi n}\frac{S_{44}D^3}{d^4}d\theta\right)^{-1} \geq \pi\left(4\int_0^{2\pi n}\frac{S_{44}D^3}{d^4_{opt}}d\theta\right)^{-1}.$$

If the fiber angle is constant over the cross-section of the spring, the compliance S_{44} does not change over the length. Substitution of the expression (8.28) for the optimal diameter of wire into the last expression reduces the stiffness requirement (8.22) to the following inequality:

$$c^* = c(D,d) \geq \frac{4\pi}{S_{44}}\left(\frac{F_c}{\pi\tau_w}\right)^{4/3}\left(\int_0^{2\pi n}D^{5/3}\,d\theta\right)^{-1}. \tag{8.29}$$

Otherwise, the expression (8.26) for the mass of spring after the substitution (8.28) results in the second inequality:

$$m \geq \frac{1}{2}\pi\rho\left(\frac{F_c}{\pi\tau_w}\right)^{2/3}\left(\int_0^{2\pi n}D^{5/3}\,d\theta\right). \tag{8.30}$$

It is well known that the inequalities of the same sign can be multiplied. The multiplication of the inequalities (8.29) and (8.30) results in a final lower boundary for spring mass:

$$m \geq m_{opt} = \frac{2\,\rho\,F_c^2}{\tau_w^2\,S_{44}c^*},$$
$$\tau_w^2\,S_{44}(\Phi) = \frac{4(s_{33}+s_{22}-2s_{23}-s_{44})\cos^2\Phi\sin^2\Phi + s_{44}}{4(F+F+4H-2N)\cos^2\Phi\sin^2\Phi + 2N}. \tag{8.31}$$

The mass of the spring is minimal if the angle φ delivers the maximum value for the function $\tau_w^2\,S_{44}(\Phi)$. The derivative of the function $\tau_w^2\,S_{44}(\Phi)$ vanishes in three stationary points:

$$\Phi_1 = 0, \qquad \Phi_2 = \pi/4, \quad \Phi_3 = \pi/2. \tag{8.32}$$

The corresponding values of the function $\tau_w^2 S_{44}(\Phi)$ in three stationary points are:

$$\tau_w^2 S_{44}(0) = \tau_w^2 S_{44}\left(\frac{\pi}{2}\right) = \frac{s_{44}}{2N}, \tag{8.33}$$

$$\tau_w^2 S_{44}\left(\frac{\pi}{4}\right) = \frac{s_{44} + (s_{33} + s_{22} - 2s_{23} - s_{44})}{2N + (F + G + 4H - 2N)}. \tag{8.34}$$

If the following inequality fulfills:

$$\frac{s_{33} + s_{22} - 2s_{23}}{F + G + 4H} > \frac{s_{44}}{2N}, \tag{8.35}$$

the mass of the spring attains its minimum if the fiber angle is equal to $\pi/4$. Otherwise, both angles 0, $\pi/2$ lead to a lower mass of the spring. The sense of this result is the following. If this inequality is fulfilled, the elastic energy, stored in stronger but stiffer fibers, is higher than the elastic energy stored in the pure matrix. Otherwise, the less stiff matrix stores more energy than fibers and the use of the pure matrix is more effective than the storage of elastic energy in fibers.

This important inequality establishes the exact lower boundary for the mass of a spring of arbitrary variable shape and variable circular cross-sections, designed to fulfill the stress condition at solid length:

$$m \geq \frac{2\,\rho\,F_c^2}{c^*}\min\left(\frac{2N}{s_{44}}, \frac{F + G + 4H}{s_{33} + s_{22} - 2s_{23}}\right). \tag{8.36}$$

Substituting the values of the shear compliance

$$S_{44} = 4(E_L^{-1} + E_T^{-1} + 2\nu_{TT}\mathrm{E}_\mathrm{T}^{-1} - G_{TL}^{-1})\sin^2\Phi\cos^2\Phi + G_{TL}^{-1}, \tag{8.37}$$

and Mises–Hill constants

$$H = G = \frac{1}{2}\sigma_L{}^{-2}, F = \sigma_T{}^{-2} - \frac{1}{2}\sigma_L{}^{-2}, 2N = \tau_{TL}^{-2}, F + G + 4H = \sigma_T{}^{-2} + 2\sigma_L{}^{-2}.$$

into the inequality for the mass, we finally get:

$$m \geq \frac{2\,\rho\,F_c^2}{c^*}\min\left(\frac{G_{TL}}{\tau_{TL}^2}, \frac{\sigma_T{}^{-2} + 2\sigma_L{}^{-2}}{E_L^{-1} + E_T^{-1} + 2\nu_{TT}\mathrm{E}_\mathrm{T}^{-1}}\right). \tag{8.38}$$

Consequently, if

$$\frac{G_{TL}}{\tau_{TL}^2} < \frac{\sigma_T{}^{-2} + 2\sigma_L{}^{-2}}{E_L^{-1} + E_T^{-1} + 2\nu_{TT}\mathrm{E}_\mathrm{T}^{-1}}. \tag{8.39}$$

then the weight optimal spring must be manufactured from a pure resin:

$$m \geq \frac{2\,\rho\,F_c^2}{c^*}\frac{G_{TL}}{\tau_{TL}^2}. \tag{8.40}$$

Otherwise, the fibrous material leads to an advantageous design:

$$m \geq \frac{2\,\rho\,F_c^2}{c^*}\frac{\sigma_T{}^{-2} + 2\sigma_L{}^{-2}}{E_L^{-1} + E_T^{-1} + 2\nu_{TT}\mathrm{E}_\mathrm{T}^{-1}}. \tag{8.41}$$

8.1.6 Axial and Twist Vibrations of Helical Springs

In this section, the axial and twist vibration of the helical anisotropic spring are studied. As we can see, if the pitch angle of the wire is substantial, the both vibrations are coupled together. For the small pitch angle the equation turn to be independent, θ. The equations in terms of shear and axial deformations lead to a coupled system consisting of second-order partial differential equations with variable coefficients. The differential equations make use of the warping deformation of a beam's cross-section made of anisotropic material, the curvatures of the rod axis, the initial twist of the cross-section and rotary inertia as the input quantities. The equilibrium equation of the element of the wire reads:

$$\frac{\partial M_T}{\partial z} = -\frac{m}{L}\left(\frac{D}{2}\right)^2 \frac{\partial^2 \varphi}{\partial t^2}, \frac{\partial F}{\partial z} = -\frac{m}{L}\frac{\partial^2 s}{\partial t^2}. \tag{8.42}$$

The equations of axial vibrations of springs follow from the expression for elastic energy of a helical anisotropic spring (8.19):

$$-\frac{\partial \varphi}{\partial z} = A^* M_T + B^* F, -\frac{\partial s}{\partial z} = B^* M_T + C^* F. \tag{8.43}$$

The substitution of Eq. (8.42) into (8.43) leads to the dynamical equations of the helical anisotropic spring:

$$\begin{aligned} \frac{\partial^2 \varphi}{\partial z^2} &= A^* \frac{m}{L}\left(\frac{D}{2}\right)^2 \frac{\partial^2 \varphi}{\partial t^2} + B^* \frac{m}{L}\frac{\partial^2 s}{\partial t^2}, \\ \frac{\partial s^2}{\partial z^2} &= B^* \frac{m}{L}\left(\frac{D}{2}\right)^2 \frac{\partial^2 \varphi}{\partial t^2} + C^* \frac{m}{L}\frac{\partial^2 s}{\partial t^2}. \end{aligned} \tag{8.44}$$

Equations (8.44) are solved by separation of variables. The functions $\varphi(z, t)$, $s(z, t)$ are the products of the functions of the spatial coordinate and time:

$$\varphi(z,t) = \widetilde{\varphi}(z)\exp(i\omega t), \qquad s(z,t) = \widetilde{s}(z)\exp(i\omega t). \tag{8.45}$$

After the substitution (8.45), the amplitudes of vibrations $\widetilde{\varphi}(z), \widetilde{s}(z)$ follow the equations:

$$\begin{aligned} -\frac{1}{\omega^2}\frac{\partial^2 \widetilde{\varphi}}{\partial z^2} &= A^* \frac{m}{L}\left(\frac{D}{2}\right)^2 \widetilde{\varphi} + B^* \frac{m}{L}\widetilde{s}\omega^2, \\ -\frac{1}{\omega^2}\frac{\partial \widetilde{s}^2}{\partial z^2} &= B^* \frac{m}{L}\left(\frac{D}{2}\right)^2 \widetilde{\varphi} + C^* \frac{m}{L}\widetilde{s}. \end{aligned} \tag{8.46}$$

The solution of the linear differential equations (8.46) with integration constants C_i, D_i, $i = 1, .., 4$ expresses as:

$$\widetilde{s} = \sum_{i=1}^{4} C_i \exp(-i\lambda z), \widetilde{\varphi} = \sum_{i=1}^{4} D_i \exp(-i\lambda z). \tag{8.47}$$

In Eq. (8.47) λ is the root of the quadric equation:

$$\begin{aligned} &\left(\frac{\lambda}{\omega}\right)^4 - \left(\frac{\lambda}{\omega}\right)^2 p + q = 0, \\ &p = \frac{m}{L}\left[A^*\left(\frac{D}{2}\right)^2 + C^*\right], \quad q = (A^* C^* - B^{*2})\left(\frac{D}{2}\frac{m}{L}\right)^2. \end{aligned} \tag{8.48}$$

The solution of Eq. (8.48) reads:

$$\left(\frac{\lambda}{\omega}\right)_{1,2,3,4} = \pm\frac{\sqrt{p \pm \sqrt{p^2 - 4q}}}{\sqrt{2}}. \tag{8.49}$$

For example, if the rotation and side movements of both ends of the spring are fixed:

$$\begin{aligned}&\widetilde{\varphi}(z = 0) = \widetilde{\varphi}(z = L) = 0,\\ &\widetilde{s}(z = 0) = \widetilde{s}(z = L) = 0.\end{aligned} \tag{8.50}$$

For the boundary conditions (8.50), the resonance equations are the following:

$$\sin \lambda_1 L = 0, \quad \sin \lambda_2 L = 0. \tag{8.51}$$

The roots of the resonance equations (8.51) are:

$$\begin{aligned}&\lambda_1 = \pi k_1/L, \quad k_1 = 0, 1, 2..,\\ &\lambda_2 = \pi k_2/L, k_2 = 0, 1, 2\ldots\end{aligned} \tag{8.52}$$

The roots (8.52) must be different, such that the integers are also different: $k_1 \neq k_2$. Finally, the fundamental frequencies of helical anisotropic spring for the boundary conditions (8.50) read:

$$\Omega_{1,2} = \frac{\pi k_{1,2}}{L}\frac{\sqrt{2}}{\sqrt{p \pm \sqrt{p^2 - 4q}}}, \qquad \lambda_{1,2} = \frac{\pi k_{1,2}}{L}\frac{\omega}{\Omega_{1,2}}. \tag{8.53}$$

Equation (8.53) takes into consideration the coupling of the axial and rotational vibrations.

For a small pitch ($\cos\alpha \cong 1$, $\sin\alpha \cong 0$), the following representations of the parameters are valid:

$$\begin{aligned}&A^* = \frac{S_{33}}{I_b}, \quad B^* = 0, \qquad C^* = \frac{S_{44}}{I_T}\left(\frac{D}{2}\right)^2,\\ &p = \left(\frac{S_{33}}{I_b} + \frac{S_{44}}{I_T}\right)\left(\frac{D}{2}\right)^2\frac{m}{L}, \quad q = \frac{S_{33}}{I_b}\frac{S_{44}}{I_T}\left(\frac{D}{2}\right)^4\left(\frac{m}{L}\right)^2.\end{aligned} \tag{8.54}$$

The fundamental frequencies of the closed-coiled anisotropic spring follow from (8.53) after the substitution (8.54):

$$\Omega_{1,2} = \frac{4\pi k_{1,2}}{mD^2}\frac{\sqrt{2}}{\sqrt{\left(\frac{S_{33}}{I_b} + \frac{S_{44}}{I_T}\right) \pm \left(\frac{S_{33}}{I_b} - \frac{S_{44}}{I_T}\right)}} = \frac{4\pi k_{1,2}}{mD^2}\begin{cases}\sqrt{I_b/S_{33}}, (T),\\ \sqrt{I_T/S_{44}}, (A).\end{cases} \tag{8.55}$$

The axial (A) and twist (T) vibrations modes of the closed-coiled anisotropic helical springs are completely decoupled.

The differential equations of motion for naturally curved and twisted elastic space beams made of anisotropic materials with non-circular cross-sections were studied in Hao et al. (2018). The equations enrich previous work on the same subject.

8.2 Conical Springs Made of Composite Material

8.2.1 Geometry of an Anisotropic Conical Spring in an Undeformed State

Conical springs are generally manufactured from spring steel and can be subjected to static loads, rarely alternating loads and dynamic loads. Conical springs can satisfy the most severe fatigue life and set loss requirements. The overwhelming majority of conical springs are made of isotropic material, steel or bronze alloys. The common theories about conical springs exclusively consider isotropic linear elastic materials.

The conical spring is well known for its configuration to nonlinearly withstand a large force with minimum deflection while storing a large amount of energy in relation to the occupied space. The load-displacement formula for an isotropic conical disk spring or a non-slotted disk spring was first developed in the articles by Timoshenko and Woinowsky-Krieger (1957) and Almen and Laszlo (1936). The cited theory is based on experimental observations according to which the cross-section of the spring merely rotates about a center point (assumed to be on the middle line of the cross-section) without undergoing an appreciable deflection. The results by Almen and Laszlo agreed with Timoshenko in regarding the radial stresses as negligible and succeeded in calculating tangential stresses and displacements of a conical spring. According to these equations, the deformation behavior of the conical spring is treated as a one-dimensional inversion of a circular ring of rectangular cross-section about an inversion center point. The resulting inverted stress condition is overlaid by a bending stress condition caused by the change in the cone angle resulting from the deflection. The cross-section of the conical spring remains rectangular so that force is always applied at the edges I and III. The cited theories consider the spring to be an elastically deformable isotropic solid disk.

Using the thin-walled shell theory, the static response of isotropic conical springs was obtained (Hübner 1982, 1984). Axisymmetric conical shells made of isotropic material under axial forces were investigated by means of calculation technique (Hübner and Emmerling 1982).

With the emergence of new construction materials (carbon fiber-reinforced plastics, boron-fiber-reinforced plastics, glass-reinforced plastics etc.) that can be regarded as anisotropic, an approximate analytical solution of the large deflection axisymmetric response of polar orthotropic thin-walled truncated conical and spherical shallow caps was presented in the paper by Dumir (1986). Donnell type equations were employed. The deflection was approximated by a one-term mode shape satisfying the boundary conditions. Galerkin's method was used to get the governing equation for the deflection at the hole.

An analytical model for representing composite wave springs is proposed in Erfanian-Naziftoosi et al. (2016). The deformation mode of the wave spring is primarily bending. The composite wave springs possess nonlinear properties under quasi-static loading. The models of the orthotropic carbon-epoxy wave springs with arbitrary geometric parameters were developed. The fatigue tests were conducted up to 5 million cycles.

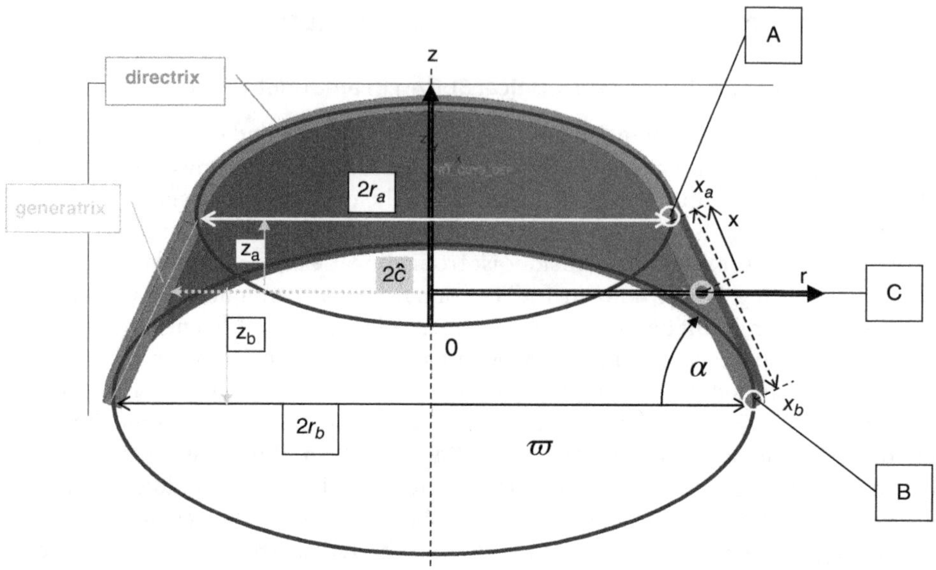

Figure 8.5 Middle surface ϖ of a free spring.

The start point of the analysis is the description of the middle surface of a conical spring (Figure 8.5). Consider a truncated conical shell made of anisotropic material. The directrix is a circle perpendicular to the line connecting its center to the apex, the motion is rotation around the fixed z-axis (Struik 1950; Pottmann and Wallner 2001). The conical frustum ϖ is generated by moving the generatrix fixed at the future apex of the cone along the directrix. The thickness t_c of the shell is assumed to be constant. The frustum ϖ serves as its middle surface in the undeformed state.

The inner radius of middle surface of free spring is r_a, the outer radius is r_b. The ratio of outer radius to inner radius reads

$$\Delta = r_b/r_a. \tag{8.56}$$

The slope angle α of the undeformed conical shell is constant. The position of the middle surface of the shell relates to a neutral plane C. In cylindrical coordinates (r, θ, z), the middle surface of a free conical shell is described, respectively, by the parametric equations

$$r = c - x\cos\alpha, z = -x\sin\alpha. \tag{8.57}$$

The distance from the point with coordinates (x, θ, y) to the axis of the undeformed cone is

$$\tilde{r} = r + y\sin\alpha = c - x\cos\alpha + y\sin\alpha, \tag{8.58}$$

where y is the normal distance from the point to the middle surface of the shell: $-t_c/2 \leq y \leq t_c/2$.

The arc length $x_b \leq x \leq x_a$ serves as a coordinate on the meridian. The arc length is measured from a certain point $\widehat{c}$, where

$$x_a = \frac{\widehat{c} - r_a}{\cos\alpha} > 0, x_b = \frac{\widehat{c} - r_b}{\cos\alpha} < 0, \tag{8.59}$$

such that

$$r_a = \widehat{c} - x_a \cos(\alpha), r_b = \widehat{c} - x_b \cos(\alpha). \tag{8.60}$$

The point c defines the parallel plane C of conical surface ϖ. The heights of the inner and outer edges of the conical surface ϖ are measured from the parallel plane C. The heights of the inner and outer edges are, correspondingly,

$$z_a = x_a \sin \alpha > 0, z_b = x_b \sin \alpha < 0. \tag{8.61}$$

The total height of the middle surface of the unloaded conical spring is

$$h_z = z_b - z_a = (\Delta - 1)\, r_a \, \tan \alpha.$$

Correspondingly, the principal radii of curvature r_1 and r_2 for the unloaded conical surface ϖ are

$$r_1^{-1} = r^{-1} \sin \, \alpha, r_2^{-1} = 0. \tag{8.62}$$

The width of the middle surface ϖ in its free state is

$$h_r = r_b - r_a = (\Delta - 1) r_a. \tag{8.63}$$

8.2.2 Curvature and Strain Deviations

The main hypothesis of the model is that the slope angle ψ of deformed conical shells changes, but also remains constant over the meridian (Figure 8.6). Namely, point $\widehat{c}$ is the inversion center point for the cross-section of the conical shell. The generatrix rotate about the point $\widehat{c}$, but remain straight and their lengths do not alter. The points of the middle surface, located on the neutral plane C, do not displace. The middle surface of the shell in the deformed state is again a frustum, Ω.

In cylindrical coordinates (r, θ, z), the middle surface Ω is given by the parametric equations

$$R = \widehat{c} - x \, \cos \psi, Z = -x \, \sin \psi. \tag{8.64}$$

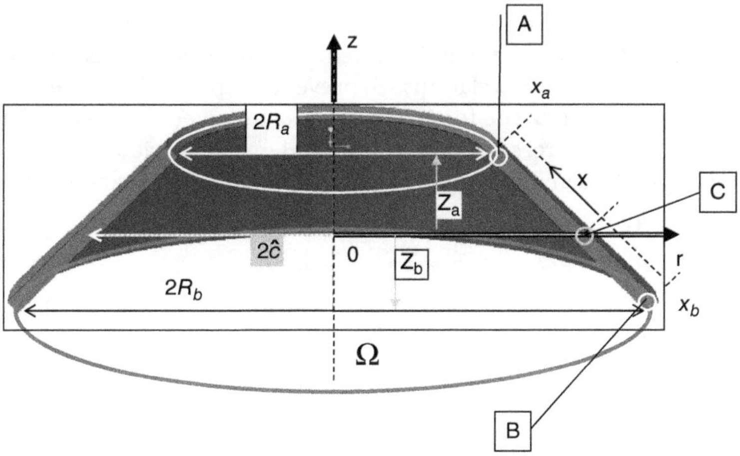

Figure 8.6 Middle surface Ω of a deformed spring.

The distance from the point of the thick-walled cone with coordinates (x, θ, y) to the axis of the deformed cone is:

$$\widetilde{R} = R + y\,\sin\psi = \widehat{c} - x\cos\psi + y\sin\psi. \tag{8.65}$$

The principal radii of curvature R_1 and R_2 for middle surface Ω are

$$R_1^{-1} = r^{-1}\sin\,\psi, R_2^{-1} = 0. \tag{8.66}$$

The inner and outer radii of Ω are equal to

$$R_a = \widehat{c} - x_a\cos\psi, R_b = \widehat{c} - x_b\cos\psi. \tag{8.67}$$

The distances from the plane C to the inner and outer edges of surface Ω read

$$Z_a = x_a\sin\psi, Z_b = x_b\sin\psi. \tag{8.68}$$

The height and width of middle surface in the deformed state Ω are, correspondingly,

$$H_z = Z_b - Z_a = r_a\frac{\sin\psi}{\cos\alpha}(\Delta - 1), \tag{8.69}$$

$$H_r = R_b - R_a = r_a\frac{\cos\psi}{\cos\alpha}(\Delta - 1). \tag{8.70}$$

The conical surface elongates during the deformation in the circumferential direction. The strain displacement relations in cylindrical coordinates (Marsden and Hughes 1994) are applied to the case of a conical shell. The following is the expression for the circumferential mid-surface strain:

$$\varepsilon_1 = \frac{R - r}{r} = \frac{\cos\alpha - \cos\psi}{\widehat{c} - x\cos\alpha}x. \tag{8.71}$$

Furthermore, the conical surface bends in the circumferential direction over the course of deformation. The circumferential curvature change is equal to the difference of curvatures in the initial and deformed states (Ventsel and Krauthammer 2001)

$$\kappa_1 = R_1^{-1} - r_1^{-1} = (\sin\,\psi - \sin\,\alpha)\cdot r^{-1}. \tag{8.72}$$

8.2.3 Thin-Walled Conical Shells Made of Anisotropic Materials

Under general orthotropic conditions, the thin-walled shell has different membrane (extensional) and bending (curvature) stiffnesses in the orthogonal directions (Vinson 1993, eq. 14.63). The integrated stress–strain and moment curvature equations for the shell are also referred to as the constitutive relations. To prevent warping of the cone, the coupling between moments and membrane forces should be eliminated. In the middle surface of the cone, the membrane forces N_1, N_2, N_{12} depend on the meridional strain ε_2, circumferential strain ε_1 and shear strain γ; see Chapter 1, and Eqs. (4.1) and (5.2):

$$\begin{aligned} N_1 &= A_{11}\varepsilon_1 + A_{12}\varepsilon_2 + A_{16}\gamma, \\ N_2 &= A_{22}\varepsilon_2 + A_{12}\varepsilon_1 + A_{26}\gamma, \\ N_{12} &= A_{26}\varepsilon_2 + A_{16}\varepsilon_1 + A_{66}\gamma. \end{aligned} \tag{8.73}$$

The moments M_1, M_2, M_{12} depend on curvature changes of the middle surface and the twist of a differential element of the middle surface due to shell bending; see Chapter 1, and Eqs. (4.1) and (5.2):

$$\begin{aligned} M_1 &= D_{11}\kappa_1 + D_{12}\kappa_2 + D_{16}\omega, \\ M_2 &= D_{22}\kappa_2 + D_{12}\kappa_1 + D_{26}\omega, \\ M_{12} &= D_{26}\kappa_2 + D_{16}\kappa_1 + D_{66}\omega. \end{aligned} \tag{8.74}$$

The constants A_{ij} are the membrane stiffness coefficients and D_{ij} are the flexural stiffness coefficients of the shell.

In the general case of coupling, the known formulas could be applied (Vinson 1993, Eqs. (14.51)–(14.54)).

When the membrane stiffness coefficients A_{16} and A_{26} are both non-zero, the stretching-shearing coupling occurs. Bending-twisting coupling comes from non-zero values of the terms of flexural stiffness coefficients D_{16} and D_{26}.

The hypotheses of the absence of meridional and twisting moments

$$M_2 = 0, \qquad M_{12} = 0, \tag{8.75}$$

delivers the expressions for curvature change in radial direction κ_2 and twist ω in terms of circumferential curvature change κ_1 only:

$$\kappa_2 = -\frac{D_{12}D_{66} - D_{26}D_{16}}{D_{22}D_{66} - D_{26}^2}\kappa_1, \omega = -\frac{D_{22}D_{16} - D_{26}D_{12}}{D_{22}D_{66} - D_{26}^2}\kappa_1. \tag{8.76}$$

The hypotheses of the absence of meridional and twisting membrane forces

$$N_2 = 0, \qquad N_{12} = 0 \tag{8.77}$$

results in the expressions for meridional strain ε_2 and shear strain γ solitary in terms of circumferential strain ε_1:

$$\varepsilon_2 = -\frac{A_{12}A_{66} - A_{26}A_{16}}{A_{22}A_{66} - A_{26}^2}\varepsilon_1, \gamma = -\frac{A_{22}A_{16} - A_{26}A_{12}}{A_{22}A_{66} - A_{26}^2}\varepsilon_1. \tag{8.78}$$

The substitution of the expressions (8.76) and (8.78) in formulas of (8.73) and (8.74) delivers the circumferential force and moment, respectively, as

$$N_1 = \frac{A_{11}A_{22}A_{66} - A_{22}A_{16}^2 - A_{11}A_{26}^2 - A_{66}A_{12}^2 + 2A_{26}A_{16}A_{12}}{A_{66}A_{22} - A_{26}^2}\varepsilon_1, \tag{8.79}$$

$$M_1 = \frac{D_{11}D_{22}D_{66} - D_{22}D_{16}^2 - D_{11}D_{26}^2 - D_{66}D_{12}^2 + 2D_{26}D_{16}D_{12}}{D_{66}D_{22} - D_{26}^2}\kappa_1. \tag{8.80}$$

8.2.4 Variation Principle

The total potential energy is:

$$\Pi = U_e + U_f. \tag{8.81}$$

Potential energy Π is the sum of the elastic strain energy, U_e, stored in the deformed body and the potential energy U_f of the applied forces:

$$U_f = \int_\alpha^\psi M d\psi, \qquad M = F_Z H_r - F_R H_z. \tag{8.82}$$

The shell is stressed by the forces in the direction of the rotation axis of the shell or by the forces in the radial direction. The total axial force acting on the edge of the upper middle surface is F_z and the total radial force is F_R.

The elastic strain energy

$$U_e = U_A + U_D, \tag{8.83}$$

is the sum of extensional energy U_A and flexural energy U_D (Vinson 1993, eq. 22.11):

$$U_A = 2\pi \int_{r_a}^{r_b} r\widetilde{W}_A dx, \qquad U_D = 2\pi \int_{r_a}^{r_b} r\widetilde{W}_D dx. \tag{8.84}$$

The functions in Eq. (8.84) are the extensional and flexural energy densities, respectively (Vinson 1993, eq. 14.34):

$$2\widetilde{W}_A \equiv A_{11}\varepsilon_1^2 + 2A_{12}\varepsilon_1\varepsilon_2 + A_{22}\varepsilon_2^2 + A_{66}\gamma^2 + 2A_{16}\varepsilon_1\gamma + 2A_{26}\varepsilon_2\gamma, \tag{8.85}$$

$$2\widetilde{W}_D \equiv D_{11}\kappa_1^2 + 2D_{12}\kappa_1\kappa_2 + D_{22}\kappa_2^2 + D_{66}\omega^2 + 2D_{16}\kappa_1\omega + 2D_{26}\kappa_2\omega. \tag{8.86}$$

For multiple layers, classical plate lamination theory leads to a set of averaged modules. For briefness, the homogeneous layer structure is assumed. In this case, the membrane stiffness coefficients A_{ij} are the linear functions of shell thickness:

$$A_{ij} = Q_{ij}t_c. \tag{8.87}$$

Here Q_{ij} are "reduced stiffness coefficients," see Chapter 1, Eq. (1.38). The flexural stiffness coefficients D_{ij} are the cubic functions of the shell thickness:

$$D_{ij} = Q_{ij}t_c{}^3/12. \tag{8.88}$$

The important constant $\widetilde{E}$ will be referred to as the effective circumferential elastic modulus:

$$\widetilde{E} = \frac{Q_{11}Q_{22}Q_{66} - Q_{22}Q_{16}{}^2 - Q_{11}Q_{26}{}^2 - Q_{66}Q_{12}{}^2 + 2Q_{26}Q_{16}Q_{12}}{Q_{66}Q_{22} - Q_{26}{}^2}. \tag{8.89}$$

With Eq. (8.89) from (8.79) and (8.80) the expressions follow:

$$N_1 = \widetilde{E}t_c\varepsilon_1, M_1 = \widetilde{E}t_c{}^3\kappa_1/12. \tag{8.90}$$

The substitution of (8.71), (8.72), (8.76) and (8.78) into the expressions (8.85) and (8.86) results in the formulas for extensional and flexural elastic energy densities:

$$2\widetilde{W}_A = \frac{A_{11}A_{22}A_{66} - A_{22}A_{16}{}^2 - A_{11}A_{26}{}^2 - A_{66}A_{12}{}^2 + 2A_{26}A_{16}A_{12}}{A_{66}A_{22} - A_{26}{}^2}\varepsilon_1^2 = \widetilde{E}t_c\varepsilon_1^2, \tag{8.91}$$

$$2\widetilde{W}_D \equiv \frac{D_{11}D_{22}D_{66} - D_{22}D_{16}{}^2 - D_{11}D_{26}{}^2 - D_{66}D_{12}{}^2 + 2D_{26}D_{16}D_{12}}{D_{66}D_{22} - D_{26}{}^2}\kappa_1^2 = \frac{\widetilde{E}t_c^3\kappa_1^2}{12}. \tag{8.92}$$

According to the principle of virtual work, external virtual work is equal to internal virtual work when equilibrated forces and stresses undergo unrelated but consistent displacements and strains (Reddy 1976). The necessary stationarity condition with respect to ψ delivers the circumferential moment:

$$M_1 = \widetilde{E}r_a^3\hat{\mu}\frac{(\cos\psi - \cos\alpha)\sin\psi}{\cos^3\alpha}\left[2(\Delta - 1) - \frac{1}{2}(\Delta^2 - 1)\right] + \widetilde{E}r_a^3\hat{\mu}^3\frac{(\sin\psi - \sin\alpha)\cos\psi}{12\cos^3\alpha}\ln\Delta, \tag{8.93}$$

with

$$\widehat{\mu} = t_c/r_a. \tag{8.94}$$

The spring is loaded by the systems of forces, which consists of two axial forces F_Z and $-F_Z$ with equal magnitudes and in opposite directions. The load on the upper edge of the spring is F_Z and the load on the lower edge of the spring is $-F_Z$. The pair of forces F_Z and $-F_Z$ with the arm H_r exerts the torque in the circumferential direction (8.93). The axial force on the shell is:

$$F_Z(c) = \frac{M_1}{H_r}. \tag{8.95}$$

The moment arm H_r is the width of middle surface in the deformed state Ω.

The substitution of (8.93) into (8.95) delivers the central formula for the axial force of the anisotropic elastic conical spring:

$$F_z = \pi \widetilde{E} {r_a}^2 \frac{(F_e\widehat{\mu} + F_f\widehat{\mu}^3)}{\cos\psi}, \tag{8.96}$$

$$F_e = \frac{2(1-\Delta) + (1+\Delta)\ln\Delta}{\ln\Delta} \cdot \frac{(\cos\psi - \cos\alpha)}{\cos^2\alpha} \sin\psi, \tag{8.97}$$

$$F_f = \frac{\ln\Delta}{6(\Delta - 1)} \cdot (\sin\alpha - \sin\psi)\cos\psi. \tag{8.98}$$

8.2.5 Structural Optimization of a Conical Spring Due to Ply Orientation

Now consider structural optimization of a conical spring made of laminated composites due to ply orientation as a design variable (Adali 1981; Banichuk 1993, 1996; Catapano et al. 2015). An extensive review of the literature can be found in the paper by Rovati and Taliercio (2003).

The objective function is strain energy. The effective circumferential elastic modulus depends upon the angle Φ between the meridian and the principal material axis (Figure 8.7). For the single lamina, the so-called transformed reduced stiffness matrix is used. The reduced stiffness coefficients q_{ij} in local material system of the solid are prescribed (Vinson 1993, eq. 14.35)

$$q_{11} = \frac{E_1}{1-\nu_{12}\nu_{21}}, q_{22} = \frac{E_2}{1-\nu_{12}\nu_{21}}, q_{12} = \frac{\nu_{21}E_1}{1-\nu_{12}\nu_{21}}, q_{66} = G_{12}, \tag{8.99}$$

E_1, E_2 are the Young's moduli in two principal directions,
ν_{12}, ν_{21} are the corresponding Poisson's ratios.

For any combination of fibers and matrix, the relation

$$\nu_{12}/E_1 = \nu_{21}/E_2 \tag{8.100}$$

is valid. The reduced stiffness coefficients q_{ij} in the local material system do alter in course of optimization.

The state of strain at each point of the solid is characterized by the given values of the three principal stresses; note that two of the stresses vanish and the remaining principal stress is the circumferential stress. The relationship between the Cartesian components

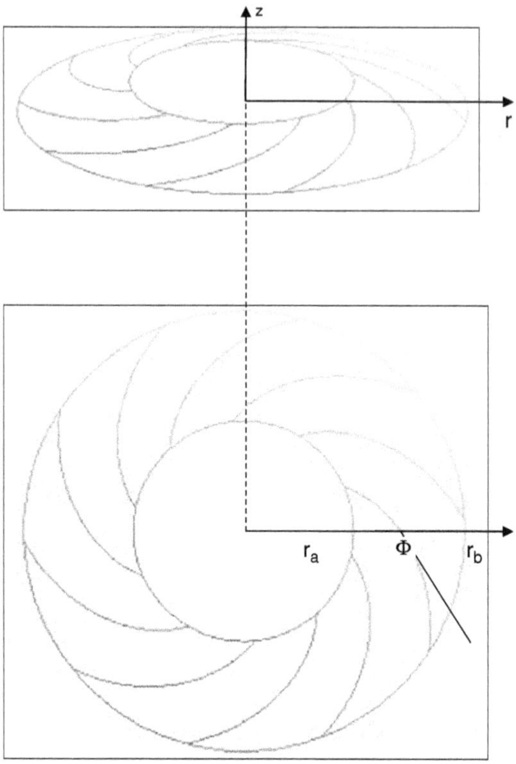

Figure 8.7 Orientation of reinforcement fibers in the anisotropic conical spring.

of the elasticity tensor Q_{ij} in the global frame, and those in the local material system q_{ij}, is given by the transformation law constants (Vinson 1993, Eq. (14.37)):

$$Q_{11} = q_{11}c^4 + 2(q_{12} + q_{66})c^2s^2 + q_{22}s^4, \tag{8.101a}$$
$$Q_{22} = q_{22}c^4 + 2(q_{12} + q_{66})c^2s^2 + q_{11}s^4, \tag{8.101b}$$
$$Q_{12} = q_{12}(c^4 + s^4) + (q_{11} + q_{22} - 4q_{66})c^2s^2, \tag{8.101c}$$
$$Q_{66} = q_{66}(c^2 - s^2)^2 + (q_{11} + q_{22} - 2q_{12})c^2s^2, \tag{8.101d}$$
$$Q_{16} = -cs \cdot (q_{22}c^2 - q_{11}s^2 - (s^2 - c^2)(q_{12} + 2q_{66})), \tag{8.101e}$$
$$Q_{26} = -cs \cdot (q_{11}c^2 - q_{22}s^2 + (s^2 - c^2)(q_{12} + 2q_{66})). \tag{8.101f}$$

Here, Φ is the angle between the meridian and the principal material axis, $c = \cos\Phi$, $s = \sin\Phi$. The angle Φ is the design variable. The effective circumferential elastic modulus

$$\widetilde{E} = \widetilde{E}(\Phi) \tag{8.102}$$

is given by the expressions (8.89) and (8.101a–f). The following inequality could be proved

$$\widetilde{E} \leq \max\left\{\frac{q_{22}q_{11} - q_{12}^2}{q_{11}}, \frac{q_{22}q_{11} - q_{12}^2}{q_{22}}\right\}. \tag{8.103}$$

8.2.6 Conical Spring Made of Orthotropic Material

In this section, the variational method is applied to shells made of special orthotropic laminates without stretching-shearing coupling:

$$A_{16} = 0, A_{26} = 0, D_{16} = 0, D_{26} = 0. \tag{8.104}$$

This kind of anisotropy demonstrates symmetric assembled laminates with equal numbers of layers of meridian angles Φ and $-\Phi$. The geometry of the spring is depicted in Figure 8.8. Consequently, the stretching-shearing coupling in the orthotropic material disappears.

With (8.73) and (8.74), the integrated stress–strain and moment curvature relations for the shell reduce to:

$$M_1 = D_{11}\kappa_1 + D_{12}\kappa_2, \tag{8.105}$$
$$M_2 = D_{22}\kappa_2 + D_{12}\kappa_1, \tag{8.106}$$
$$M_{12} = 0. \tag{8.107}$$
$$N_1 = A_{11}\varepsilon_1 + A_{12}\varepsilon_2, \tag{8.108}$$
$$N_2 = A_{22}\varepsilon_2 + A_{12}\varepsilon_1, \tag{8.109}$$
$$N_{12} = 0. \tag{8.110}$$

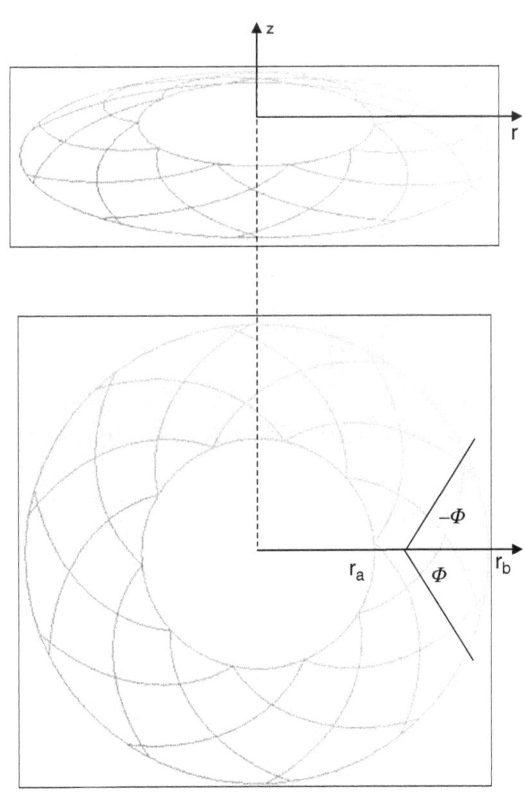

Figure 8.8 Orientation of two families of reinforcement fibers in the orthotropic conical spring.

The hypothesis of the absence of meridional forces $N_2 = 0$ and moments $M_2 = 0$ is supposed. This hypothesis immediately leads to the expressions for meridional curvature change κ_2 and twist ω, respectively:

$$\kappa_2 = -(D_{12}/D_{22})\kappa_1, \omega = 0. \tag{8.111}$$

The hypothesis delivers the meridional deformation ε_2 and shear γ, respectively, as:

$$\varepsilon_2 = -(A_{12}/A_{22})\varepsilon_1, \gamma = 0. \tag{8.112}$$

The substitution of the expressions (8.112) into (8.108) delivers the circumferential force N_1 as a function of circumferential strain ε_1 only:

$$N_1 = (A_{11}A_{22} - A_{12}{}^2)\varepsilon_1/A_{22}. \tag{8.113}$$

The substitution of the expressions (8.111) in (8.105) delivers the circumferential moment M_1 as a function of solely circumferential curvature change κ_1:

$$M_1 = (D_{11}D_{22} - D_{12}{}^2)\kappa_1/D_{22}. \tag{8.114}$$

The elastic strain energy is the sum of extensional part $\overline{U}_e$ and flexural part $\overline{U}_f$:

$$U_e = U_A + U_D. \tag{8.115}$$

For the extensional part and flexural part, the following expressions are valid in the case of an orthotropic material:

$$U_A = 2\pi \int_{r_a}^{r_b} r\overline{W}_A dx, U_D = 2\pi \int_{r_a}^{r_b} r\overline{W}_D dx. \tag{8.116}$$

The extensional energy density and flexural energy density in (8.116) for an orthotropic material are, respectively:

$$2\overline{W}_A \equiv A_{11}\varepsilon_1^2 + 2A_{12}\varepsilon_1\varepsilon_2 + A_{22}\varepsilon_2^2, \tag{8.117}$$

$$2\overline{W}_D \equiv D_{11}\kappa_1^2 + 2D_{12}\kappa_1\kappa_2 + D_{22}\kappa_2^2. \tag{8.118}$$

The stretching-shearing coupling orthotropic material disappears due to Eq. (8.104) and $Q_{16} = 0$, $Q_{26} = 0$. Accordingly, from (8.89) follows the quotient:

$$\overline{E} = \frac{Q_{22}Q_{11} - Q_{12}^2}{Q_{11}}. \tag{8.119}$$

The constant $\overline{E}$ serves as the effective circumferential elastic modulus for an orthotropic material. With Eq. (8.119) the formulas for the extensional energy density follow:

$$2\overline{W}_A = \frac{A_{22}A_{11} - A_{12}{}^2}{A_{22}}\varepsilon_1^2 \equiv \overline{E}t_c\varepsilon_1^2, \tag{8.120}$$

and flexural energy density:

$$2\overline{W}_D \equiv \frac{D_{22}D_{11} - D_{12}{}^2}{D_{22}}\kappa_1^2 \equiv \frac{\overline{E}t_c{}^3\kappa_1^2}{12}. \tag{8.121}$$

Comparison of (8.89) and (8.119), (8.91) and (8.92) with (8.120) and (8.121) leads to the following relation:

$$\overline{U}/\tilde{U} = \overline{E}/\tilde{E}. \tag{8.122}$$

Therefore, the axial force of spring made of orthotropic material relates to reaction force of spring made of isotropic material with elasticity modulus E as:

$$F_{z.orth} = \pi \overline{E} r_a{}^2 \frac{(F_e \widehat{\mu} + F_f \widehat{\mu}^3)}{\cos \psi}. \quad (8.123)$$

8.2.7 Bounds for Stiffness of a Spring Made of Orthotropic Material

The numerator of (8.89) reads:

$$Q_{22}Q_{11} - Q_{12}^2 = \frac{1}{2}(q_{11}q_{22} + 2q_{66}q_{12} + q_{66}q_{11} + q_{66}q_{22} - q_{12}^2)$$
$$- \frac{1}{2}(-q_{11}q_{22} + 2q_{66}q_{12} + q_{66}q_{11} + q_{66}q_{22} + q_{12}^2)\cos 4\Phi. \quad (8.124)$$

If $q_{11}<q_{22}$, then the exact inequality holds:

$$\overline{E}_{\min} \leq \overline{E} \leq \overline{E}_{\max}, \quad (8.125)$$

where the maximal and minimal values of effective circumferential elastic modulus are:

$$\Phi_1 = 0 : \overline{E}_{\max} = \frac{q_{22}q_{11} - q_{12}^2}{q_{11}}, \quad (8.126)$$

$$\Phi_2 = \pi/2 : \overline{E}_{\min} = \frac{q_{22}q_{11} - q_{12}^2}{q_{22}}. \quad (8.127)$$

It follows from (3.69) that the maximum of the effective circumferential elastic modulus is attained if the upmost stiffness matches the circumferential direction.

Equation (8.127) expresses the fact that minimal stiffness of the effective circumferential elastic modulus is achieved if the direction of upmost modulus of the orthotropic lamina coincides with the meridional direction.

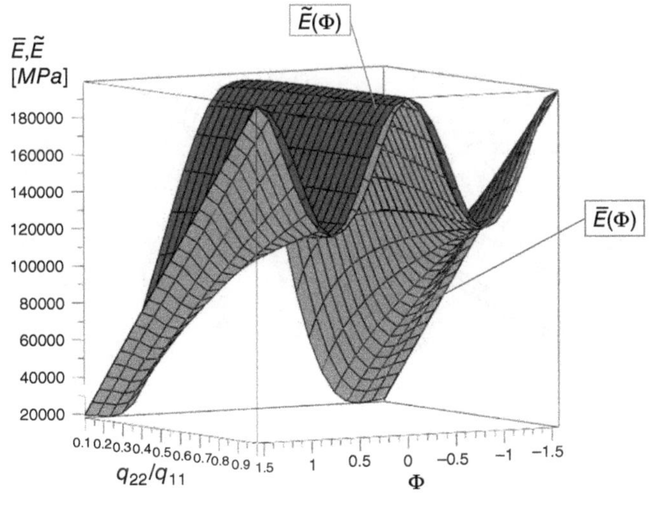

Figure 8.9 Dependence effective circumferential modules $\overline{E}, \tilde{E}$ upon meridian angle Φ for different ratios q_{22}/q_{11}.

The graph of functions $\widetilde{E}(\Phi), \overline{E}(\Phi)$ as the function of the angle Φ between the meridian and the principal material axis is shown in Figure 8.9.

8.3 Alternative Concepts for Chassis Springs Made of Composites

US20040256829A1 (2003-06-20, https://www.uspto.gov) Suspension system having a composite beam:

> A suspension system having a composite beam. In a described embodiment, a suspension system for a vehicle having a frame is provided. The suspension system includes an axle and a beam interconnected between the vehicle frame and the axle. The beam has opposite ends, an elongated body extending between the opposite ends, and a metal end connection at one of the opposite ends. The body is made of a composite material (Figure 8.10).

EP 0459220 B1 (1993, https://www.epo.org) Ring-shaped spring made out of fiber composites:

> In the case of a ring-shaped spring made out of fiber composites for vibration insulation of machines and installations, having at least two concentric wound bodies of fibers impregnated with synthetic resin, as well as slip sheets wound in between the wound bodies, to achieve characteristic values with long-term stability for the dynamic rigidity and damping of such a spring the invention provides

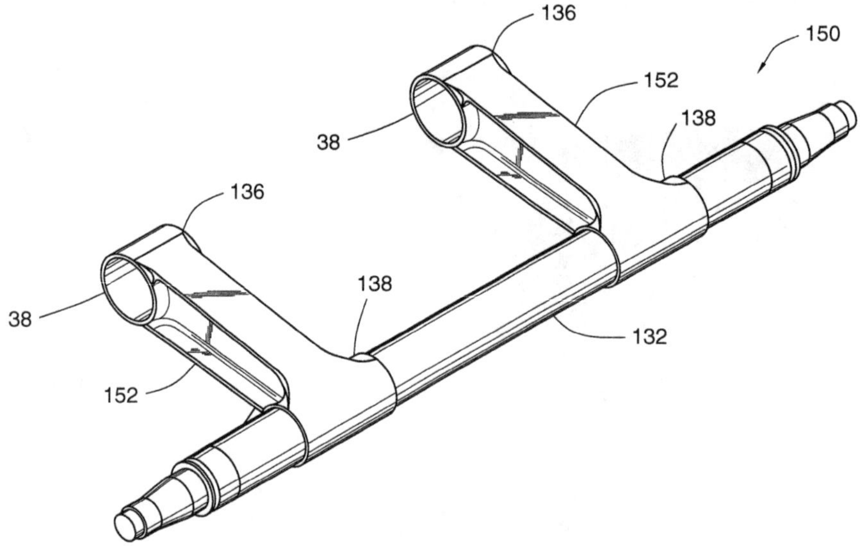

Figure 8.10 US20040256829A1 (2003) "Suspension system having a composite beam."

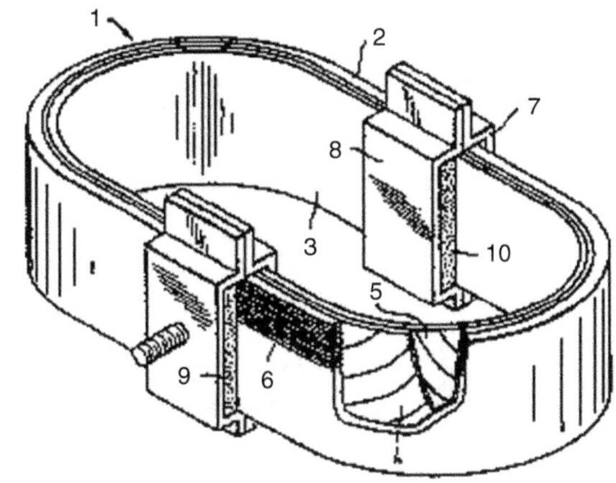

Figure 8.11 EP0459220B1 (1993): "Ring-shaped spring made out of fiber composites."

that two slip sheets (4, 5) are in each case wound in between two wound bodies (2, 3) in such a way that they respectively adhere to the laminate on the wound body side, but are designed to slip freely on the sheet sides lying against each other (Figure 8.11).

US4801019 (1989, https://www.uspto.gov) Shock Absorbing Unit Assisted by Fiberglass Reinforced Spring:

A shock absorbing spring device (A,C) is illustrated which includes a plurality of elongated, slender spring elements (B). The device may also include a central shock absorbing unit (10). Spring elements (B) comprise fiber reinforcing strands (32) embedded in the matrix material (31). Spring elements (B) have a contoured profile which includes a thickened medial bend (38) and a reverse bend (44) formed at ends (34, 36). In this manner, spring elements (B) deflect outwardly from their longitudinal axis in a reliable manner when ends (34) and (36) are loaded with compression. In one embodiment of the device (A) spring elements (B) are arranged to deflect radially with respect to longitudinal axis (37) of the device. In another embodiment (C) medial portion (38) of spring elements (B) deflect circumferentially in a row about longitudinal axis (43) of the device. Preferably, the cross-section of spring elements (B) varies from a maximum thickness at medial portion (38) to a minimum thickness at ends (34, 36) defining cantilevered arms (34a, 36a) which deflect in as medial portion (38) deflects outwardly. A second row (50) of spring elements (B′) may be included for overload capacity (Figure 8.12).

DE 102009029300A1 (2009, https://www.epo.org):

'Plastic spring for a motor vehicle undercarriage: The spring (1) has a corrugated tubular body (3) comprising two longitudinal ends, where a connection surface (5) is arranged at one of the longitudinal ends of the corrugated tubular body provided as a tubular body to engage in a cup element of one of supporting structural

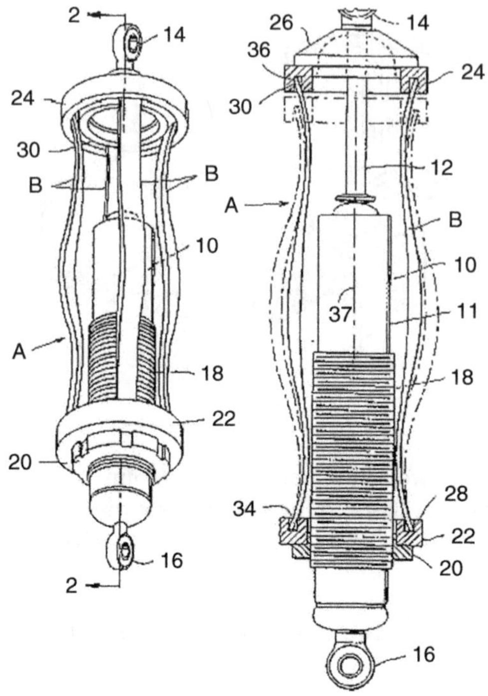

Figure 8.12 US4801019 (1989): "Shock Absorbing Unit."

component parts (9, 11). Another connection surface is arranged at the other longitudinal end of the corrugated tubular body. A displacement mass is filled in the cup element, where the tubular body is provided for displacing movement within the cup element to axially adjust preloading of a plastic spring (Figure 8.13).

DE 102012202625A1 (2013, https://www.epo.org) Plastic composite spring for vehicle suspension and apparatus and method for manufacturing the same:

> Disclosed is a corrugated plastic composite spring for a vehicle suspension and an apparatus and method for manufacturing the same. The apparatus includes a corrugated extrusion part, a braiding part, and a pultrusion part. The corrugated extrusion part forms a preform having a hollow corrugated structure. The braiding part weaves a three-dimensional woven fabric on the preform. The pultrusion part impregnates the three-dimensional woven fabric with thermosetting resin (Figure 8.14).

US20030222385 (2003) Composite wave ring spring:

> A compression spring comprising a plurality of composite wave rings stacked and attached to one another in crest to crest configuration. Each of the wave rings has a plurality of alternating peaks and troughs wherein one peak and one adjacent trough define a wave having a wave height measured axially from the peak to the trough and the wave heights vary such that the compression spring provides a varying spring rate as the compression spring is compressed (Figure 8.15).

Figure 8.13 DE102009029300A1 (2009): "Plastic spring for a motor vehicle undercarriage."

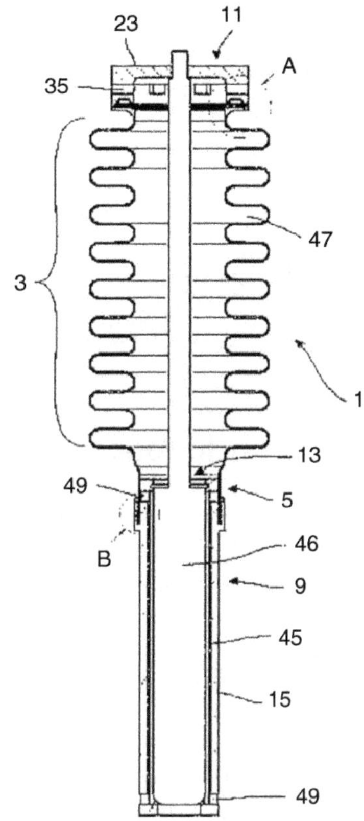

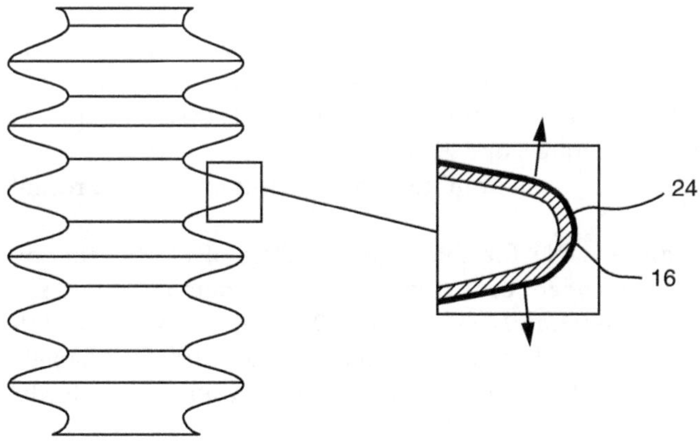

Figure 8.14 DE102012202625A1 (2013): "Plastic composite spring."

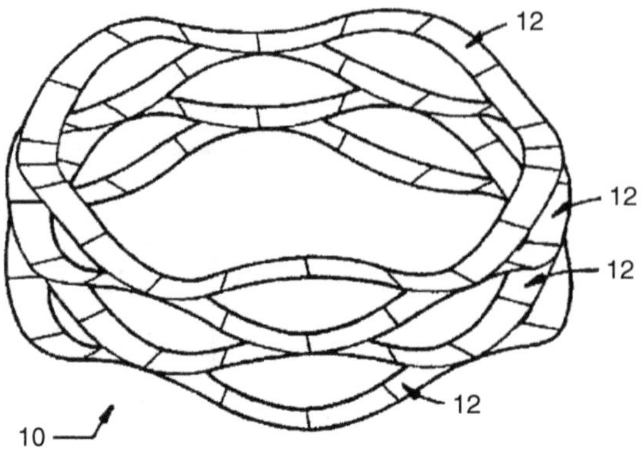

Figure 8.15 US20030222385 (2003) "Composite wave ring spring."

8.4 Conclusions

This chapter showed that the optimal wire shape, determined from the certain equal stress condition, guarantees the lowest possible mass of the spring. This mass depends only on the ultimate allowable stress for the spring material, the load at full stroke and the spring stiffness. This is an important milestone for comparison of different spring designs and spring materials. As the density and shear module are almost the same for all spring steels, the spring quality parameter can serve as the benchmarking property for spring design.

The minimal mass of the springs in parallel or in series assemblies has a single spring with the highest specific elastic energy density of material.

Further, the conical spring was investigated using variation methods. For derivation of the static response of the conical spring the equations of thin-walled and moderately thick-walled isotropic and anisotropic shells were applied. The developed equations were based on common assumptions of the theory of conical springs. The main assumption was the rigidity of the cross-section of the truncated conical shell. In the actual study, the nonlinearities due to large displacements were accounted for without the usual simplifications. The closed form solutions for spring forces and stresses in springs of rectangular cross-section with moderate slope were delivered.

Finally, some innovative solutions for the automotive spring components and recent patent applications are exposed.

The following studies are indispensable for the understanding of elastic energy-harvesting elements and are recommended for further reading and detailed reference: Ashby (2010); Norris (2005); Cowin (1994, 1997); Pedersen (1989, 1990); Rovati and Taliercio (2003); Seregin and Troitskii (1981); Sgarra and Vianello (1997a, 1997b); Vianello (1996a); Vianello (1996b); Koiter (1960); Chen and Wei (2005); Vasiliev and Gurdal (1999); EN 13906-1:2013-11 (2013); EN 13906-2:2013-09 (2013); EN 13906-3:2014-06 (2014); SAE International (1996); Meissner et al. (2015); Kruzelecki (1990); Ashby and Bréchet (2003); Lipton and Chen (2004); Chen and Lipton (2007); Lipton (1998); Blasques et al. (2011); Sonmez (2017); Ganguli (2013); Qiao et al. (1998); Del Piero (1979); Ogden (1997); Yu and Hao (2013a); Yu and Hao (2013b); Yu and Hao (2011); Xu et al. (2010); Hao and Guan (2016); Sapountzakis and Mokos (2006); Yildirim (1999); ANSYS (2015); Atluri and Pian (1972); Blom et al. (2009); Catapano et al. (2014); Lellep and Polikarpus (2014); Libai and Simmonds (1988); Meissner and Schorcht (2013); Pedersen and Pedersen (2011); Saini et al. (2007); SAE (1996); Shen and Fang (2007); Sofiyev (2003); Sofiyev and Kuruoglu (2013); Schremmer (1973); Tavares (1996); Ekelöw (2014); Schürmann (2015); Scharr (2015); Kobelev et al. (2015); Müller (2015); Müller et al. (2013); Manjunatha and Abdul Budan (2012); Petrich et al. (2017); Neubrand (2014); David (2015); Hofmeister (2016); Schürmann (2016) and Petrich et al. (2017a).

References

Adali, S. (1981). Optimization of a thin-walled, anisotropic curved Bar for maximum torsional stiffness. *J. Struct. Mech.* 9 (4): 389–413.

Almen, J.O. and Laszlo, A. (1936). The uniform section disc spring. *Trans. ASME* 58 (4): 305–314.

ANSYS (2015), http://www.ansys.com, ANSYS, Inc. Southpointe, 2600 ANSYS Drive, Canonsburg, PA, 15317c.

Ashby, M.F. (2010). *Materials Selection in Mechanical Design*, 4e. Burlington: Elsevier.

Ashby, M.F. and Bréchet, Y.J.M. (2003). Designing hybrid materials. *Acta Mater.* 51 (19): 5801–5821.

Atluri, S. and Pian, T. (1972). Theoretical formulation of finite-element methods in linear-elastic analysis of general shells. *J. Struct. Mech.* 1 (1): 1–41.

Banichuk, N.V. (1993). Shape design sensitivity analysis for optimization problems with local and global functionals. *Mech. Struct. Mach.* 21 (3): 375–397.

Banichuk, N.V. (1996). Optimization of anisotropic properties for continuum bodies and structural elements using spectral methods of tensor analysis. *Mech. Struct. Mach.* 24 (1): 71–87.

Blasques, J.P.A.A., Stolpe, M., Berggreen, C., Branner, K. (2011). Optimal Design of Laminated Composite Beams. Kgs. Lyngby, Denmark: Technical University of Denmark (DTU). DCAMM Special Report; No. S134.

Blom, A.W., Tatting, B.F., Hol, J., and Gürdala, Z. (2009). Fiber path definitions for elastically tailored conical shells. *Composites Part B* 40 (1): 77–84.

Catapano, A., Desmorat, B., and Vannucci, P. (2015). Stiffness and strength optimization of the anisotropy distribution for laminated structures. *J. Optim. Theory Appl.*, Springer Verlag, https://doi.org/10.1007/s10957-014-0693-5.

Chen, T. and Lipton, R. (2007). Bounds for the torsional rigidity of shafts with arbitrary cross-sections containing cylindrically orthotropic fibres or coated fibres. *Proc. R. Soc. A* 463: 3291–3309.

Chen, T. and Wei, C.-J. (2005). Saint-Venant torsion of anisotropic shafts: theoretical frameworks, extremal bounds and affine transformations. *Q. J. Mech. Appl. Math.* 58 (2): 269–287.

Chiu, C.-H., Chung-Li, H., Tsai, H.-S., and Lee, W.-P. (2007). An experimental investigation into the mechanical behaviors of helical composite springs. *Compos. Struct.* 77 (3): 331–340.

Choi, B.-L. and Choi, B.-H. (2015). Numerical method for optimizing design variables of carbon-fiber-reinforced epoxy composite coil springs. *Composites Part B* 82: 42–49.

Cowin, S.C. (1994). Optimization of the strain energy density in linear anisotropic elasticity. *J. Elast.* 34: 45–68.

Cowin, S.C. (1997). Remarks on coaxiality of strain and stress in anisotropic elasticity. *J. Elast.* 47: 83–84.

Dumir, P.C. (1986). Nonlinear axisymmetric response of orthotropic thin truncated conical and spherical caps. *Acta Mechanica* 60: 121–132.

Ehrich F. (2015). GFRP coil springs for use in vehicle suspensions, Federn im Fahrzeugbau, Deutscher Verband für Materialforschung und-prüfung e.V. pp. 11–18.

Ekanthappa, J., Shiva Shankar, G.S., Amith, B.M., and Gagan, M. (2016). Fabrication and experimentation of FRP helical spring. *IOP Conf. Ser.: Mater. Sci. Eng.* 149: 012098.

Ekelöw J. (2014). Design and Manufacturing of Thin Composite Tape Springs, KTH Royal Institute of Technology, Degree Project, In Lightweight Structures, Second Level, Stockholm, Sweden 2014.

EN 13906-1:2013-11 (2013). Cylindrical helical springs made from round wire and bar – Calculation and design – Part 1: Compression springs; German version DIN EN 13906-1:2013.

EN 13906-2:2013-09 (2013). Cylindrical helical springs made from round wire and bar – Calculation and design – Part 2: Extension springs; German version DIN EN 13906-2:2013.

EN 13906-3:2014-06 (2014). Cylindrical helical springs made from round wire and bar – Calculation and design – Part 3: Torsion springs; German version DIN EN 13906-3:2014.

Erfanian-Naziftoosi, H.R., Shams, S.S., and Elhajjar, R. (2016). Composite wave springs: theory and design. *Mater. Des.* 95: 48–53.

Ganguli, R. (2013). Optimal design of composite structures: a historical review. *J. Indian Inst. Sci.* 93: 557–570.

Hao, Y. and Guan, J. (2016). Parametric study on the influence of warping deformation upon natural frequencies of die springs. *Shock Vib.* 2016, Article ID 6231676: 1–10.

Hao, Y., He, W., and Shi, Y. (2018). Differential equations of motion for naturally curved and twisted composite space beams. *Shock Vib.* 2018, Article ID 5015807: 1–11.

Hofmeister K. (2016). Anforderungen an GFK-Fahrwerks-Federn aus OEM-Sicht. *"VDFI Lecture event Federn aus faserverstärkten Kunststoffen";* University Siegen, 20.04.2016.

Hübner, W. (1982). Deformationen und Spannungen bei Tellerfedern. *Konstruktion* 34: 387–392.

Hübner, W. (1984). Large deformations of elastic conical shells. In: *Flexible Shells* (ed. E.L. Axelrad and F.A. Emmerling), 257–270. Berlin: Springer Verlag.

Hübner, W. and Emmerling, F.A. (1982). Axialsymmetrische große Deformationen einer elastischen Kegelschale. *ZAMM* 62: 408–411.

Jadhav, G.B. and Gawande, V. (2015). Review on development and analysis of helical spring with combination of conventional and composite materials. *Int. J. Eng. Res. Gen. Sci.* 3 (2): 19–21.

Kara, Y. (2017). A review: fiber reinforced polymer composite helical springs. *J. Mater. Sci. Nanotechnol.,* 5 (1): 101. https://doi.org/10.15744/2348-9812.5.101.

Kobelev V., Alfes M., Müller D. (2015). Meandering axle springs from fibre-reinforced plastic for car suspension, *Federn im Fahrzeugbau, Deutscher Verband für Materialforschung und-prüfung e.V.* pp. 139–148.

Koiter, W.T. (1960). General theorems of the theory of elastic-plastic media. In: *Progress in Solid Mechanics,* vol. 1 (ed. L.N. Sneddon and R. Hill), 167–244. The Netherlands: North-Holland.

Kotaro, W., Tamura, M., Yamaya, K., and Kunor, T. (2001). Development of a new type suspension spring for rally cars. *J. Mater. Process. Technol.* 1 (111): 132–134.

Kruzelecki, J. (1990). Optimal design of helical springs. *Mechanika teoretyczna i stosowana* 1–2: 28.

Kumbhar, A., Aitavade, E., Patil, S.A., and Kumbhar, V. (2016). Analysis and experimental validation for behavior (compression, tensile, fatigue etc.) of composite material helical compressive spring used for four-wheeler suspension system. *Int. J. Innov. Eng. Res. Tech.* 3 (5): 1–15.

Lellep, J. and Polikarpus, J. (2014). Optimization of anisotropic circular plates. In: *Recent Advances in Mechanical Engineering* (ed. M.V. Shitikova, L. Vladareanu and C. Guarnaccia), Proceedings of the 5th European Conference of Mechanical Engineering (ECME' 14) Florence, Italy, November 22–24, 2014, 40–45. Athens: WSEAS Press. ISBN: 978-960-474-402-2.

Libai, A. and Simmonds, J.G. (1988). *The Nonlinear Theory of Elastic Shells, One Spatial Dimension*. London: Academic Press.

Lipton, R. (1998). Optimal fiber configurations for maximum torsional rigidity. *Arch. Ration. Mech. Anal.*, 144: 79–106.

Lipton, R. and Chen, T. (2004). Bounds and extremal configurations for the torsional rigidity of coated fiber reinforced shafts. *SIAM J. Appl. Math.* 65 (1): 299–315.

Manjunatha, T.S. and Abdul Budan, D. (2012). Manufacturing and experimentation of composite helical springs for automotive suspension. *Int. J. Mech. Eng. Rob. Res.* 1 (2): 229–241.

Marsden, J. and Hughes, T. (1994). *Mathematical Foundations of Elasticity*. New York: Dover.

Meissner, M. and Schorcht, H.-J. (2013). *Metallfedern: Grundlagen, Werkstoffe*. Berlin: Berechnung und Gestaltung Springer-Verlag.

Meissner, M., Schorcht, H.-J., and Kletzin, U. (2015). *Metallfedern*. Berlin Heidelberg: Springer-Verlag. ISBN: 978-3-642-39123-1.

Müller D. (2015) *Beitrag zur Entwicklung von Achsfedern aus Glasfaser-Kunststoff-Verbund – Methoden und Konzepte*, (Contribution to the development of axle springs made of glass-fibre plastic composites – methods and concepts). PhD Thesis, Department of Mechanical Engineering at the Technical University of Darmstadt.

Müller D., Wagner K., Brandt, R. (2013)Faserverbundwerkstoffe in Fahrwerk – Reduktion von ungefederten Massen. *4. Dresdner Werkstoffsymposium*. University of Siegen. Available at http://www.mb.uni-siegen.de/lwf/downloadbereich/131119_faserverbundwerkstoffe_im_fahrwerk_reduktion_von_ungefederten_massen.pdf.

Neubrand, J. (2014). Possibilities of coil springs and fiber-reinforced suspension parts. In: *Encyclopedia of Automotive Engineering* (ed. D. Crolla) Online © 2014, 1–11. Wiley.

Norris, A.N. (2005). Optimal orientation of anisotropic solids. *Q. J. Mech. Appl. Math.* 59 (1): 29–53.

Ogden, R.W. (1997). *Non-linear Elastic Deformations*. Mineola: Dover Publications Inc.

Pedersen, P. (1989). On optimal orientation of orthotropic materials. *Struct. Optim.* 1: 101–106.

Pedersen, P. (1990). Bounds on elastic energy in solids of orthotropic materials. *Struct. Optim.* 2: 55–62.

Pedersen, N.L. and Pedersen, P. (2011). Stiffness and design for strength of trapezoidal Belleville springs. *J. Strain Anal. Eng. Des.* 46: 825–836.

Petrich, M., Schrodin, B., Llimpe, J., and Kletzin, U. (2017a). Grundlagen der Auslegung von Federn aus faserverstärkten Kunststoffen. In: *Ilmenauer Federntag 2017* (ed. U. Kletzin), 5–16. Ilmenau: Verlag ISLE. ISBN: 978-3-938843-89-5.

Petrich M., Schrodin B., Kletzin U. (2017b) Material Selection Method for Composite Springs, *59th Ilmenau Scientific Colloquium*, Technische Universität Ilmenau, 11–15 September 2017 URN: urn:nbn:de:gbv:ilm1-2017iwk-140:5, https://www.db-thueringen.de/servlets/MCRFileNodeServlet/dbt_derivate_00039377/ilm1-2017iwk-140.pdf.

Petrich M., Llimpe J., Titze S., Kletzin U. (2017c) New concept for composite compression springs, *9th International Congress of Spring Industry, E.S.F., Taormina, Italy – 29.09.2017*, http://www.esf-springs.com/wp-content/uploads/2017/10/01-Petrich-Kletzin-New-Concept-for-Composite-Compression-Springs.pdf, Retrieved, 2018-Nov-3.

Del Piero, G. (1979). Some properties of the set of fourth-order tensors, with application to elasticity. *J. Elast.* 9: 245–261.

Pottmann, H. and Wallner, J. (2001). *Computational Line Geometry*. Berlin: Springer-Verlag.

Qiao, P., Davalos, J.F., and Barbero, E.J. (1998). Design optimization of fiber reinforced plastic composite shapes. *J. Compos. Mater.* 32 (2): 177–196.

Reddy, J.N. (1976). On complementary variational principles for the linear theory of plates. *J. Struct. Mech.* 4 (4): 417–436.

Reid, S.R., Wong, W.H., Lau, K.J., and Tse, P.C. (2002). Spring stiffnesses of composite circular springs with extended flat contact surfaces under unidirectional line-loading and surface-loading configurations. *Compos. Struct.* 55: 367–386, PII: S02 63-8223(01)00168-4.

Reid, S.R., Tse, P.C., Wong, W.H., and Lau, K.J. (2004). Large deflections of composite circular springs with extended flat contact surfaces. *Compos. Struct.* 63: 253–260. https://doi.org/10.1016/S0263-8223(03)00172-7.

Rovati, M. and Taliercio, A. (2003). Stationarity of the strain energy density for some classes of anisotropic solids. *Int. J. Solids Struct.* 40: 6043–6075.

SAE (1996). *Spring Design Manual*. Warrendale, PA: Society of Automotive Engineers: Part 5, SAE, HS-158.

Saini, P.K., Kumar, P., and Tandon, P. (2007). Design and analysis of radially tapered disc springs with parabolically varying thickness. *Proc. Inst. Mech. Eng. C* 221 (2): 151–158. https://doi.org/10.1243/0954406JMES114.

Sapountzakis, E.J. and Mokos, V.G. (2006). Dynamic analysis of 3-D beam elements including warping and shear deformation effects. *Int. J. Solids Struct.* 43 (22–23): 6707–6726.

Sardou M.A., Djomseu P. (2000). "Light and low cost composite compression C springs for vehicle suspension." *No. 2000-01-0100. SAE Technical Paper*, 2000.

Scharr G. (2015). Springs made of fiber reinforced plastics, *DVM-Bericht 1682. Federn im Fahrzeugbau*, Deutscher Verband für Materialforschung und -prüfung e. V. pp. 129–138.

Schremmer, G. (1973). The slotted conical disc spring. *Trans. ASME J. Manuf. Sci. Eng.* 95 (3): 765–770. https://doi.org/10.1115/1.3438223.

Schürmann H. (2015). The material and design capabilities of fibre-polymer-composites in spring application, *DVM-Bericht 1682. Federn im Fahrzeugbau*, Deutscher Verband für Materialforschung und-prüfung e.V. pp. 119–128.

Schürmann H. (2016). "Zum werkstofflichen und konstruktiven Potenzial von Faserverbunden beim Einsatz in Federn"; *VDFI Lecture event "Federn aus faserverstärkten Kunststoffen"*, University of Siegen, 20.04.2016.

Seregin, G.A. and Troitskii, V.A. (1981). On the best position of elastic symmetry planes in an orthotropic body. *J. Appl. Math. Mech.* 45: 139–142.

Sgarra, C. and Vianello, M. (1997a). Directions of coaxiality between pure strain and stress in linear elasticity. *J. Elast.* 46: 263–265.

Sgarra, C. and Vianello, M. (1997b). Rotations which make strain and stress coaxial. *J. Elast.* 47: 217–224.

Shen, W. and Fang, W. (2007). Design of a friction clutch using dual Belleville structures. *ASME J. Mech. Des.* 129 (9): 986–990.

Sofiyev, A.H. (2003). The buckling of an orthotropic composite truncated conical shell with continuously varying thickness subject to a time dependent external pressure. *Composites Part B* 34 (3): 227–233.

Sofiyev, A.H. and Kuruoglu, N. (2013). Buckling analysis of nonhomogeneous orthotropic thin-walled truncated conical shells in large deformation. *Thin-Walled Struct.* 62 (1): 131–141.

Sonmez, F.O. (2017). Optimum design of composite structures: a literature survey (1969–2009). *J. Reinf. Plast. Compos.* 36 (1): 3–39.

Struik, D.J. (1950). *Lectures on Classical Differential Geometry*. Cambridge, MA: Addison-Wesley.

Tavares, S.A. (1996). Thin conical shells with constant thickness and under axisymmetric load. *Comput. Struct.* 60 (6): 895–921.

Teodorescu, P.P. (2013). *Treatise on Classical Elasticity, Theory and Related Problems*. Springer.

Timoshenko, S. and Woinowsky-Krieger, S. (1957). *Theory of Plate and Shell*, 2e. New York: McGraw Hill.

Vasiliev, V.V. and Gurdal, Z. (1999). *Optimal Design: Theory and Applications to Materials and Structures*, 1e. CRC Press.

Ventsel, E. and Krauthammer, T. (2001). *Thin Plates and Shells, Theory, Analysis, and Applications*. Basel: Marcel Dekker AG.

Vianello, M. (1996a). Coaxiality of strain and stress in anisotropic linear elasticity. *J. Elast.* 42: 283–289.

Vianello, M. (1996b). Optimization of the stored energy and coaxiality of strain and stress in finite elasticity. *J. Elast.* 44: 193–202.

Vinson, J.R. (1993). *The Behavior of Shells Composed of Isotropic and Composite Materials*. Netherlands: Springer,.

Wong, W.H., Tse, P.C., Lu, K.J., and Ng, Y.F. (2004). Spring constant of fibre-reinforced plastics circular springs embedded with nickel–titanium alloy wire. *Compos. Struct.* 65: 319–328. https://doi.org/10.1016/j.compstruct.2003.11.006.

Xu, R.Q., He, J.S., and Chen, W.Q. (2010). Saint-Venant torsion of orthotropic bars with inhomogeneous rectangular cross section. *Compos. Struct.* 92 (6): 1449–1457.
Yildirim, V. (1999). Governing equations of initially twisted elastic space rods made of laminated composite materials. *Int. J. Eng. Sci.* 37 (8): 1007–1035.
Yu, A.M. and Hao, Y. (2011). Free vibration analysis of cylindrical helical springs with noncircular cross-sections. *J. Sound Vib.* 330 (11): 2628–2639.
Yu, A.-M. and Hao, Y. (2013a). Warping effect in free vibration analysis of unidirectional composite non-cylindrical helical springs. *Meccanica* 48 (10): 2453–2465.
Yu, A.M. and Hao, Y. (2013b). Effect of warping on natural frequencies of symmetric cross-ply laminated composite non-cylindrical helical springs. *Int. J. Mech. Sci.* 74: 65–72.

9

Equivalent Beams of Helical Anisotropic Springs

9.1 Helical Compression Springs Made of Composite Materials

9.1.1 Statics of the Equivalent Beam for an Anisotropic Spring

For the substitution of helical compression springs, a filamentary-wounded composite helical compression spring was primarily developed. With a circular cross-section, a weight reduction of around 50% compared to a steel compression is realistic. The fibers are wound over a thin fiber core in a +45° and −45° arrangement until the finished cross-section is produced. The cross-section is then wrapped around the negative mold. The blanks must harden in the autoclave under high pressure and temperature. The finished spring can simply be removed from the negative mold. Frequently, the so-called "lost form" principle is used, whereby the negative form is no longer used after one occasion. An advantage of the winding process is the hardening of the springs in the autoclave, which allows a large number of springs to harden in parallel.

On the actual compression spring (helical or meander-shaped) and the hypothetical beam, the equal force and moment act. Under these conditions, the axial s and lateral s_Q displacement of a spring element (coil or meander) correspond to the displacements of the hypothetical beam. Therefore, the hypothetical beam is also referred to as the equivalent beam. An equivalent beam deforms exactly like a compression spring under the same load (Figure 9.1). Analysis of space-curved beams and, particularly, a helical spring is based on Kirchhoff's equations (Kirchhoff 1859; Reissner 1981; Yildrim 2016). As shown in Kruzelecki and Zyczkowski (1990), Skoczeń and Skrzypek (1992) and Chassie et al. (1997), equivalent beam equations are considerably easier to handle than Kirchhoff's equations.

In this chapter, we apply the Timoshenko-type equivalent beam equations for the helical composite spring. The motivation for the Timoshenko-type equivalent beam is the following. The common Bernoulli's hypothesis for equivalent beams (Grammel 1924) or (Dick 1942) leads to significantly higher flexural stiffness of a spring. There is the Timoshenko-type model that provides realistic evaluation of side stiffness. The Timoshenko-type model, in comparison to the Bernoulli-type, describes two effects: shearing force effect and rotary motion effect.

The centerline of the equivalent to the helical composite spring beam is of variable length L and coincides with the axis of the free spring. In a static case, the equivalent

Design and Analysis of Composite Structures for Automotive Applications: Chassis and Drivetrain, First Edition. Vladimir Kobelev.

Companion website: www.wiley.com/go/kobelev/automotive_suspensions

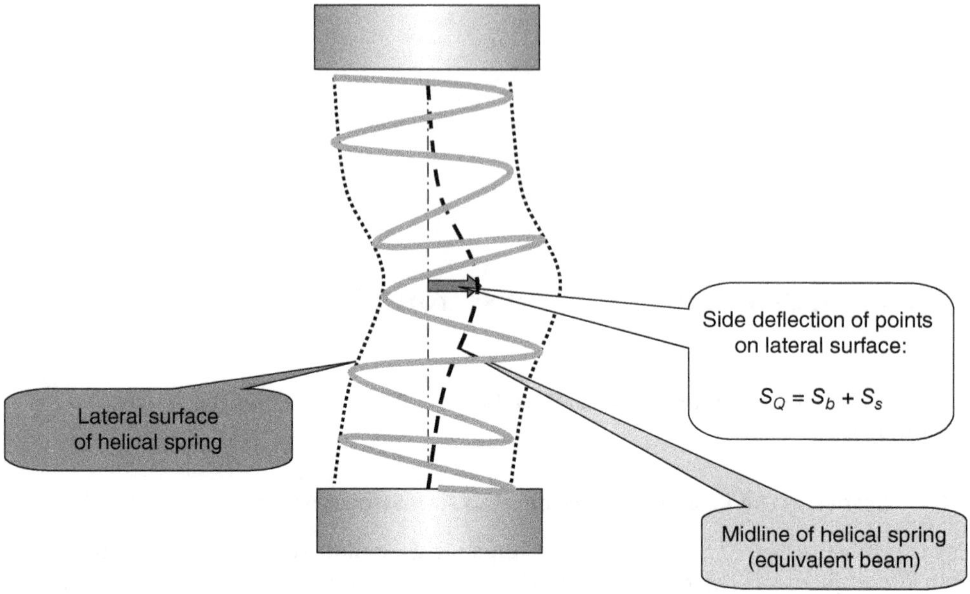

Figure 9.1 Lateral surface and equivalent beam of the composite spring.

beam is exposed to an axial load F. Consider an element of the spring that is loaded with the lateral force f_Q and the bending moment m_B. The lateral force and bending moment cause flexure of the initially straight axis of the equivalent beam. The inclination angle of the curved axis in the linear approximation reads:

$$\varphi_Q = s'_Q. \quad ()' = \partial/\partial z.$$

The transversal displacement is the sum of two partial displacements (Figure 9.2):

$$s_Q = s_b + s_s. \tag{9.1}$$

The partial displacements are:

s_s displacement due to the shear force,
s_b displacement due to bending moment.

For anisotropic helical springs, the equivalent shear, bending and axial compression rigidities of the equivalent beam are listed in Table 9.1. We use brackets for rigidities of the equivalent beam. The formulas represent the averaged stiffness of the hypothetical equivalent beam. In Table 9.1, mass per unit length $\langle m \rangle$ is a mass related to the unit length of the axis of the spring. The stiffness of spring wire in the case of bending in the normal direction $C_{33}I_n$ is assumed to be different from the stiffness of wire $C_{33}I_b$ by bending in the direction of binormal. A is the area of the wire cross-section, $C_{44}I_T$ is the twist stiffness of wire with respect to its axis, d is the diameter of round wire, D is the mean diameter of coil, n is the number of active coils and ρ is the density of the material. For the circular wire of diameter d:

$$I_T = 2I_b = 2I_n = \pi d^4/32.$$

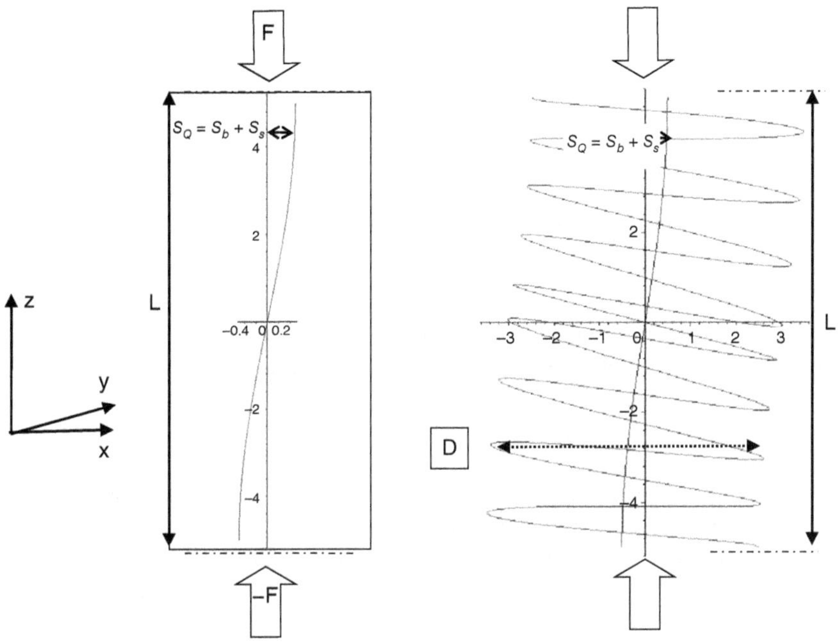

Figure 9.2 Composite spring (right) and its equivalent beam (left).

Table 9.1 Effective stiffnesses for shear, bending and compression and mass per unit length of a composite helical spring.

	Shear stiffness $\langle GA \rangle$	Bending stiffness $\langle EI_b \rangle$	Axial stiffness $\langle EA \rangle \equiv c_F L$	Mass per unit length $\langle m \rangle$
Arbitrary wire	$\frac{8\,C_{33} I_b}{\pi\, n_a\, D^3} L$	$\frac{2C_{33}\, I_n}{\pi\, n_a D\left(1 + \frac{C_{33} I_n}{C_{44} I_T}\right)} L$	$\frac{4\,C_{44} I_T}{\pi\, n_a\, D^3} L$	$\frac{\rho\, \pi\, a\, n_a D}{L}$
Round wire	$\frac{d^4\, C_{33}}{8\, n_a\, D^3} L$	$\frac{C_{33} d^4}{32\pi\, n_a\, D\left(1 + \frac{C_{33}}{2C_{44}}\right)} L$	$\frac{d^4 C_{44}}{8\, n_a\, D^3} L$	$\frac{\rho\, \pi^2\, d^2\, n_a\, D}{4L}$

The static deformation of an equivalent beam with the equivalent bending stiffness $\langle EI_b \rangle$ and shear stiffness $\langle GA \rangle$ under the action of the torque m_B and external force f_Q is described by the Timoshenko-type equations:

$$M_B = \langle EI_b \rangle (s_Q - s_s)'' + F \cdot s_Q, \tag{9.2}$$

$$Q = \langle GA \rangle s_s' - F(s_Q - s_s)', \tag{9.3}$$

$$Q + M_B' = m_B, \tag{9.4}$$

$$Q' = f_Q. \tag{9.5}$$

where:

Q is the shear force
M_B is the bending moment
m_B the external torque per unit length (in the y-direction)
f_Q the external load in the transverse direction per unit length (in the x-direction)

9.1.2 Dynamics of an Equivalent Beam for an Anisotropic Spring

The vibrations of helical composite springs were studied in Yildrim (2001) and Calim (2009). The free vibration (Yildrim 2016) problem of helical springs is modeled theoretically as continuous systems considering rotary inertia, shear and axial deformation effects. The equivalent beam model for the helical springs, made of isotropic materials, was rigorously studied in Michalczyk (2015, 2017). In these articles, different boundary conditions were investigated. The presented results (Michalczyk 2015, 2017) are also applicable for composite springs.

The Timoshenko-type equations of an equivalent beam in the dynamic case follow from the static equation with appropriate replacements for moment and lateral force. In addition to its lateral motion, the coils of the spring rotate slightly over the horizontal y-axis. The bending moment m_B is the product of static moment $\langle m\rangle D^2/8$ and the angular acceleration $\ddot{s}'_Q \equiv \partial^3 s_Q/\partial z\, \partial t^2$. The dynamical bending moment and the shear force, correspondingly, are:

$$m_B = \langle m\rangle D^2 \ddot{s}'_Q/8, \tag{9.6}$$

$$f_Q = \langle m\rangle \ddot{s}_Q. \tag{9.7}$$

Neglecting damping of the material, dynamical equations follow after the substitution of (9.6) and (9.7) into Eqs. (9.2)–(9.5):

$$M_B = \langle EI_b\rangle (s_Q - s_s)'' + F \cdot s_Q, \tag{9.8}$$

$$Q = \langle GA\rangle s'_s - F(s_Q - s_s)', \tag{9.9}$$

$$Q + M'_B = \langle m\rangle D^2 \ddot{s}'_Q/8, \tag{9.10}$$

$$Q' = \langle m\rangle \ddot{s}_Q. \tag{9.11}$$

The unknowns of four partial differential Eqs. (9.8)–(9.11) are the functions M_B, Q, s_Q, s_s. In terms of s_Q, s_s these equations reduce to partial differential equations

$$\langle GA\rangle s'_s - F(s_Q - s_s)' + [\langle EI_b\rangle (s_Q - s_s)'' + F \cdot s_Q]' = \langle m\rangle D^2 \ddot{s}'_Q/8,$$

$$[\langle GA\rangle s'_s - F(s_Q - s_s)']' = \langle m\rangle \ddot{s}_Q.$$

These equations reduce to one differential equation for the total lateral deformation $s_Q(z, t)$ of a pre-stressed compression spring:

$$\frac{\langle EI_B\rangle\langle GA\rangle}{\langle GA\rangle + F} s''''_Q - \langle m\rangle \left(\frac{\langle EI_b\rangle}{\langle GA\rangle + F} + \frac{D^2}{8} \right) \ddot{s}''_Q + \langle m\rangle \ddot{s}_Q + F s''_Q = 0. \tag{9.12}$$

Equation (9.12) is a linear partial differential equation of fourth order in coordinates and second order in time.

For the following, we need the expressions for bending moment and shear force as functions of total lateral deformation s_Q:

$$s'_s = \frac{Q + Fs'_Q}{\langle GA\rangle + F},$$

$$M_B = \frac{\langle EI_b\rangle\langle GA\rangle}{\langle GA\rangle + F}s''_Q - \frac{\langle EI_b\rangle\langle m\rangle}{\langle GA\rangle + F}\ddot{s}_Q + F\cdot s_Q,$$

$$Q = \left[\langle m\rangle\frac{D^2}{8} - \frac{\langle EI_b\rangle\langle m\rangle}{\langle GA\rangle + F}\right]\ddot{s}'_Q - \frac{\langle EI_b\rangle\langle GA\rangle}{\langle GA\rangle + F}s'''_Q - F\cdot s'_Q.$$

The clamping conditions are determined by static conditions on the spring ends. Two boundary conditions must be satisfied at each end (Table 9.3). The clamping (boundary) conditions relate to the actual, compressed length of the spring L, which shortens under the action of force F.

For example, the calculations for moment-free, hinged ends are explained in Table 9.3 (*case 5*):

$$s_Q|_{z=0} = 0, \tag{9.13}$$

$$\left[\frac{\langle EI_b\rangle\langle GA\rangle}{\langle GA\rangle + F}s''_Q - \frac{\langle EI_b\rangle\langle m\rangle}{\langle GA\rangle + F}\ddot{s}_Q + F\cdot s_Q\right]_{z=0} = 0, \tag{9.14}$$

$$s_Q|_{z=L} = 0, \tag{9.15}$$

$$\left[\frac{\langle EI_b\rangle\langle GA\rangle}{\langle GA\rangle + F}s''_Q - \frac{\langle EI_b\rangle\langle m\rangle}{\langle GA\rangle + F}\ddot{s}_Q + F\cdot s_Q\right]_{z=L} = 0. \tag{9.16}$$

The initial values for displacement and velocity are assumed:

$$s_Q(z, t = 0) = s_o(z), \tag{9.17}$$

$$\dot{s}_Q(z, t = 0) = v_o(z) \text{ for } 0 < z < L, \tag{9.18}$$

$$s_o|_{z=0} = s_o|_{z=L} = 0, \tag{9.19}$$

$$v_o|_{z=0} = v_o|_{z=L} = 0. \tag{9.20}$$

Here $s_{\cdot o}(z)$ is the initial transverse deflection and $v_o(z)$ the initial transverse velocity in the time $t = 0$. The zero boundary conditions for the displacement and for moment at each end section are to be satisfied:

$$s_Q|_{z=0} = s_Q|_{z=L} = 0, s''_Q|_{z=0} = s''_Q|_{z=L} = 0. \tag{9.21}$$

The other common types of the boundary conditions are shown in Table 9.2.

Table 9.2 Conditions at the ends of the spring in terms of shear force, moment and total transversal displacement.

Type of clamping		Condition 1		Condition 2	
1. Fixed	Clamped end deflection	Zero deflection	$s_Q = 0$	Zero slope	$s'_Q = 0$
2. Free	Free end shear force	Zero force $Q = 0$	$s'''_Q = 0$	Zero moment $M_B = 0$	$s''_Q = 0$
3. Hinged	Moment-free end (ball joint)	Zero deflection	$s_Q = 0$	Zero moment $M_B = 0$	$s''_Q = 0$
4. Sliding	Lateral force-free, sliding end	Zero slope	$s'_Q = 0$	Zero force $Q = 0$	$s'''_Q = 0$

9.2 Transverse Vibrations of a Composite Spring

9.2.1 Separation of Variables

By separation of variables, the function $s_Q(z,t)$ is presented as a product of the function $Z(z)$ of the spatial coordinate and the function $T(t)$ of time:

$$s_Q(z,t) = Z(z)T(t). \tag{9.22}$$

The substitution of (9.22) in Eq. (9.12) and after separation of variables, the following condition is established:

$$\frac{1}{\langle m\rangle}\frac{\frac{\langle EI_b\rangle\langle GA\rangle}{\langle GA\rangle+F}Z''''(z)+F\,Z''(z)}{\left(\frac{\langle EI_b\rangle}{\langle GA\rangle+F}+\frac{D^2}{8}\right)Z''(z)-Z(z)} = -\frac{\ddot{T}}{T(t)} = \omega^2. \tag{9.23}$$

This condition can be satisfied only for a constant ω^2. Equation (9.23) is separated into two ordinary differential equations:

(a) The eigenvalue problem:

$$\frac{\langle EI_b\rangle\langle GA\rangle}{\langle GA\rangle+F}Z''''(z)+F\,Z''(z) = \omega^2\langle m\rangle\left[\left(\frac{\langle EI_b\rangle}{\langle GA\rangle+F}+\frac{D^2}{8}\right)Z''(z)-Z(z)\right]. \tag{9.24}$$

(b) The evolution equation:

$$\ddot{T}+\omega^2 T = 0. \tag{9.25}$$

For solution of the eigenvalue problem (a) the representation is used:

$$Z(z) = C\exp(\Lambda z). \tag{9.26}$$

With this substitution (9.26), the Eq. (9.24) leads to an algebraic equation for Λ:

$$\Lambda^4\frac{\langle EI_b\rangle\langle GA\rangle}{\langle GA\rangle+F}+\Lambda^2 F = \omega^2\langle m\rangle\left[\left(\frac{\langle EI_b\rangle}{\langle GA\rangle+F}+\frac{D^2}{8}\right)\Lambda^2-1\right]. \tag{9.27}$$

The quartic Eq. (9.27) has four solutions:

$$\Lambda = i\beta_1, \Lambda = -i\beta_1, \Lambda = i\beta_2, \Lambda = -i\beta_2. \tag{9.28}$$

The solution to the problem (a) is the linear combination of the partial solutions:

$$Z(z) = C_1\sin(\beta_1 z)+C_2\cos(\beta_1 z)+C_3\sin(\beta_2 z)+C_4\cos(\beta_2 z), \tag{9.29}$$

C_i, $i = 1, .., 4$ are the integration constants.

On both sides the conditions of momentless hinged ends are used (Table 9.3, *case 5*). Displacement, caused by the lateral force, is supposedly omitted at the spring ends. The conditions for bending moment at the hinged ends are:

$$M_B|_{z=0} = 0, M_B|_{z=L} = 0.$$

These conditions for bending moment in terms of functions $Z(z)$, $T(t)$ read:

$$\frac{\langle EI_b\rangle\langle GA\rangle}{\langle GA\rangle+F}T(t)Z''|_{z=0} = 0, \frac{\langle EI_b\rangle\langle GA\rangle}{\langle GA\rangle+F}T(t)Z''|_{z=L} = 0.$$

Table 9.3 Conditions at the ends of the spring for different loading.

End conditions and characteristic equation	Euler column	Helical spring
$s_1 = \sin(\beta_1 L)$, $s_2 = \sin(\beta_2 L)$, $c_1 = \cos(\beta_1 L)$, $c_2 = \cos(\beta_2 L)$		
1: One end is fixed, the other slides without rotation: $s_Q(0) = 0, s'_Q(0) = 0,$ $s'''_Q(L) = 0, s'_Q(L) = 0.$ Characteristic equation: $s_1 s_2(\beta_1^4 + \beta_2^4) = (1 - c_1 c_2)\beta_1\beta_2(\beta_1^2 + \beta_2^2).$	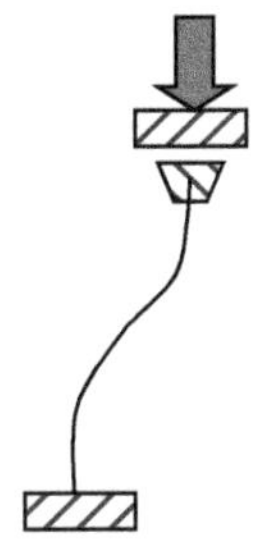	
2: Both ends are clamped: $s_Q(0) = 0, s'_Q(0) = 0,$ $s_Q(L) = 0, s'_Q(L) = 0.$ Characteristic equation: $s_1 s_2(\beta_1^2 + \beta_2^2) = 2(1 - c_1 c_2)\beta_1\beta_2.$		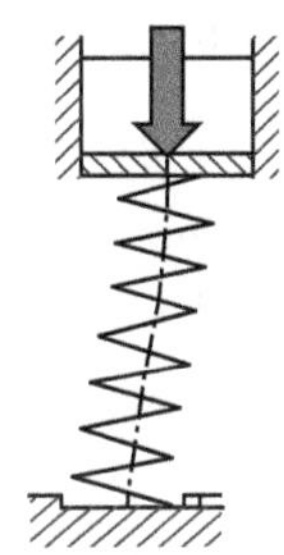
3: One end is clamped, the deflection and rotation of other is free, load direction is fixed: $s_Q(0) = 0, s'_Q(0) = 0,$ $s'''_Q(L) = 0, s''_Q(L) = 0.$ Characteristic equation: $\beta_1\beta_2(\beta_1^4 + \beta_2^4) = s_1 s_2 \beta_1^2\beta_2^2(\beta_1^2 + \beta_2^2) + 2c_1 c_2 \beta_1^3\beta_2^3.$		
4: One end is clamped, the other rotates free and side deflection is fixed: $s_Q(0) = 0, s'_Q(0) = 0,$ $s''_Q(L) = 0, s'_Q(L) = 0.$ Characteristic equation: $2s_1 s_2 \beta_1^2\beta_2^2 = (1 - c_1 c_2)\beta_1\beta_2(\beta_1^2 + \beta_2^2).$		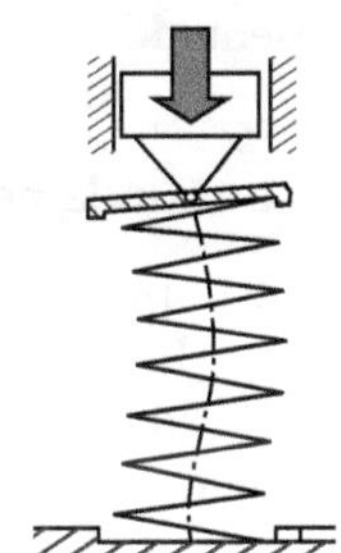

(*continued*)

Table 9.3 (Continued)

End conditions and characteristic equation	Euler column	Helical spring
5: Both ends rotate free and side deflection is restricted: $s_Q(0) = 0, s''_Q(0) = 0,$ $s_Q(L) = 0, s''_Q(L) = 0.$ Characteristic equation: $s_1 s_2 (\beta_1^2 - \beta_2^2)^2 = 0.$	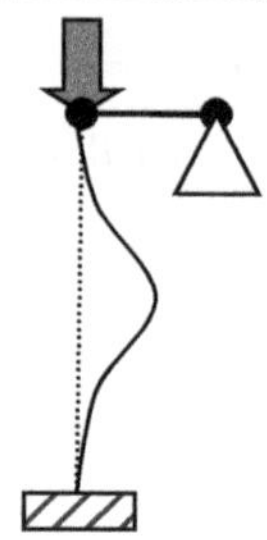	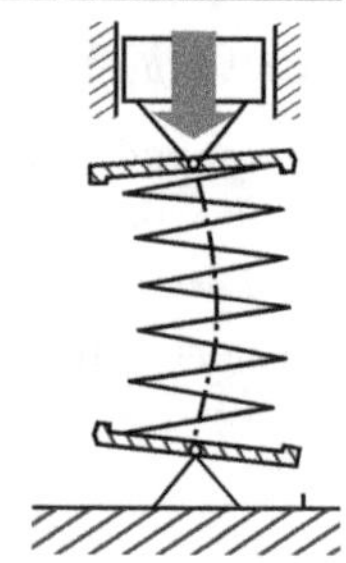

The boundary conditions for the hinged end read in this case:

$$Z(0) = Z''(0) = Z(L) = Z''(L) = 0. \tag{9.30}$$

This simplification is corrected in the final paragraph.

The use of (9.29) into (9.30) gives the following system of homogeneous equations:

$$\begin{bmatrix} 0 & 1 & 0 & 1 \\ 0 & -\beta_1^2 & 0 & -\beta_2^2 \\ \sin\beta_1 L & \cos\beta_1 L & \sin\beta_2 L & \cos\beta_2 L \\ -\beta_1^2 \sin\beta_1 L & -\beta_1^2 \cos\beta_1 L & -\beta_1^2 \sin\beta_2 L & -\beta_1^2 \cos\beta_2 L \end{bmatrix} \begin{bmatrix} C_1 \\ C_2 \\ C_3 \\ C_4 \end{bmatrix} = \begin{bmatrix} 0 \\ 0 \\ 0 \\ 0 \end{bmatrix}. \tag{9.31}$$

The homogeneous system of Eq. (9.31) only leads to non-trivial solutions if its determinant is zero:

$$(\beta_1^2 - \beta_2^2)^2 \sin(\beta_1 L)\sin(\beta_2 L) = 0. \tag{9.32}$$

The condition (9.32) is satisfied if:

$$\beta_1(k) = \beta_2(k) = \pi\frac{k}{L}. \text{ with } k = 1, 2, 3 \ldots \tag{9.33}$$

Other practically interesting cases are summarized in Table 9.3. In this table, the boundary conditions on both ends and the corresponding characteristic equations are presented.

9.2.2 Fundamental Frequencies of Transversal Vibrations

For clamping conditions, the characteristic equations from Table 9.3 should be solved for $\beta_{1,2}$ by means of numerical methods. Substitution of the roots $\beta_{1,2}$ of the characteristic equations into (9.27), and then $\Lambda = i\beta_{1,2}$ into (9.28) leads to the circular natural frequency:

$$\omega_k{}^2 = \frac{\beta_{1,2}^2(k)\left[\beta_{1,2}^2(k)\frac{\langle EI_b\rangle\langle GA\rangle}{\langle GA\rangle + F} - F\right]}{\langle m\rangle\left[\left(\frac{\langle EI_b\rangle}{\langle GA\rangle + F} + \frac{D^2}{8}\right)\beta_{1,2}^2(k) + 1\right]}, \text{with } k = 1, 2, 3 \ldots$$

In the case of moment-free support (9.30), $\beta_{1,2}$ are resolved explicitly and the natural frequencies are:

$$\omega_k^{\,2} = \frac{\left(\frac{\pi k}{L}\right)^2 \left[\left(\frac{\pi k}{L}\right)^2 \frac{\langle EI_b\rangle\langle GA\rangle}{\langle GA\rangle+F} - F\right]}{\langle m\rangle \left[\left(\frac{\langle EI_b\rangle}{\langle GA\rangle+F} + \frac{D^2}{8}\right)\left(\frac{\pi k}{L}\right)^2 + 1\right]} \quad \text{with } k = 1, 2, 3\ldots \tag{9.34}$$

Effective constants and mass of the cylindrical helical spring made of anisotropic material for shear, bending and compression are displayed in the Table 9.1. The results will now be presented in terms of two dimensionless parameters. The first parameter is the slenderness ratio of the free spring:

$$\xi = L_0/D.$$

The second parameter is the dimensionless length:

$$\mu = L/L_0.$$

The product of dimensionless length and slenderness ratio does not depend upon the stroke of the spring:

$$\xi\mu = L/D.$$

The dimensionless deflection μ characterizes the degree of compression of the spring. For a spring that is free of load: $\mu = 1$. The completely squeezed to the ideal plane form of spring corresponds to $\mu = 0$.

The axial spring force is a function of the actual spring length L. The axial spring force can be also expressed as the function of dimensionless deflection:

$$F = c_0(L_0 - L) \equiv \langle EA\rangle(\mu^{-1} - 1). \tag{9.35}$$

For the springs with circular cross-section the fundamental frequency of an order $k = 1, 2, ..$ reads:

$$\omega_k(\mu) = \frac{d\,k}{n_w D^2}\sqrt{\frac{2C_{44}\mu}{\rho}}\sqrt{\frac{(\mu-1)4\xi^2 C_{44} + (2\xi^2(\mu-1) + \pi k^2)C_{33}}{(2\pi^2k^2 + 16\xi^2\mu^2)C_{44} + (\pi^2k^2 + 4\pi k^2\mu + 8\xi^2\mu^2)C_{33}}}. \tag{9.36}$$

Remarkably, the fundamental frequency of the linear spring is not constant, but depends on dimensionless length deflection μ. The free spring in its initial length has the frequency $\omega_k^{\,0} = \omega_k(\mu = 1)$. After the substitution $\mu = 1$, the fundamental frequency for the same spring in its undeformed length is:

$$\omega_k^{\,0} = \frac{d\,k}{n_w D^2}\sqrt{\frac{2C_{44}}{\rho}}\sqrt{\frac{\pi k^2 C_{33}}{(2\pi^2k^2 + 16\xi^2)C_{44} + (\pi^2k^2 + 4\pi k^2 + 8\xi^2)C_{33}}}. \tag{9.37}$$

The formulas (9.36) and (9.37) account for slenderness ratio ξ as a parameter. After inserting the related rigidities in (9.34), the natural frequency ω_k could be obtained as a function of the dimensionless length μ and slenderness ratio ξ.

The dimensionless fundamental frequency of transverse oscillations of the order k is:

$$\Omega_k(\mu) = \omega_k/\omega_k^0 \equiv \omega_k(\mu)/\omega_k(\mu = 1). \tag{9.38}$$

9.2.3 Transverse Vibrations of a Symmetrically Stacked Helical Spring

The simplest estimation could be performed for a spring, which is fabricated by symmetric stacking [$\pm$ Φ] to the wire axis of the spring. The fiber contents of the first and second family of fibers are, correspondingly, V_{f1} and V_{f2}. The total fiber content is $V_f = V_{f1} + V_{f2}$. In this case, the modulus of the matrix is much lower than the modulus of fibers. In the local coordinates of the spring wire, the elastic constants are:

$$C_{33} = E_f[V_{f1}\cos^4(\Phi) + V_{f2}\sin^4(\Phi)],\ C_{44} = E_f(V_{f1} + V_{f2})\sin^2(\Phi)\cos^2(\Phi). \tag{9.39}$$

If the spiral angles are equal to [$\pm 45^\circ$], the elastic constants will be:

$$C_{33} = C_{44} = E_f V_f/4. \tag{9.40}$$

Accordingly, for the symmetric fabricated spring the fundamental frequency of the order k reads:

$$\omega_k(\mu) = \frac{d\,k}{n_w D^2}\sqrt{\frac{E_f V_f \mu}{2\rho}}\sqrt{\frac{6\xi^2(\mu-1)+\pi k^2}{4\pi k^2\mu + 3\pi^2 k^2 + 24\xi^2\mu^2}}. \tag{9.41}$$

The dimensionless fundamental frequency of the order for a symmetric fabricated spring [$\pm 45^\circ$] is equal to:

$$\Omega_k(\mu) = \sqrt{\mu\frac{6\xi^2(\mu-1)+\pi k^2}{\pi k^2}\,\frac{4\pi k^2 + 3\pi^2 k^2 + 24\xi^2}{4\pi k^2\mu + 3\pi^2 k^2 + 24\xi^2\mu^2}}. \tag{9.42}$$

The contour plot of a dimensionless fundamental frequency is demonstrated in Figure 9.3. There are two regions on the contour plot. The first region is bounded from

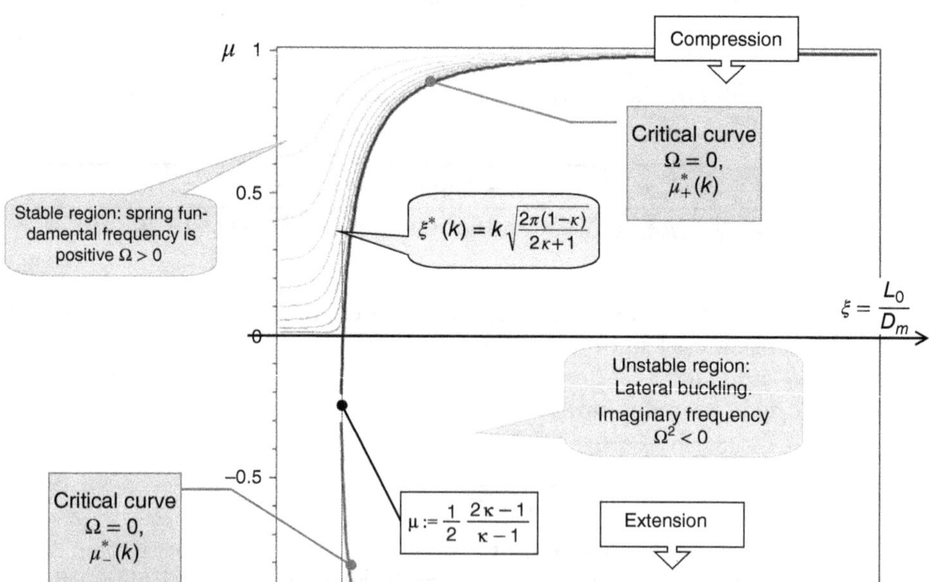

Figure 9.3 Contour plot of first fundamental frequency and critical loads as function of degree of slenderness ξ during compression $\mu_1^*(1)$ of helical composite spring and its lengthening $\mu_1^*(2)$ from the flattened state.

above by the line $\mu = 1$. The fundamental frequency of the spring is given along the line $\mu = 1$ by the expression (9.37) for springs free of load. If the spring is shorter, that is, $\mu < 1$, the frequency reduces. For relatively "thick" springs with a low value of slenderness, the springs are stable until the flat state $\mu = 0$. For modest slenderness, the spring will be unstable. The higher the slenderness is, the higher the reduction in frequency and also the more readily buckling happens.

9.3 Side Buckling of a Helical Composite Spring

9.3.1 Buckling Under Axial Force

The simplest way to demonstrate the essence of the dynamic stability method consists of its application to the traditional concept of the equivalent beam. The annulling of the fundamental frequency of transversal vibrations indicates a divergence-type instability of the corresponding vibration mode.

If the dimensionless fundamental frequency $\Omega_k(\mu)$ vanishes, lateral buckling of the spring happens. The fundamental frequency fades, if:

$$\Omega_k(\mu) = 0. \tag{9.43}$$

Equation (9.43) could be resolved for the critical spring compression $\mu^*(k)$. At that moment, the degeneration of the transversal vibration mode to buckling mode occurs. The matching mode shape with $k = 1$ corresponds to the buckling of the spring with moment-free, simply supported, hinged ends. The mode $k = 2$ matches approximately to the buckling of the spring with clamped ends.

Substitution of the expression for frequency (9.36) into the condition (9.43) leads to the following quadratic equation for critical dimensionless deflection μ as a function of slenderness ratio of the spring ξ:

$$\mu^2 + p\mu + q = 0,$$

$$p = \frac{1 - 2\kappa}{\kappa - 1}, \quad q = -\frac{\pi k^2 - 4\xi^2 k^2 - 2\xi^2 \kappa}{2\xi^2(2\kappa + 1)(\kappa - 1)}, \quad \kappa = \frac{C_{44}}{C_{33}}. \tag{9.44}$$

For each mode shape the quadratic equation (9.44) has two roots $\mu_+^*(k)$, $\mu_-^*(k)$. If the spring is compressed from its free length $\mu = 1$, buckling occurs when μ reaching the critical deflection $\mu_+^*(k)$ from free length. If the spring is unloaded from its flat state $\mu = 0$, buckling occurs at the moment of critical deflection $\mu_-^*(k)$ from the flattened state.

There is the unstable region between these two critical states:

$$\mu_-^*(k) < \mu < \mu_+^*(k), \tag{9.45}$$

$$\mu_\pm^*(k) = \frac{2\kappa - 1}{2(\kappa - 1)} \pm \frac{\sqrt{2\pi k^2(\kappa - 1) + \xi^2(2\kappa + 1)}}{2\sqrt{2\kappa + 1}(\kappa - 1)\xi}.$$

These two critical states exist only if the expression under square root is positive. This happens as long as the following condition is satisfied:

$$\xi \geq \xi^*(k) = k\sqrt{\frac{2\pi(1 - \kappa)}{2\kappa + 1}}.$$

Otherwise, the spring is always stable over the compression range, if:

$$\xi < \xi^*(k) = k\sqrt{\frac{2\pi(1 - \kappa)}{2\kappa + 1}}. \tag{9.46}$$

In this case, no buckling of the spring occurs and the spring remains straight over the full compression range $0 < \mu < 1$. With the successive compression of spring, the dimensionless natural frequency gradually reduces. The shorter the spring, the lower the fundamental frequency. In this region, the force action line of the spring drifts from the center line of the spring. On the left region of the plot $\xi < \xi^*(1)$ the dimensionless natural frequency gradually reduces until the spring flattens. On the right-hand region of the plot $\xi > \xi^*(1)$ the frequency sharply reduces to zero and the spring buckles when its critical length is achieved $\mu = \mu_+^*(1)$. To put it briefly, the spring loses its stability and transversal buckling of spring happens. The spring behaves similarly when being unloaded from its flattened shape. Consider the spring in its flattened state. The spring is not stable in this position for all values of the slenderness ratio. When expanding from its flattened position, the frequency of spring drops rapidly and the spring buckles when $\mu = \mu_-^*(1)$. This situation has minor importance for practical applications.

The angle between the direction of fiber and the normal of the wire cross-section is Φ. This angle is assumed to be constant along the wire length, as shown on the Figures 8.2 and 8.3 in Chapter 8. The elastic modulus depends on the angle Φ:

$$C_{33} = c_{11}\cos^4\Phi + 2(c_{12} + c_{66})\cos^2\Phi\sin^2\Phi + c_{22}\sin^4\Phi,$$
$$C_{44} = (c_{11} + c_{22} - 2c_{12})\cos^2\Phi\sin^2\Phi + c_{66}(\cos^2\Phi - \sin^2\Phi)^2. \tag{9.47}$$

The frequency is calculated using the values in Table 1.4, Chapter 1 and Eqs. (2.6) and (2.7) in Chapter 2. Figure 9.3 demonstrates the first fundamental frequency and critical compression for a symmetric fabricated spring with two families of fibers with an angle to the axis of the wire of $[\pm 45^\circ]$.

9.3.2 Simplified Formulas for Buckling of a Symmetrically Stacked Helical Spring

For simplification, both elastic constants are assumed to be approximately equal:

$$C_{33} = c_{11}\cos^4\Phi + c_{22}\sin^4\Phi, C_{44} = (c_{11} + c_{22})\cos^2\Phi\sin^2\Phi. \tag{9.48}$$

For the symmetric fabricated spring with two families of fibers with an angle to the axis of the wire of $[\pm 45^\circ]$, Eq. (9.43) reduces to a linear algebraic equation:

$$6\xi^2\mu - \pi k^2 - 6\xi^2 = 0. \tag{9.49}$$

From Eq. (9.49) it follows that the spring is longer than the critical compression length:

$$\mu > \mu_{crit} = 1 - \frac{\pi k^2}{6\xi^2}, \tag{9.50}$$

and possesses the positive natural frequency. The frequency reduces gradually with the compression of the spring from $\mu = 1$ to $\mu = \mu_{crit}$. Thus, above the critical curve $\mu > \mu_{crit}$ the spring is stable. When the spring is compressed and the dimensionless deflection reaches the value μ_{crit}, the frequency vanishes and the side bucking occurs following the mode in the order of k. In the general case, there is a tiny region that describes the steady lengthening of the spring from the flat state. This stable region disappears for a symmetric stacked spring. Figure 9.4 demonstrates the single critical compression that corresponds to the first fundamental form.

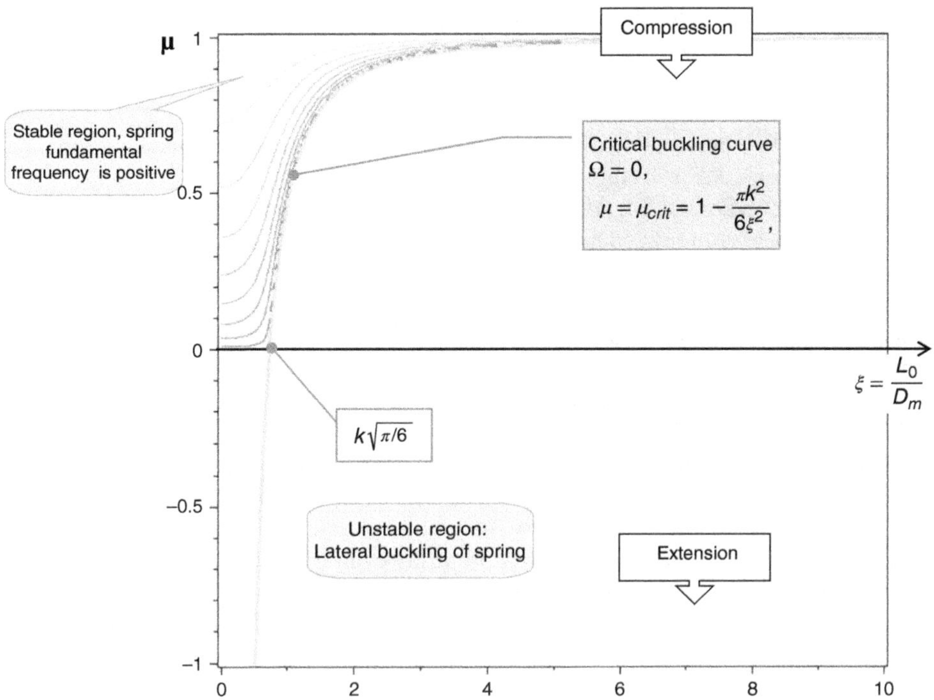

Figure 9.4 Critical loads as function of the degree of slenderness ξ during compression of symmetrical fabricated helical composite springs.

9.4 Conclusions

The bending, torsion and compression properties of a helical spring with an anisotropic wire were investigated in this chapter. The three-dimensional helical form of the spring was replaced by an imitation in the form of an imaginal, one-dimensional elastic rod. Mechanical properties of the original helical spring and its imaginal substitution were shown to be identical. Therefore, the imaginal replacement was referred to as an equivalent beam. Because of the simplicity of the geometric form and governing equations of the equivalent beam, the mathematical examination of the equivalent beam was shown to be unpretentious compared to study of the three-dimensional spatial curved beam. The exact solution of static and dynamic problems together with transition to buckling and snap-in due to axial load were presented.

References

Calim, F.F. (2009). Dynamic analysis of composite coil springs of arbitrary shape. *Composites Part B* 40: 741–757.

Chassie, G.G., Becker, L.E., and Cleghorn, W.L. (1997). On the buckling of helical springs under combined compression and torsion. *Int. J. Mech. Sci.* 39 (6): 697–704, ISSN 0020-7403, https://doi.org/10.1016/S0020-7403(96)00070-7.

Dick, J. (1942). On transverse vibrations of a helical spring with pinned ends and no axial load. *Phil. Mag. Ser.* 7 (33): 222, 513–519.

Grammel, R. (1924). Die Knickung von Schraubenfedern. *Z. Angew. Math. Mech.* 4: 384–389.

Kirchhoff, G. (1859). Über das Gleichgewicht und die Bewegung eines unendlich dünnen elastischen Stabes. *J. Reine u. Angew. Math.* 56: 285–313.

Kruzelecki, J. and Zyczkowski, M. (1990). On the concept of an equivalent beam in the stability problem of compressed helical springs. *Ing. Arch.* 60: 367–377.

Michalczyk, K. (2015). Analysis of lateral vibrations of the axially loaded helical spring. *J. Theor. App. Mech.* 53 (3): 745–755, Warsaw 2015.

Michalczyk, K. (2017). Natural transverse vibrations of helical springs in sections covered with elastic coatings. *Bull. Pol. Acad. Sci., Tech. Sci.* 65 (6): 949–959.

Reissner, E. (1981). On finite deformations of space-curved beams. *Z. Angew. Math. Phys.* 32: 734. https://doi.org/10.1007/BF00946983.

Skoczeń, B. and Skrzypek, J. (1992). Application of the equivalent beam concept to the stability of axially compressed bellows. *Int. J. Mech. Sci.* 34 (11): 901–916, ISSN 0020-7403, https://doi.org/10.1016/0020-7403(92)90020-H.

Yildrim, V. (2001). Free vibration of uniaxial composite cylindrical helical springs with circular section. *J. Sound Vib.* 239: 321–333.

Yildrim, V. (2016). Axial static load dependence free vibration analysis of helical springs based on the theory of spatially curved bars. *Lat. Am. J. Solids Struct.* 13 (2016): 2852–2875.

10

Composite Leaf Springs

10.1 Longitudinally Mounted Leaf Springs for Solid Axles

10.1.1 Predominantly Bending-Loaded Leaf Springs

The leaf type of spring is currently the most widely developed concept of the composite springs. This type of spring has been used for a long time, but industrial application started a decade ago. Leaf springs made of steel with a wheel-guiding function have long been used extensively for the transportation of vehicles. Analogously, leaf springs, made of glass fiber-reinforced plastic, are thin strips that exert a spring force when bent. There are several types of leaf spring with different mechanical properties and suspension functions. A brief comparison of different concepts for leaf springs is presented in Table 10.1 (Kobelev et al. 2015). At a low density of 1.9 g cm^{-3}, excellent fatigue resistance and low longitudinal modulus of elasticity of 40 GPa, the fiberglass permits high elastic strain and low spring rate that improves vehicle comfort (Kelly and Zweben 2000, Vol. 2, *Polymer Matrix Composites*).

Fiber failure during fatigue is commonly not critical. Distinct fiber breaks are stopped again and again by the ductile matrix. The destruction of fibers remains localized and cannot grow abruptly through the body of the spring. In the case of bending springs, failure progress is limited to the boundary layers of springs and can easily be visually detected. Both components of the composite – the glass fibers and epoxy resin matrix – are principally corrosion-resistant. Contact with automotive-specific substances such as petrol, brake fluid, neutral cleaner and distilled water provoke no significant corrosion of the critical structural components.

Temperatures in the range of −40–100 °C are tolerable without difficulty. Both during test drives on sheets and on the water, and in field trials over a longer period of time, it has been shown that fiberglass impact from stones is tolerated. Protective measures such as pressing on of plastic coverings or coatings with a high-toughness rubber compound are superfluous in the majority of cases.

Because fiberglass leaf springs are usually manufactured using precisely shaped steel tools, the dispersion of the mechanical parameters of composite springs can be kept low. By replacing the steel material with fiberglass, a weight reduction in the region of 60% can be achieved.

Compared to other suspension devices, however, leaf springs, which can be produced very cost-effectively, have a low level of suspension comfort and only a limited

Design and Analysis of Composite Structures for Automotive Applications: Chassis and Drivetrain, First Edition. Vladimir Kobelev.

Companion website: www.wiley.com/go/kobelev/automotive_suspensions

Table 10.1 Comparison of different concepts for leaf springs (Kobelev et al. 2015).

	Leaf spring with constant cross-section	Single layer elliptic leaf spring	Multileaf elliptical spring	O-elliptical leaf spring	Leaf-tension springs
Spring rate	Constant		Linear-continuous or multilinear progressive	Bilinear progressive	Linear-quadratic progressive
Non-linear spring rate adjustment	–		By changing the cross-sections		Adjustable by geometry variation
Material utilization	~50%		~100%		~60%
Reaction forces in direction of travel and reaction moments			–		Depend on spring travel, progressive
Spring shape	Flat		Predominantly flat		Spatial
Cross-section	Constant		Variable cross-section for optimum utilization of the material		

improvement in driving safety, so they are currently usually only used in commercial vehicles. Leaf springs for motor vehicles are usually provided in a curved shape and are often designed in such a way that the unloaded leaf spring in a curved state is bent by a given load into a substantially flat or flat shape.

A single leaf with variable thickness and width for a constant cross-sectional area of unidirectional fiberglass was designed, fabricated and tested in the study by Al-Qureshi (2001). Shiva Shankar and Vijayarangan (2006) composed a "low cost mono composite leaf spring" with bonded end joints. Goudah et al. (2006) manufactured elliptical springs using woven roving composites and investigated controlling their failure by using their strength in the principal direction instead of shear.

The paper by Qian et al. (2017) suggested the fatigue law interpretation of a parabolic composite leaf spring. The proposed fatigue law was confirmed by fatigue test. The experimental evaluation of the proposed ply scheme design method demonstrated the improvement in fatigue life of composite leaf springs. A report about the introduction of the new leaf springs by automotive manufacturers was presented by Pawsey (2015).

10.1.2 Moments and Forces of Leaf Springs in a Pure Bending State

The bending-dominated anisotropic spring in automotive suspensions is the leaf spring. This type of spring locates, for example, the solid drive axle in the Hotchkiss drive (Gillespie 1992. The leaf composite spring is a simply supported beam type used in light tracks. The eyes are provided for attaching the spring to the car body (Figure 10.1b). The spring is fixed to the wheel axle by means of a central clamp. The chamber is the amount of bend that is given to the spring from the central line, passing through the eyes. The wheel exercises the force F_B on the spring, and support reactions at the two eyes of the

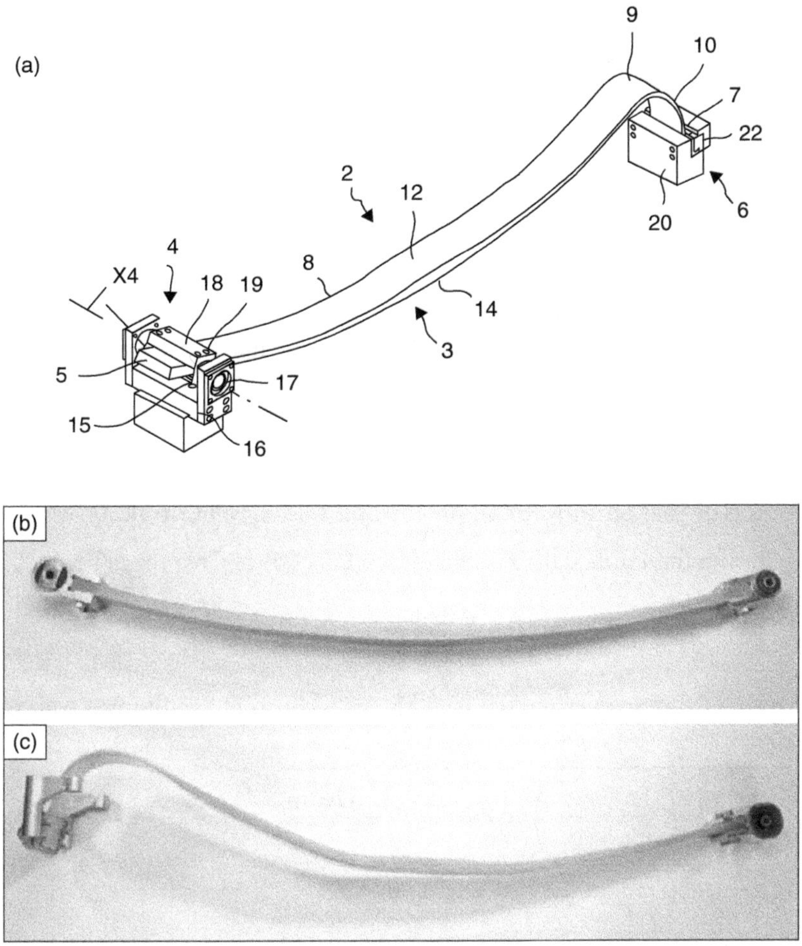

Figure 10.1 Leaf-tension composite spring (a, c) and conventional leaf spring (b). Source: (a) WO2015011181A1 (2015); (b and c) Stimpfl (2016).

spring come from the carriage. The free length of the horizontal leaf spring is L_0. In this section the elongation of the leaf spring is neglected and all equations relate to the free length of the spring. The force F_B leads to the bending moment in the cross-section of the spring:

$$M_B = M_B(x). \tag{10.1}$$

The bending moment (10.1) depends on the position x along the axis of the spring. For example, the bending moment in the cross-section of the simply supported spring with the force F_B in its middle section reads:

$$M_B(x) = F_B \begin{cases} L_0/2 + x, & -L_0/2 \le x < 0, \\ L_0/2 - x, & 0 \le x \le L_0/2. \end{cases} \tag{10.2}$$

The stored elastic energy is given by:

$$U_e = \frac{1}{2}\int_{L_0} \frac{S_{33}{M_B}^2}{I_b}dx = \frac{{F_B}^2}{2c}. \tag{10.3}$$

This expression delivers the general expression for spring rate of a leaf composite spring:

$$\frac{1}{c} = \frac{1}{{F_B}^2}\int_{L_0} \frac{S_{33}{M_B}^2}{I_b}dx. \tag{10.4}$$

The volume of material with a variable cross-section is given by the integral of the cross-sectional area $a = a(x)$ over the spring length, so that the mass of the spring is:

$$m = \rho \int_{L_0} a dx. \tag{10.5}$$

10.1.3 Optimization of Leaf Springs for an Anisotropic Mises–Hill Criterion

The Mises–Hill criterion for a uniaxial stress-state (see Eq. (2.4) of Chapter 2) reads:

$$\frac{\sin^4\Phi}{\widetilde{\sigma}_T^2} + \left(\frac{2}{\widetilde{\tau}_{LT}^2} - \frac{1}{\widetilde{\sigma}_L^2}\right)\cos^2\Phi \sin^2\Phi + \frac{\cos^4\Phi}{\widetilde{\sigma}_L^2} \leq {\sigma_B}^{-2}.$$

To protect the spring from inadvertent damage, the stress at solid height must be less than the critical stress:

$$\sigma_w = 1/\sqrt{\frac{\sin^4\Phi}{\widetilde{\sigma}_T^2} + \left(\frac{2}{\widetilde{\tau}_{LT}^2} - \frac{1}{\widetilde{\sigma}_L^2}\right)\cos^2\Phi \sin^2\Phi + \frac{\cos^4\Phi}{\widetilde{\sigma}_L^2}}. \tag{10.6}$$

The restriction (10.6), applied on the bending stress:

$$\sigma \equiv \frac{M_B}{w_B} \leq \sigma_w(\Phi). \tag{10.7}$$

The axes of the rectangular cross-section are:

$$b_s = b_s(x), \quad t_s = t_s(x).$$

The geometric characteristics of the cross-section are the following:

$$w_b = \frac{{t_s}^2 b_s}{6}, \quad I_b = \frac{b_s {t_s}^3}{12}, a = b_s t_s.$$

It is possible to minimize the mass of the spring (10.5) assuming the spring rate $c = c^*$ and the force F_B are prescribed and the stress restriction (10.7) is fulfilled.

The following estimation for the height of the cross-section follows from (10.7):

$$t_s(x) \geq t_{so}(x), t_{so}(x) = \sqrt{\frac{6M_B(x)}{b_s \sigma_w}}. \tag{10.8}$$

The lowest mass possesses a uniform stress beam, which is known as a parabolic spring. For the given moment distribution, the profile height of the uniform stress beam is $t_{so}(x)$. The substitution of the height from (10.8) into Eq. (10.5) leads to the inequality for spring rate:

$$\frac{1}{c^*} \leq \frac{1}{F_B^{\;2}} \int_{L_0} \frac{S_{33} M_B^{\;2}}{\frac{b_s}{12}\left(\frac{6M_B}{b\sigma_w}\right)^{3/2}} dx = \frac{1}{\frac{F_B^{\;2}}{12}\left(\frac{6}{\sigma_w}\right)^{3/2}} \int_{L_0} S_{33}\sqrt{b_s M_B}\, dx. \tag{10.9}$$

Similarly, the substitution (10.8) into expression (10.6) for the spring mass delivers:

$$\rho \int_{L_0} \sqrt{\frac{6BM_B}{\sigma_w}}\, dx \leq \rho \int_{L_0} t_s b_s dx = m. \tag{10.10}$$

The multiplication of inequalities (10.9) and (10.10) leads to an inequality:

$$\frac{\sqrt{6}}{c^* \sqrt{\sigma_w}} \rho \int_{L_0} \sqrt{b_s M_B}\, dx \leq \frac{m}{\frac{F_B^{\;2}}{12}\left(\frac{6}{\sigma_w}\right)^{3/2}} \int_{L_0} S_{33}\sqrt{b_s M_B}\, dx. \tag{10.11}$$

The integrals of the positive function $\sqrt{b_s M_B}$ are there on both sides of the inequality (10.11) and could be shortened in the case of a constant compliance S_{33} along the axis of the spring. From the inequality (10.11) the exact lower mass of the spring with a rectangular cross-section reads:

$$m \geq \frac{3\rho}{c^*} \frac{F_B^{\;2}}{S_{33}\sigma_w^{\;2}}. \tag{10.12}$$

For springs with a circular cross-section of the diameter $d = d(x)$, the geometric characteristics of the cross-section are the following:

$$w_b = \frac{\pi d^3}{32}, \quad I_b = \frac{\pi d^4}{64}, a = \frac{\pi d^2}{4}.$$

From (10.7) the estimation of the diameter of the cross-section follows as:

$$d \geq \sqrt[3]{\frac{32 M_B}{\pi \sigma_w}}. \tag{10.13}$$

The substitution of the wire diameter from (10.13) into Eq. (10.5) leads to the inequality for the spring rate:

$$\frac{64\pi^{1/3}\sigma_w^{\;4/3}}{32^{4/3}} \int_{L_0} S_{33} M_B^{\;2/3} dx \geq \frac{F_B^{\;2}}{c^*}. \tag{10.14}$$

The substitution (10.13) into the expression (10.6) for the spring mass reads:

$$m \geq \frac{\rho\pi}{4} \int_{L_0} \left(\frac{32 M_B}{\pi \sigma_w}\right)^{2/3} dx. \tag{10.15}$$

The multiplication of inequalities (10.9) and (10.10) leads to an inequality for the spring mass with a circular cross-section:

$$m \geq \frac{4\rho}{c^*} \frac{F_B^{\;2}}{S_{33}\sigma_w^{\;2}}. \tag{10.16}$$

The lowest mass of the circular wire is one-third greater than the lowest mass of the rectangular wire. This relation is valid for all fully stress-designed springs for an arbitrary moment M_B along the spring axes if the restrictions for allowable stress and bending stiffness are satisfied.

The stacking design of spring consists of two families of unidirectional layers, symmetric arranged in angles Φ and $-\Phi$ such that the material is transversally orthotropic respective to the axis of the spring. The compliance module in the Eq. (10.16) depends on the orientation angle of fibers to the axis of the spring:

$$S_{33} = E_L^{-1}\sin^4\Phi + (-2\nu_{LT}E_L^{-1} + G_{TL}^{-1})\sin^2\Phi\cos^2\Phi + E_T^{-1}\cos^4\Phi. \tag{10.17}$$

According to the Mises–Hill criterion, Eq. (2.4) in Chapter 2, the ultimate axial stress is the function of the orientation angles:

$$\sigma_w = 1/\sqrt{\frac{\sin^4\Phi}{\widetilde{\sigma}_T^2} + \left(\frac{2}{\widetilde{\tau}_{LT}^2} - \frac{1}{\widetilde{\sigma}_L^2}\right)\cos^2\Phi\sin^2\Phi + \frac{\cos^4\Phi}{\widetilde{\sigma}_L^2}}. \tag{10.18}$$

With these two functions, the factor in the denominator of Eqs. (10.12) and (10.16) is a function of the orientation angle Φ:

$$S_{33}{\sigma_w}^2(\varphi) = \frac{E_L^{-1}\sin^4\Phi + (-2\nu_{LT}E_L^{-1} + G_{TL}^{-1})\sin^2\Phi\cos^2\Phi + E_T^{-1}\cos^4\Phi}{\sqrt{\frac{\sin^4\Phi}{\widetilde{\sigma}_T^2} + \left(\frac{2}{\widetilde{\tau}_{LT}^2} - \frac{1}{\widetilde{\sigma}_L^2}\right)\cos^2\Phi\sin^2\Phi + \frac{\cos^4\Phi}{\widetilde{\sigma}_L^2}}}. \tag{10.19}$$

The mass of the spring is minimal if the angle Φ delivers the maximum value for the function $\sigma_w^2\, S_{33}(\Phi)$.

The derivative of the function $\sigma_w^2\, S_{33}(\Phi)$ vanishes in three stationary points:

$$\Phi_1 = 0, \Phi_3 = \pi/2,$$

$$\Phi_2 = \arccos\sqrt{\frac{2E_L\sigma_T^2 - E_T\sigma_L^2 + \sqrt{(E_L\sigma_T - E_T\sigma_L)^2 + E_LE_T\sigma_T^4}}{3E_L\sigma_T^2 - E_T(2\sigma_L^2 + \sigma_T^2)}}. \tag{10.20}$$

Correspondingly, the function $\sigma_w^2\, S_{33}(\Phi)$ assumes the following values in the stationary points:

$$\sigma_w^2\, S_{33}(0) = \frac{\sigma_L^2}{E_L}, \sigma_w^2\, S_{33}\left(\frac{\pi}{2}\right) = \frac{\sigma_T^2}{E_T}, \tag{10.21}$$

$$\sigma_w^2\, S_{33}(\Phi_2) \approx \frac{\sigma_L^2}{E_L}\frac{\varepsilon^2(4\chi^2 - \chi\varepsilon^4 - \varepsilon^6)}{4\chi^3}, \chi = \frac{E_T}{E_L}, \varepsilon = \frac{\sigma_T}{\sigma_L}. \tag{10.22}$$

To explain the obtained results, compare the masses of the spring for three suspicious stationary points. If the following inequality

$$\frac{\sigma_L^2}{E_L} > \frac{\sigma_T^2}{E_T}. \tag{10.23}$$

is satisfied, the mass of the spring attains its minimum if the fiber angle is equal to $\Phi_1 = 0$ and the fibers are optimally oriented along the axis of the spring.

Otherwise, the angle $\Phi_3 = \pi/2$ leads to a lower mass of spring. The sense of this result is the following. If this inequality is fulfilled, the elastic energy, stored in stronger but stiffer fibers, is higher than the elastic energy stored in pure matrix. Otherwise, the less stiff matrix stores more energy as fibers and the use of pure matrix is more effective than the storage of elastic energy in fibers. Consequently, the mass-optimal spring must be manufactured from a pure resin.

The intermediate case of an inclined orientation Φ_2 of the fibers to the axis of the spring, always leads to the minimum of the function $\sigma_w^2 S_{33}(\Phi_2)$ and consequently to the highest mass of the spring. This case is ineffective from a technological viewpoint.

This important inequality establishes the exact lower boundary for the mass of a spring of arbitrary variable shape and variable circular cross-sections, designed to fulfill the stress condition at full load:

$$m \geq \frac{2\,\rho\,F_B^2}{c^*} \min\left(\frac{E_L}{\sigma_L^2}, \frac{E_T}{\sigma_T^2}\right). \tag{10.24}$$

Consequently, if

$$\frac{E_L}{\sigma_L^2} > \frac{E_T}{\sigma_T^2}. \tag{10.25}$$

then the mass of the optimal spring is:

$$m \geq \frac{2\,\rho\,F_B^2}{c^*} \frac{E_T}{\sigma_T^2}. \tag{10.26}$$

If the inequality (10.25) is violated, the fibrous material leads to an optimal design with the mass:

$$m \geq \frac{2\,\rho\,F_B^2}{c^*} \frac{E_L}{\sigma_L^2}. \tag{10.27}$$

10.2 Leaf-Tension Springs

10.2.1 Combined Bending and Tension of a Spring

The development of a glass fiber-reinforced plastic leaf spring resulted in the leaf-tension spring (WO2011110611A3 2011, WO2015011181A1, 2015). The leaf-tension leaf spring comprises a first end portion (5), a spring portion (8), a bendable portion (10) and a second end portion (6) (Figure 10.1). This figure displays the leaf-tension spring (a) and (c) and the corresponding conventional leaf spring (b).

The leaf-tension spring provides an efficient solution to the task of producing a leaf spring with the appropriate suspension properties for motor vehicles. In addition to the pure suspension of the vehicle, the leaf-tension spring also performs wheel-guiding tasks. In addition to a standard leaf spring, the leaf-tension spring has a simple, torque-resistant connection at both ends.

At the same time, the leaf-tension spring takes over another feeding point, which saves a fastening component as the spring does not require a revolving bearing in some cases. Additionally, the spring force response curve can be continuously adjusted by the suitable variations of the curvature and thickness of the spring body.

A leaf-tension spring is a meander-shaped leaf spring made of a fiber-reinforced plastic material that can absorb any lateral guiding forces and transfer them to a leaf spring

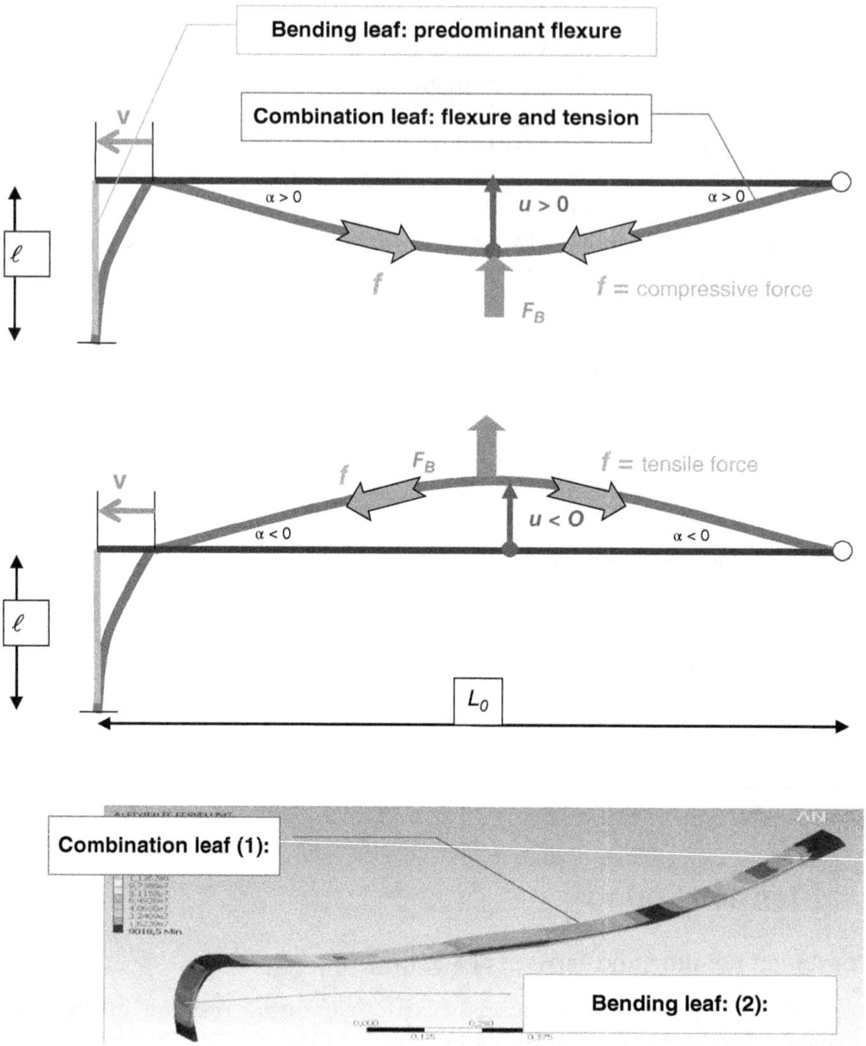

Figure 10.2 Leaf-tension composite spring.

receiving device (Figure 10.2). The longitudinal force due to braking stretches the front half part of the combination leaf (1). A leaf-tension spring has a bending joint section (2), which can compensate for a change in length of a spring section of the leaf spring when the spring section is deformed due to a load. In an unloaded state, the bending joint section and the leaf spring are arranged to be approximately flat and at an angle relative to each other. The bending joint section is shorter than the leaf spring and approximately vertically aligned.

The torque-resistant and displacement-resistant fixing of the two ends of the leaf spring, for example on a motor vehicle, not only generates a bending load transversely to the longitudinal direction of the leaf spring in the event of load-related deformation of the leaf spring, but also causes a tensile load in the longitudinal direction of the leaf spring. This tensile load is caused by a change in length of the spring section deformed transversely to the longitudinal direction and can be compensated for by an adapted deformation of the bending joint section.

In the transition section connecting the two spring legs, the leaf spring shows a curved course according to this. Due to the alignment of the two spring legs at an angle to each other, lengthening of one spring leg can be compensated for by bending the other spring leg.

10.2.2 Forces and Rates of Leaf-Tension Springs

A simple model is used for dimensioning the leaf-tension spring. A vertical, straight bending joint section can be easily formed on the horizontal, also straight leaf spring by the fact that both springs are arranged perpendicular to each other (Kobelev et al. 2015). The cross-sections of both legs are constant. The free length of the horizontal leaf spring is L_0. To simplify matters, it is assumed that the leaf spring bends in the middle. After the vertical deformation u, both halves of the leaf spring are no longer horizontal, but slightly inclined with the inclination angle:

$$\alpha = \pm \arctan(2u/L).$$

As a result, the length of the spring after deformation is slightly shorter than the free length:

$$L = 2\sqrt{(L_0/2)^2 - u^2}. \tag{10.28}$$

Shortening the horizontal leaf spring causes the vertical section of the bending joint to bend in the horizontal direction:

$$V = L_0 - L \approx 2u^2/L_0^2. \tag{10.29}$$

The horizontal bending force in the bending joint section generates a compressive force. The force in the leaf spring is:

$$f = 3E_L I_{b2} v/l^3. \tag{10.30}$$

The compressive force f in the two halves of the leaf spring is slightly inclined to the horizontal with the corresponding inclination angles α. The spring force supporting the vehicle in the vertical direction F_B is the sum of two components F_b, F_s. The first component F_s is the vertical projection of the sum of both compressive forces. The second component F_b results from the bending of the leaf spring:

$$F_B = F_b + F_s, F_s = 2f \sin\alpha, F_b = 48E_L I_{b1} u/L^3. \tag{10.31}$$

The first component F_s has a cubic dependence on u:

$$F_B \cong \frac{48}{L_0^3} E_L I_{b1} u + \left(\frac{288}{L_0^5} I_{b1} + \frac{24}{L_0^2 l^3} I_{b2} \right) E_L u^3. \tag{10.32}$$

The second component F_b depends in the first approximation linearly on the vertical deformation u.

The force-displacement characteristic of such a leaf spring can therefore initially exhibit a flat curve under low load and predominant bending load, which results in a comfortably soft suspension. Subsequently, under higher loads, a disproportionately increasing proportion can be caused by the compressive load so that the force-displacement characteristic rises increasingly steeply and a progressive characteristic curve is present.

Thanks to the vertical displacement of the pivot points of the wishbone and the blades, the different displacement of the pivot points in the direction of travel, and thus initially a displacement of the camber in relation to the deflection, is achieved with correspondingly differently selected path curves of the pivot points. However, this also achieves a displacement of the wheel carrier in the direction of travel, which causes the desired change in the toe-in due to the articulated connection of the trailing arm with the vehicle.

The model (Kobelev et al. 2015) also allows one to evaluate the stresses that occur over the course of deformation. Stresses on the lower and upper sides of the bending joint section are, correspondingly:

$$\sigma_1 = \pm\frac{F_b L}{2w_{b1}} + \frac{F_s}{a_1} \cong \pm\frac{24}{L_0^2}\frac{E_L I_{b1}}{w_{b1}}u + \left(\pm\frac{96}{L_0^4}\frac{I_{b1}}{w_{b1}} + \frac{24}{L_0^2 l^3}\frac{I_{b2}}{a_1}\right)E_L u^3, \tag{10.33}$$

$$\sigma_2 = \pm\frac{f \cdot l}{w_{b2}} \cong \pm\frac{6}{L_0 l^2}\frac{E_L I_{b2}}{w_{b2}}u^2. \tag{10.34}$$

In Eqs. (10.33) and (10.34), the moments and cross-sectional areas:

$$I_{b1} = b_1 t_1^3/12, w_{1b} = b_1 t_1^2/6, a_1 = b_1 t_1,$$
$$I_{b2} = b_2 t_2^3/12, w_{b2} = b_2 t_2^2/6, a_2 = b_2 t_2,$$

the parameters b_i and t_i are the corresponding widths and thicknesses of both elastic energy-harvesting components of the spring.

10.3 Transversally Mounted Leaf Springs

10.3.1 Axle Concepts of Transverse Leaf Springs

The transversal leaf spring can be integrated into today's designs for driving comfort as well as for driving dynamics of important independent wheel suspensions as an alternative to axle springs on the front and rear axle. The composite material is appropriate for manufacturing this type of leaf spring, since the load of these springs is bending.

As the name already indicates, a transversal leaf spring is integrated into the wheel suspension transverse to the direction of travel of the car. A positive aspect of the transversal leaf spring is the high level of functional integration. It can be used for stabilizing stiffness and for a vertical stiffness of the axis. This means that axle components, such as handlebars or stabilizers, are no longer required. The transversal leaf spring can also improve the elastic-kinematics of the axis, providing additional lateral stiffness. The transversely positioned leaf spring is not harmful from the crash safety viewpoint due to lack of penetration into the passenger compartment.

On the other hand, vehicles with transverse springs have some disadvantages. The installation space of the spring the spring lies transversely in the center of the axle in the direction of travel, colliding with the powertrain components. Rubber mountings can also transmit moments and local forces on the vehicle frame and cause overload of structural components of the car body.

The transversal leaf spring is mainly combined with a four-point bearing. This type of packaging provides some advantages for driving dynamics. The spring can undertake wheel guiding and stabilizing tasks. As a result of this functional integration, the other components that perform these tasks can be entirely eliminated or their mass or volume can be squeezed.

There are two different mounting types of hanging for the transversal spring. The first, now outdated configuration is the three-point bearing, with a middle bearing and two outer attachment points attached to the wheel carrier or handlebars. The second configuration with four bearing points is the typical application. The four-point bearing has two inner and two outer connection points. The position of the inner bearing is located on the axle beam, in rare cases also on the body or frame of the vehicle. The outer bearings are positioned at a direct distance from the wheel on the wheel carrier or wishbone. In this design, the transversal leaf spring does not influence the suspension kinematics and can be wheel-guided or non-wheel-guided. In particular, a wheel-guiding transversal leaf spring made of glass fiber-reinforced plastic presents a great design challenge, especially due to the high transverse forces. The actual examples of the transverse mounted composite leaf springs are displayed in Figures 10.3 and 10.4.

The spring performs two tasks with the four-point bearing. It undertakes the vertical load on the axis on one hand, and roll compensation on the other.

When both wheels move symmetrically in the same direction, the spring forms a C-shape and opposes the vertical force that resists the vehicle weight on the axis (Figure 10.5a). The bending moment in the middle area between two inner bushings of the spring is constant. Thus, it makes sense to keep the cross-section in the middle area constant.

In the other case, the movement of both wheels of an axis is opposite, for example, when swaying while cornering (Figure 10.5b). Cornering is accompanied by rolling of the vehicle. The spring bends to an S-shape. In this case, one wheel of the axle rebounds

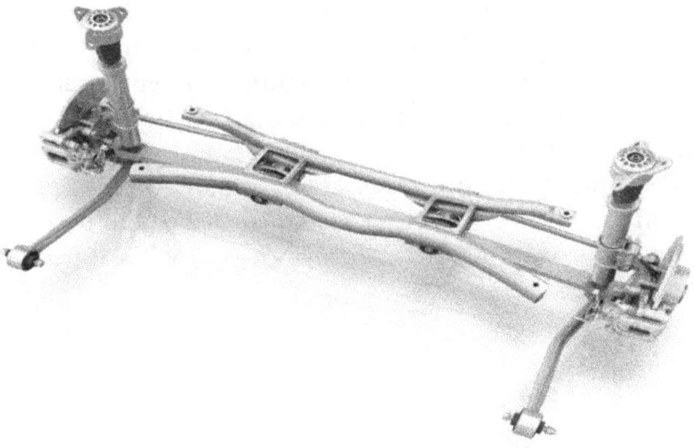

Figure 10.3 ZF lightweight axle CLA. Source: Courtesy of ZF Friedrichshafen AG.

Figure 10.4 ZF CLA system (ZF Friedrichshafen AG, Press Information 2015). Source: Courtesy of ZF Friedrichshafen AG.

and the other one jounces. Usually, the rolling of a vehicle should be kept as low as possible to enable comfortable driving behavior. The dynamic loads on both ends of the spring are also exactly opposite, thereby achieving the moment along the longitudinal axis of the vehicle, x_f. The moment reduces the roll movement of the vehicle and has a stabilizing effect. The torsion bar can thus be made lighter or even in some cases be completely avoided.

Ideally, both load cases should be combined and an optimized design developed so that the desired stroke and roll spring rate is achieved. For this purpose, simple analysis formulas are displayed in the next section.

10.3.2 Analysis of a Transverse Leaf Spring

The wheel-guided transverse leaf spring axle possesses a slightly different moment distribution over the spring length (WO2008125076A1 2007). In this design, the distance between the force appliance points is $L_a = \kappa L_0$, $0 < \kappa < 1$ (Figure 10.5). For briefness, assume that the spring-to-wheel ratio is equal to one that is the spring transmits both end forces immediately to both wheels on the axle.

First, consider the symmetric loading of the transversely mounted leaf spring (Figure 10.5a). Both end loads in the vertical direction are equal to F_v. The bending moment along the spring length is mirror symmetric $M_v(y) = M_v(-y)$ with respect to the middle plane of the vehicle $y = 0$:

$$M_v(y) = F_v \begin{cases} (L_0 - L_a)/2, & 0 \le y \le L_a/2, \\ L_0/2 - y, & L_a/2 \le y \le L_0/2. \end{cases} \tag{10.35}$$

The stored elastic energy in half of the spring is given by:

$$U_v = \frac{1}{2} \int_{L_0/2} \frac{S_{33} M_v^{\,2}}{I_b} dy = \frac{F_v^{\,2}}{2c_v} = \frac{u_v F_v}{2}. \tag{10.36}$$

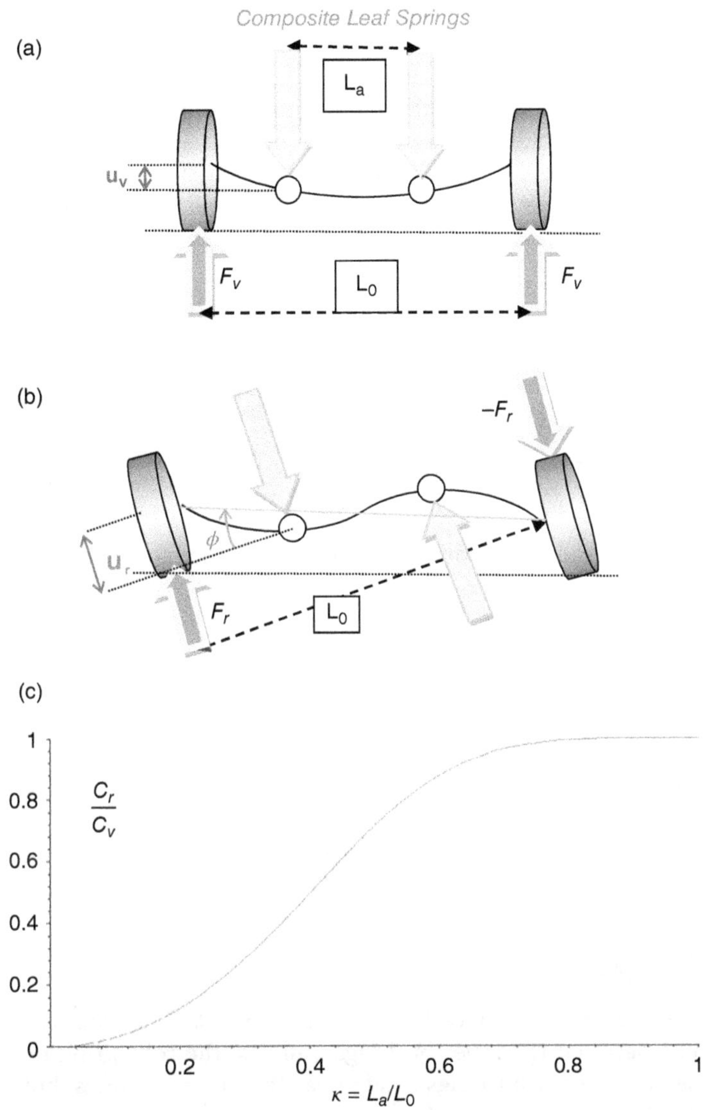

Figure 10.5 Ratio of roll spring rate to vertical spring rate.

The value u_v is the vertical stroke of the ends of spring. This expression delivers the general expression for spring rate of the leaf composite spring for one wheel:

$$\frac{1}{c_v} = \frac{1}{F_v^{\,2}} \int_{L_0/2} \frac{S_{33} M_s^{\,2}}{I_b} dy. \tag{10.37}$$

With the formulas (10.35) and (10.36), the vertical rate of the spring $c_v = F_v/u_v$ [N/m] and the vertical rate of the axis $R_v = 2(L_0/B)^2 F_v/u_v$ [N/m] are, correspondingly:

$$c_v = \frac{1}{3 - 6\kappa + 6\kappa^2 - 2\kappa^3} \frac{24 I_b}{S_{33} L_0^{\,3}}, R_v = 2\left(\frac{L_0}{B}\right)^2 c_v, \tag{10.38}$$

B = rear track width = thread (is approximately the twist-beam length)

Second, consider the asymmetric loading of the transversely mounted leaf spring (Figure 10.5b). Both end loads are equal in magnitude F_r, but oppositely directed. Both forces induce the roll moment in the driving direction of the vehicle: $M_{RA} = F_r L_0$, similar to Eq. (7.61) in Chapter 7. The bending moment in the spring along its length is the point symmetric function $M_r(y) = -M_r(-y)$ with respect to the middle plane of the vehicle $y = 0$:

$$M_r(y) = F_r \begin{cases} (L_0 - L_a)y/L_a, & 0 \le y \le L_a/2, \\ L_0/2 - y, & L_a/2 \le y \le L_0/2. \end{cases} \tag{10.39}$$

The stored elastic energy in half of the spring is given by:

$$U_r = \frac{1}{2} \int_{L_0/2} \frac{S_{33} M_r^{\,2}}{I_b} dy = \frac{F_r^{\,2}}{2c_r} = \frac{F_r u_r}{2}. \tag{10.40}$$

This expression delivers the general expression for asymmetric spring rate of leaf composite spring per wheel:

$$\frac{1}{c_r} = \frac{1}{F_r^{\,2}} \int_{L_0/2} \frac{S_{33} M_r^{\,2}}{I_b} dy. \tag{10.41}$$

With the formulas (10.39) and (10.41), the asymmetric spring rate per wheel delivers:

$$c_r = \frac{\kappa^2}{1 - 2\kappa + \kappa^2 + 2\kappa^3 - \kappa^4} \frac{24 I_b}{S_{33} L_0^{\,3}}. \tag{10.42}$$

The ratio of the asymmetric spring rate to the vertical spring rate is solely the function of the factor $\kappa = L_0/L_f$:

$$\frac{c_r}{c_v} = \kappa^2 \frac{3 - 6\kappa + 6\kappa^2 - 2\kappa^3}{1 - 2\kappa + \kappa^2 + 2\kappa^3 - \kappa^4}. \tag{10.43}$$

The graph of ratio of roll spring rate to vertical spring rate (10.43) is shown in Figure 10.5c. If the distance between the inner bushings nullifies, the roll spring rate vanishes. In the other case, if the inner bushings locate near the outer bushings, both the roll and vertical spring rates increase, becoming nearly equal to each other.

According to (10.40), the relative stroke of the both ends of the spring is equal to $u_r = F_r/c_r$. For the small stroke the roll angle results in $\varphi = u_r/(L_0/2)$. Thus, $F_r = c_r L_o \varphi/2$.

The roll moment M_{RA} (Nm) and the roll spring rate R_r (Nm rad^{-1}) of the axle, correspondingly, are:

$$M_{RA} = F_r L_0 = c_r L_0^2 \varphi/2, R_r \equiv M_{RA}/\varphi = c_r L_0^2/2. \tag{10.44}$$

The ratio of the roll spring rate to the vertical spring rate of the axis is:

$$\frac{R_r}{R_v} = \left(\frac{2}{B}\right)^2 \frac{c_r}{c_v} = \kappa^2 \frac{3 - 6\kappa + 6\kappa^2 - 2\kappa^3}{1 - 2\kappa + \kappa^2 + 2\kappa^3 - \kappa^4} \left(\frac{2}{B}\right)^2. \tag{10.45}$$

10.3.3 Examples and Patents for Transversely Mounted Leaf Springs

The idea of a transverse leaf spring made of steel and, in particular, of fiberglass is not new. Patent applications for the transversely mounted composite leaf spring have been around since 1993.

(EP0660005A1, 1993-12-24, https://www.epo.org*): Composite material leaf spring for motor vehicle suspension*:

> Suspension leaf spring (1) made of composite material transversely connecting two wheels of one and the same motor vehicle axle, which leaf spring has at least one external reinforcement (6) based on unidirectional fibers (7) distributed between a central zone (2) and two end zones (3, 4) that are inclined toward that central zone (2), characterized in that fibers (7) of the reinforcement are mutually parallel and follow the shape of the leaf spring so that the lines tangential to the fibers (7) at the various points thereof form an angle of between 0° and 60° with the axis of the axle (9) of the vehicle (Figure 10.6).

The identified patents could be divided into several categories:

- Cross leaf springs from fiberglass
- Cross leaf spring bearings
- Axle concepts with transverse leaf spring

The patent (DE102015108484A1 2015) was applied for a cross leaf spring made of fiberglass. The invention concerns a transverse leaf spring and an axle arrangement for a motor vehicle. Due to its shape, the transverse leaf spring represents an optimum of possible spring rate and installation space. In the end areas, the spring is curved around the longitudinal direction of the vehicle and in the middle area it is curved around the vertical direction of the vehicle. Above all, the curvature in the middle area of the transverse leaf spring is a novelty, since the spring usually has a straight beam.

In the "Axle concepts with transverse leaf spring" category, three similar patents were applied for: (DE102014215871A1 2014, https://www.epo.org), (DE102013218055A1 2013, https://www.epo.org), (DE102013213674A1 2013, https://www.epo.org). All three represent different types of connections of the middle bearings to the vehicle axle beam. The patent (DE102014215871A1 2014, https://www.epo.org) concerns a wheel suspension for motor vehicles and focuses in particular on a driven axle. The characteristic feature of this concept is a transverse leaf spring bent forward in the

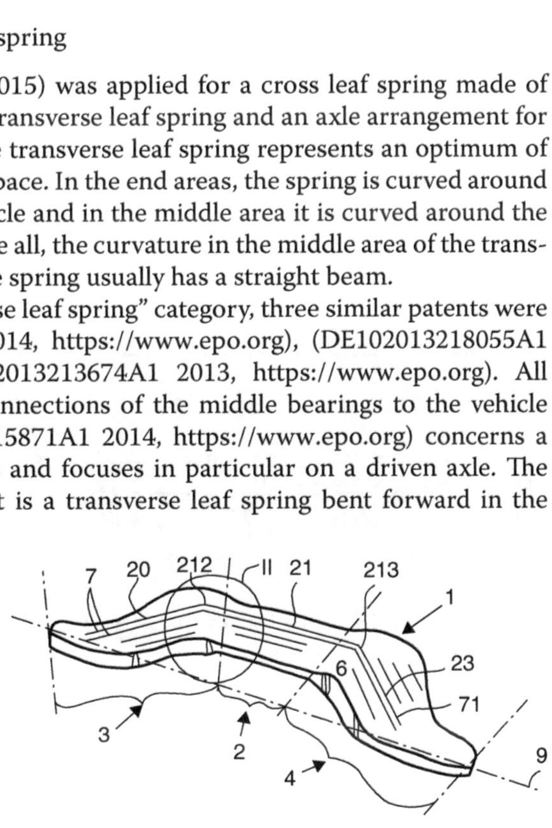

Figure 10.6 EP0660005A1 (1993) "Composite material leaf spring for motor vehicle suspension" (www.epo.org).

vertical direction of the vehicle. The patent (DE102011083248A1 2011) differs from the other bearing connections with regard to the functional integration of the transverse leaf spring. This invention also contains the outer and inner bearing connections, which are designed for a wheel-guiding cross leaf spring. The patent (DE102013218055A1 2013) includes a swiveling middle bearing connection. This measure places only a minor load on the rubber bearing due to the fact that the bearings are hardly twisted any more.

The category "Axle concepts with transverse leaf spring" represents transverse leaf springs. The patent (DE102013210338A1 2013) uses different types of multi-link suspension in its designs. A special feature is the patent (DE102010056388A1 2010). With this axle concept, the spring rate and the standing height of the axle can be adjusted via the middle bearings, which are adjustable in the transverse direction of the vehicle. This design thus offers a high degree of variability, compared to the previously discussed patent (DE102015108484A1 2015). Another invention with the patent number DE102014215871A1 (2014) contains an axle with the wheel-guiding transverse leaf spring.

DE102013003958A1 (2013, www.epo.org*) Storage arrangement of a transverse leaf spring to a structure of a vehicle as well as vehicle with such a bearing assembly*:

> The invention relates to a bearing assembly (10) at least one transverse leaf spring (16) to a structure (14) of a vehicle, wherein the at least one extending in the vehicle transverse direction transverse leaf spring (16) to the structure (14) on at least one storage racks (21) via a bearing means is mounted (22), comprising at least one adjusting element (34) by means of which a extending in the vehicle vertical direction distance (a) between the at least one bearing point (21) and the

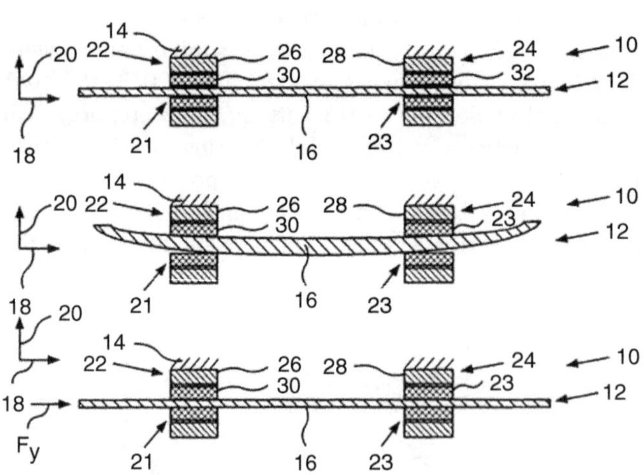

Figure 10.7 DE102013003958A1 (2013) Storage arrangement of a transverse leaf spring to a structure of a vehicle as well as vehicle with such a bearing assembly (www.epo.org).

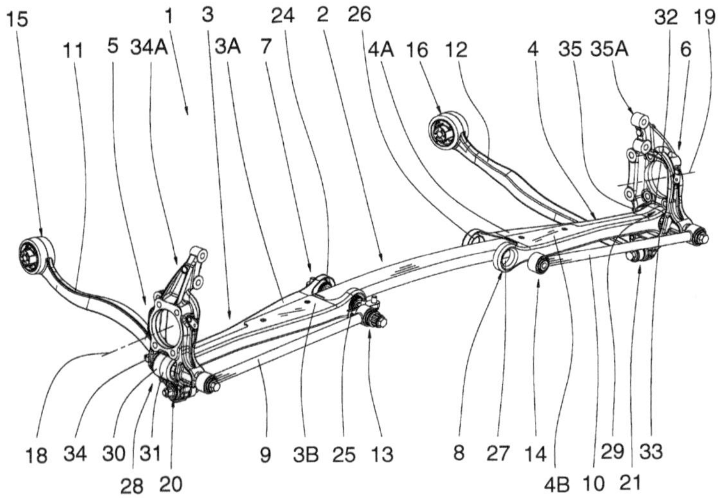

Figure 10.8 DE102014215871 (2014) Vehicle axle with an axis extending in the in-stalled position in the vehicle transverse direction and wheel-guiding transverse leaf spring (www.epo.org).

assembly (14) is adjustable, wherein the bearing means (22) at least one the adjustment member (34) in the circumferential direction comprises at least partially surrounding the rubber element (30), via which the transverse leaf spring (16) is mounted on the structure (14), and a vehicle having such a bearing assembly (10) (Figure 10.7).

DE102014215871 (2014, www.epo.org*) Vehicle axle with an axis extending in the installed position in the vehicle transverse direction and wheel-guiding transverse leaf spring:*

It is a vehicle axle (1) extending with an in the installed position in the vehicle transverse direction (y) and wheel-guiding transverse leaf spring (2) above, which control arm (3, 4) having wheel carriers (5, 6) and on the transverse link (3, 4), engageable with a vehicle body in operative connection. According to the invention, the transverse leaf spring (2) is in the vehicle transverse direction (y) each between the connecting portions between the suspension arms (3, 4) and the wheel carriers (5, 6) and the operative connections between the cross arms (3, 4) and the vehicle body (with the transverse links 3, 4) (Figure 10.8).

US2016/0347139A1 (2016, https://www.uspto.gov*) Transverse leaf spring for a motor vehicle as well as axle assembly with a transverse leaf spring*:

In a transverse leaf spring 2 and an axle assembly 1 for a motor vehicle, transverse leaf spring 2 is curved about the longitudinal direction X of the motor vehicle in the end regions 11 and is curved about the vertical direction Z of the motor vehicle in a central region 10. As a result, optimum potential spring rates and the associated installation space requirement is achieved (Figure 10.9).

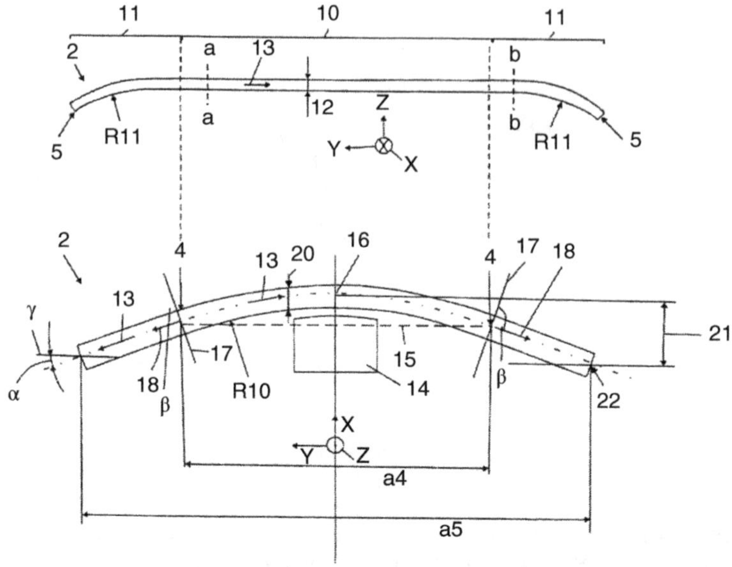

Figure 10.9 US2016/0347139A1 (2016) Transverse leaf spring for a motor vehicle as well as axle assembly with a transverse leaf spring (https://www.uspto.gov).

10.4 Conclusions

Composite leaf springs of different types were reviewed in Chapter 10. The dominant stress in all types of leaf spring is stress due to bending moments.

The first type of leaf spring discussed was the longitudinally mounted leaf spring. This spring carries the weight of the vehicle and withstands the oscillating vertical load and severe moments during breaking. The leaf spring is responsible for the cornering behavior of the vehicle because the vertical loads deform the spring and push the solid axis of the vehicle along the longitudinal axis. If the length travel of the solid axis differs on the right and left wheel, the compliance steer of the dead axis occurs. This steer can cause the oversteering of the vehicle and needs to be profoundly controlled. The side forces affect the lateral shift of the axis, which is highly undesirable. The side shift must be eliminated, regularly with the additional links (Watt or four-link arrangement), or sliding tube or splined half shafts.

The second type of leaf spring discussed was the so-called leaf-tension spring, which exploits the longitudinal forces in the spring. The longitudinal forces arise in the spring in course of bending with fixed endpoints. The length of spring is fixed, and the vertical flexural deflection provokes reaction forces in the length direction. Because these reaction forces depend nearly quadratically upon vertical travel, the vertical spring force is progressive so that the spring rate increases with vertical travel. This feature is favorable for the vertical dynamics of the vehicle, because it allows nearly constant vertical frequency of the axis over the load range of the vehicle.

The third type of leaf spring discussed was the transversally mounted leaf members. These springs are traditionally mounted on four hinge points. Depending on the position of the hinge points, the ratio of lateral-symmetric to mirror-symmetric spring rates differs. The lateral-symmetric spring rate affects the vertical wheel forces and consequently determines the vertical spring rate of the vehicle. The mirror-symmetric spring rate affects the difference between vertical wheel forces on the left and right wheel and consequently changes the roll spring rate of the vehicle.

For further reading, the following studies on other concepts of leaf springs made of composite materials are recommended: Götte (1989), Vijayarangan and Rajendran (2001), Neubrand (2014), Mallick (1988), Sancakkar and Gratton (1997), Morris (1986), Shokrieh and Rezaei (2003), Dai and Chang (2004), Fruhmann et al. (2010), Hurter et al. (2014), Simmonds (2018), Abu Talib et al. (2010) and Petrich et al. (2017). The cited works also explain the details of manufacturing and specific calculation methods for the mentioned composite springs.

References

Abu Talib, A.R., Goudah, A.A.G., Lah, N.C., and Golestaneh, A.F. (2010). Developing a composite based elliptic spring for automotive applications. *Mater. Des.* 31: 475–484.

Al-Qureshi, H.A. (2001). Automobile leaf springs from composite materials. *J. Mater. Process. Technol.* 108: 58–61.

Dai, G. and Chang, S. (2004). Novel applications of composite structures to robots, machine tools and automobiles. *Compos. Struct.* 66: 17–39.

DE102010056388A1 (2010) Vehicle with a leaf spring member for the suspension of the vehicle, European Patent Office, Espacenet https://www.epo.org.

DE102011083248A1 (2011) Bearing means for a transverse leaf spring, European Patent Office, https://www.epo.org.

DE102013210338A1 (2013) Mehrlenker-hinterachse für ein Fahrzeug, European Patent Office, Espacenet https://www.epo.org.

DE102013003958A1 (2013); Vehicle with a leaf spring member for the suspension of the vehicle, European Patent Office, Espacenet https://www.epo.org.

DE102013213674A1 (2013) Wheel suspension for a motor vehicle having a curved transverse leaf spring, European Patent Office, https://www.epo.org

DE102013218055A1 (2013) Storage device of a transverse leaf spring which is mounted in the region of a vehicle axle of a vehicle, European Patent Office, Espacenet https://www.epo.org

DE102014215871 (2014) Vehicle axle with an axis extending in the installed position in the vehicle transverse direction and wheel-guiding transverse leaf spring, European Patent Office, Espacenet https://www.epo.org

DE102015108484A1 (2015) Transverse leaf spring for a motor vehicle and axle with a transverse leaf spring, European Patent Office, Espacenet https://www.epo.org.

EP0660005A1 (1993) Composite material leaf spring for motor vehicle suspension; European Patent Office, Espacenet https://www.epo.org.

Fruhmann G., Elbers C., Vossel A., Stretz K. (2010). Achskonzeptstudie mit radführender Querblattfeder. 19. Aachener Kolloqium Fahrzeug-und Motorentechnik.

Gillespie, T.D. (1992). *Fundamentals of Vehicle Dynamics*. Warrendale, PA: Society of Automotive Engineers, 495 p.

Götte, T. (1989). *Zur Gestaltung und Dimensionierung von Lkw-Blattfedern aus Glasfaser-Kunststoff*. Düsseldorf: VDI-Verlag GmbH.

Goudah, G., Mahdi, E., Abu Talib, A.R. et al. (2006). Automobile compression composite elliptic spring. *Int. J. Eng. Technol.* 3: 139–147.

Hurter W.S., van Rensburg N.J., Madyira D.M. (2014) Static Analysis of Advanced Composites for the Optimal Design of an Experimental Lightweight Solar Vehicle Suspension System, ASME 2014 International Mechanical Engineering Congress & Exposition November 14–20, 2014 – Montreal, Canada.

Kelly, A. and Zweben, C. (eds.) (2000). *Comprehensive Composite Materials*. Pergamon: Elsevier Ltd.

Kobelev V., Alfes M., Müller D. (2015) Meandering axle springs from fibre-reinforced plastic for car suspension, April 2015, DOI: 10.13140/RG.2.1.5099.9843 Conference: Federn im Fahrzeugbau DVM-Tag, Berlin.

Mallick, P.K. (1988). Static mechanical performance of composite elliptic springs. *Trans. ASME J. Eng. Mater. Technol.*, 109: 22–26.

Morris, C.J. (1986). Composite integrated rear suspension. *Compos. Struct.* 5: 233–242.

Neubrand, J. (2014). *Possibilities of Coil Springs and Fiber-Reinforced Suspension Parts*, Encyclopedia of Automotive Engineering. Wiley.

Pawsey C. (2015) *Lightweight Materials Reducing Weight of Suspension Systems*, https://www.automotive-iq.com/chassis-systems/articles/lightweight-materials-reducing-weight-suspension-systems.

Petrich M., Schrodin B., Kletzin U. (2017) Material Selection Method for Composite Springs, *59th Ilmenau Scientific Colloquium*, Technische Universität Ilmenau, 11–15 September 2017. URN: urn:nbn:de:gbv:ilm1-2017iwk-140:5.

Qian, C., Shi, W., Chen, Z. et al. (2017). Fatigue reliability design of composite leaf springs based on ply scheme optimization. *Compos. Struct.* 168: 40–46.

Sancakkar, E. and Gratton, M. (1997). Design, analysis and optimization of composite leaf springs for light vehicle applications. *Compos. Struct.* 44: 195–204.

Shiva Shankar, G.S. and Vijayarangan, S. (2006). Mono composite leaf spring for light weight vehicle – design, end joint analysis and testing. *Mater. Sci. (Medziagotyra)* 12: 220–225.

Shokrieh, M. and Rezaei, D. (2003). Analysis and optimization of a composite leaf spring. *Compos. Struct.* 60: 317–325.

Simmonds, H. (2018). *CLASS Research Project, Composite Lightweight Automotive Suspension System*. UK: University of Warwick. https://warwick.ac.uk/fac/sci/wmg/research/engineering_materials/composites/projects/poster_for_wmg_website.pdf.

Stimpfl J. (2016) "Leichtbau in faserverstärktem Kunststoff für Fahrwerksfedern im Automobil," *VDFI Lecture event "Federn aus faserverstärkten Kunststoffen,"* University Siegen, 20.04.2016.

US 2016/0347139 A1 (2016) Transverse leaf spring for a motor vehicle as well as axle assembly with a transverse leaf spring, US Patent and Trademark Office, https://www.uspto.gov.

Vijayarangan, S. and Rajendran, I. (2001). Optimal design of a composite leaf spring using genetic algorithm. *Comput. Struct.* 79: 1121–1129.

WO2008125076A1 (2007) Leaf spring made of a fiber-plastic composite and force transmission element therefore, European Patent Office, Espacenet, https://www.epo.org.

WO2011110611A3 (2011). Leaf spring for motor vehicles, European Patent Office, Espacenet, https://www.epo.org.

WO2015011181A1 (2015) Leaf spring arrangement for motor vehicles, European Patent Office, Espacenet, https://www.epo.org.

ZF CLA system (2015) Press Information, ZF Friedrichshafen AG, Friedrichshafen.

ZF lightweight axle CLA (2015) Press Information, ZF Friedrichshafen AG, Friedrichshafen.

11

Meander-Shaped Springs

11.1 Meander-Shaped Compression Springs for Automotive Suspensions

11.1.1 Bending Stress State of Corrugated Springs

The meander compression spring – meander spring or corrugated spring – is a promising design variant for the compression axle springs made of composite material (EP2082903B1 2009) (Figures 11.1 and 11.2). Several known patented variants of meander springs are displayed in Section 11.6. The meander spring exhibits a design that is more suitable for composite materials than the helical spring. The torsional stresses in meander springs are negligible and bending stresses dominate, which can be better compensated for by the reinforcement fibers. However, this geometry also has some natural disadvantages. For example, there are two bending directions and a relatively high block length of meander spring. In contrast to the meander spring, the composite helical spring has a lower block length, because the coils dive into each other. Thus, for isotropic materials such as steel, the design of the helical spring is more appropriate. Similar to leaf springs, the meander spring is manufactured with unidirectional prepreg and this is another reason why it is pertinent for composite materials. The main stress on the meander spring occurs in the direction of tension of the fibers. Fiber composite components are very strongly anisotropic, that is, direction-dependent structures. This ensures the best utilization of fiber properties, since the highest modulus of elasticity is in the tensile direction of the fibers. Further advantages of meander spring are the theoretical lightweight construction potential with nearly 50% weight saving compared to a steel helical spring and suitability for various axle concepts.

The invention (EP2082903B1 2009) relates to a flat spring made of unidirectional fiber-reinforced material. A spring band is a single wave of reversal regions and intermediate portions around a longitudinal axis of the meander. The longitudinal axis corresponds to the direction of the force. The thickness of the material in the reversal regions increases in comparison to the thickness of the connecting intermediate regions. Accordingly, in the reversal regions of the meander, the resistance moment of the spring band increases. As a result, the stresses in all cross-sections remain in essence equal along the meander. In other words, the design of the meander spring is fully stressed, so that the uniform stress dominates in the optimally designed spring and the best possible utilization of the material is assured. Accordingly, the lowest possible

Design and Analysis of Composite Structures for Automotive Applications: Chassis and Drivetrain, First Edition. Vladimir Kobelev.

Companion website: www.wiley.com/go/kobelev/automotive_suspensions

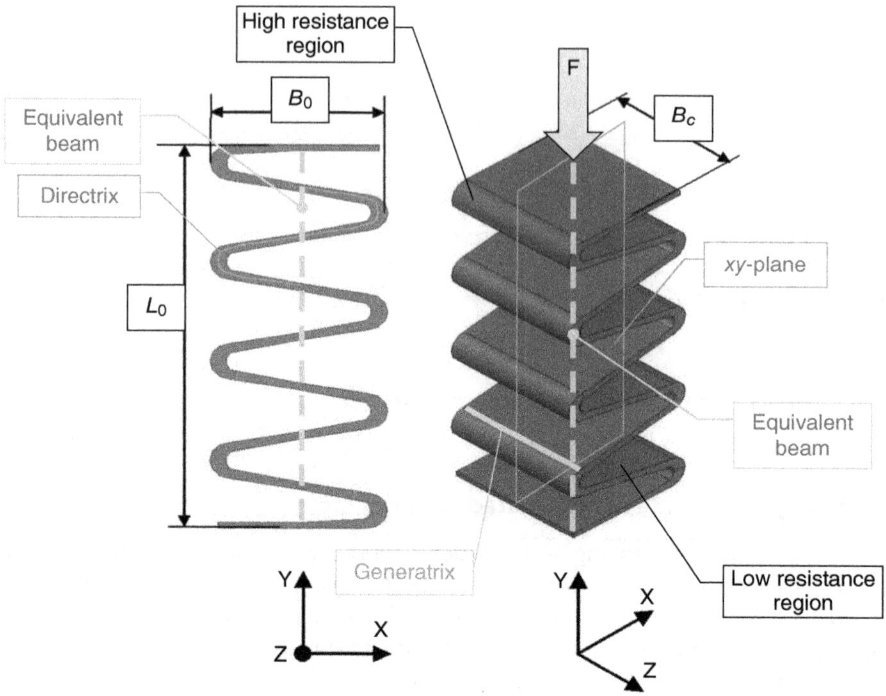

Figure 11.1 Elements of a meander spring.

weight for the fixed maximal force and required spring rate will be attained. One design variant assumes a constant width over the length of the band. The area of cross-section varies over the meander. This design is preferable from the viewpoint of packaging and production aspects.

Another variant of the design supposes the effective cross-sectional area of the spring band to be constant over the length of meander surface. At the same time, the thickness of the band might be variable. Thus, the width varies along the length of the spring band, but their product remains constant. With this design, the resistance moment is customized to the bending moment in each section of the band. Consequently, the fully stressed design will be achieved even for a constant area of the cross-section of the spring. With this design option, the mass of material will be minimal.

The middle surface of the meander springs is the so-named "ruled surface." A surface is referred to as ruled (or a scroll surface) if through every point of the surface there is a straight line that lies on it (Pottmann and Wallner 2001). A ruled surface is a curved surface that can be generated by the continuous motion of a straight line in space along a space curve called a directrix. This straight line is called a generatrix, or ruling, of the surface. The meander-shaped neutral line (directrix) of the spring is in the *xy* plane (Figure 11.1). The direction of the generatrix is parallel to the *z*-axis.

Specifically, the middle surface of the meander spring is the developable surface (Struik 1950). A developable surface is a surface that can be, at least locally, unrolled onto a flat plane without stretching or tearing it. If a developable surface lies in three-dimensional Euclidean space and is complete, then it is necessarily ruled. This

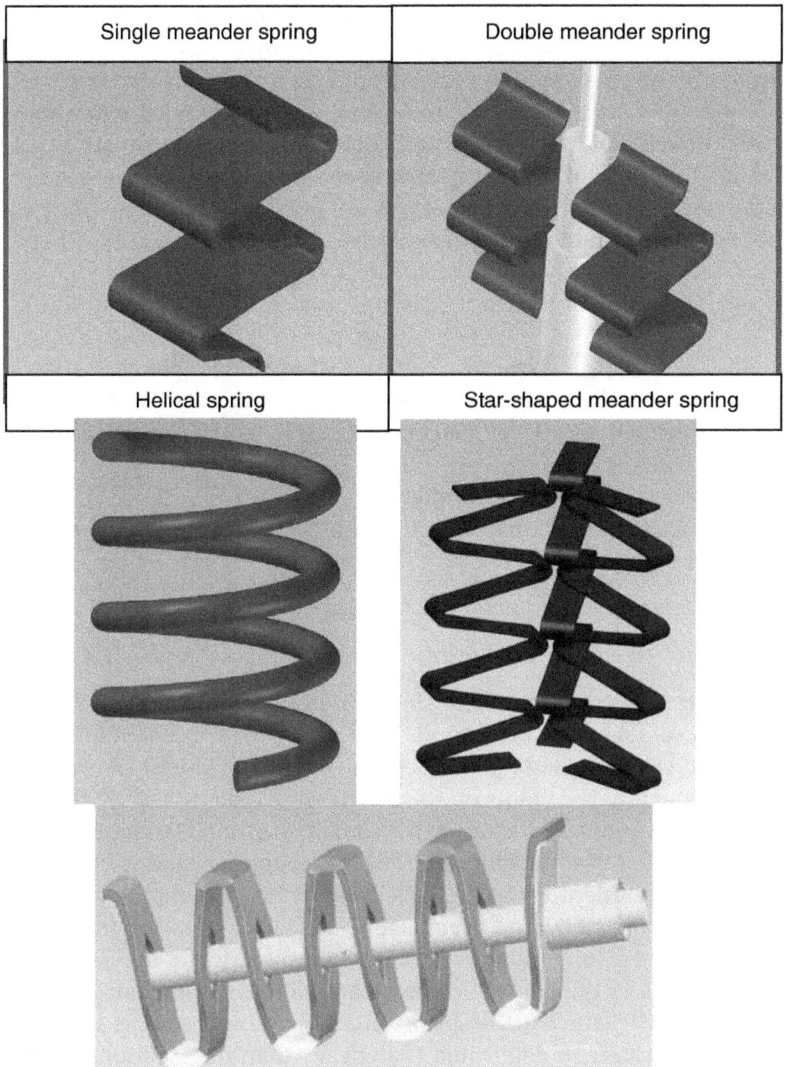

Figure 11.2 Different concepts of meander springs and a helical spring.

property is of primary importance for the manufacturability of the spatial layered composite elements and, particularly, the spring elements.

The meander spring, made of an orthotropic material, is modeled for mechanical analysis as a thin shell. The midpoints of the neutral surface of the thin shell form the directrix of the ruled surface. The shell with periodically variable thickness replaces the solid elastic spring body and simplifies its mechanical analysis. The force and moment resultants (also called membrane forces, shear forces and bending moment) determine the stress state in each cross-section of the spring. The resultant forces and resultant moments vary from point to point of the band, and the maximal bending moment is attained on the turning points of meander on the utmost distance from the axis of force. In the other width direction of the shell, the stress is nearly constant. Further simplification is achieved if the meander spring is represented by the curved beam with the varying cross-section. The cross-section of the beam varies along the directrix.

Stress resultants are defined as integrals of stress over the thickness of a structural element. The integrals are weighted by integer powers of the thickness coordinate. Stress resultants are so defined to represent the effect of stress as a membrane force, a bending moment on a meander beam. The dominated stress state of the shell is caused by bending moments. The bending moment in the spring cause normal stresses in the curved band.

The elliptic springs behave similarly to the meander springs with one meander element. The analysis and simulation of the elliptic springs was performed in Mahdi et al. (2006).

Another design, similar to the mechanical viewpoint of meander springs, shows a flexure-stressed spring, the so-called "circular spring" (Figures 8.1 and 8.11). The methods of analysis of circular springs were developed by Reid et al. (2002), Wong et al. (2004) and Reid et al. (2004).

In the same way, the springs, exposed in the cited patents (Figures 8.12–8.15), are all dominantly flexure-loaded. Thus, from mechanical viewpoint these elements are analogous to the meander spring.

11.1.2 "Equivalent Beam" of a Meander Spring

For simplified analysis, the thin-shell model of the meander spring is replaced by an equivalent beam. The equivalent beam corresponds to the hypothetical center line of the spring, which is attributed with longitudinal and bending features (Figure 11.1). Instead of describing the displacement of the points on the meander-shaped thin-shell, the displacement of the point of the equivalent beam at the same y-position is used. The axial stiffness and transverse stiffness of the spring and the corresponding axial stiffness and transverse stiffness of the equivalent beam are equal. The forces and moments acting on the spring and on the equivalent beam are equal as well. Under these conditions the axial and lateral displacements in longitudinal and transverse direction of the spring and corresponding displacements of the equivalent beam are all the same. Consequently, the equivalent beam deforms exactly like the spring under the same load. The solutions for the equivalent beam are easier to determine analytically than the solutions for the dimensional thin-shell models of the corrugated spring. For validity of the results, the numerous stiffness constants of the equivalent beam must be correctly evaluated.

11.1.3 Axial and Lateral Stiffness of Corrugated Springs

The free length of the meander-shaped compression spring is L_0. The compressed length of the spring is L. The axial stiffness of the equivalent beam is:

$$\langle ES\rangle = c\,L. \tag{11.1}$$

The axial spring rate of the spring is c. The equivalent beam possesses different bending stiffness in two orthogonal planes $\langle EI_i\rangle$, $i = 1, 2$. Two equations

$$M_i = \langle EI_i\rangle d^2 s_i/dy^2, \text{and } \ i = 1, 2 \tag{11.2}$$

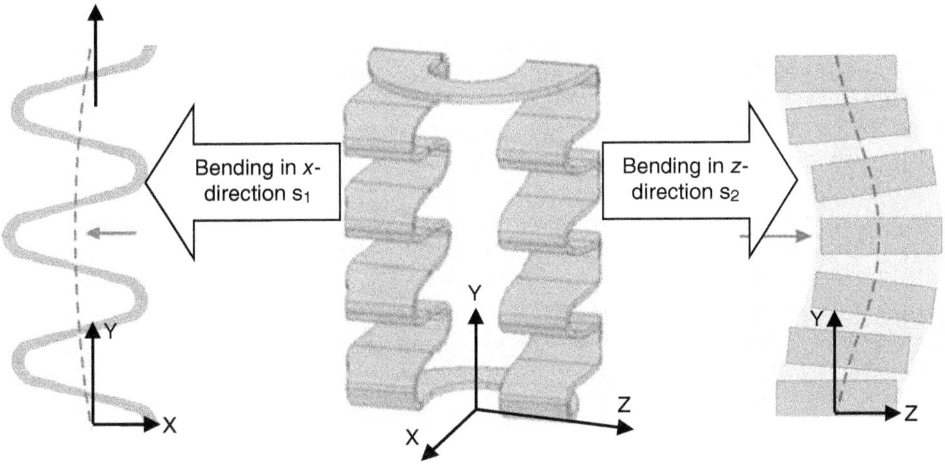

Figure 11.3 Anisotropic behavior of a meander compression spring: different stiffness in longitudinal and transverse directions.

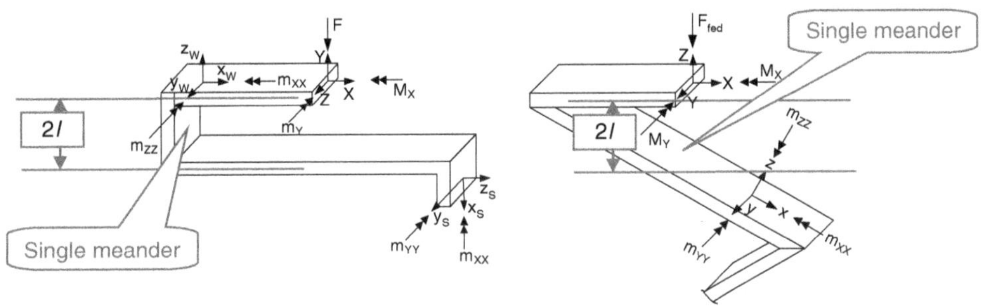

Figure 11.4 Representation of the continuous meander spring and its replacement models (rectangular and triangular spring).

describe the bending of the equivalent beam under the effect of the moments M_i in transverse and longitudinal directions, respectively. The lateral displacements of the axes of the equivalent beam in corresponding longitudinal and transverse directions are s_i (Figures 11.3 and 11.4).

11.1.4 Effective Spring Constants of Meander and Coil Springs for Bending and Compression

For evaluation of the elastic behavior for a meander spring, the stiffness coefficients of the equivalent beam are required. The models (triangular and rectangular spring, Figures 11.5 and 11.6) provide the effective constants of rectangular meander springs. The rough estimations for the coefficients of bending and axial compression are summarized in Table 11.1. The corresponding coefficients are also given for a helical spring with circular anisotropic wire. For simplification, the thickness of the material t_c is taken to be constant. The detailed derivations of the coefficients for bending and axial compression are performed in the following section.

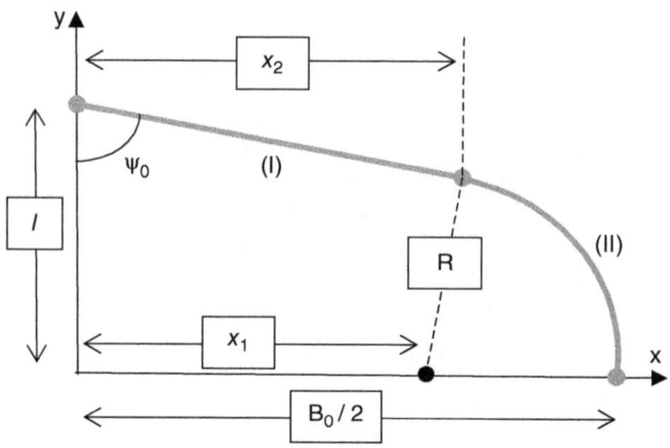

Figure 11.5 Directrix of a quarter of the neutral surface of a multiarc-profiled meander spring.

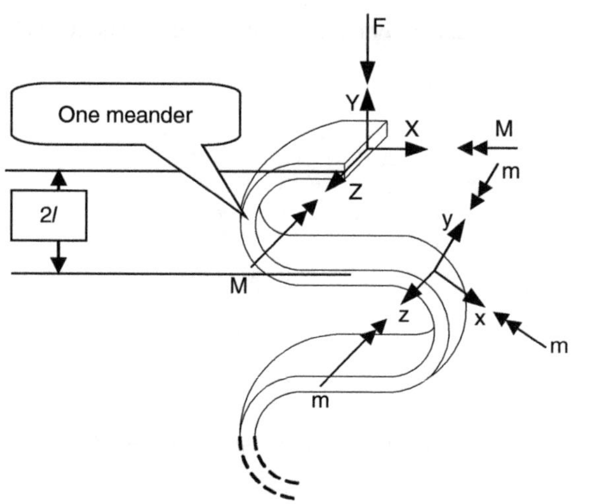

Figure 11.6 Moments and forces in the meander.

The influences of the fiber orientation angle on the stiffness and strength of the meander springs and the leaf springs (Chapter 10) are identical, because both spring types are pure bending members R. Accordingly, the estimations for the stored elastic energy and strength of the leaf springs from Section 10.1.2 are applicable for the meander springs without modifications.

11.2 Multiarc-Profiled Spring Under Axial Compressive Load

11.2.1 Multiarc Meander Spring With Constant Cross-Section

One of the common designs of the meander spring is the so-called multiarc design. Consider a quarter of the spring of height l. The total free length of the spring with n_w full meanders is equal to $L_0 = 2n_w l$. In the following consideration, linear elastic deformation is assumed. The shortening of the spring is neglected, such that $L_0 \cong L$.

The directrix of neutral surface of each meander is composed of one straight segment and one circular arc of the radius R. The straight segment is referred to as (I) and the

Table 11.1 Effective spring constants of meander and coil springs for bending and axial compression.

Spring type	Transversal bending	Axial compression
Meander spring	$\langle EI_1\rangle \approx \dfrac{B_c t_c^3 E_L L}{8n_w(l+B_0)}$	$c \approx \dfrac{B_c t_c^3 E_L}{2(B_0/2)^2 n_w(3l+B_0)}$
	$\langle EI_2\rangle = \dfrac{B_c t_c^3 G_{TL} L}{2n_w(l+B_0)}$	
Helical anisotropic spring with diameter d of wire	$\langle EI_b\rangle = \dfrac{E_L d^4 L}{32(2+\nu)\, n_w D_m}$	$c = \dfrac{G_{TL} d^4}{8\, n_w D_m^3}$

circular arc segment as (II). The width of the meander spring in the y-direction B_c is constant. In other words, the length of the ruler or generator is B_c.

For a multiarc-profiled spring, the appropriate parameterization for the directrix of the neutral surface uses x as an independent variable. The inclination angle of the straight segment to the y-axis of the spring is ψ_0. The total width of the neutral surface in the x-direction is a given constant B_0. With these parameters, the x-coordinates of the key point of the arc x_1 and of the tangent point of the arc and straight x_2 segment are R:

$$x_1 = \frac{1}{2}B_0 - R,\;\; x_2 = \frac{1}{2}B_0 - R + R\cos\psi_0, \kappa = \sqrt{\frac{R}{B_0}} \le \frac{1}{\sqrt{2}}. \tag{11.3}$$

The height of quarter of the meander is

$$l = R\sin\psi_0 + l_2/\tan\psi_0.$$

The along the y-axis the vertical force F acts. The bending moment along the neutral surface of the meander is:

$$M_Z = \begin{cases} F\cdot\left(\frac{1}{2}B_0 - x\right), & region\, I\, :\, 0 < x \le x_2, \\ F\cdot R\cdot(1-\cos\varphi), & region\, II\, :\, x_2 < x \le \frac{1}{2}B_0. \end{cases} \tag{11.4}$$

The number of meanders is an integer and is specified: n_w. Under the action of external loads, the length of the spring changes to $L = 2n_w l$.

For the beginning, both the thickness t_c and the width B_c of the cross-section are considered to be constant. With this assumption the area of a cross-section, its section modulus and its area moment of inertia are, correspondingly:

$$a = B_{cR}t_{cR}, \qquad w_b = B_{cR}\, t_{cR}^{\,2}/6, \qquad I_b = B_{cR}\, t_{cR}^{\,3}/12.$$

The mass of the quarter of one meander for a spring with a constant profile shape is:

$$\frac{m_{cR}}{4n_w} = \rho\int_0^{x_2}\frac{B_{cR}t_{cR}}{\sin\psi_0}dx + R\rho\int_0^{\psi_0} B_{cR}t_{cR}d\varphi, \tag{11.5}$$

where material density is ρ. The spring consists of a total of n_w meanders, or, correspondingly, of entirely of $4n_w$ quarters. Accordingly, the complete mass of the spring is m_{cR}.

For simplicity, the Euler–Bernoulli theory of slender beams, in which major assumption is that “plane sections remain plane,” is applied. In other words, any deformation due to shear across the section is not accounted for (no shear deformation). With the Euler–Bernoulli theory the elastic energy due to bending of one-quarter of the spring reads:

$$\frac{U_{cR}}{4n_w} = \frac{1}{2}\int_0^{x_2} \frac{M_Z^2(I)}{E_L I_b \sin\psi_0} dx + \frac{R}{2}\int_0^{\psi_0} \frac{M_Z^2(II)}{E_L I_b} d\varphi. \tag{11.6}$$

The modulus of elasticity in the direction of the fibers is E_L. The maximal applied axial load of the spring is set to $F_{\max}$. The thickness of the material t_{cR} must be determined from the stress requirement. Namely, the stress due to bending under the action of the moment $F_{\max}\mathrm{B}_0/2$ in the outmost cross-section fibers does not exceed the ultimate value σ_0:

$$t_{cR} = \sqrt{\frac{3F_{\max}\mathrm{B}_0}{\sigma_0 \mathrm{B}_{cR}}}. \tag{11.7}$$

Substitution of the expressions for the bending moment (11.4) and the area moment of inertia into Eq. (11.6) leads to the expression for the stored elastic energy in the spring.

For brevity, the special case $\psi_0 = \pi/2$ will be explained in detail. The total elastic energy reads:

$$U_{cR} = \frac{n_w\sqrt{\mathrm{B}_{cR}}}{E_L}\left(\frac{\mathrm{B}_0\sigma_0}{3F_{\max}}\right)^{3/2} \cdot \frac{F^2}{g_{cR}(\kappa)}. \tag{11.8}$$

The auxiliary function on the right side of Eq. (11.8) is:

$$\frac{1}{g_{cR}(\kappa)} = 1 - 56\kappa^6 + 18\kappa^6\,\pi.$$

The mass of the spring follows with (11.7) from Eq. (11.5)

$$\frac{m_{cR}}{4n_w} = \frac{\rho}{2}\sqrt{\frac{3F_{\max}\,\mathrm{B}_{cR}\mathrm{B}_0^{\,3}}{\sigma_0}} \cdot (1 - 2\kappa^2 + \kappa^2\pi). \tag{11.9}$$

From Eq. (11.8) the expression for the spring rate (spring constant) follows:

$$c_{cR} = \left(\frac{d^2U_{cR}}{d^2F}\right)^{-1} = \frac{E_L\, g_{cR}(\kappa)}{2n_w\sqrt{\mathrm{B}_{cR}}}\left(\frac{3F_{\max}}{\mathrm{B}_0\sigma_0}\right)^{3/2} = \frac{\langle EA\rangle_{cR}}{L}. \tag{11.10}$$

In Eq. (11.10) $\langle EA\rangle_{cR}$ is the effective axial stiffness of the multiarc spring with the constant thickness of the layer. For design characterization, the spring rate is set:

$$c_{cR} = c^*,$$

where c^* is a given spring constant of the dimension $[N/m]$.

From the requirement of the prescribed spring rate $c_{cR} = c^*$, it is possible to determine the so-far undefined width of the spring. The width of the spring is the following:

$$\mathrm{B}_{cR} = \left(\frac{g_{cR}(\kappa)E_L}{2c^* n_w}\right)^2 \left(\frac{3F_{\max}}{\sigma_0\mathrm{B}_0}\right)^3. \tag{11.11}$$

Substitution of the expression for the width (11.11) into Eq. (11.9) delivers the formula for the mass of a meander spring. The mass of the spring reads in terms of the predefined geometrical parameters, the spring rate and allowable ultimate stress in the material:

$$m_{cR} = \frac{9F_{\max}^2 \rho E_L}{\sigma_0^2 c^*} \frac{1 - 2\kappa^2 + k^2\pi}{1 - 56\kappa^6 + 18k^6\pi}. \tag{11.12}$$

The total stored elastic energy of the spring with the spring constant c^* appears:

$$U_{cR} = \frac{F^2}{2c^*}. \tag{11.13}$$

The effective axial stiffness of the equivalent beam is equal to $\langle EA \rangle_R$.

In a multiarc design with a constant thickness of material, the stresses in the outmost fibers vary from point to point. The stresses attain their maximum in the distant cross-sections $x = \mathrm{B}_0/2$. On the axis of the spring $x = 0$ the bending moments vanish and the bending stresses disappear. Thus, as considered in this section, the multiarc design with a constant material thickness leads to a variable stress field and correspondingly heavy spring. Mass reduction is achieved with the full stress design as considered in the following section.

11.2.2 Multiarc Meander Spring With Optimal Cross-Section

A fully stressed state is reached when the material over the whole surface of the spring is utilized to its full-strength capacity. This design in engineering is considered a design optimum. This section examines optimality of the fully stressed design of a multiarc meander spring.

The width of the fully stressed spring B_{oR} is assumed again to be constant. The height of the cross-section $t_{oR}(x)$, or the thickness of the material is found from the condition of constancy of the stresses over the outer surfaces of the meander. With this assumption, the area of a cross-section, its section modulus and its area moment of inertia, correspondingly, are:

$$a(x) = \mathrm{B}_{oR} t_{oR}(x), \qquad w_b(x) = \mathrm{B}_{oR} t_{oR}^2(x)^2/6, \qquad I_b = \mathrm{B}_{oR} t_{oR}^3(x)/12.$$

The optimal height of the cross-section along the directrix is:

$$t_{oR}(x) = \sqrt{\frac{6F_{\max}}{\sigma_0 \mathrm{B}_{oR}}} \cdot \begin{cases} \sqrt{\frac{1}{2}\mathrm{B}_0 - x}, & region\, I: \quad 0 < x \le x_2, \\ \sqrt{R \cdot (1 - \cos\varphi)}, & region\, II: \quad x_2 < x \le \frac{1}{2}\mathrm{B}_0. \end{cases} \tag{11.14}$$

The mass of the quarter of one meander for a spring with optimal thickness is:

$$\frac{m_{oR}}{4n_w} = \rho \int_0^{x_2} \frac{\mathrm{B}_{oR} t_{oR}(x)}{\sin\psi_0} dx + R\rho \int_0^{\psi_0} \mathrm{B}_{oR} t_{oR}(\varphi) d\varphi. \tag{11.15}$$

Substitution of the expressions for the bending moment (11.4) and the area moment of inertia into Eq. (11.6) leads to the expression for stored elastic energy in the spring. The total deformation energy of the fully stressed spring reads:

$$\frac{U_{oR}}{4n_w} = \frac{1}{2} \int_0^{x_2} \frac{M_Z{}^2(I)}{E_L I_b(x) \sin\psi_0} dx + \frac{R}{2} \int_0^{\psi_0} \frac{M_Z{}^2(II)}{E_L I_b(\varphi)} d\varphi. \tag{11.16}$$

For the special case $\psi_0 = \pi/2$ the total deformation energy and the mass of the spring, correspondingly, are:

$$U_{oR} = \frac{n_w\sqrt{B_{oR}}}{E_L}\left(\frac{\sigma_0 B_0}{3F_{max}}\right)^{3/2} \cdot \frac{F^2}{g_{oR}(\kappa)}, \tag{11.17}$$

$$\frac{m_{oR}}{4n_w} = \rho\sqrt{\frac{F_{max}B_{oR}B_0{}^3}{3\sigma_0}\frac{1}{g_{oR}(\kappa)}}. \tag{11.18}$$

The auxiliary function on the right side of the Eq. (11.18) is:

$$\frac{1}{g_{oR}(\kappa)} = 2(1 - 8\sqrt{2}\kappa^3 + 12\kappa^3).$$

From the expression (11.17), the spring constant of the optimal spring results:

$$c_{oR} = \left(\frac{d^2 U_{oR}}{d^2 F}\right)^{-1} = \frac{E_L g_{oR}(\kappa)}{2n_w\sqrt{B_{oR}}}\left(\frac{3F_{max}}{B_0\sigma_0}\right)^{3/2} = \frac{\langle EA\rangle_{oR}}{L}. \tag{11.19}$$

In Eq. (11.19) $\langle EA\rangle_{oR}$ is the effective axial stiffness of the multiarc spring with optimal thickness of the layer. The spring rate is equal to the identical constant: $c_{oR} = c^*$. The width of the spring that assures the desired spring rate c^*, reads:

$$B_{oR} = \left(\frac{g_{oR}(\kappa)E_L}{2c^* n_w}\right)^2\left(\frac{3F_{max}}{\sigma_0 B_0}\right)^3. \tag{11.20}$$

Accordingly, the substitution of the width of the spring (11.20) into (11.17) delivers the value of stored elastic energy under the load F:

$$U_{oR} = \frac{F^2}{2c^*}. \tag{11.21}$$

The optimal mass of the spring after the substitution (11.20) into (11.18):

$$m_{oR} = \frac{3F_{max}^2\rho E_L}{\sigma_0^2 c^*} \cdot \frac{1 - 8\kappa^3\sqrt{2} + 12k^3}{\sqrt{1 - 16\kappa^3\sqrt{2} - 192k^6\sqrt{2} + 24\kappa^3 + 272\kappa^6}}. \tag{11.22}$$

11.2.3 Comparison of Masses for Fixed Spring Rate and Stress

The stored elastic energies of the fully stressed spring and of the spring with a constant cross-section are equal. The ratio of mass of optimal spring to the mass of the spring with the constant cross-section is (Figure 11.7):

$$\frac{m_{oR}}{m_{cR}} = 4f_R, f_R(\kappa) = \frac{1}{12}\frac{1 - 56\kappa^6 + 18k^6\pi}{1 - 2\kappa^2 + k^2\pi}. \tag{11.23}$$

The ratio of masses is limited:

$$\frac{1}{3} \leq \frac{m_{oR}}{m_{cR}} \leq \frac{3\pi - 8}{2\pi} \text{ for } 0 \leq \kappa \leq 1/\sqrt{2}. \tag{11.24}$$

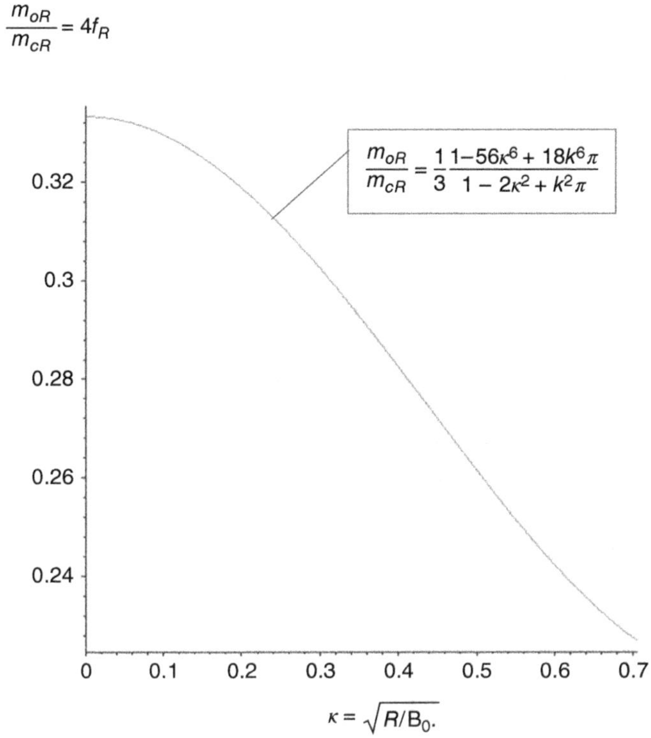

Figure 11.7 Mass ratio of an optimal spring to the mass of a spring with a constant cross-section, see Eq. (11.23).

11.3 Sinusoidal Spring Under Compressive Axial Load

11.3.1 Sinusoidal Meander Spring With Constant Cross-Section

Another important design of meander spring is referred to as a "sinusoidal spring." Its neutral surface is also a developable surface. Consider a quarter of a spring of the height l. The total length of the free spring with n_w single meanders is equal to $L_0 = 2n_w l$. The equation for the directrix of the neutral surface of the spring reads:

$$x = \frac{B_0}{2}\cos\left(\frac{\pi y}{l}\right) \cdot \hat{\kappa} = \frac{l}{B_0}, 0 < \hat{\kappa} < \infty. \tag{11.25}$$

For the sinusoidal spring, the appropriate parameterization of the neutral surface uses y as an independent variable.

$$M_Z = \frac{FB_0}{2}\cos Y, \qquad Y = \frac{\pi y}{l}. \tag{11.26}$$

To begin, both the thickness t_c and the width B_{cS} of the cross-section of the sinusoidal spring are taken to be constant. With this guess, the area of a cross-section, its section modulus and its area moment of inertia, correspondingly, are:

$$a = B_{cS}\, t_{cS}, \qquad w_b = B_{cS}\, t_{cS}{}^2/6, \qquad I_b = B_{cS}\, t_{cS}{}^3/12.$$

The constant thickness t_{cS} of the material determines from the condition that the stress due to bending under the action of the moment $F_{\max}B_0/2$ in the outmost points does not exceed an ultimate value σ_0. The mass of a quarter of the sinusoidal meander is:

$$\frac{m_{cS}}{4n_w} = \rho\int_0^{l/2} B_{cS}t_{cS}\sqrt{1+\left(\frac{dx}{dy}\right)^2}\,dy =$$

$$= \rho\sqrt{\frac{3F_{\max}B_{cS}{B_0}^3}{\sigma_0}}\,\frac{\sqrt{4\hat{\kappa}^2+\pi^2}}{2\pi}\cdot\mathbf{E}\left(\frac{\pi}{\sqrt{4\hat{\kappa}^2+\pi^2}}\right), \qquad (11.27)$$

where **E** is the Legendre's complete elliptic integral of the second kind (NIST 2010, section 19-1-1).

Accordingly, the calculation of the stored elastic energy in a quarter of the sinusoidal spring leads to the expression:

$$\frac{U_{cS}}{4n_w} = \frac{1}{2}\int_0^{l/2}\frac{12}{E_L B_{cS}{t_{cS}}^3}{M_Z}^2\sqrt{1+\left(\frac{dx}{dy}\right)^2}\,dy =$$

$$= \frac{F^2}{36\pi^3E_L}\left(\frac{3\sigma_0B_0}{F_{\max}}\right)^{3/2}\sqrt{B_{cS}}\sqrt{4\hat{\kappa}^2+\pi^2}\cdot\varpi(\hat{\kappa}). \qquad (11.28)$$

The dimensionless function $\varpi(\hat{\kappa}) = \varpi(l/B_0)$ reads:

$$\varpi(\hat{\kappa}) = (\pi^2-4\hat{\kappa}^2)\cdot\mathbf{E}\left(\frac{\pi}{\sqrt{4\hat{\kappa}^2+\pi^2}}\right) + 4\cdot\mathbf{K}\left(\frac{\pi}{\sqrt{4\hat{\kappa}^2+\pi^2}}\right).$$

where **K** is the Legendre's complete elliptic integral of the first kind (NIST 2010, section 19-1-1).

From Eq. (11.28) the spring rate (spring constant) is evaluated as follows:

$$c_{cS} = \left(\frac{d^2U_{cS}}{d^2F}\right)^{-1} = \frac{E_Lg_{cS}(\hat{\kappa})}{2n_w\sqrt{B_{cS}}}\left(\frac{3F_{\max}}{B_0\sigma_0}\right)^{3/2} = \frac{\langle EA\rangle_{cS}}{L}. \qquad (11.29)$$

The dimensionless auxiliary function in Eq. (11.29) reads:

$$\frac{1}{g_{cS}(\hat{\kappa})} = \frac{\varpi(\hat{\kappa})\sqrt{4\hat{\kappa}^2+\pi^2}}{\pi^3}.$$

In Eq. (11.29), $\langle EA\rangle_{cS}$ is the effective axial stiffness of a sinusoidal spring with a constant layer thickness. Solving the equation $c_{cS} = c^*$ with respect to the yet unknown width of the spring, one gets the formula for this value:

$$B_{cS} = \left(\frac{g_{cS}(\hat{\kappa})E_L}{2c^*n_w}\right)^2\left(\frac{3F_{\max}}{\sigma_0B_0}\right)^3. \qquad (11.30)$$

The substitution of (11.30) into (11.27) delivers the final expression for the mass of a sinusoidal meander spring:

$$m_{cS} = \frac{3F_{\max}^2E_L\rho}{\sigma_0^2c^*}\frac{1}{4f_S(\hat{\kappa})}, \quad f_S(\hat{\kappa}) = \frac{\varpi(\hat{\kappa})}{12\pi^2\mathbf{E}\left(\dfrac{\pi}{\sqrt{4\hat{\kappa}^2+\pi^2}}\right)}. \qquad (11.31)$$

The substitution of (11.30) into (11.28) once again gives the apparent expression for the stored energy of a sinusoidal meander spring:

$$U_{cS} = \frac{F^2}{2c^*}. \tag{11.32}$$

11.3.2 Sinusoidal Meander Spring With Optimal Cross-Section

The full stress design of the sinusoidal meander spring provides the minimal achievable mass for springs with a variable cross-section height. This section examines optimality of the fully stressed design of the sinusoidal meander spring. For a constant width of cross-section B_{oS}, the optimal height of the cross-section is:

$$t_{oS}(y) = \sqrt{\frac{3F_{\max}B_0}{\sigma_0 B_{oS}}\cos\frac{\pi y}{l}}. \tag{11.33}$$

With Eq. (11.33) the mass of the spring is:

$$\frac{m_{oS}}{4n_w} = \rho\int_0^{l/2} B_{oS}t_{oS}(y)\sqrt{1+\left(\frac{dx}{dy}\right)^2}dy = \frac{\rho}{2}\sqrt{\frac{F_{\max}B_{oS}{B_0}^3}{3\sigma_0}}\cdot\frac{1}{g_{oS}(\hat{\kappa})}. \tag{11.34}$$

The dimensionless auxiliary function in Eq. (11.29) is defined as follows:

$$\frac{1}{g_{oS}(\hat{\kappa})} = \frac{3\Gamma(3/4)}{\pi^{5/2}}\hat{\kappa}^2\, G\left(\left(\frac{2\hat{\kappa}}{\pi}\right)^2; \left[\left[\frac{1}{2}\right], \left[\frac{3}{4}\right]\right]; [[0,-1],[]]\right).$$

The higher transcendental function $G(\xi; [[a_1], [a_2]]; [[b_1, b_2], [b_3]])$ is the Meijer G-function (NIST 2010, section 16.17). The evaluation of the integral (11.34) is based on the formula by Gradshteyn and Ryzhik (2014, eq. 3.671):

$$\int_0^{\pi/2}\sin^\alpha\theta\cos^\beta\theta\sqrt{1-k^2\sin^2\theta}d\theta =$$

$$\frac{1}{2}\mathbf{B}\left(\frac{\alpha+1}{2}, \frac{\beta+1}{2}\right)\mathbf{F}\left(\frac{\alpha+1}{2}, -\frac{1}{2}, \frac{\alpha+\beta+2}{2}; k^2\right), \alpha > -1, \beta > -1, |k| < 1.$$

In this expression, two transcendental functions are used: **F** is the hypergeometric function and **B** is the beta-function (NIST 2010).

The stored elastic energy is equal to:

$$\frac{U_{oS}}{4n_w} = \frac{1}{2}\int_0^{l/2}\frac{12}{E_L B_{oS}{t_{oS}}^3(y)}{M_Z}^2\sqrt{1+\left(\frac{dx}{dy}\right)^2}dy.$$

After the substitution of (11.33) into this expression and evaluation of the integral, the stored energy of the optimal sinusoidal spring reads:

$$U_{oS} = \frac{n_w\sqrt{B_{oS}}F^2}{E_L g_{oS}(\hat{\kappa})}\left(\frac{\sigma_0 B_0}{3F_{\max}}\right)^{3/2}. \tag{11.35}$$

For the spring rate, the following expression comes:

$$c_{oS} = \left(\frac{d^2U_{oS}}{d^2F}\right)^{-1} = \frac{E_L g_{oS}(\hat{\kappa})}{2n_w\sqrt{B_{oS}}}\left(\frac{3F_{\max}}{\sigma_0 B_0}\right)^{3/2} = \frac{\langle EA\rangle_{oS}}{L}. \tag{11.36}$$

In Eq. (11.29) $\langle EA \rangle_{cS}$ is the effective axial stiffness of the sinusoidal spring with an optimal layer thickness. After the equating the spring rate (11.36) and the prescribed spring constant c^*, the solution for the width of the spring is:

$$\mathrm{B}_{oS} = \left(\frac{g_{oS}(\hat{\kappa})E_L}{2c^* n_w}\right)^2 \left(\frac{3F_{\max}}{\sigma_0 \mathrm{B}_0}\right)^3. \tag{11.37}$$

Using the expression (11.37) in the formulas (11.34) and (11.35), the final statements for the mass and stored elastic energy of the optimal sinusoidal meander spring are, correspondingly:

$$m_{oS} = \frac{3F_{\max}^2 \rho E_L}{\sigma_0^2 c^*}, \tag{11.38}$$

$$U_{oS} = \frac{F^2}{2c^*}. \tag{11.39}$$

11.3.3 Comparison of Masses for Fixed Spring Rate and Stress

The compassion of the stored energies of the optimal and constant thickness springs demonstrates the equality (Figure 11.8):

$$\frac{U_{oS}}{U_{cS}} \equiv 1 \quad . \tag{11.40}$$

The masses of the optimal (11.40) and constant thickness springs (11.31) relate to each other as follows:

$$\frac{m_{oS}}{m_{cS}} = 4f_S(\hat{\kappa}). \tag{11.41}$$

The mass ratio of optimal to the constant thickness sinusoidal meander springs is limited by the values:

$$\frac{1}{3} < \frac{m_{oS}}{m_{cS}} < \frac{1}{2}, \left.\frac{m_{oS}}{m_{cS}}\right|_{\hat{\kappa}=0} = \frac{1}{3}, \left.\frac{m_{oS}}{m_{cS}}\right|_{\hat{\kappa}\to\infty} = \frac{1}{2}. \tag{11.42}$$

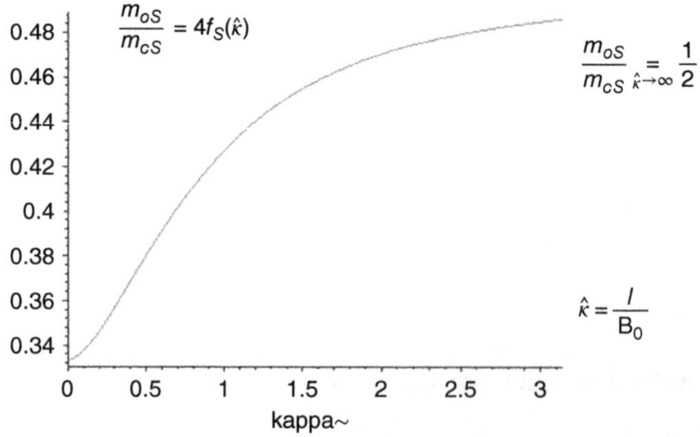

Figure 11.8 Mass ratio of constant to optimal sinusoidal meander springs, see Eq. (11.39).

11.4 Bending Stiffness of Meander Spring With a Constant Cross-Section

11.4.1 Bending Stiffness of a Multiarc Meander Spring With a Constant Cross-Section

The evaluation of bending stiffness of the meander spring is necessary for calculation of lateral stability and buckling loads of the spring. For evaluation of bending stiffness, the meander spring is considered to be in a state of pure bending. The bending moment is constant over the directrix of the middle surface:

$$M_Z = M_0 = const. \tag{11.43}$$

The stored elastic energy for the pure bending for the multiarc spring with a constant thickness of material equals:

$$\frac{U_{bR}}{4n_w} = \frac{1}{2}\int_0^{x_2} \frac{12{M_Z}^2}{E_L I_b \sin\psi_0} dx + \frac{R}{2}\int_0^{\psi_0} \frac{{M_Z}^2}{E_L I_b} d\varphi, \tag{11.44a}$$

$$U_{bR} = \frac{6M_0^2}{{B_0}^2 c^*} \frac{1 - 2\kappa^2 + k^2\pi}{1 - 56\kappa^6 + 18k^6\pi} \equiv \frac{M_0^2}{{B_0}^2 c^*} \cdot \frac{1}{2f_R(\kappa)}. \tag{11.44b}$$

The spring rate for pure bending possesses the dimension [Nm rad^{-1}] and is expressed in the course of the formula:

$$c_{bR} = \left(\frac{d^2 U_{bR}}{d^2 M_0}\right)^{-1} = \frac{\langle EI_b \rangle_R}{L}. \tag{11.45}$$

The effective bending stiffness of the multiarc with constant layer thickness is $\langle EI_b \rangle_R$ in Eq. (11.45).

With the expression (11.29) for the axial spring rate, the spring rate for pure bending is:

$$c_{bR} = f_R(\kappa) c^* B_0^2. \tag{11.46}$$

From (11.46) the effective bending constant results:

$$\langle EI_b \rangle_R = c_{bR} L = f_R(\kappa) c^* B_0^2 L. \tag{11.47}$$

Notably, the ratio of bending stiffness to axial stiffness relates to the ratio of optimal mass to the mass of the multiarc spring with a constant material thickness as:

$$\frac{c_{bR}}{c_{cR} B_0^2} \equiv \frac{m_{oR}}{4m_{cR}} \equiv f_R(\kappa). \tag{11.48}$$

11.4.2 Bending Stiffness of a Sinusoidal Meander Spring with a Constant Cross-Section

The same method is appropriate for evaluation of the pure bending stiffness of a sinusoidal meander spring with constant material thickness. The stored elastic energy for pure bending is:

$$\frac{U_{bS}}{4n_w} = \frac{1}{2}\int_0^{l/2} \frac{12}{E_L B_{cS} {t_{cS}}^3} {M_Z}^2 \sqrt{1 + \left(\frac{dx}{dy}\right)^2} dy, \tag{11.49a}$$

$$U_{bS} = \frac{6M_0^2\pi^2}{{B_0}^2 c^* \varpi(\hat{\kappa})} \cdot \mathrm{E}\left(\frac{\pi}{\sqrt{4\hat{\kappa}^2+\pi^2}}\right) \equiv \frac{M_0^2}{{B_0}^2 c^*} \cdot \frac{1}{2f_S(\hat{\kappa})}. \tag{11.49b}$$

The bending spring rate follows from (11.49a,11.49b) as:

$$c_{bS} = \left(\frac{d^2 U_{bS}}{d^2 M_0}\right)^{-1} = \frac{\langle EI_b\rangle_S}{L}. \tag{11.50}$$

In Eq. (11.50), the value $\langle EI_b\rangle_S$ is the effective bending stiffness of the sinusoidal with a constant layer thickness. The integration (11.49a,11.49b) leads to the expression of bending stiffness in terms of the axial stiffness of the spring

$$c_{bS} = f_S(\hat{\kappa}) \cdot c^* B_0^2. \tag{11.51}$$

The ratio of bending stiffness to axial stiffness relates to the ratio of optimal mass to the mass of a sinusoidal spring with a constant material thickness:

$$\frac{c_{bS}}{c_{cS}{B_0}^2} \equiv \frac{m_{oS}}{4m_{cS}} \equiv f_S(\hat{\kappa}). \tag{11.52}$$

Remarkably, both formulas (11.48) and (11.52) express the identity of the ratios of masses and ratios of bending stiffness for multiarc and sinusoidal meander springs.

Summarizing these results, the effective axial and bending stiffness of an effective rod with a total length $L = 2n_w l$ could be represented as follows for the multiarc meander spring and for sinusoidal meander spring, correspondingly:

$$c_{cR} = \frac{\langle EA\rangle_{cR}}{L} \equiv c^*, c_{bR} = \frac{\langle EI_b\rangle_R}{L} \equiv c^* f_R {B_0}^2. \tag{11.53}$$

$$c_{cS} = \frac{\langle EA\rangle_{cS}}{L} = c^*, c_{bS} = \frac{\langle EI_b\rangle_S}{L} \equiv c^* f_S {B_0}^2. \tag{11.54}$$

The design formulas for the multiarc and sinusoidal springs with constant material thickness are recapitulated in Table 11.2. The expressions for effective axial stiffness and spring rates of multiarc and sinusoidal meander spring with constant and optimal material thickness are listed in Table 11.3. The dimensionless functions for multiarc and sinusoidal meander spring with constant and optimal material thickness are summarized in Table 11.4.

11.5 Stability of Corrugated Springs

11.5.1 Euler's Buckling of an Axially Compressed Rod

Buckling refers to the loss of stability up to the sudden and violent failure of straight beams or beams under the effect of compression forces whose line of action lies in the beam axis. The loss of stability, according to Euler, manifests itself in rapidly increasing deformations of the bar or beam from a certain load (buckling load), namely with a lateral deviation of the rod or beam axis. When calculating the buckling load of a member according to Euler, the reduction of the member caused by the load may always be neglected:

$$F_{krit} = EI_b\left(\frac{k_{cl}\pi}{L_0}\right)^2 \tag{11.55}$$

Table 11.2 Design formulas for axial and bending stiffness of multiarc and sinusoidal meander spring with constant material thickness.

	Multiarc meander spring	**Sinusoidal meander spring**
Effective compression stiffness	$\langle EA\rangle_{cR} = \dfrac{E_L l}{\sqrt{\mathrm{B}_c}}\left(\dfrac{3F_{\max}}{\mathrm{B}_0\sigma_0}\right)^{3/2} g_{cR}$	$\langle EA\rangle_{cS} = \dfrac{E_L l}{\sqrt{\mathrm{B}_c}}\left(\dfrac{3F_{\max}}{\mathrm{B}_0\sigma_0}\right)^{3/2} g_{cS}$
	$g_{cR}(\kappa) = \dfrac{1}{1-56\kappa^6+18\kappa^6\,\pi}$	$g_{cS}(\hat{\kappa}) = \dfrac{\pi^3}{\varpi(\hat{\kappa})\sqrt{4\hat{\kappa}^2+\pi^2}}$
Effective bending stiffness	$\langle EI_b\rangle_R = f_R \cdot \langle EA\rangle_{cR}\mathrm{B}_0^2$	$\langle EI_b\rangle_S = f_S \cdot \langle EA\rangle_{cS}\mathrm{B}_0^2$
	$f_R(\kappa) = \dfrac{1}{12}\dfrac{1-56\kappa^6+18k^6\pi}{1-2\kappa^2+k^2\pi}$	$f_S(\hat{\kappa}) = \dfrac{\varpi(\hat{\kappa})}{12\pi^2\mathbf{E}\left(\dfrac{\pi}{\sqrt{4\hat{\kappa}^2+\pi^2}}\right)}$
Width of the cross-section	$\mathrm{B}_{cR} = \dfrac{g_{cR}^2(\kappa)}{4}\cdot\dfrac{E_L{}^2}{c^{*2}n_w{}^2}\left(\dfrac{3F_{\max}}{\sigma_0\mathrm{B}_0}\right)^3$	$\mathrm{B}_{cS} = \dfrac{g_{cS}^2(\kappa)}{4}\cdot\dfrac{E_L{}^2}{c^{*2}n_w{}^2}\left(\dfrac{3F_{\max}}{\sigma_0\mathrm{B}_0}\right)^3$
Parameters	$\kappa = \sqrt{\dfrac{R}{\mathrm{B}_0}} \le \dfrac{1}{\sqrt{2}}$	$\hat{\kappa} = \dfrac{l}{\mathrm{B}_0}, \quad 0 < \hat{\kappa} < \infty$
	$\varpi(\hat{\kappa}) = (\pi^2 - 4\hat{\kappa}^2)\cdot\mathbf{E}\left(\dfrac{\pi}{\sqrt{4\hat{\kappa}^2+\pi^2}}\right) + 4\cdot\mathbf{K}\left(\dfrac{\pi}{\sqrt{4\hat{\kappa}^2+\pi^2}}\right)$	

Table 11.3 Effective axial stiffness and spring rates of multiarc and sinusoidal meander spring with constant and optimal material thickness.

$c_{cS} = K_c\dfrac{g_{cS}(\hat{\kappa})}{\sqrt{\mathrm{B}_{cS}}} = \dfrac{\langle EA\rangle_{cS}}{L}$	$B_{cS} = K_B g_{cS}^2(\hat{\kappa})$
$c_{oS} = K_c\dfrac{g_{oS}(\hat{\kappa})}{\sqrt{\mathrm{B}_{oS}}} = \dfrac{\langle EA\rangle_{oS}}{L}$	$B_{oS} = K_B g_{oS}^2(\hat{\kappa})$
$c_{cR} = K_c\dfrac{g_{cR}(\kappa)}{\sqrt{\mathrm{B}_{cR}}} = \dfrac{\langle EA\rangle_{cR}}{L}$	$B_{cR} = K_B g_{cR}^2(\kappa)$
$c_{oR} = K_c\dfrac{g_{oR}(\kappa)}{\sqrt{\mathrm{B}_{oR}}} = \dfrac{\langle EA\rangle_{oR}}{L}$	$B_{oR} = K_B g_{oR}^2(\kappa)$
$K_c = \dfrac{E_L}{2n_w}\left(\dfrac{3F_{\max}}{\sigma_0\mathrm{B}_0}\right)^{3/2}$	$K_B = \dfrac{E_L{}^2}{4c^{*2}n_w{}^2}\left(\dfrac{3F_{\max}}{\sigma_0\mathrm{B}_0}\right)^3$

Here are:

- EI_b the bending stiffness of the rod,
- I_b the moment of inertia of the cross-section of the straight bar around the beige or buckling axis,
- L_0 the initial or free length of the undeformed rod,
- k_{cl} the coefficient dependent on clamping conditions.

Slender compression springs also tend to buckling. The buckling resistance of a spring depends on the ratio of length to diameter, the specific load and the type of bearing. When applying Euler's methods to a meander spring, two special features should be

Table 11.4 Dimensionless functions for multiarc and sinusoidal meander spring with constant and optimal material thickness.

$\frac{1}{g_{cS}(\hat{\kappa})}$	$\frac{\varpi(\hat{\kappa})\sqrt{4\hat{\kappa}^2+\pi^2}}{\pi^3}$
$\frac{1}{g_{oS}(\hat{\kappa})}$	$\frac{3\Gamma(3/4)}{\pi^{5/2}}\hat{\kappa}^2\, G\left(\left(\frac{2\hat{\kappa}}{\pi}\right)^2; \left[\left[\frac{1}{2}\right], \left[\frac{3}{4}\right]\right]; [[0,-1],[]]\right)$
$\frac{1}{g_{cR}(\kappa)}$	$1-56\kappa^6+18\kappa^6\,\pi$
$\frac{1}{g_{oR}(\kappa)}$	$2(1-8\sqrt{2}\kappa^3+12\kappa^3)$

considered. First, the axial stiffness of the springs is much lower than the axial stiffness of the rods. For calculation according to Euler, it is assumed that the rod is shortened only negligibly under the effect of the compressive force. A similar method was applied to the study of buckling of helical springs in Chapter 9.

11.5.2 Side Buckling of Meander Springs

The springs are considerably softer in comparison to straight rods and shorten significantly under compressive force. On the other hand, the Euler equation uses two equivalent bending stiffnesses of the spring for longitudinal $\langle E_L I_1\rangle$ and for transverse directions $\langle E_L I_2\rangle$ instead of one bending stiffness of the rod EI_b. The bending stiffness of the springs EI_b depends on the current, shortened length of the spring and is unequal in the longitudinal and transverse directions. If we use the correct stiffness and length of the spring in the Euler formula, we obtain an expression of the critical buckling force of a spring in its compressed state:

$$F_{krit} = \min\left[\langle E_L I_1\rangle, \langle E_L I_2\rangle\right](k_{cl}\pi/L)^2. \tag{11.56}$$

In the formula (11.56) we use the free length L_0 on one hand, and the actual compressed length of the spring L depending on the buckling force F_{crit} on the other. To calculate the buckling force of the spring, the length shortened under the effect of the force should be used in (11.56) instead of the original length. The shortened length of the spring depends of force with the spring rate c^*:

$$L = L_0 - F_{crit}/c^*. \tag{11.57}$$

Taking this dependence into account in Eq. (11.57) leads to the main equation for calculating the critical buckling force F_{crit}. The buckling behavior of a meander spring in the transverse direction depends on the material used or the ratio between elastic modulus E_L and shear modulus G_{TL}. The buckling in the longitudinal direction, however, as well as axial deflection behavior, depends only on geometrical pare meters. The meander spring exhibits an unfavorable stability behavior in relation to the helical spring. The storage of the meander spring plays a decisive role. This influences the storage coefficient, which significantly changes the buckling behavior. A meander spring clamped on both sides in a wide range of freedom of geometric design gilts is kink-resistant.

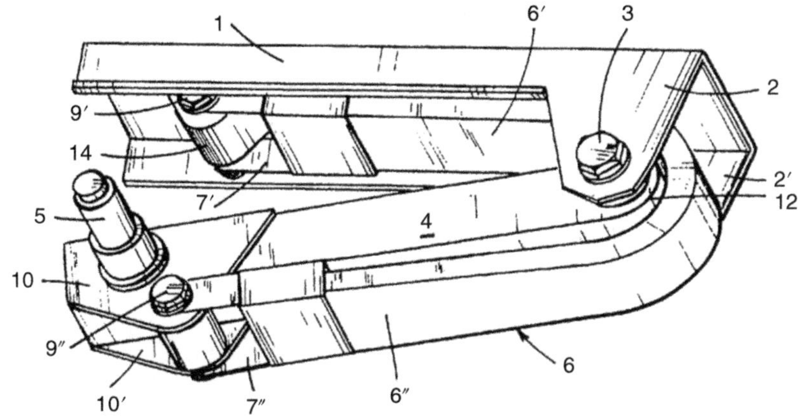

Figure 11.9 WO1994018019A1 (1993-02-15) Suspension for the wheel of a vehicle.

11.6 Patents for Chassis Springs Made of Composites in Meandering Form

This section displays the short compilation of the relevant patents and patent applications for meander-shaped springs in an automotive chassis.

WO1994018019A1 (1993-02-15, https://www.epo.org*) Suspension for the wheel of a vehicle such as a trailer and set of such suspensions*:

> Suspension of the type in which a resilient device opposes the reaction forces applied to the wheel where it contacts the ground. The device (6) consists of a resilient U-shaped leaf made of a composite material containing fibers which are continuous from one end of the blade to the other and are embedded in a binder, the two so shaped arms (6′, 6′) of the leaf being mechanically coupled to the vehicle chassis and the wheel's suspension arm (4), respectively (Figure 11.9).

WO85/00207A1 (1985, https://www.epo.org*) Springs for high specific energy storage:*

> A spring element (12) comprises a strip of fiber-reinforced plastics material formed as a zig-zag consisting of a plurality of limbs lying in substantially the same plane and joined at their ends by reflex portions (14). The strip includes at least a proportion of fibers aligned in its generally longitudinal sense, sufficient to spread stress concentrations away from the reflex portions, in use. The ratio of the limb length to the radius of the reflex portions is desirably in the range 0.2 to 200, preferably 0.5 to 30, and this leads to an extremely compact spring. Several elements (12) can be assembled together, possibly in different planes (e.g. Fig. 6) to produce a spring assembly whose spring rates in different modes of deflection are easily controlled. The spring elements (12) may include elastomer either as blocks (17) in the reflex portions (14) or as layers interleaved with fiber-reinforced plastic layers (e.g. Fig. 7). The elastomer and plastics materials can be cured in mutual contact, to provide cross-linking between them (Figure 11.10).

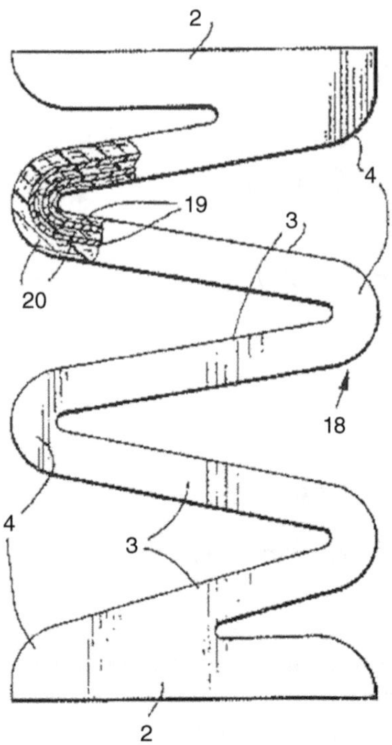

Figure 11.10 WO8500207(A1) Springs for high specific energy storage.

WO8700252A1 (1985, https://www.epo.org*) spring assemblies*:

This invention relates to springs each comprising a strip of fiber-reinforced plastics material in the form of a zig-zag, having a plurality of limbs connected by reflex portions, the longitudinal center line of the strip lying substantially in a single plane. Such springs, which will hereafter be referred to as springs of the kind specified, are disclosed in International – Patent Application Publication No. O85/00207. More particularly, the invention relates to assemblies wherein such springs can be mounted for use in, for example, vehicle suspensions.
Experience with springs made of composite, fiber-reinforced plastics, material has shown that one of the most difficult aspects of design associated with such springs is in respect of the means by which the springs are connected to the apparatus or equipment where they are to be used. This is the case whatever the configuration of the springs, including springs of the kind specified. It is the broad object of the present invention to provide assemblies wherein such connection of springs of the kind specified is effectively made. It is an object of a further aspect of the invention to provide an assembly of springs of the kind specified which is advantageous for use in vehicle suspensions.
According to the invention, we provide an assembly comprising a spring of the kind specified and a mounting at an end thereof comprising means affording a socket in which a limb portion of the spring is held. The limb which is held in the socket may comprise one of said zig-zag limbs of the spring, extending generally transversely of the spring considered as a whole, or may be a limb

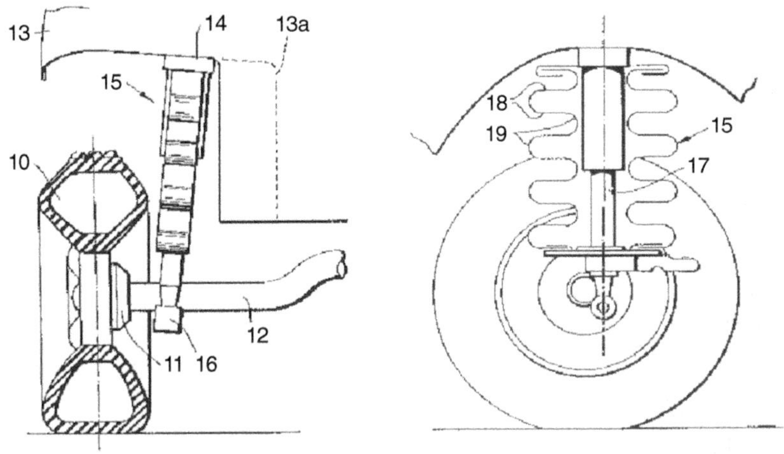

Figure 11.11 WO8700252: Spring assemblies.

portion extending generally longitudinally of the spring considered as a whole (Figure 11.11).

DE3641108 C2 (1988, https://www.epo.org*) Spring element*:

The invention relates to a spring element which is apposite for receiving linear forces on pressure or train. For this purpose, cylindrical coil springs made of spring steel wire with different wire thickness and pitch are commonly used.
In these known cylindrical coil springs, it is a disadvantage that to achieve a certain spring force or a certain spring travel relatively much space is needed because the cylinder interior consists of mostly unused cavity…
It is an object of the invention to provide a spring element which has small dimensions with a comparatively high travel and high spring force and for the spring connection given by the spring wire restrictions are not given.
To solve this problem, the spring element consists of a plurality of consecutively formed from flat material, contiguous, angular, or U-shaped edgings whose openings point in opposite directions (Figure 11.12).

US4927124A (1990, https://www.uspto.gov*) Spring Assemblies*:

This invention relates to springs comprising a strip of fiber-reinforced plastics material of zig-zag configuration, having a plurality of limbs connected by reflex portions and the longitudinal center line of the strip lying substantially in a single plane. Such springs, which will hereafter be referred to as springs of the kind specified, are disclosed in International Patent Application Publication No. W085/00207. More particularly, the invention relates to arrangements by which such springs can be mounted for use.
Springs of the kind specified have been proposed for use in motor vehicle suspension systems, where the properties of such springs can be advantageous. As compared with steel coil springs, currently widely used in vehicle suspensions,

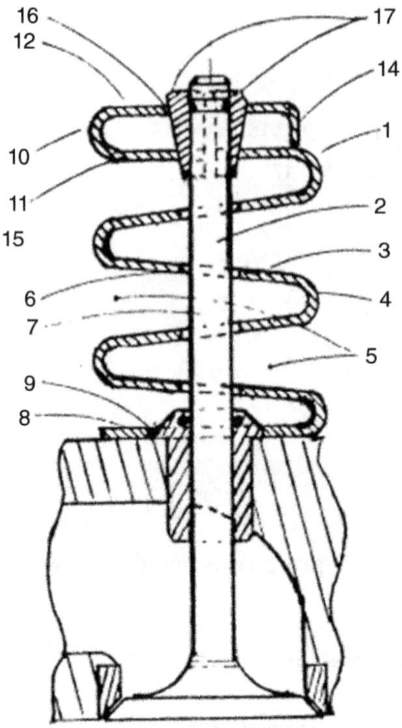

Figure 11.12 DE3641108C2: Spring element.

springs of the kind specified can lead to a reduction in weight and certain installation advantages resulting from the reduced space occupied by the springs. A spring of the kind specified is used as a compression spring, loaded between mounting components secured to opposite end limbs of the spring. When the spring is loaded, the limbs thereof are moved closer together (Figure 11.13).

DE19962026A1 (2001, https://www.epo.org*) Spring/suspension device for motor vehicles has corrugated spring element to accommodate bending loads*:

The spring/suspension device incorporates a corrugated spring element (3), which when fitted extends mainly parallel to the longitudinal vehicle direction, and accommodates bending loads. The spring element is held secured against buckling on upper and lower connection ends of a damper element (2). The spring element consists of metallic and/or fiber-reinforced or non-reinforced plastic (Figure 11.14).

DE102012111252A1 (2012, https://www.epo.org*) Spring of suspension for vehicle:*

A spring member of a suspension for a vehicle may include a spring provided to be supported between a lower seat disposed in the suspension and a car body to provide an elastic force, wherein an inner part of the spring is formed in a shape of a hollow corrugated tube (rotationally symmetric meander spring) (Figure 11.15).

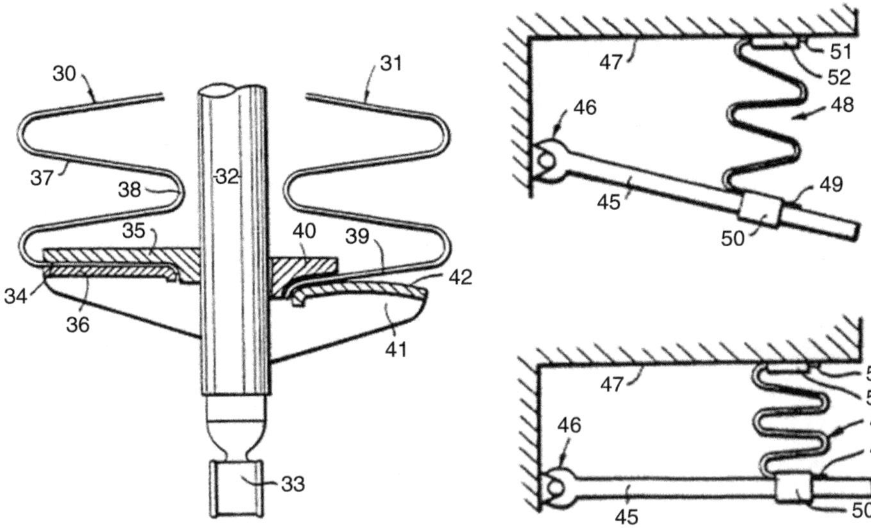

Figure 11.13 US4927124 Spring assemblies.

Figure 11.14 DE19962026A1 (2001): Spring/suspension device.

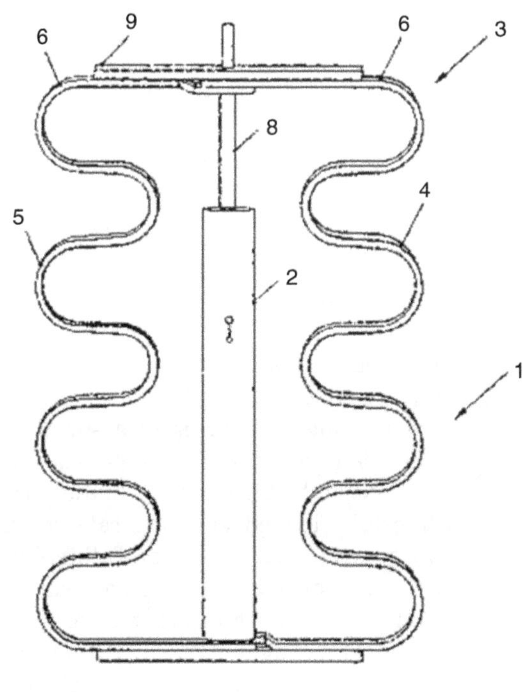

US2007267792A1 (2007, https://www.uspto.gov*) Sigma-Springs for Suspension Systems:*

> Sigma-spring can be used as a passive vehicle suspension system of built-in damping, a passive vehicle suspension system with hydraulic damper, a semi-active vehicle suspension system of two opposite Sigma-springs of built-in damping,

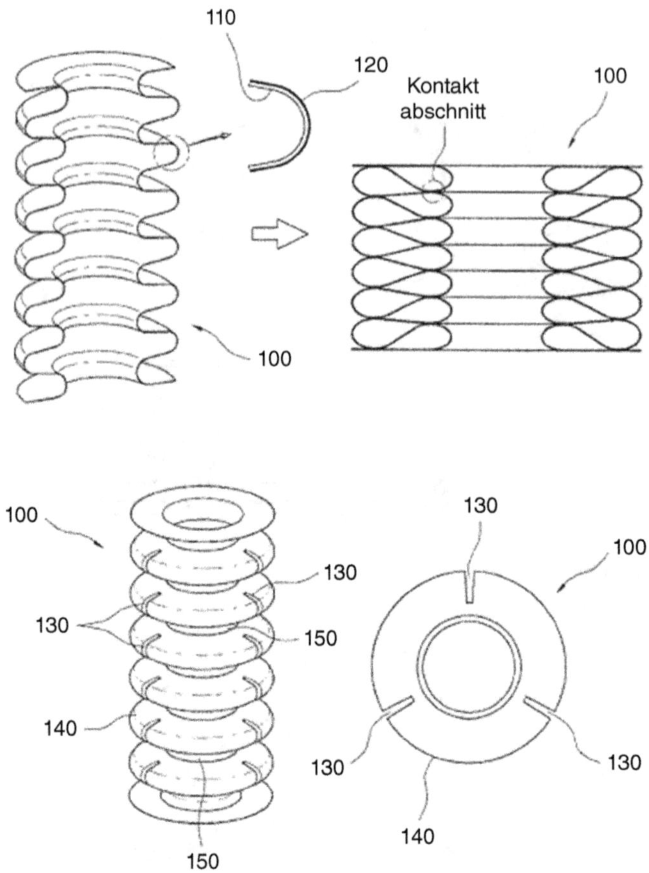

Figure 11.15 DE102012111252A1 (2012) Spring of suspension for a vehicle.

or a semi-active vehicle suspension system of two opposite Sigma-springs (10, 11) penetrated at line of external loading by hydraulic damper (12). The Sigma-spring is a spring of a special Sigma-shape above which mass can be suspended vertically while under static or dynamic loading conditions. The outer two arms of the Sigma-shape of the spring are inclined such that to be horizontal while fully loaded ensuring safe compression pattern. The Sigma-shape has two opposite sets of turns of different sizes one of small radius of curvature at the side of line of the applied external load and the opposite one has large radius of curvature in order to maximize spring vertical deflection capability. Thickness throughout the Sigma-spring developed length is graduated in order to minimize induced stresses, weight, and cost. Stiffness of the Sigma-spring can be nonlinear in order to provide improved vibration isolation and can be adjusted for same compact space allowance through increasing or decreasing its number of turns. The Sigma-spring can be made of Polymeric Matrix Composite of resin strengthened at nano-scale by nanometer-sized powder E-glass fibers. Moreover, the Sigma-spring can be made of Metallic Matrix Composite or of whole metallic monolithic material (Figure 11.16).

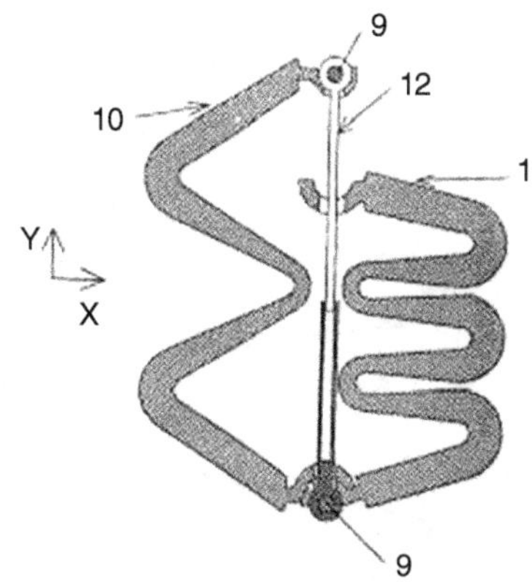

Figure 11.16 US2007267792A1 (2007) Sigma-springs for suspension systems.

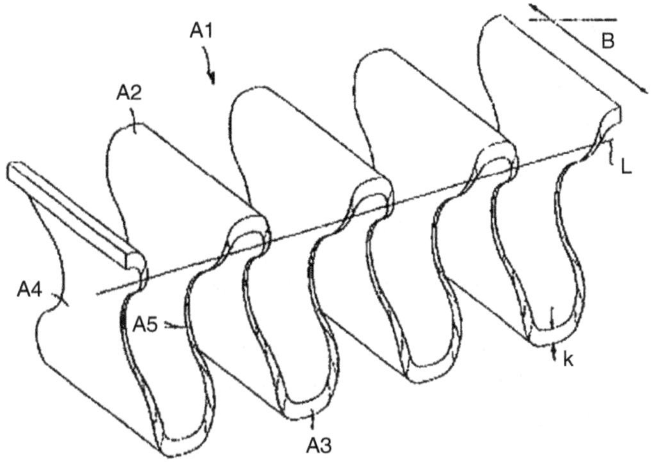

Figure 11.17 DE102008006411 (2009) Vehicle spring made from composite material.

DE102008006411 (2009, https://www.epo.org*) Vehicle spring made from composite material:*

> Band spring 11, 21, 31 made of fiber composite material having an undulating course, wherein a spring band as a single wave of reversal regions 12, 13 and intermediate portions 14, 15 meanders about a longitudinal center line L, which substantially corresponds to the direction of introduction of force K, wherein in each case an increased resistance moment of the spring band is provided in the reversal regions 12, 13 of the wave train (Figure 11.17).

DE102008057463 (2008, https://www.epo.org*): Spring, particularly drawing and pressure spring, is made from fiber-reinforced material and has two end sections, where twisting sections are arranged between end sections:*

> Spring (1) consists of a fiber composite material, the two end portions (2, 6) and arranged in between the turn portions (4, 5), wherein the end portions (2, 6) and

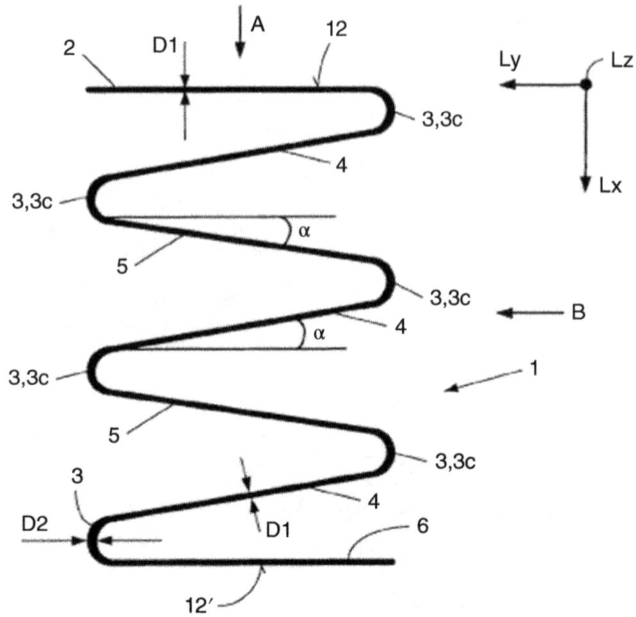

Figure 11.18 DE102008057463 (2008) Spring, particularly drawing and pressure spring.

the turn portions (4, 5) via deflection sections (3) integral with each other are connected, and in which all the sections (2, 3, 4, 5, 6) consist of a meander-shaped folded web (8) of a fiber composite material, characterized in that the fiber volume fraction of the turn (3) is smaller than the fiber volume fraction of the respective adjacent sections (2, 4, 5, 6) (Figure 11.18).

11.7 Conclusions

Compared to conventional cold-formed metal springs, springs made of composite materials have disadvantages due to low cycle rates due to slow manufacturing processes. In addition, each spring requires a special press mold. This reduces design variability. As a result, it is difficult to fine-tune wheel suspensions for optimum chassis tuning. On the other hand, the substitution of leaf springs by composite springs allows a great weight saving. Another advantage of the composite spring is that the installation space changes only slightly compared to conventional springs, since both have the same shape in principle. Therefore, a good and fast substitution of steel springs is possible. Furthermore, the unsprung masses of a vehicle are reduced, which causes a reduced vibration excitation of the body to the same extent. The reduced vibration amplitudes subsequently lead to lower loads, which can reduce the weight on the body itself as the stresses occurring are lower. The percentage weight saving of the springs is estimated to be between 40 and 70%, depending on the design. The negative effects of the unsprung mass on the wheel on fast-running vehicles are noticeably reduced. However, since helical springs are much smaller components, the absolute weight bonus is lower compared to steel springs. Furthermore, customers' installation space restrictions place high demands on the design and stability of the composite meander spring, as this is a rectangular spring base. Despite these difficulties, weight reduction is imaginable due to the high material utilization of the geometry.

References

DE102008006411 (2009). Vehicle spring made from composite material. European Patent Office, Espacenet, https://www.epo.org.

DE102008057463 (2008). Spring, particularly drawing and pressure spring. European Patent Office, Espacenet, https://www.epo.org.

DE102012111252A1 (2012). Spring of suspension for vehicle. European Patent Office, Espacenet, https://www.epo.org.

DE19962026A1 (2001). Spring/suspension device for motor vehicles has corrugated spring element to accommodate bending loads. European Patent Office, Espacenet, https://www.epo.org.

DE3641108C2 (1988). Spring element, European Patent Office, Espacenet, https://www.epo.org.

EP2082903B1 (2009). Vehicle spring made from composite material. European Patent Office, Espacenet, https://www.epo.org.

Gradshteyn, I.S. and Ryzhik, I.M. (2014). *Table of Integrals, Series, and Products*, 8e. Elsevier Inc.

Mahdi, E., Alkoles, O.M.S., Goudah, G. et al. (2006). Light composite elliptic springs for vehicle suspension. *Compos. Struct.* 75: 24–28. https://doi.org/10.1016/j.compstruct.2006.04.082.

NIST (2010). *Handbook of Mathematical Functions*. Cambridge: Cambridge University Press.

Pottmann, H. and Wallner, J. (2001). *Computational Line Geometry*. Berlin: Springer-Verlag.

Reid, S.R., Wong, W.H., Lau, K.J., and Tse, P.C. (2002). Spring stiffnesses of composite circular springs with extended flat contact surfaces under unidirectional line-loading and surface-loading configurations. *Compos. Struct.* 55: 367–386.

Reid, S.R., Tse, P.C., Wong, W.H., and Lau, K.J. (2004). Large deflections of composite circular springs with extended flat contact surfaces. *Compos. Struct.* 63: 253–260. https://doi.org/10.1016/S0263-8223(03)00172-7.

Struik, D.J. (1950). *Lectures on Classical Differential Geometry*. Cambridge, MA: Addison-Wesley.

US 2007267792A1 (2007). Sigma-springs for suspension systems, The United States Patent and Trademark Office (USPTO) https://www.uspto.gov.

US4927124A (1990). Spring assemblies, The United States Patent and Trademark Office (USPTO), https://www.uspto.gov.

Wong, W.H., Tse, P.C., Lau, K.J., and Ng, G.Y.F. (2004). Spring constant of fibre-reinforced plastics circular springs embedded with nickel–titanium alloy wire. *Compos. Struct.* 65: 319–328. https://doi.org/10.1016/j.compstruct.2003.11.006.

WO1994018019A1 (1993-02-15). Suspension for the wheel of a vehicle, European Patent Office, Espacenet, https://www.epo.org.

WO85/00207A1 (1985). Springs for high specific energy storage: European Patent Office, Espacenet, https://www.epo.org.

WO8700252A1 (1985). Spring assemblies, European Patent Office, Espacenet, https://www.epo.org.

12

Hereditary Mechanics of Composite Springs and Driveshafts

12.1 Elements of Hereditary Mechanics of Composite Materials

12.1.1 Mechanisms of Time-Dependent Deformation of Composites

The stiffness and strength of a composite material forms the basis of the mechanical properties of a laminate. In general anisotropy and spatial stress states, the generalized Hooke's law is used to establish the relationship between stress and strain. However, the material law of linear elasticity law describes the dependence between the stress and strain of a material only at considerably low temperatures. With increasing temperatures, pure elastic, time-independent deformation is complemented by different hereditary effects, such as viscoelasticity (Kerner 1956). Different deformation processes (e.g. quasi-static loading, short-term and long-term creep, stress relaxation and cyclic loading) can be described by equations with a definite set of parameters (Rabotnov 1980). This approach is known as hereditary mechanics.[1]

Composite materials with polymer matrix demonstrate pronounced time-dependent behavior. Therefore, to make a prediction of long-term performance of structural members, hereditary effects must be accounted for. Principally, there are two distinct physical mechanisms of hereditary behavior of composite materials. First, there is the "true creep" in the matrix material, because the matrix exhibits distinct viscoelastic behavior, typical for polymer materials (Tschoegl 1990). For temperatures closer to the glass transition temperature, deformation of polymers is accompanied by homogeneous viscous creep, with strain rate proportional to the applied stress. Second, there is "micro-fracture-induced" creep due to dilute volume local breakage of fibers and matrix. The damage depends on loading history and accumulates over time. Dilute damage reduces the mean stiffness of the composite material and finally leads to time-dependent elongation of the material in the direction of load application (Herakovich and Hyer 1986; Gao 1994).

1 Hereditary mechanics: "The field of mechanics involving boundary conditions extending over continuous intervals of space and time and demanding integrals for their representation. For example, in the application of stress to a deformable elastic medium, the final strain at any instant depends not only on the stress at that instant but on the whole previous stress to which the medium has been exposed." (Van Nostrand's Scientific Encyclopedia 2005).

Design and Analysis of Composite Structures for Automotive Applications: Chassis and Drivetrain, First Edition. Vladimir Kobelev.

Companion website: www.wiley.com/go/kobelev/automotive_suspensions

Generally speaking, both mechanisms overlap and the total creep in the composite material is caused by both effects. In composite materials, which are reinforced with continuous brittle fibers, fiber breakage due to axial tension sporadically occurs. The axial stress in broken fibers changes from zero on a ruptured cross-section to an average stress of intact fibers. The increase of axial stress happens along the scale of the plastic zone length l_p, as shown in Chapter 3, Eq. (3.37). In the vicinity of fiber rupture, the shear stress in the matrix transfers the axial stress in a fiber to the neighboring intact fibers. In this region, strong creep accompanies the pure elastic shear deformation in the polymer matrix. The fibers slide along their interface surfaces and matrix-induced creep occurs. The creep deformation of cylindrical cell containing a broken fiber by assuming the perfect bonding at fiber-matrix separation surface was studied by Lifshitz and Rotem (1970). The cited article found the linear viscoelastic solution that describes this mechanism by means of a time-dependent integral transform. The power law creep of matrix equations were established by Mason et al. (1992). The mechanisms of matrix-induced creep were intensively studied using analytical and experimental methods (Du and McMeeking 1995; Miyake et al. 1998; Ohno et al. 2001). The investigation of a composite under creep loads applied in the transverse direction was performed by Kruch et al. (2001). The model was based on a micromechanical approach performed on a hexagonal unit cell representative of the internal microstructure.

This chapter investigates the hereditary effects in a phenomenological way. Namely, the time-dependent deformation law of the material and its coefficients will be presumed, for example, acquired from the experiments. The phenomenological, anisotropic hereditary laws will be used for the analysis of some types of composite springs and torque-loaded shafts.

12.1.2 Linear Viscoelasticity of Composites

In the simplest linear formulation, the equations of the Hooke's law can be converted to the viscoelastic formulations in order to characterize the composite, since the properties of the material are frequency-dependent (Gibson 2016). The independent constants in the elastic formulations are no longer constants in viscoelastic formulations. Instead, they are frequency-dependent variables that depend upon the load frequency ω. If a prescribed strain ε is applied to a linearly viscoelastic material, the stress-strain relationship can be derived as (Findley et al. 1989):

$$\sigma(t) = \int_0^t E(t-\vartheta)\dot{\varepsilon}(\vartheta)d\vartheta, \tag{12.1}$$

where $\sigma(t)$ is time-dependent stress; $E(t)$ is relaxation modulus; $\dot{\varepsilon}(\vartheta)$ is prescribed strain rate; and $\vartheta < t$ is a dummy variable.

The Laplace transform of the function $f(t)$ is:

$$\bar{f}(\vartheta) = \mathcal{L}\{f(t)\} = \int_0^\infty \exp(-\vartheta t)f(t)dt. \tag{12.2}$$

Applying the Laplace transform to equations of Hooke's law yields:

$$\bar{\boldsymbol{\sigma}}(\vartheta) = \vartheta\bar{\mathbf{C}}(\vartheta)\cdot\boldsymbol{\varepsilon}(\vartheta). \tag{12.3}$$

According to the correspondence principle, elastic solutions can be converted into viscoelastic solutions by taking the Laplace transform of the time-dependent stress (or force) and strain (or displacement) and taking the Carson transform and Poisson's ratio (Graham 1968; Park and Schapery 1997; Lakes and Wineman 2006).

12.1.3 Nonlinear Creep Mechanics of Anisotropic Materials

However, it is well known that realistic models of hereditary mechanics of composites require nonlinear formulations. The strain rate depends on creep time on one hand, and on applied stress in the material on the other. Moreover, the parameters of nonlinear creep laws highly depend on temperature (McKeen 2009). A summary of creep laws for isotropic and anisotropic engineering materials is provided in Naumenko and Altenbach (2007). The constitutive models and solution methods for creep problems are discussed in Kassner (2008). The monograph by Guedes (2011) discourses the fundamental concepts of viscoelastic behavior of polymers and polymeric composites as well as respective modeling for both linear and nonlinear behavior. The nonlinear viscoelastic behavior of polymers and polymeric composites and different predicting models and methods along with specific applications are presented.

When a composite is subjected to a destructive compressive load, it experiences a sequence of three deformation stages: primary, secondary and tertiary stages. Most of the research in the literature has focused on plastic deformation in the primary and secondary stages, or the prediction of the flow number, which is the initiation of the tertiary stage. Common material models adequately describe the primary and secondary creep stage from constant stress uniaxial tests. These phenomenological models represent an attempt to approximate experimentally acquired data in the simplest formula. From the mechanical viewpoint, the phenomenological models must characterize stress, time and temperature dependences. Among others, the most widespread creep constitutive model is the Norton–Bailey law. This law provides a power law relationship between creep rate and stress.

The main objectives of this paragraph are to provide a mechanistic characterization method for damage modeling of a composite in the primary stage and to provide a method to determine anisotropic damage densities used in effective stress during nonlinear anisotropic viscous-plastic modeling.

The creep component of strain rate is defined by a material specific creep law. The models are based on the time- and strain-hardening constitutive equations for stress that depend on time, t.[2] For the spring element, a typical temperature is assumed to be prescribed. Therefore, the parameters of the material law correspond to a certain temperature. In this chapter, we adopt, following the common procedure (Hillier 1973; Betten 2008; Altenbach 2003), the general expression creep rates of an anisotropic material:

$$\dot{\varepsilon}_{ij} = \lambda \frac{\partial f(\sigma_{ij})}{\partial \sigma_{ij}}, \lambda > 0. \tag{12.4}$$

An anisotropic creep function $f(\sigma_{ij})$ is a homogeneous function of stress that satisfies the conditions:

$$\sum_{1 \le i,j \le 3} \sigma_{ij} \frac{\partial f(\sigma_{ij})}{\partial \sigma_{ij}} = f(\sigma_{ij}), \tag{12.5}$$

2 For material thickness the symbol t_c is used. The symbol t is reserved for the time variable.

$$\dot{\varepsilon}_{ij}\sigma_{ij} = \lambda f \geq 0. \tag{12.6}$$

The "over-dot" denotes the time derivative:

$$\dot{e}_{ij} = de_{ij}/dt, \dot{\varepsilon}_{ij} = d\varepsilon_{ij}/dt. \tag{12.7}$$

Hereafter, the incompressible creep behavior of a material ($\dot{\varepsilon}_{kk} = 0$) is presumed. In Eq. (12.7), the following notations are used:

$$\dot{e}_{ij} = \dot{\varepsilon}_{ij} - \frac{1}{3}\dot{\varepsilon}_{kk}\delta_{ij} \tag{12.8}$$

is the deviatoric component of creep strain rate;

$$s_{ij} = \sigma_{ij} - \frac{1}{3}\sigma_{kk}\delta_{ij} \tag{12.9}$$

is the deviatoric component of stress. The Mises–Hill creep potential in an isotropic case is the scalar function of the second invariant of a stress tensor. The second invariant of the stress tensor is identically equal to the second invariant of the deviator of a stress tensor. Accordingly, the creep potential is the function of the second invariant of the deviator of a stress tensor:

$$f = f(s_{ij}s_{ji}/2). \tag{12.10}$$

In the anisotropic case, the creep potential is the scalar function of the positive-definite quadratic form of components of a stress tensor (Tsvelodub 1991; Banshchikova et al. 2016):

$$f = f(K_{ijkl}\sigma_{ij}\sigma_{kl}/2), K_{ijkl} = K_{jikl} = K_{ijlk} = K_{jilk}. \tag{12.11}$$

From the conditions of creep incompressibility, the equations follow:

$$\begin{aligned} -2K_{1212} &= K_{1111} + K_{2222} - K_{3333}, \\ -2K_{2233} &= K_{2222} + K_{3333} - K_{1111}. \\ -2K_{3311} &= K_{3333} + K_{1111} - K_{2222}. \end{aligned} \tag{12.12}$$

If the material possesses creep orthotropic in a certain coordinate system, the number of non-vanishing coefficients is reduced. In a coordinate system, which is associated with the direction of fibers of a uniaxial fiber-reinforced material, the non-zero coefficients are:

$$\begin{aligned} &K_{1111} = \widehat{G} + \widehat{H}, K_{2222} = \widehat{H} + \widehat{F}, K_{3333} = \widehat{F} + \widehat{G}, \\ &K_{1122} = -\widehat{H}, K_{2233} = -\widehat{F}, K_{3311} = -\widehat{G}, \\ &K_{1212} = \widehat{N}, K_{2323} = \widehat{L}, K_{3131} = \widehat{M}. \end{aligned} \tag{12.13}$$

The expression (12.11) for the anisotropic creep potential transforms to a quadratic form that looks similar to the expression of the Mises–Hill failure criterion:

$$\begin{aligned} &f = f(\sigma_H), \\ &\sigma_H^2 = \widehat{F}(\sigma_{22} - \sigma_{33})^2 + \widehat{G}(\sigma_{33} - \sigma_{11})^2 \\ &\quad + \widehat{H}(\sigma_{11} - \sigma_{22})^2 + 2\widehat{L}\sigma_{23}^2 + 2\widehat{M}\sigma_{31}^2 + 2\widehat{N}\sigma_{12}^2. \end{aligned} \tag{12.14}$$

The value σ_H will be used as the equivalent anisotropic creep stress. In an isotropic case, this value reduces to $\sigma_H^2 = s_{ij}s_{ji}/2$. The coefficients of the Mises–Hill quadratic form must be evaluated in uniaxial experiments for creep.

The advantage of this form of the anisotropic creep potential is the following. If the coordinate system varies, the coefficients for anisotropic creep potential in the altered coordinate system transform as the components of the fourth rank tensor **K**. The transformation law of tensor **K** is identical to the transformation law for the coefficients of the compatibility tensor **S** and of the Mises–Hill failure criterion. These transformation laws were reported in Chapters 1 and 2.

We derive the expressions for strain rate for uniaxial and shear stress states for the certain representations of the stress function.

12.1.4 Anisotropic Norton–Bailey Law

Consider the anisotropic Norton–Bailey law (Robinson et al. 2003a,b). The stress function of the anisotropic creep potential in this case reads:

$$f(\sigma_H, t) = \overline{\varepsilon} \cdot (\sigma_H/\overline{\sigma})^{\xi+1}, \tag{12.15}$$

where $\overline{\varepsilon}$, $\overline{\sigma}$ and ξ are the experimental constants.

Consider two cases of the stress state. First, for the uniaxial stress state there is only non-vanishing component of stress tensor

$$\sigma_{11} = \sigma, \sigma_H = \sqrt{\widehat{G} + \widehat{H}}\ \sigma. \tag{12.16}$$

Correspondingly, the non-vanishing components of strain rate are

$$\dot{\varepsilon} \equiv \dot{\varepsilon}_{11} = -2\dot{\varepsilon}_{22} = -2\dot{\varepsilon}_{33}, \tag{12.17}$$

where

$$\dot{\varepsilon} = \overline{\varepsilon} \cdot (\sigma_H/\overline{\sigma})^{\xi+1} = \overline{\varepsilon} \cdot (\sqrt{\widehat{G} + \widehat{H}}\ \sigma/\overline{\sigma})^{\xi+1}. \tag{12.18}$$

For brevity of equations, we introduce the material constant

$$c_\sigma = \overline{\sigma}^{-\xi-1}(\widehat{G} + \widehat{H})^{\frac{\xi+1}{2}}\overline{\varepsilon}. \tag{12.19}$$

With this constant, the dependence of uniaxial strain rate upon stress reads

$$\dot{\varepsilon} = c_\sigma \sigma^{\xi+1}. \tag{12.20}$$

Second, for the pure shear stress state

$$\sigma_{12} = \sigma_{21} = \tau, \sigma_H = \sqrt{2\widehat{N}}\sigma \tag{12.21}$$

the non-vanishing components of deformation rate are

$$\dot{\varepsilon}_{12} = \dot{\varepsilon}_{21} \equiv \dot{\gamma}/2 = \overline{\gamma} \cdot (\sqrt{2\widehat{N}}\ \tau/\overline{\tau})^{\xi+1}/2, \tag{12.22}$$

where $\overline{\gamma} = \sqrt{3\overline{\varepsilon}}$ and $\overline{\tau} = \overline{\sigma}/\sqrt{3}$.

The anisotropic creep constant for shear strain reads:

$$c_\tau = \overline{\tau}^{-\xi-1}\overline{\gamma}(2\widehat{N})^{\frac{\xi+1}{2}}. \tag{12.23}$$

The anisotropic Norton–Bailey creep law for pure shear deformation reduces to

$$\dot{\gamma} = c_\tau \tau^{\xi+1}. \tag{12.24}$$

There is a simple relation between the constants in Eqs. (12.19) and (12.23):

$$c_\tau = 3^{\xi/2+1} c_\sigma. \tag{12.25}$$

12.2 Creep and Relaxation of Twisted Composite Shafts

12.2.1 Constitutive Equations for Relaxation in Torsion of Anisotropic Shafts

The deformation of a body during relaxation does not alter, but the stress gradually reduces. Consider the relaxation problem for a driveshaft with circular cross-section under the constant twist. Let $\tau(r, t)$ be the shear stress in the cross-section of shaft. The total shear strain in any instant of the time is $\gamma(r, t)$, the sum of the elastic and the creep components of shear strain:

$$\gamma = \gamma_e + \gamma_c. \tag{12.26}$$

The creep component of shear strain is $\gamma_c(r, t)$. The elastic component of shear strain is

$$\gamma_e = \tau / C_{44}. \tag{12.27}$$

First, in this chapter, we investigate the creep for the total deformation that remains constant in time. Thus, we consider the total strain $\gamma_0(r)$ as function of radius only. However, the elastic and the creep components of strain are the functions as well of radius and of time, such that:

$$\gamma(r, t) = \gamma_e(r, t) + \gamma_c(r, t) \equiv \gamma_0(r). \tag{12.28}$$

The time differentiation of (12.26) leads to the differential equation for elastic and creep strain rates:

$$\dot{\gamma}(r, t) = \dot{\gamma}_e(r, t) + \dot{\gamma}_c(r, t) \equiv 0, \tag{12.29}$$

where the dot denotes the time derivative.

The differentiation of the Eq. (12.27) over time delivers the elastic component of strain rate

$$\dot{\gamma}_e = \dot{\tau} / C_{44}. \tag{12.30}$$

12.2.2 Torque Relaxation for an Anisotropic Norton–Bailey Law

First, we assume the Norton–Bailey law for the state of shear stress:

$$\dot{\gamma}_c(r, t) = c_\tau \tau^{\xi+1}. \tag{12.31}$$

The substitution of material law (12.31) into Eq. (12.29) results in the ordinary non-linear differential equation of the first order for total shear stress $\tau(r, t)$:

$$\dot{\tau} / C_{44} + c_\tau \tau^{\xi+1} = 0. \tag{12.32}$$

The initial condition for the Eq. (12.32) presumes the pure elastic shear stress in the initial moment $t = 0$:

$$\tau(\rho, t = 0) = \tau_0(\rho). \tag{12.33}$$

The shear stresses in the moment $\mathbf{t} = 0$ for the driveshaft with circular cross-section of the diameter $d = 2r$ are

$$\tau_0(\rho) = C_{44}\theta\rho, 0 < \rho < r,$$

where θ is the twist angle per unit of length. The torque at the moment $t = 0$ is

$$M_T^0 = C_{44}I_T\theta.$$

The solution of the ordinary differential Eq. (12.32) with initial condition (12.33) delivers the shear stress over the cross-section of the twisted shaft as the function of time and radius:

$$\tau(\rho, t) = [\tau_0^{-\xi}(\rho) + c_\tau\, C_{44}\, \xi t\,]^{-1/\xi}. \tag{12.34}$$

The couple as the function of time is

$$M_T(t) = 2\pi \int_0^r r^2\tau(\rho, t)d\rho. \tag{12.35}$$

With the expression for total shear stress (12.35), we can calculate the couple

$$M_T(t) = 2\pi C_{66}\theta \int_0^r \rho^2[\rho^{-\xi} + c_\tau\, C_{44}{}^{\xi+1}\theta^\xi \xi t]^{-1/\xi} d\rho. \tag{12.36}$$

The integral could be expressed in terms of hypergeometric function (Kobelev 2014):

$$\frac{M_T(t)}{M_T^0} = \Phi_T(t) \equiv {}_2\mathbf{F}_1\left(\frac{4}{\xi}, \frac{1}{\xi}; \frac{4+\xi}{\xi}; -c_\tau\, \theta^\xi C_{44}{}^{\xi+1}\, \xi r^\xi t\right) \tag{12.37}$$

12.3 Creep and Relaxation of Composite Helical Coiled Springs

12.3.1 Compression and Tension Composite Springs

Helical springs are made up of a wire coiled in the form of a helix and are primarily intended for compressive or tensile loads. The cross-section of the wire from which the spring is made may be circular, square or rectangular. Compression and tension springs are made of an anisotropic elastic wire. Consequently, one essential task of creep analysis is the derivation of the exact closed-form expressions for torsion and bending creep for anisotropic materials. The major stresses produced in conical and volute springs are also shear stresses due to twisting. The stress in the circular wire varies from zero on the axis of the wire to its maximum value on the surface.

Consider a helical composite spring with mean helix diameter D, wire diameter d and active coils number n. The force $F_z(t)$ applied to a spring that causes a deflection s is

$$F_z(t) = 2M_T(t)/D. \tag{12.38}$$

The relaxation of helical composite springs occurs when the compressed length of the spring remains constant. The spring force $F_z(t)$ reduces over time. During the relaxation the twist angle per unit length remains constant:

$$\theta = 2s \cdot (\pi n D^2)^{-1}. \tag{12.39}$$

The spring force at initial moment $t = 0$ due to pure elastic deformation is:

$$F_z^0 = 2M_T^0/D = C_{44}d^4s/8nD^3, C_{44} = G_{TL}. \tag{12.40}$$

Alternatively, the creep of helical composite springs occurs when the spring force remains constant. In this situation, the compressed length continuously reduces over time and the deflection increases.

12.3.2 Relaxation of Helical Composite Springs

For the solution of the relaxation problem, we apply the results of the relaxation problem for a twisted driveshaft with the circular cross-section for helical composite springs. If the material of the helical composite spring obeys the Norton–Bailey law, the spring load as a function of time is (Kobelev 2014):

$$F_z(t) = {}_2\mathbf{F}_1\left(\frac{4}{\xi}, \frac{1}{\xi}; \frac{4+\xi}{\xi}; -\frac{c_\tau\, d^\xi \theta^\xi C_{44}{}^{\xi+1}\, \xi\, t}{2^\xi}\right) F_z^0, \tag{12.41}$$

With the constant value for θ, the expression (12.41) yields

$$F_z(t) = {}_2\mathbf{F}_1\left(\frac{4}{\xi}, \frac{1}{\xi}; \frac{4+\xi}{\xi}; -c_\tau\, C_{44}\xi\overline{\tau}t\right) F_z^0. \tag{12.42}$$

The further simplification expressions for spring load as the function of time is achieved for certain exponents in the Norton–Bailey law:

$$F_z(t) = \frac{4}{3}\frac{(1 + c_\tau C_{44}t^4\overline{\tau}^4)^{3/4} - 1}{c_\tau C_{44}t^4\overline{\tau}^4} F_z^0 \text{ for } \xi = 4, \tag{12.43}$$

and

$$F_z(t) = \frac{2}{3}\frac{1 + (c_\tau C_{44}t^2\overline{\tau}^2 - \varsigma)\sqrt{1 + 2c_\tau C_{44}t^2\overline{\tau}^2}}{c_\tau{}^2 C_{44}{}^2 t^4\overline{\tau}^4} F_z^0 \text{ for } \xi = 2. \tag{12.44}$$

The relaxation of spring load over time demonstrates the dimensionless relaxation function:

$$\frac{F_z(t)}{F_z^0} = \Phi_H(t) \equiv {}_2\mathbf{F}_1\left(\frac{4}{\xi}, \frac{1}{\xi}; \frac{4+\xi}{\xi}; -c_\tau\, C_{44}\, \xi\, \overline{\tau}t\right). \tag{12.45}$$

12.3.3 Creep of Helical Composite Compression Springs

During the creep deformation of anisotropic springs, the spring force remains constant over time. Correspondingly, the spring length reduces with time. The length decrease rate over time is $\dot{s}$. The shear strain rate varies with time:

$$\dot{\gamma}(t, \rho) = \dot{\theta}\rho = 2\dot{s}\rho/\pi n D^2 \text{ for } 0 < \rho < d/2. \tag{12.46}$$

According to the anisotropic Norton–Bailey law, the shear stress due to creep is:

$$\tau = \left(\frac{\dot{\gamma}}{c_\tau}\right)^{\frac{1}{\xi+1}} = \left(\frac{2\dot{s}\rho}{c_\tau \pi n D^2}\right)^{\frac{1}{\xi+1}}. \tag{12.47}$$

Performing the integration over the area of the wire, we get the moment due to creep:

$$M_T^0 = 2\pi \int_0^r \rho^2 \tau(\rho, t) dr = \frac{\pi}{4}\frac{\xi+1}{4+3\xi}\left(\frac{2d^{4+3\xi}\dot{s}}{\pi D^2 n c_\tau}\right)^{\frac{1}{1+\xi}}. \tag{12.48}$$

Solution of this equation delivers the spring length reduction rate for constant spring force F_z^0 as the function of time:

$$\dot{s} = \frac{\pi D^2 n c_\tau}{2d^{4+3\xi}}\left(\frac{4M_T^0}{\pi}\frac{4+3\xi}{\xi+1}\right)^{\xi+1} = \frac{\pi D^2 n c_\tau}{2d^{4+3\xi}}\left(\frac{4DF_z^0}{2\pi}\frac{4+3\xi}{\xi+1}\right)^{\xi+1}. \tag{12.49}$$

12.4 Creep and Relaxation of Composite Springs in a State of Pure Bending

12.4.1 Constitutive Equations for Bending Relaxation

In this section, we study the relaxation in the state of pure bending of a leaf or meander spring of rectangular cross-section ($b_c \times t_c$). In the applied Euler–Bernoulli theory of slender beams subjected to a bending moment M_B, a major assumption is that the plane cross-sections remain plane during the relaxation process. The deformations due to shear across the section are not accounted for. During the relaxation, the curvature of the neutral axis of spring κ remains constant over time, such that the bending moment M_B continuously decreases. This case describes the relaxation of bending stress, assuming that the flexure deformation of leaf spring does not alter in time.

Let $\sigma(z, t)$ is normal stress in the beam in the direction of the beam axis. The total strain in any instant of the time is $\varepsilon(z, t)$; the sum of the elastic and the creep components of the strain is:

$$\varepsilon = \varepsilon_e + \varepsilon_c. \tag{12.50}$$

The elastic component of strain is

$$\varepsilon_e = \sigma / C_{33}, \tag{12.51}$$

where $\varepsilon_c(z, t)$ is the creep partial strain. Consider creep as a constant in time total strain:

$$\varepsilon(z, t) = \varepsilon_e(z, t) + \varepsilon_c(z, t) \equiv \varepsilon_0(z). \tag{12.52}$$

The normal strain $\varepsilon_0(z) = \varepsilon(z, t = 0)$ is the function of normal coordinate, but remains constant over time. The time differentiation of (12.52) leads to

$$\dot{\varepsilon} = \dot{\varepsilon}_e + \dot{\varepsilon}_c \equiv 0. \tag{12.53}$$

12.4.2 Relaxation of the Bending Moment for the Anisotropic Norton–Bailey Law

The anisotropic Norton–Bailey law for a uniaxial state of stress reads

$$\dot{\varepsilon}_c(z,t) = c_\sigma \sigma^{\xi+1}, \tag{12.54}$$

The substitution of material laws results in the ordinary differential equation for uniaxial stress

$$\dot{\sigma}/C_{33} + c_\sigma \sigma^{\xi+1} = 0. \tag{12.55}$$

The initial condition for Eq. (12.55) delivers the pure elastic shear stress in the initial moment

$$\sigma(z, t = 0) = \sigma_0(z). \tag{12.56}$$

For pure elastic bending, the following initial distribution of stresses over the cross-section of a leaf spring is valid:

$$\sigma_0(z) = C_{33}\kappa z, C_{33} = E_L.$$

where κ is the bending curvature, which is presumed to be constant over time and z is the perpendicular distance to the neutral axis. The solution of the ordinary differential Eq. (12.55) with initial condition (12.56) is

$$\sigma(z,t) = [\sigma_0^{-\xi}(z) + c_\sigma\, C_{33}\, \xi\, t]^{-1/\xi}. \tag{12.57}$$

The bending moment for the rectangular cross-section of width b_c and height t_c is a function of time.

$$M_B(t) = b_c \int_{-t_c/2}^{t_c/2} z\sigma(z,t)dz. \tag{12.58}$$

With the expression (12.57) we can calculate from (12.58):

$$M_B(t) = 2C_{33}b_c\kappa \int_0^{t_c/2} z \cdot \left[\frac{1}{z^\xi} + c_\sigma\, \kappa^\xi C_{33}{}^{\xi+1}\, \xi t\right]^{-1/\xi} dz. \tag{12.59}$$

The integral in (12.59) could be expressed in terms of hypergeometric function (Kobelev 2014):

$$\frac{M_B(t)}{M_B^0} = \Phi_B(t) \equiv {}_2\mathrm{F}_1\left(\frac{3}{\xi}, \frac{1}{\xi}; \frac{3+\xi}{\xi}; -c_\sigma\, \kappa^\xi C_{33}{}^{\xi+1}\, \xi\, t\left(\frac{t_c}{2}\right)^\xi\right), \tag{12.60}$$

where $M_B^0 = C_{33}I_B\kappa = \kappa\, C_{33}b_c t_c{}^3/12$ is the elastic bending moment at the time $t = 0$.

12.4.3 Creep in a State of Bending

On the contrary, during the bending creep deformation the moment M_B^0 remains constant over time. The curvature $\kappa = \kappa(t)$ also continually increases. The elongation rate of the strip, which is located on the perpendicular distance z to the neutral axis, is:

$$\dot{\varepsilon}(t,z) = \dot{\kappa} z \text{ for } -t_c/2 < z < t_c/2. \tag{12.61}$$

According to Norton–Bailey law the shear stress due to creep is:

$$\sigma_c = \left(\frac{\dot{\varepsilon}}{c_\sigma}\right)^{\frac{1}{\xi+1}} = \left(\frac{\dot{\kappa}z}{c_\sigma}\right)^{\frac{1}{\xi+1}}. \tag{12.62}$$

Performing the integration (12.62) over the area of the wire, we get the bending moment due to creep (Kobelev 2014):

$$M_B^0 = b_c \int_{-t_c/2}^{t_c/2} z\sigma_c(z,t)dz = \frac{b_c t_c^{\,2}}{2}\frac{\xi+1}{3+2\xi}\left(\frac{t_c\dot{\kappa}}{2c_\sigma}\right)^{\frac{1}{1+\xi}}. \tag{12.63}$$

The bending moment M_B^0 remains constant over time. From Eq. (12.63) the curvature time rate resolves as

$$\dot{\kappa} = \frac{2c_\sigma}{t_c}\left(\frac{2M_B^0}{b_c t_c^{\,2}}\frac{3+2\xi}{\xi+1}\right)^{\xi+1}. \tag{12.64}$$

The rate of flexure $\dot{\kappa}$ remains constant such that the bending radius linearly increases over time.

The results of this section are applicable for anisotropic springs that are overwhelmingly stressed by bending loads, like leaf springs and helical springs, when loaded by the axial twist moment. The formulas of this section could be instantly applied to estimate the effects of creep and relaxation of such springs.

12.5 Conclusions

The structures examined comprised a beam in bending, a driveshaft in torsion and helical composite springs. The closed-form solutions demonstrated the basic characteristics of nonlinear creep in anisotropic material. Bending and torsion dominate most spring elements and driveshafts. Following the procedure, we established closed-form solutions for creep and relaxation in twisted driveshafts, helical and leaf springs made of composite materials.

For further reading, the reader is directed to Movahedi-Rad et al. (2016), Monfared et al. (2012), Carmai and Dunne (2001), Kawai (2001), Abdolavahid et al. (2016), Vinogradov and Schumacher (2001), Elanchezhian et al. (2014), Daniel (2014), Daniel et al. (2011), Sevenois and Van Paepegem (2015), Kaminski et al. (2015), Brinson and Gates (1994), Naumenko (2006), Chou (1992), Mises (1928), Hill (1950) and Timoshenko (1948).

References

Abdolavahid M.-R., Keller T., Vassilopoulos A.P. (2016) Creep-fatigue interaction in composite materials, in *Eccm17 – 17th European Conference on Composite Materials*, Munich, Germany, June 26–30, 2016.

Altenbach, H. (2003). Topical problems and applications of creep theory. *Int. Appl. Mech.* 39 (6): 631–655.

Banshchikova, I.A., Petrov, D.M., and Tsvelodub, I.Y. (2016). Torsion of circular rods at anisotropic creep. *J. Phys. Conf. Ser.* 722: 012004.

Betten, J. (2008). *Creep Mechanics*, 367. Berlin Heidelberg: Springer-Verlag.

Brinson L.C., Gates T.S. (1994) Effects of Physical Aging on Long-Term Creep of Polymers and Polymer Matrix Composites, *NASA Technical Memorandum 109081*, Langley Research Center, Hampton, Virginia 23681-0001.

Carmai, J. and Dunne, F.P.E. (2001). Micromechanical models for creep in the consolidation of composites. In: *IUTAM Symposium on Creep in Structures* (ed. S. Murakami and N. Ohno), 463–468. Kluwer Academic Publishers.

Chou, T.-W. (1992). *Microstructural Design of Fiber Composites*. Cambridge: Cambridge University Press.

Daniel, I.M. (2014). Failure of composite materials under multi-axial static and dynamic loading international symposium on dynamic response and failure of composite materials, DRaF2014. *Procedia Eng.* 88 (2014): 10–17.

Daniel, I.M., Werner, B.T., and Fenner, J.S. (2011). Strain-rate-dependent failure criteria for composites. *Compos. Sci. Technol.* 71: 357–364.

Du, Z.-Z. and McMeeking, R.M. (1995). Creep models for metal matrix composites with long brittle fibers. *J. Mech. Phys. Solids* 43: 701–726.

Elanchezhian C., Vijaya Ramnath B., Hemalatha J. (2014) Mechanical behaviour of glass and carbon fiber reinforced composites at varying strain rates and temperatures, in: *3rd International Conference on Materials Processing and characterisation (ICMPC 2014), Proc. Mat. Sci.*, 6, 1405–1418.

Findley, W.N., Lai, J.S., and Onaran, K. (1989). *Creep and Relaxation of Nonlinear Viscoelastic Materials*. New York: Dover Publications.

Gao Z. (1994) Fatigue stiffness reduction and failure of composite laminates, in: *Proceedings of the International Mechanical Engineering Congress, Chicago, IL, ASME DE-79*, pp. 37–42.

Gibson, R.L. (2016). *Principles of Composite Material Mechanics*. Boca Raton: CRC Press.

Graham, G.A.C. (1968). The correspondence principle of linear viscoelasticity theory for mixed boundary value problems involving time-dependent boundary regions. *Q. Appl. Math.* 26: 167–174.

Guedes, R.M. (ed.) (2011). *Creep and Fatigue in Polymer Matrix Composites*. Philadelphia, PA, USA: Woodhead Publishing Limited.

Herakovich, C.T. and Hyer, M.W. (1986). Damage-induced property changes in composites subjected to cyclic thermal loading. *Eng. Fract. Mech.* 25 (5/6): 779–791.

Hereditary Mechanics (2005) In: *Van Nostrand's Scientific Encyclopedia*, G.D. Considine (ed.). doi:https://doi.org/10.1002/0471743984.vse3896.

Hill, R. (1950). *Mathematical Theory of Plasticity*. Oxford: Clarendon Press.

Hillier, M.J. (1973). On anisotropic creep. *Nucl. Eng. Des.* 25 (3): 438–440.

Kaminski, M., Laurin, F., Maire, J.F. et al. (2015). Fatigue damage modeling of composite structures: the onera viewpoint. *Aerospace Lab.* 1–12. https://doi.org/10.12762/2015.AL09-06.

Kassner, M. (2008). *Fundamentals of Creep in Metals and Alloys*, 2e. Elsevier Science, 295 p.

Kawai, M. (2001). Off-axis creep behavior of unidirectional polymer matrix composites at high temperature. In: *IUTAM Symposium on Creep in Structures* (ed. S. Murakami and N. Ohno), 464–478. Kluwer Academic Publishers.

Kerner, E.H. (1956). The elastic and thermo-elastic properties of composite media. *Proc. Phys. Soc. Sect. B* 69 (8): 808.

Kobelev, V. (2014). Relaxation and creep in twist and flexure. *Multidiscipline Model. Mater. Struct.* 10 (3): 304–327. https://doi.org/10.1108/MMMS-11-2013-0067.

Kruch, S., Chaboche, J.L., and Carrere, N. (2001). Micromechanics based creep damage analysis of unidirectional metal matrix composites. In: *IUTAM Symposium on Creep in Structures* (ed. S. Murakami and N. Ohno), 453–462. Kluwer Academic Publishers.

Lakes, R.S. and Wineman, A. (2006). On Poisson's ratio in linearly viscoelastic solids. *J. Elast.* 85: 45. https://doi.org/10.1007/s10659-006-9070-4.

Lifshitz, J.M. and Rotem, A. (1970). Time-dependent longitudinal strength of unidirectional fibrous composites. *Fibre Sci. Tech.* 3: 1–20.

Mason, D.O., Hui, C.-Y., and Phoenix, S.L. (1992). Stress profiles around a fiber break in a composite with a nonlinear, power law creeping matrix. *Int. J. Solids Struct.*, 29: 2829–2854.

McKeen, L.W. (2009). *The Effect of Creep and Other Tme Related Factors on Plastics and Elastomers.* Oxford: Elsevier Inc.

von Mises, R. (1928). Mechanik der plastischen Formänderung von Kristallen. *Zeits. f. Angew. Math. u. Mech.* 8: 161.

Miyake, T., Yamakawa, T., and Ohno, N. (1998). Measurement of stress relaxation in broken fibers embedded in epoxy using Raman spectroscopy. *J. Mater. Sci.* 33: 5177–5183.

Monfared, V., Mondali, M., and Abedian, A. (2012). Steady-state creep analysis of polymer matrix composites using complex variable method. *Proc. Inst. Mech. Eng. C: J. Mech. Eng. Sci.* 227 (10): 2182–2194.

Movahedi-Rad A., Keller T., Vassilopoulos A.P. (2016) Creep-Fatigue Interaction. In: *Composite Materials, Eccm17 – 17th European Conference on Composite Materials,* Munich, Germany, June 26–30, 2016.

Naumenko K. (2006) *Modeling of High-Temperature Creep for Structural Analysis Applications,* Habilitation thesis, Martin-Luther University of Halle-Wittenberg.

Naumenko, K. and Altenbach, H. (2007). *Modelling of Creep for Structural Analysis.* Berlin: Springer.

Ohno, N., Ando, T., Miyake, T., and Biwa, S. (2001). Effects of matrix creep on fiber stress profiles in unidirectional composites: mesoscopic analysis based on a variational method. In: *IUTAM Symposium on Creep in Structures* (ed. S. Murakami and N. Ohno), 439–452. Kluwer Academic Publishers.

Park, S.W. and Schapery, R.A. (1997). A viscoelastic constitutive model for particulate composites with growing damage. *Int. J. Solids Struct.* 34: 931–947.

Rabotnov, Y.N. (1980). *Elements of Hereditary Solid Mechanics.* Moscow: Mir Publishers.

Robinson, D.N., Binienda, W.K., and Ruggles, M.B. (2003a). Creep of polymer matrix composites. I: Norton/Bailey creep law for transverse isotropy. *J. Eng. Mech. ASCE* 129 (3): 310–317.

Robinson, D.N., Binienda, W.K., and Ruggles, M.B. (2003b). Creep of polymer matrix composites. II: Monkman–Grant failure relationship for transverse isotropy. *J. Eng. Mech. ASCE* 129 (3): 318–323.

Sevenois, R.D.B. and Van Paepegem, W. (2015). Fatigue damage modeling techniques for textile composites: review and comparison with unidirectional composite modeling techniques. *Appl. Mech. Rev. ASME* 67 (2): 020802–020802-12.

Timoshenko, S. (1948). *Strength of Materials.* Toronto, NY, London: Van Nostrand.

Tschoegl, N.W. (1990). Viscoelasticity. In: *The Encyclopedia of Physics* (ed. R.M. Besancon). Springer Science+Business Media, LLC.

Tsvelodub, I.Y. (1991). *Stability Postulate and its Applications in the Creep Theory of Metallic Materials*. Novosibirsk: Institute of Hydrodynamics.

Vinogradov, A.M. and Schumacher, S. (2001). Cyclic creep of polymers and polymer-matrix composites. *Mech. Compos. Mater.* 37 (1): 29–34.

Appendix A

Mechanical Properties of Composites

A.1 Fibers

A.1.1 Glass Fibers

Glass fiber was the first kind of fiber that was widely used as a reinforcement fiber. The so-called E-glass, with over 90% of all glass fibers used for reinforced plastics has the greatest share. Glass fiber has an isotropic behavior. The modulus of elasticity is approximately 70 GPa, as large as that of aluminum and about a third of the modulus of elasticity for steel. The tensile strength is nearly 2.4 GPa and clearly below that of carbon fibers. This value is definitely higher than the strengths of aluminum and most steel alloys. The density of glass fibers is roughly 2600 kg m^{-3}.

A.1.2 Carbon Fibers

Carbon fibers are currently the most widely spread reinforcement fibers for high quality parts. The density of carbon fibers is a relatively low 1600–2000 kg m^{-3}. Carbon fibers are highly anisotropic, like aramid fibers. The thermal expansion coefficient is also negative as it is for aramid fibers. The carbon fibers possess a high fatigue endurance limit. The dynamic characteristics of laminates based on carbon fibers for reinforcement dominate the dynamic characteristics of such materials as steel or aluminum. These fibers possess exceptional chemical and thermal properties and good resistance to corrosion. Strength, stiffness and thermal conductivity substantially differ depending on the type of fiber.

A.1.3 Aramid Fibers

Aramid fibers are characterized by the high-strength, low density, high impact and fatigue strength, good heat resistance and dimensional stability, good vibration damping and high-energy absorption capacity. The density of the aramid fibers, derived from aromatic polyamide, is lower (1450 kg m^{-3}) than the density of carbon fibers. The tensile strength is approximately 2700 MPa. Consequently, the aramid fibers possess remarkably high specific, density-related, strength. The application scope is chiefly ballistic and industrial protective devices. Aramid fiber plastic composites are

Design and Analysis of Composite Structures for Automotive Applications: Chassis and Drivetrain, First Edition. Vladimir Kobelev.

Companion website: www.wiley.com/go/kobelev/automotive_suspensions

Table A.1 Mechanical properties of fibers.

Mechanical properties of fibers				
	E-Glass	Carbon HT	Aramid HM	Basalt
Density (kg m^{-3})	2600	1780	1450	2750
Tensile strength (MPa)	3400	3400	2800	4840
Modulus of elasticity, along the fibers (GPa)	73	235	100	89
Modulus of elasticity, transverse to the fibers (GPa)	73	15	5.4	
Elongation at break (%)	3.5–4	1.4–1.5	2.8	3.15
Poisson ratio	0.18			
Coefficient of thermal expansion, along the fibers (10^{-6} K^{-1})	5	−0.1	−3.5	
Coefficient of thermal expansion, transverse to the fibers (10^{-6} K^{-1})	5	10	17	
Thermal conductivity (W m^{-1} K)	1	17	0.04	
Resistivity (Ω m)	10^{13}	10^{-5}–10^{-6}	10^{13}	
Dielectric constant (10^6 Hz)	5.6–6.7			
Degradation temperature (°C)	>400		550	1450
Moisture absorption, at 20 °C/65% relative air humidity (%)	0.1	0.1	3.5	
Ultimate axial stress σ_{fb} (GPa)	2.55–2.9	3.3–3.5	2.8	2.52

exceptionally suited for tensile-stressed lightweight components. However, the fibers are highly anisotropic and pressure sensitive in the direction of the fibers. This feature poses a problem for the design of components made of aramid reinforced plastic, which are predominantly subject to bending or compressive stresses. Aramid fibers feature an outstanding chemical resistance and low thermal conductivity. Starting at a temperature of approximately 400 °C, while the fibers do not melt they do begin to carbonize. Similar to carbon fibers, they also have a negative thermal expansion coefficient. Their environmental field of application is limited because the aramid fibers absorb moisture. Aramid fibers react to high-energy radiation with a clear reduction in strength, but this feature is gratefully irrelevant to automotive applications.

A.2 Physical Properties of Resin

The task of the resin (or matrix) is to fix the fibers in place and to protect them from environmental effects, such as chemicals or moisture. Regarding the mechanical properties of the fiber-reinforced plastic, the purpose of the matrix is to transfer the forces to the fibers or between the fibers and to bear them with respect to compressive stress. The widespread matrix materials are:

- Thermosetting plastics: epoxy resin (EP), polyurethane resin (PU), unsaturated polyester (UP), vinyl ester resin (VE), cyanate ester resin (CE) and phenolic resin (PF) (Kelly and Zweben 2000, Vol. 2), polymer matrix composites.
- Thermoplastics: polymethyl-methylacrylate (PMMA), polyamide (PA), polyetheretherketone (PEEK), polyphenyl sulfide (PPS), (Kelly and Zweben 2000, Vol. 2), polymer matrix composites.
- Metals: Aluminum alloys, sometimes magnesium or copper alloys (Kelly and Zweben 2000, Volume 3), metal matrix composites (Lenke et al. 2001).
- Ceramics (Kelly and Zweben 2000, Vol. 4), carbon/carbon, cement and ceramic matrix composites.

Table A.2 Mechanical properties of the non-reinforced matrix resins.[a)]

Mechanical properties of the non-reinforced matrix resins			
	Polyester	**Epoxy**	**Vinyl ester**
Density ($kg\,m^{-3}$)	1120–1250	1100–1250	1070
Fracture elongation[b)] (%)	<3	6–8	3.5–7
Processing temperature[c)] (°C)	20–180	20–230	20–175
Operating temperature (°C)	50–160	45–230	100–150
Tensile strength (MPa)	50–70	70–90	75–85
E-modulus (GPa)	3.5–4.7	2.8–3.6	3.4–3.5
Flexural strength (MPa)	60–120	140–160	125–135
Coefficient of thermal expansion (°μK)		60–70	
Heat conductivity ($°K\,W\,m^{-1}$)		0.52	
Surface resistance (Ω)		10^{13}	
Specific volume resistance (Ω , m)		10^{13}	
Viscosity[d)] (mPa s)		400 (HT2) 700 (L, L285) 900 (L20)	
Fracture toughness K ($MNm^{-3/2}$)	5	4	

a) Faserverbund-Werkstoffdaten, suter-kunststoffe AG, info@swiss-composite.ch • www.swiss-composite.ch.

b) The elongation at break of laminating resins must preferably be equal to or greater than the elongation at break of the reinforcing fibers, so that in the event of a load no component failure occurs due to fractures and cracks in the resin.

c) The highest value describes the maximum temperature required during hot hardening. This heat curing is only necessary for some tempering resins with a correspondingly high temperature resistance. The cold-curing resins, most of which are available, achieve almost their full strength at room temperature.

d) www.chemie.fu-berlin.de/fb/fachdid/kunststoffe/glas.htm; www.kunstharzputze.de/abc/kap3/g04.htm.

A.3 Laminates

A.3.1 Unidirectional Fiber-Reinforced Composite Material

Unidirectional plies are pertinent for high-grade fiber composite components, which require high-strength and also high-stiffness components. The unidirectional layers have a very low thread waviness, which has an advantageous effect on the strength and stiffness. The drawback of unidirectional laminas is the reduced cohesion between the individual layers, which can result in delamination and separation of laminates.

A.3.2 Fabric

There are several types of bonds with different technical properties. The frequently used types are the plain weave and twill weave. A plain weave primarily has points of intersection per square unit, which give it the known stability of shape. Because of the high shape retention, the drapability of this fabric on physical contours is limited; therefore, they do not slip and are firm. The twill weave has a better drapability, and is therefore easier to insert in sharply curved areas.

A.3.3 Non-Woven Fabric

Non-woven fabrics consist of two to four unidirectional plies of varied fiber orientation that are placed on top of each other. The fibers of the individual plies are secured by knitting threads or bonding adhesives. Field of application, advantages and disadvantages of non-woven fabrics are identical to those of unidirectional laminates. The advantage over unidirectional laminas is that with one non-woven ply there is an equal coverage to multiple fiber directions

Table A.3 Mechanical properties of fabric laminate.

Mechanical properties of fabric laminates			
	Glass	**Carbon**	**Aramid**
Fiber volume fraction (%)	43	50	50
Tensile strength, Bidirectional fabrics[a)] (DIN EN 61) (MPa)	330–400	560–650	460–540
Tensile strength, Unidirectional fabrics[b)] (DIN EN 61) (MPa)	590–680	950–1100	790–900
Tensile modulus, bidirectional fabrics (DIN EN 61) (GPa)	19–21	52–58	22–27
Tensile modulus, unidirectional fabrics (DIN EN 61) (GPa)	24–35	90–100	44–47
Compressive strength, bidirectional fabrics (DIN 534554) (MPa)	310–440	450–520	130–165
Compressive strength, unidirectional fabrics (DIN 534554) (MPa)	480–600	600–800	180–190
Heat conductivity (°K W m^{-1})	1	17	0.04–0.05

a) Bidirectional fabric (warp and weft are equal). The higher values are obtained with twill and sateen at the optimal laminate quality.

b) Unidirectional fabric. Values depend on proportion of warp to weft thread.

References

Kelly, A. and Zweben, C. (eds.) (2000). *Comprehensive Composite Materials*. Oxford: Pergamon Elsevier Ltd.

Lenke, I., Richter, G., and Rogowski, R. (2001). Ceramic engineering with performs for locally reinforced light metal components. In: *Ceramic (Hrsg.) Materials and Components for Engines* (ed. J. Heinrich and F. Aldinger), 383–386. Weinheim: Wiley-VCH Verlag GmbH.

Appendix B

Anisotropic Elasticity

B.1 Elastic Orthotropic Body

In linear elasticity theory, only small changes in shape are considered, and a macroscopic-
homogeneous, linear-elastic material is also assumed. Elastic deformations disappear completely after release. For orthotropic bodies the generalized Hooke's law applies with nine effective elasticity constants:

$$\varepsilon_{xx} = \frac{1}{E_1}\sigma_{xx} - \frac{\upsilon_{21}}{E_2}\sigma_{yy} - \frac{\upsilon_{32}}{E_3}\sigma_{zz}, \quad \varepsilon_{yy} = \frac{1}{E_2}\sigma_{yy} - \frac{\upsilon_{12}}{E_1}\sigma_{xx} - \frac{\upsilon_{32}}{E_3}\sigma_{zz},$$
$$\varepsilon_{zz} = \frac{1}{E_3}\sigma_{zz} - \frac{\upsilon_{13}}{E_1}\sigma_{xx} - \frac{\upsilon_{23}}{E_2}\sigma_{yy}, \quad \gamma_{yz} = \gamma_{zy} = \frac{1}{G_{23}}\tau_{yz} = \frac{1}{G_{32}}\tau_{zy}$$
$$\gamma_{zx} = \gamma_{xz} = \frac{1}{G_{31}}\tau_{zx} = \frac{1}{G_{13}}\tau_{xz}, \quad \gamma_{xy} = \gamma_{yx} = \frac{1}{G_{12}}\tau_{xy} = \frac{1}{G_{21}}\tau_{yx}. \tag{B.1}$$

E_1, E_2, E_3 modules of elasticity,
$G_{23} = G_{32},\ G_{13} = G_{31},\ G_{12} = G_{21}$ shear modules,
$\nu_{12},\ \nu_{21},\ \nu_{23},\ \nu_{32},\ \nu_{13},\ \nu_{31}$ Poisson numbers.

There is a relationship:

$$E_1\upsilon_{21} = E_2\upsilon_{12}, \quad E_2\upsilon_{32} = E_3\upsilon_{23}, \quad E_3\upsilon_{13} = E_1\upsilon_{31}. \tag{B.2}$$

The linear-elastic material behavior with respect to the associated material system can be characterized by nine elasticity constants C_{ij}, or the equivalent by nine stiffness constants S_{ij}:

$$\varepsilon_{xx} = S_{11}\sigma_{xx} + S_{12}\sigma_{yy} + S_{13}\sigma_{zz}, \quad \varepsilon_{yy} = S_{21}\sigma_{xx} + S_{22}\sigma_{yy} + S_{23}\sigma_{zz}$$
$$\varepsilon_{zz} = S_{31}\sigma_{xx} + S_{32}\sigma_{yy} + S_{33}\sigma_{zz}, \quad \gamma_{yz} = \gamma_{zy} = S_{44}\tau_{yz} = S_{44}\tau_{zy},$$
$$\gamma_{zx} = \gamma_{xz} = S_{55}\tau_{zx} = S_{55}\tau_{xz}, \quad \gamma_{xy} = \gamma_{yx} = S_{66}\tau_{xy} = S_{66}\tau_{yx}. \tag{B.3}$$

The inverse of (B.3) is:

$$\sigma_{xx} = C_{11}\varepsilon_{xx} + C_{12}\varepsilon_{yy} + C_{13}\varepsilon_{zz}, \quad \sigma_{yy} = C_{21}\varepsilon_{xx} + C_{22}\varepsilon_{yy} + C_{23}\varepsilon_{zz}$$
$$\sigma_{zz} = C_{31}\varepsilon_{xx} + C_{32}\varepsilon_{yy} + C_{33}\varepsilon_{zz}, \quad \tau_{yz} = \tau_{zy} = C_{44}\gamma_{yz} = C_{44}\gamma_{zy},$$
$$\tau_{zx} = \tau_{xz} = C_{55}\gamma_{zx} = C_{55}\gamma_{xz}, \quad \tau_{xy} = \tau_{yx} = C_{66}\gamma_{xy} = C_{66}\gamma_{yx}. \tag{B.4}$$

Design and Analysis of Composite Structures for Automotive Applications: Chassis and Drivetrain, First Edition. Vladimir Kobelev.

Companion website: www.wiley.com/go/kobelev/automotive_suspensions

The relations between the elasticities C_{ij} and stiffnesses S_{ij} read:

$$
\begin{aligned}
&S_{11} = E_1^{-1}, S_{22} = E_2^{-1}, S_{33} = E_3^{-1}, S_{12} = S_{21} = -\upsilon_{21}E_2^{-1} = -\upsilon_{12}E_1^{-1}\\
&S_{23} = S_{32} = -\upsilon_{32}E_3^{-1} = -\upsilon_{23}E_2^{-1}, S_{31} = S_{13} = -\upsilon_{13}E_1^{-1} = -\upsilon_{31}E_3^{-1},\\
&S_{44} = G_{23}^{-1} = G_{32}^{-1}, S_{55} = G_{31}^{-1} = G_{13}^{-1}, S_{66} = G_{12}^{-1} = G_{21}^{-1},
\end{aligned}
\tag{B.5}
$$

and

$$
\begin{aligned}
&C_{11} = \frac{E_1(\upsilon_{23}\upsilon_{32} - 1)}{\Delta}, \quad C_{22} = \frac{E_2(\upsilon_{13}\upsilon_{31} - 1)}{\Delta}, \quad C_{33} = \frac{E_3(\upsilon_{12}\upsilon_{21} - 1)}{\Delta},\\
&C_{12} = C_{21} = -E_1(\upsilon_{21}+\upsilon_{23}\upsilon_{31}) = -E_2(\upsilon_{12}+\upsilon_{13}\upsilon_{32}),\\
&C_{31} = C_{13} = -E_3(\upsilon_{13}+\upsilon_{12}\upsilon_{23}) = -E_1(\upsilon_{31}+\upsilon_{32}\upsilon_{21}),\\
&C_{44} = G_{23} = G_{32}, \quad C_{55} = G_{31} = G_{13}, \quad C_{66} = G_{12} = G_{21},
\end{aligned}
\tag{B.6}
$$

$$\Delta = \upsilon_{32}\upsilon_{23} + \upsilon_{13}\upsilon_{31} + \upsilon_{12}\upsilon_{21} + \upsilon_{12}\upsilon_{23}\upsilon_{31} + \upsilon_{13}\upsilon_{32}\upsilon_{21} - 1.$$

The generalized Navier equations result from the equilibrium conditions (for $X = Y = Z = 0$):

$$
\begin{aligned}
&C_{44}\frac{\partial^2 u}{\partial x^2} + C_{66}\frac{\partial^2 u}{\partial y^2} + C_{55}\frac{\partial^2 u}{\partial z^2}\\
&\quad + \frac{\partial}{\partial x}\left[(C_{11} - C_{44})\frac{\partial u}{\partial x} + (C_{12} + C_{66})\frac{\partial v}{\partial y} + (C_{13} + C_{55})\frac{\partial w}{\partial z}\right] = 0,\\
&C_{66}\frac{\partial^2 v}{\partial x^2} + C_{55}\frac{\partial^2 v}{\partial y^2} + C_{44}\frac{\partial^2 v}{\partial z^2}\\
&\quad + \frac{\partial}{\partial y}\left[(C_{12} + C_{66})\frac{\partial u}{\partial x} + (C_{22} - C_{55})\frac{\partial v}{\partial y} + (C_{23} + C_{44})\frac{\partial w}{\partial z}\right] = 0,\\
&C_{55}\frac{\partial^2 w}{\partial x^2} + C_{44}\frac{\partial^2 w}{\partial y^2} + C_{66}\frac{\partial^2 w}{\partial z^2}\\
&\quad + \frac{\partial}{\partial z}\left[(C_{13} + C_{55})\frac{\partial u}{\partial x} + (C_{23} + C_{44})\frac{\partial v}{\partial y} + (C_{33} - C_{66})\frac{\partial w}{\partial z}\right] = 0
\end{aligned}
\tag{B.7}
$$

Transformation formulas for stresses in course of rotation φ with respect to the z-axis are:

$$
\begin{aligned}
&\sigma_{11} = \sigma_{xx}\cos^2\Phi + \sigma_{yy}\sin^2\Phi + 2\tau_{xy}\cos\Phi\sin\Phi,\\
&\sigma_{22} = \sigma_{xx}\sin^2\Phi + \sigma_{yy}\cos^2\Phi - 2\tau_{xy}\cos\Phi\sin\Phi,\\
&\tau_{12} = (\sigma_{yy} - \sigma_{xx})\sin\Phi\cos\Phi + \tau_{xy}(\cos^2\Phi - \sin^2\Phi),\\
&\tau_{13} = \tau_{xz}\cos\Phi + \tau_{yz}\sin\Phi, \tau_{23} = -\tau_{xz}\sin\Phi + \tau_{yz}\cos\Phi.
\end{aligned}
\tag{B.8}
$$

In Eq. (B.8) the components of stresses in the global coordinate system, which is associated with the geometry of the structural element, are $\sigma_{xx}, \ldots, \tau_{yz}$ and Φ is the ply angle or fiber inclination angle.

B.2 Distortion Energy and Supplementary Energy

The forces that deform a solid body do the work as the points of force application displace. According to the first law of thermodynamics, this work of "external forces" is connected with an increase in "internal" energy. The internal energy is called distortion energy or elastic energy.

The elastic energy density is

$$W_E(\sigma,\varepsilon) = \frac{1}{2}(\sigma_{xx}\varepsilon_{xx} + \sigma_{yy}\varepsilon_{yy} + \sigma_{zz}\varepsilon_{zz} + \tau_{yz}\gamma_{yz} + \tau_{zx}\gamma_{zx} + \tau_{xy}\gamma_{xy}). \tag{B.9}$$

The energy density of linear-elastic materials reads:

$$W_\varepsilon(\varepsilon) = \frac{1}{2}(C_{11}\varepsilon_{xx}{}^2 + 2C_{12}\varepsilon_{xx}\varepsilon_{yy} + 2C_{13}\varepsilon_{xx}\varepsilon_{zz} + \ldots + C_{66}\gamma_{xy}{}^2), \tag{B.10}$$

$$W_\sigma(\sigma) = \frac{1}{2}(S_{11}\sigma_{xx}{}^2 + 2S_{12}\sigma_{xx}\sigma_{yy} + 2S_{13}\sigma_{xx}\sigma_{zz} + \ldots + S_{66}\tau_{xy}{}^2). \tag{B.11}$$

The potential energy (or total potential) for a body with volume V is

$$U_\varepsilon = \int_V [W_\varepsilon(\varepsilon) - f_x u - f_y v - f_z w]\, dV - \int_{\Gamma_\sigma} [F_x u + F_y v + F_z w]\, d\Gamma. \tag{B.12}$$

The supplementary energy (or total supplementary potential) for a body of volume V is:

$$U_\sigma = \int_V W_\sigma(\sigma)\, dV - \int_{\Gamma_u} [t_x U + t_y V + t_z W]\, d\Gamma. \tag{B.13}$$

B.3 Plane Elasticity Problems

B.3.1 Plane Strain State

The relations between stresses and strains distinguish two special cases, the plane strain state and the plane stress state. In the plane strain case:

$$\varepsilon_{zz} = \gamma_{xz} = \gamma_{yz} = 0, \quad \sigma_{zz} \neq 0.$$

The stress–strain relations in the plane strain case are:

$$\begin{aligned} &\varepsilon_{xx} = q_{11}\sigma_{xx} + q_{12}\sigma_{yy}, \quad \varepsilon_{yy} = q_{21}\sigma_{xx} + q_{22}\sigma_{yy} \\ &\gamma_{xy} = \gamma_{yx} = q_{66}\tau_{xy} = q_{66}\tau_{yx}, \end{aligned} \tag{B.14}$$

where

$$q_{ij} = S_{ij} - \frac{S_{i3}S_{j3}}{S_{33}}. \tag{B.15}$$

The plane strain state can be realized in a long cylinder of constant cross section with fixed ends under a constant line load along the lateral surface.

B.3.2 Plane Stress State

In the plane stress state, we have:

$$\sigma_{zz} = \tau_{xz} = \tau_{yz} = 0, \varepsilon_{zz} \neq 0.$$

From (B.8) it follows that:

$$\begin{aligned} &\sigma_{xx} = Q_{11}\varepsilon_{xx} + Q_{12}\varepsilon_{yy} \quad \sigma_{yy} = Q_{21}\varepsilon_{xx} + Q_{22}\varepsilon_{yy}, \\ &\tau_{xy} = \tau_{yx} = Q_{66}\gamma_{xy} = Q_{66}\gamma_{yx}, \end{aligned} \tag{B.16}$$

where

$$Q_{ij} = C_{ij} - \frac{C_{i3}C_{j3}}{C_{33}}. \tag{B.17}$$

B.4 Generalized Airy Stress Function

B.4.1 Plane Stress State

The equilibrium conditions (B.20) for the plane stress state are identically fulfilled by the scalar function F(x, y) (the Airy stress function):

$$\sigma_{xx} = \frac{\partial^2 \mathrm{F}}{\partial y^2}, \quad \sigma_{yy} = \frac{\partial^2 \mathrm{F}}{\partial x^2}, \quad \tau_{xy} = -\frac{\partial^2 \mathrm{F}}{\partial x \partial y}.$$

From the two-dimensional compatibility condition follows the equation for the Airy stress function F(x, y) for the plane stress state:

$$S_{11}\frac{\partial^4 \mathrm{F}}{\partial y^4} + (2S_{12} + S_{66})\frac{\partial^4 \mathrm{F}}{\partial x^2 \partial y^2} + S_{22}\frac{\partial^4 \mathrm{F}}{\partial x^4} = 0. \tag{B.18}$$

B.4.2 Plane Strain State

The generalized Airy equation for plane strain state is:

$$q_{11}\frac{\partial^4 \mathrm{F}}{\partial y^4} + (2q_{12} + q_{66})\frac{\partial^4 \mathrm{F}}{\partial x^2 \partial y^2} + q_{22}\frac{\partial^4 \mathrm{F}}{\partial x^4} = 0. \tag{B.19}$$

B.4.3 Rotationally Symmetric Elasticity Problems

For another important class of rotationally symmetric spatial elasticity problems, the displacements and stresses depend only on two cylindrical coordinates (r, z). The basic equations for this result from the general relationships for the three-dimensional continuum.

For this purpose, set:

$$u_r = u(r, z), \quad u_\theta = 0, \quad w = w(r, z) \tag{B.20}$$

Between the distortion components

$$\varepsilon_r = \frac{\partial u}{\partial r}, \varepsilon_\theta = \frac{u}{r}, \varepsilon_z = \frac{\partial w}{\partial z}, \quad \gamma_{r\,z} = \frac{\partial u}{\partial z} + \frac{\partial w}{\partial r}, \gamma_{z\,\theta} = \gamma_{\theta\,r} = 0$$

there are two compatibility conditions:

$$\varepsilon_r - \frac{\partial}{\partial r}(r\,\varepsilon_\theta) = 0, \quad \frac{\partial^2 \varepsilon_r}{\partial z^2} + \frac{\partial^2 \varepsilon_z}{\partial r^2} - \frac{\partial^2 \gamma_{r\,z}}{\partial r\,\partial z} = 0. \tag{B.21}$$

The equilibrium conditions are

$$\frac{\partial \sigma_r}{\partial r} + \frac{\partial \tau_{rz}}{\partial z} + \frac{\sigma_r - \sigma_\theta}{r} = 0, \tag{B.22}$$

$$\frac{\partial \sigma_z}{\partial z} + \frac{\partial \tau_{rz}}{\partial r} + \frac{\tau_{rz}}{r} = 0 \tag{B.23}$$

For transversal-isotropic homogeneous bodies, the generalized Hooke's law applies:

$$\begin{aligned}
&\varepsilon_r = S_{11}\sigma_r + S_{12}\sigma_\theta + S_{13}\sigma_z, \varepsilon_\theta = S_{21}\sigma_r + S_{11}\sigma_\theta + S_{13}\sigma_z,\\
&\varepsilon_z = S_{31}(\sigma_r + \sigma_\theta) + S_{33}\sigma_z, \gamma_{rz} = \gamma_{zr} = S_{44}\tau_{rz} = S_{44}\tau_{zr},\\
&\gamma_{z\theta} = \gamma_{\theta z} = S_{44}\tau_{z\theta} = S_{44}\tau_{\theta z}, \quad \gamma_{r\theta} = \gamma_{\theta r} = 2(S_{11} - S_{12})\tau_{r\theta}.
\end{aligned} \tag{B.24}$$

Appendix C

Integral Transforms in Elasticity

C.1 One-Dimensional Integral Transform

A general integral transform of the function $\varphi(x)$ is defined by:

$$J\{\varphi(x), x \to \xi\} = \varphi^*(\xi) = \int_a^b \varphi(x)K(\xi, x)dx, \tag{C.1}$$

where $K(\xi, x)$ is the transformation kernel.

Complex Fourier transform:

$$\mathfrak{F}\{\varphi(x), x \to \xi\} = \varphi^*(\xi) = \frac{1}{\sqrt{2\pi}} \int_{-\infty}^{\infty} \varphi(x)e^{i\xi x}dx. \tag{C.2}$$

Fourier-SINUS- and Fourier-COSINUS transforms:

$$\mathfrak{F}_S\{\varphi(x), x \to \xi\} = \varphi^*(\xi) = \sqrt{\frac{2}{\pi}} \int_{-\infty}^{\infty} \varphi(x)\sin(\xi\, x)dx, \tag{C.3}$$

$$\mathfrak{F}_C\{\varphi(x), x \to \xi\} = \varphi^*(\xi) = \sqrt{\frac{2}{\pi}} \int_{-\infty}^{\infty} \varphi(x)\cos(\xi\, x)dx. \tag{C.4}$$

The inversion of Fourier transforms:

$$\mathfrak{F}^{-1}\{\varphi^*(\xi), \xi \to x\} = \varphi(x) = \frac{1}{\sqrt{2\pi}} \int_{-\infty}^{\infty} \varphi^*(\xi)e^{-i\,\xi\,x}d\xi, \tag{C.5}$$

$$\mathfrak{F}_S{}^{-1}\{\varphi^*(\xi), \xi \to x\} = \varphi(x) = \sqrt{\frac{1}{2\pi}} \int_{-\infty}^{\infty} \varphi^*(\xi)\sin(\xi\, x)\, d\xi, \tag{C.6}$$

$$\mathfrak{F}_C{}^{-1}\{\varphi^*(\xi), \xi \to x\} = \varphi(x) = \sqrt{\frac{1}{2\pi}} \int_{-\infty}^{\infty} \varphi^*(\xi)\cos(\xi\, x)\, d\xi. \tag{C.7}$$

The transform of the derivative reads

$$\mathfrak{F}\left\{\frac{d^n}{dx^n}\varphi(x), x \to \xi\right\} = (\varphi^{(n)}(\xi))^* = (-i\xi)^n\varphi^*(\xi). \tag{C.8}$$

Design and Analysis of Composite Structures for Automotive Applications: Chassis and Drivetrain, First Edition. Vladimir Kobelev.

Companion website: www.wiley.com/go/kobelev/automotive_suspensions

C.2 Two-Dimensional Fourier Transform

Two-dimensional Fourier transform of the function $\varphi(x, y)$:

$$\mathfrak{F}\{\varphi(x, y), x \to \xi\} = \varphi^*(\xi, y) = \frac{1}{\sqrt{2\pi}} \int_{-\infty}^{\infty} \varphi(x, y) e^{i\xi x} dx, \tag{C.9}$$

$$\mathfrak{F}\{\varphi(x, y), y \to \eta\} = \varphi^*(x, \eta) = \frac{1}{\sqrt{2\pi}} \int_{-\infty}^{\infty} \varphi(x, y) e^{iy\eta} dy, \tag{C.10}$$

$$\mathfrak{F}\{\varphi(x, y), x \to \xi, y \to \eta\} = \varphi^*(\xi, \eta) = \frac{1}{\sqrt{2\pi}} \int_{-\infty}^{\infty} \int_{-\infty}^{\infty} \varphi(x, y) e^{i(y\eta + x\xi)} dx\, dy. \tag{C.11}$$

The corresponding retransformation formulas are:

$$\mathfrak{F}\{\varphi^*(\xi, y), \xi \to x\} = \varphi(x, y) = \frac{1}{\sqrt{2\pi}} \int_{-\infty}^{\infty} \varphi^*(\xi, y) e^{-i\,\xi\, x} d\xi, \tag{C.12}$$

$$\mathfrak{F}\{\varphi^*(x, \eta), \eta \to y\} = \varphi(x, y) = \frac{1}{\sqrt{2\pi}} \int_{-\infty}^{\infty} \varphi^*(x, \eta) e^{-i\,\eta\, y} d\eta, \tag{C.13}$$

$$\mathfrak{F}\{\varphi^*(\xi, \eta), \xi \to x, \eta \to y\} = \varphi(x, y) = \frac{1}{\sqrt{2\pi}} \int_{-\infty}^{\infty} \int_{-\infty}^{\infty} \varphi^*(\xi, \eta) e^{-i(y\,\eta + x\,\xi)} d\eta\, d\xi \tag{C.14}$$

C.3 Potential Functions for Plane Elasticity Problems

For plane elasticity problems of an orthotropic medium, the displacements, distortions and stresses depend only on two coordinates in the plane. These equations can be represented as follows:

$$\mathbf{L}^E[\mathrm{F}] = 0, \tag{C.15}$$

where a generalized biharmonic operator for plane stress state reads:

$$\mathbf{L}^E[\mathrm{F}] = C_{11} \frac{\partial^4 \mathrm{F}}{\partial y^4} + (2C_{12} + C_{66}) \frac{\partial^4 \mathrm{F}}{\partial x^2 \partial y^2} + C_{22} \frac{\partial^4 \mathrm{F}}{\partial x^4} = 0$$

For a plane strain state, the coefficients should be recalculated according to formulas from Appendix B.

The strains after the substitution read:

$$\frac{\partial u}{\partial x} = C_{11} \frac{\partial^2 \mathrm{F}}{\partial y^2} + C_{12} \frac{\partial^2 \mathrm{F}}{\partial x^2},$$

$$\frac{\partial v}{\partial y} = C_{21} \frac{\partial^2 \mathrm{F}}{\partial y^2} + C_{22} \frac{\partial^2 \mathrm{F}}{\partial x^2}$$

$$\frac{\partial u}{\partial y} + \frac{\partial v}{\partial x} = -C_{66} \frac{\partial^2 \mathrm{F}}{\partial x\, \partial y}. \tag{C.16}$$

The polynomial $\mathbf{L}^A[\mu] = C_{11}\mu^4 + (2C_{12} + C_{66})\mu^2 + c_{22}$ Is the characteristic polynomial to (C.15), and the corresponding equation is:

$$\mathbf{L}^A[\mu] = 0. \tag{C.17}$$

The roots of (C.17) μ_1, μ_2, μ_3 and μ_4 are imaginary:

$$\mu_1 = -\mu_3 = ip_1, \quad \mu_2 = -\mu_4 = ip_2, \tag{C.18}$$

$$p_{1,2}{}^2 = \frac{2C_{12} + C_{66} \pm \sqrt{(2C_{11} + C_{66})^2 - 4C_{11}C_{22}}}{2C_{11}} > 0.$$

From the positivity of $\mathbf{L}^A[\mu]$ for real argument, the condition follows:

$$(2C_{11} + C_{66})^2 - 4C_{11}C_{22} > 0$$

The substitution:

$$\Delta_1^E = \frac{\partial^2}{\partial y^2} + p_1^2\frac{\partial^2}{\partial x^2}, \quad \Delta_2^E = \frac{\partial^2}{\partial y^2} + p_2^2\frac{\partial^2}{\partial x^2}$$

leads to factorization of the differential Eq. (C.17):

$$\mathbf{L}^E[\mathrm{F}] = \Delta_1^E\Delta_1^E\mathrm{F} = 0. \tag{C.19}$$

The characteristic polynomial of the factorized differential Eq. (C.19) is:

$$\mathbf{L}^A[\mu] = C_{11}(\mu - ip_1)(\mu + ip_1)(\mu - ip_2)(\mu + ip_2) = 0. \tag{C.20}$$

By comparing (C.17) with (C.20) it can be concluded that:

$$p_1^2p_2^2 = C_{22}/C_{11}, \quad p_1^2 + p_2^2 = (2C_{12} + C_{66})/C_{11} \tag{C.21}$$

For the isotropic body $p_1 = p_2 = 1$ and $\mu_1 = \mu_3 = i$, $\mu_2 = \mu_4 = -i$.

The general representation of a real solution (C.19) using two complex analytical functions $\mathbf{U}_1(z_1)$ and $\mathbf{U}_2(z_2)$ the complex variables $z_1 = x + \mu_1 y$, $z_2 = x + \mu_2 y$ is:

$$\mathrm{F}(x, y) = 2\mathrm{Re}[\mathbf{U}_1(z_1) + \mathbf{U}_2(z_2)]. \tag{C.22}$$

The plane problem of the elasticity theory of the orthotropic body is thus attributed to the determination of both analytical functions $\mathbf{U}_1(z_1)$ and $\mathbf{U}_2(z_2)$.

The relationship between the stress and displacement components and the complex stress functions applies:

$$\sigma_{xx} = 2\mathrm{Re}\,[\mu_1^2\Phi'(z_1) + \mu_2^2\Psi'(z_2)], \quad \sigma_{yy} = 2\mathrm{Re}\,[\Phi'(z_1) + \Psi'(z_2)]$$

$$\tau_{xy} = -2\mathrm{Re}\,[\mu_1\Phi'(z_1) + \mu_2\Psi'(z_2)],$$

$$u = 2\mathrm{Re}\,[P_1\Phi(z_1) + P_2\Psi(z_2)], \quad v = 2\mathrm{Re}\,[Q_1\Phi(z_1) + Q_2\Psi(z_2)].$$

In these equations

$$P_i = C_{11}\mu_i^2 + C_{12} - C_{66}\mu_i, \quad Q_i = C_{12}\mu_i^2 - C_{66} + C_{22}/\mu_i,$$

$$\Phi(z_1) = \frac{d\mathbf{U}(z_1)}{dz_1}, \quad \Psi(z_1) = \frac{d\mathbf{U}(z_2)}{dz_2}.$$

The boundary condition plays an important role in determining the two stress $\mathbf{U}_1(z_1)$, $\mathbf{U}_2(z_2)$ functions. In a complex representation for the first boundary value problem, this reads as follows:

$$(1 - p_1)\Phi(z_1) + (1 - p_2)\Psi(z_2) +$$
$$(1 + p_1)\overline{\Phi}(\overline{z}_1) + (1 + p_2)\overline{\Psi}(\overline{z}_2) = f_1 + if_2$$

where

$$f_1(l) + i f_2(l) = i \int_{l_0}^{l} (F_x + iF_y)\, ds.$$

It is useful to represent the function F(x, y) as the sum of two functions:

$$\mathrm{F}(x, y) = \psi_1(x, y) + \psi_2(x, y). \tag{C.23}$$

The functions $\psi_1(x, y)$, $\psi_2(x, y)$ are the solutions of the following equations:

$$\Delta_1^E \psi_1 = \frac{\partial^2 \psi_1}{\partial y^2} + p_1^2 \frac{\partial^2 \psi_1}{\partial x^2} = 0, \quad \Delta_1^E \psi_1 = \frac{\partial^2 \psi_1}{\partial y^2} + p_1^2 \frac{\partial^2 \psi_1}{\partial x^2} = 0 \tag{C.24}$$

Inserting (B.27) in (B.22) and (C.16) returns:

$$\begin{aligned}
\sigma_{yy} &= -\frac{1}{p_1^2}\frac{\partial^2 \psi_1}{\partial y^2} - \frac{1}{p_2^2}\frac{\partial^2 \psi_2}{\partial y^2}, \\
\tau_{xy} &= -\frac{\partial^2 \psi_1}{\partial x \partial y} - \frac{\partial^2 \psi_2}{\partial x \partial y}, \\
v &= \delta_1 \frac{\partial \psi_1}{\partial x} + \delta_2 \frac{\partial \psi_2}{\partial y},
\end{aligned} \tag{C.25}$$

with

$$\delta_1 = C_{21} + C_{22} p_1^{-2}, \quad \delta_2 = C_{21} + C_{22} p_2^{-2}.$$

C.4 Rotationally Symmetric, Spatial Elasticity Problems

For another important class of rotationally symmetric, spatial elasticity problem, the displacements and stresses depend only on two cylindrical coordinates (r, z). The basic equations for this result from the general relationships for the three-dimensional continuum. The equilibrium conditions (B.22–B.23) and the compatibility conditions (B.21) are determined by the following approach:

$$\begin{aligned}
\sigma_r &= -\frac{\partial}{\partial z}\left(\frac{\partial^2 F}{\partial r^2} + \frac{b}{r}\frac{\partial F}{\partial r} + a\frac{\partial^2 F}{\partial z^2}\right), \\
\sigma_\theta &= -\frac{\partial}{\partial z}\left(b\frac{\partial^2 F}{\partial r^2} + \frac{1}{r}\frac{\partial F}{\partial r} + a\frac{\partial^2 F}{\partial z^2}\right), \\
\sigma_z &= \frac{\partial}{\partial z}\left(c\frac{\partial^2 F}{\partial r^2} + \frac{c}{r}\frac{\partial F}{\partial r} + d\frac{\partial^2 F}{\partial z^2}\right), \\
\tau_{rz} &= \frac{\partial}{\partial r}\left(\frac{\partial^2 F}{\partial r^2} + \frac{1}{r}\frac{\partial F}{\partial r} + a\frac{\partial^2 F}{\partial z^2}\right)
\end{aligned} \tag{C.26}$$

The equilibrium equations are satisfied by of the scalar stress function $F(r, z)$. The coefficients read:

$$\begin{aligned}
a &= \frac{S_{13}(S_{11} - S_{12})}{D}, \quad b = \frac{S_{13}(S_{13} + S_{44}) - S_{12}S_{33}}{D}, \\
c &= \frac{S_{13}(S_{11} - S_{12}) + S_{11}S_{44}}{D}, \quad d = \frac{{S_{11}}^2 - {S_{12}}^2}{D}, \\
D &= S_{11}S_{33} - {S_{13}}^2.
\end{aligned} \tag{C.27}$$

From the second equilibrium condition (B.21) follows the basic equation of the rotationally symmetric, spatial elasticity problem:

$$\mathbf{L}^R[F] = \left(\frac{\partial^2}{\partial r^2} + \frac{1}{r}\frac{\partial}{\partial r}\right)\left(\frac{\partial^2 F}{\partial r^2} + \frac{1}{r}\frac{\partial F}{\partial r} + a\frac{\partial^2 F}{\partial z^2}\right)$$
$$+ \frac{\partial^2}{\partial z^2}\left(c\frac{\partial^2 F}{\partial r^2} + \frac{c}{r}\frac{\partial F}{\partial r} + d\frac{\partial^2 F}{\partial z^2}\right) = 0. \tag{C.28}$$

The substitution

$$\Delta_1^R = \frac{\partial^2}{\partial z^2} + q_1^2 \frac{1}{r}\frac{\partial}{\partial r}\left(r\frac{\partial}{\partial r}\right), \tag{C.29}$$

$$\Delta_2^R = \frac{\partial^2}{\partial z^2} + q_2^2 \frac{1}{r}\frac{\partial}{\partial r}\left(r\frac{\partial}{\partial r}\right) \tag{C.30}$$

with

$$\begin{cases} 2\,d\,q_1^2 = a + c + \sqrt{(a+c)^2 - 4d} \\ 2\,d\,q_2^2 = a + c - \sqrt{(a+c)^2 - 4d} \end{cases} \tag{C.31}$$

Leads to the following representation of (C.28):

$$\mathbf{L}^R[F] = d\,\Delta_1^R \Delta_2^R F = 0. \tag{C.32}$$

It is useful to represent the stress function $F(r, z)$ as an indefinite integral of the sum of two functions $\varphi_1(r, z)$ $\varphi_2(r, z)$:

$$F(r, z) = \int_0^z [\varphi_1(r, Z) + \varphi_2(r, Z)]\, dZ. \tag{C.33}$$

The functions fulfill the equations

$$\Delta_1^R \varphi_1 = 0, \quad \Delta_2^R \varphi_2 = 0. \tag{C.34}$$

Substitution of (C.33) into (C.26) provides:

$$\sigma_{zz} = \alpha_1 \frac{\partial^2 \varphi_1}{\partial z^2} + \alpha_2 \frac{\partial^2 \varphi_2}{\partial z^2},$$

$$\tau_{rz} = \beta_1 \frac{\partial^2 \varphi_1}{\partial r\, \partial z} + \beta_2 \frac{\partial^2 \varphi_2}{\partial r\, \partial z},$$

$$w = \gamma_1 \frac{\partial \varphi_1}{\partial z} + \gamma_2 \frac{\partial \varphi_2}{\partial z} \tag{C.35}$$

with

$$\alpha_i = a - c\omega_i^2, \quad \beta_i = a - \omega_i^2,$$

$$\omega_i = q_i^{-1}, \quad \gamma_i = \frac{S_{13}(1 - 2aq_i^2 + b) + S_{33}(aq_i^2 - c)}{q_i^2}.$$

This transformation is similar to the transformation (C.25) used in the plane elasticity problem.

C.5 Application of the Fourier Transformation to Plane Elasticity Problems

The application of the Fourier transformation with respect to one of the variables allows the partial differential slip for the Airy stress function to be reduced to a common differential equation. With the Fourier transform

$$G(\xi, y) = \mathfrak{F}\{\ \mathbf{L}^E[F(x, y)], x \rightarrow \xi\ \} \tag{C.36}$$

follows the equation:

$$\int_{-\infty}^{\infty} \mathbf{L}^E[F(x, y)]e^{i\xi x}dx =$$
$$\left[C_{11}\frac{\partial^4}{\partial y^4} + (2C_{12} + C_{66})\xi^2\frac{\partial^2}{\partial y^2} + C_{22}\xi^4\right]\int_{-\infty}^{\infty} F(x, y)e^{i\xi x}dx.$$

Thus, from the partial differential Eq. (C.19) follows the ordinary differential equation for $G(\xi, y)$:

$$C_{11}\frac{d^4G}{dy^4} + (2C_{12} + C_{66})\xi^2\frac{d^2G}{dy^2} + C_{22}\xi^4G = 0. \tag{C.37}$$

The solution of (C.37) reads:

$$G(\xi, y) = A(\xi)e^{-p_1\xi y} + B(\xi)e^{-p_2\xi y}$$
$$+ C(\xi)e^{p_1\xi y} + D(\xi)e^{p_2\xi y} \tag{C.38}$$

The integration constants $A(\xi)$, $B(\xi)$, $C(\xi)$ and $D(\xi)$ are determined from the boundary conditions of the problem. In the case of a semi-infinite medium in which all stresses and displacements decay for $y \rightarrow \infty$, the integration constants $C(\xi)$ and $D(\xi)$ disappear.

With the retransformation of (C.36) we finally get an integral representation for the Airy stress function:

$$F(x, y) = \frac{1}{\sqrt{2\pi}}\int_{-\infty}^{\infty} G(\xi, y)e^{-i\xi x}dx. \tag{C.39}$$

The stress and displacement components can also be expressed by the function $G(\xi, y)$ and its derivatives. It follows:

$$\sigma_{xx}(x, y) = \frac{1}{\sqrt{2\pi}}\int_{-\infty}^{\infty} \frac{d^2}{dy^2}G(\xi, y)e^{-i\xi x}dx,$$
$$\sigma_{yy}(x, y) = \frac{1}{\sqrt{2\pi}}\int_{-\infty}^{\infty} \xi^2G(\xi, y)e^{-i\xi x}dx,$$
$$\tau_{xy}(x, y) = \frac{1}{\sqrt{2\pi}}\int_{-\infty}^{\infty} \xi\frac{d}{dy}G(\xi, y)e^{-i\xi x}dx \tag{C.40}$$

and

$$u(x,y) = \frac{1}{\sqrt{2\pi}} \int_{-\infty}^{\infty} \frac{i}{\xi} \left[C_{11} \frac{d^2}{dy^2} G(\xi, y) - C_{12} \xi G(\xi, y) \right] e^{-i\xi x} dx,$$

$$v(x,y) = \frac{1}{\sqrt{2\pi}} \int_{-\infty}^{\infty} \frac{1}{\xi^2} \left[C_{11} \frac{d^3}{dy^3} G(\xi, y) - (C_{12} + C_{66}) \xi^2 \frac{d}{dy} G(\xi, y) \right] e^{-i\xi x} dx. \tag{C.41}$$

C.6 Application of the Hankel Transformation to Spatial, Rotation-Symmetric Elasticity Problems

The application of the Hankel transformation with respect to one of the variables allows the partial differential equation for the stress function $F(r, z)$ to be reduced to an ordinary differential equation. This partial differential equation is obtained by introducing the Hankel transformation of zero order with respect to the coordinate r to a common differential equation in z, in which the role of the coordinate r is represented by a parameter ξ:

$$G(\xi, z) = \mathrm{H}_0\{ \mathbf{L}^R[F(r,z)],\ r \to \xi \} = \int_0^{\infty} F(r,z) r \mathrm{J}_0(\xi r) dr$$

The procedure shows parallels to the plane problem when the Fourier transformations used there are replaced by the Hankel transformation.

For the function $F(r, z)$, $0 < r < \infty$ the formula is valid:

$$\int_0^{\infty} \mathbf{L}^R[F(r,z)] r \mathrm{J}_0(\xi r) dr = \left[d\frac{d^4}{dz^4} - (a+c)\xi^2 \frac{d^2}{dz^2} + \xi^4 \right] \int_0^{\infty} F(r,z) r \mathrm{J}_0(\xi r) dr. \tag{C.42}$$

From the partial differential Eq. (C.28) for $F(r, z)$ follows the ordinary differential equation for $G(\xi, z)$:

$$\left[d\frac{d^4}{dz^4} - (a+c)\xi^2 \frac{d^2}{dz^2} + \xi^4 \right] G(\xi, z) = 0 \tag{C.43}$$

The integration of (C.43) provides:

$$G(\xi, z) = A(\xi) e^{-q_1 \xi z} + B(\xi) e^{-q_2 \xi z} + C(\xi) e^{q_1 \xi z} + D(\xi) e^{q_2 \xi z}. \tag{C.44}$$

Index

Design and Analysis of Composite Structures for Automotive Applications: Chassis and Drivetrain, First Edition. Vladimir Kobelev.

Companion website: www.wiley.com/go/kobelev/automotive_suspensions

d

e

f

www.ingramcontent.com/pod-product-compliance
Lightning Source LLC
LaVergne TN
LVHW060439031026
849429LV00005B/27